Pili and Flagella

Current Research and Future Trends

Edited by Ken F. Jarrell

Department of Microbiology and Immunology
Queen's University
Kingston, ON
Canada

Caister Academic Press

Caister Academic Press
Norfolk, UK

www.caister.com

British Library Cataloguing-in-Publication Data
A catalogue record for this book is available from the British Library

ISBN: 978-1-904455-48-6

Printed and bound in Great Britain

Cover: Background images by Mindy Lim and Lisa Craig. Front cover image from Figure 5.3. Back cover image from Figure 12.1.

Contents

Contributors

Shin-Ichi Aizawa
Department of Life Sciences
Prefectural University of Hiroshima
Hiroshima
Japan

aizawa@pu-hiroshima.ac.jp

Phillip Aldridge
Centre for Bacterial Cell Biology
Newcastle University
Newcastle upon Tyne
UK

p.d.aldridge@ncl.ac.uk

Lena Anton
General Microbiology
Department of Biological and Environmental Sciences
Faculty of Biosciences
University of Helsinki
Finland

lena.anton@helsinki.fi

David F. Blair
Department of Biology
University of Utah
Salt Lake City, UT
USA

blair@bioscience.utah.edu

Jonathon Brown
Centre for Bacterial Cell Biology
Newcastle University
Newcastle upon Tyne
UK

Lisa Craig
Department of Molecular Biology and Biochemistry
Simon Fraser University
Burnaby, BC
Canada

licraig@sfu.ca

Asis Das
Department of Molecular, Microbial, and Structural Biology
University of Connecticut Health Center
Farmington, CT
USA

adas2nso2.uchc.edu

Michael S. Donnenberg
University of Maryland School of Medicine
Division of Infectious Diseases
Baltimore, MD
USA

mdonnenb@umaryland.edu

Alexandra Faulds-Pain
Centre for Bacterial Cell Biology
Newcastle University
Newcastle upon Tyne
UK

Andrew H. Gaspar
Department of Molecular, Microbial, and Structural Biology
University of Connecticut Health Center
Farmington, CT
USA

agaspar@student.uchc.edu

Ken F. Jarrell
Department of Microbiology and Immunology
Queen's University
Kingston, ON
Canada

jarrellk@queensu.ca

Michael Koomey
Department of Molecular Biosciences and Centre for Molecular Biology and Neuroscience
University of Oslo
Oslo
Norway

johnk@imbv.uio.no

Riikka Kylväjä
General Microbiology
Department of Biological and Environmental Sciences
Faculty of Biosciences
University of Helsinki
Finland

rikka.kylvaja@helsinki.fi

Renyi Liu
Department of Botany and Plant Sciences
University of California
Riverside, CA
USA

renyi.liu@ucr.edu

Susan M. Logan
Institute for Biological Sciences
National Research Council
Ottawa, ON
Canada

susan.logan@nrc-cnrc.gc.ca

Katariina Majander
General Microbiology
Department of Biological and Environmental Sciences
Faculty of Biosciences
University of Helsinki
Finland

katariina.majander@helsinki.fi

Anjali Mandlik
Department of Molecular, Microbial, and Structural Biology
University of Connecticut Health Center
Farmington, CT
USA

anjali.mandlik@gmail.com

Susana Merino
Departamento de Microbiología
Facultad de Biología
Universidad de Barcelona
Barcelona
Spain

smerino@ub.edu

Ekaterina Milgotina
University of Maryland School of Medicine
Division of Infectious Diseases
Baltimore, MD
USA

emilg001@umaryland.edu

Arunima Mishra
Department of Mirobiology and Molecular Genetics
University of Texas Health Science Center
Houston, TX
USA

arunima.mishra@uth.tmc.edu

Han Remaut
Institute of Structural and Molecular Biology
University College London and Birkbeck College
London
UK

Present address:
Structural and Molecular Microbiology
Department of Molecular and Cellular Interactions
Vrije Universiteit Brussel and VIB
Brussels
Belgium

han.remaut@vib-vub.be

Fadel A. Samatey
Trans-membrane Trafficking Unit
Okinawa Institute of Science and Technology
Onna, Kunigami
Okinawa
Japan

f.a.samatey@oist.jp

Anu Swaminathan
Department of Molecular, Microbial, and Structural Biology
University of Connecticut Health Center
Farmington, CT
USA

anu_swamina@yahoo.co.in

Juan M. Tomás
Departamento de Microbiología
Facultad de Biología
Universidad de Barcelona
Barcelona
Spain

jtomas@ub.edu

Hung Ton-That
Department of Mirobiology and Molecular Genetics
University of Texas Health Science Center
Houston, TX
USA

ton-that.hung@uth.tmc.edu

David J. VanDyke
Department of Microbiology and Immunology
Queen's University
Kingston, ON
Canada

dv569372@dal.ca

Gabriel Waksman
Institute of Structural and Molecular Biology
University College London and Birkbeck College
London
UK

g.waksman@bbk.ac.uk
g.waksman@ucl.ac.uk

Benita Westerlund-Wikström
General Microbiology
Department of Biological and Environmental Sciences
Faculty of Biosciences
University of Helsinki
Finland

benita.westerlund@helsinki.fi

John Wu
Department of Microbiology and Immunology
Queen's University
Kingston, ON
Canada

3jcow@queensu.ca

Preface

As Editor of *Pili and Flagella: Current Research and Future Trends*, I am absolutely delighted with the outcome and feel very privileged to have coordinated the efforts of such fine scientists and writers that resulted in this exciting compilation. Within the broad areas of pili and flagella are topics which highlight the beauty of the various appendages, their elegant control systems and the variety of mechanisms that allow cells to assemble these complex organelles, the bulk of which lies outside the cytoplasmic membrane on the cell surface. Study of these surface organelles has, among other advances, expanded our knowledge of regulation mechanisms, protein export, and detailed analysis of the structure and assembly has created new potential targets for antibacterial agents and biotechnology applications. Four different pili classes are highlighted in the book, each with unique features and assembly mechanisms. Lisa Craig delves into the structure of various Type IV pili and shows how such structural information has lead to an understanding of their assembly mechanism and provided insight about how they perform their diverse functions. Michael Koomey follows with an in depth review and analysis of Type IV pili assembly and function, relying on various Type IVa systems but focusing on the *Pseudomonas aeruginosa* and *Neiserria gonorrhoeae* pili. It contains an informative historical perspective and a critical analysis of some puzzling emerging systems. Ekaterina Milgotina and Michael Donnenberg finish off Type IV pili systems with a chapter devoted to Type IVb systems, concentrating on the bundle forming pili systems of enteropathogenic *Escherichia coli*, with particularly enlightening discussions of protein interactions in the inner and outer membrane subassembly components. Han Remaut and Gabriel Waksman discuss the entirely different structure, function and biogenesis of pili assembled by a chaperone/usher mechanism. The mechanism of periplasmic chaperone-mediated delivery of pilus subunits to the outer membrane usher is beautifully presented and the chapter ends with a segue from the molecular characterization of the assembly mechanism into how this knowledge can be used to target the pili in antibacterial therapies. Hung Ton-That's group ends the pili section of the book with a thorough examination of the unusual pili in Gram positive organisms that are dependent on both pilin specific and housekeeping sortases to covalently link the structure and anchor it to the cell wall.

The flagella portion of the book begins with a thought-provoking chapter by Shin-Ichi Aizawa who asks the simple question "What is essential for flagellar assembly?" He shows by consideration of the data accumulated on a number of flagellar mutants that, while many proteins may be required for efficient flagellation, they can be dispensable for assembling a functional flagellum. The difficult task of presenting the coordination of flagellar gene expression with flagella assembly is expertly handled by Phillip Aldridge and his group. They discuss the conservation of flagellar regulatory systems and also the divergence of these networks in organisms that lie outside the *E. coli/Salmonella* paradigm and include recent work, for example, on the role of cyclic di-GMP. In chapter 8, David Blair discusses the flagella rotary motor and its component rotor and stator before closing the chapter with a chemically detailed model that incorporates the latest structural information. He stresses the need for information on motors in species other than *E. coli* and *Salmonella*. Flagellum structure is presented by Fadel Samatey who interweaves historical data with the most recent information obtained from a combination of X-ray diffraction, various electron microscopy techniques, and three dimensional imaging tools to produce detailed models of the filament and hook in *Salmonella*. Susan Logan discusses the structure and assembly of glycans that can be O-linked or N-linked to flagellins in a wide variety of bacteria and archaea, respectively, as well as the effects that disruption in glycosylation have on flagella assembly and virulence. Susana Merino and Juan Tomás present the unusual cases where bacterial species have two flagella systems, including a lateral one. They discuss the separate regulation of these systems, their roles in virulence as well as their evolutionary origin and bring to light their possible more common distribution, based on comparative bioinformatics. Benita Westerlund-Wikström and colleagues summarize the use of the bacterial flagellum for surface display and expression, both in basic research and applied areas, including vaccine development and diagnostics. Studies using either the flagellin or the capping protein for surface display, and double or even triple display techniques expressing different peptides along the length and at the tip of the flagellum are discussed. The fascinating topic of the evolution of the flagellar system is succinctly discussed by Renyi Liu, wherein the role of various evolutionary forces including gene duplication, horizontal gene transfer and even gene deletion have all been identified in the evolution of the structure from a secretion system to the present day motility structure. The book closes with an article by my group on the unusual flagella and pili of Archaea. In the archaeal domain, flagella are more similar to type IV pili systems in composition, structure and likely assembly while research on archaeal pili, though scant, also points to novel features not observed in any of the pili systems of Bacteria.

This book brings together the efforts of 14 world experts discussing their research specialties. Their excitement for the field shines through and readers will enjoy the clarity and insight provided in all the chapters. Not only are the chapters comprehensive and current but they close with a section on future trends that point out the pressing questions that are likely to be addressed in the near future. My personal thanks go to all the authors that contributed to this book. They made my job as editor an easy and enjoyable one.

Ken F. Jarrell
March 2009

Type IV Pilus Structure

1

Lisa Craig

Abstract

The Type IV pili are architectural marvels of biology. These helical arrays of thousands of copies of a single pilin subunit are extremely thin and flexible yet remarkably strong, and they possess a diverse array of functions. Type IV pili are essential for host colonization and virulence for many Gram-negative bacteria, and may also play a role in pathogenesis for some Gram-positive bacteria. High-resolution structures of pilin subunits have been determined by X-ray crystallography and nuclear magnetic resonance spectroscopy both as full-length proteins and as soluble fragments. These structures have been used to generate computational models of pilus filament assemblies, guided by biophysical parameters extracted from fibre diffraction data and electron microscopy image analysis, and by a dense literature of biological data. Most recently, cryo-electron microscopy has provided an intermediate resolution structure of a pilus filament. The pilin structures and filament models have been instrumental in advancing our understanding of the molecular mechanisms driving pilus assembly and the role of Type IV pili in key bacterial functions such as immune evasion, microcolony formation and DNA uptake. In very general terms the structural data point to a shared subunit structure and filament architecture for all Type IV pili, but a comprehensive, atomic-level understanding of these filaments and their biological processes will require additional higher resolution filament structures as well as new structural, genetic and biochemical data on the many components of the pilus assembly apparatus.

Introduction

The molecular structures of the Type IV pili (T4P) are of great interest because of their intriguing macromolecular design and the central role they play in bacterial virulence. The T4P are typically a few microns in length but can be as long as 20 μm (Fig. 1.1). They are extremely flexible and thin, measuring 50–90 Å in diameter, yet they have impressive mechanical strength, being able to withstand stress forces of ≥100 piconewtons, which is in the order of forces required to unfold proteins (Maier *et al.*, 2002). These pili have a remarkable array of biological functions that are central to the survival of many Gram-negative bacteria and to their success as pathogens. These functions include immune evasion, microcolony formation and phage attachment. T4P comprise thousands of copies of the pilin subunit, a 15- to 20-kDa protein. Pilin from different bacterial species share amino acid sequence conservation in their hydrophobic N-terminal segment but are much less similar in their C-terminal two-thirds, all of which possess a cysteine pair. T4P are distributed in a polar or peritrichous manner on the bacterial surface and some are capable of retraction, which further broadens their functional repertoire to include bacterial motility, DNA uptake and phage transduction, and in some cases may help to facilitate secretion of proteases and colonization factors. Their prominent display on the bacterial surface and their importance in virulence make these filaments prime targets for the host immune system as well as attractive targets for vaccines and antibacterial therapeutics. Knowledge of the pilus

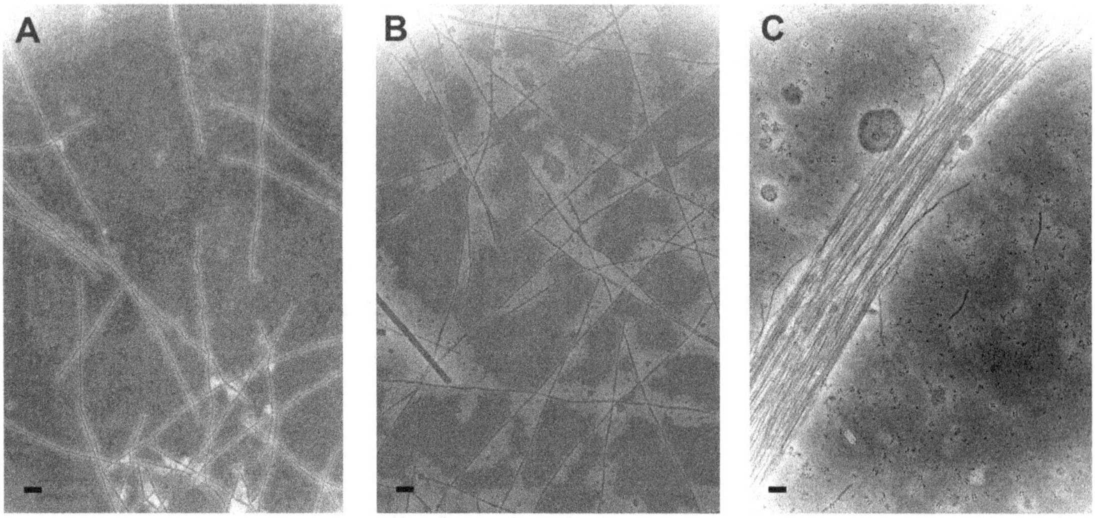

Figure 1.1 Negative-stain electron micrographs of Type IV pili. **A** *Neisseria gonorrhoeae* GC pili. **B** *Pseudomonas aeruginosa* PAK pili. **C** *Vibrio cholerae* toxin coregulated pili (TCP). Scale bar is 50 nm.

structure, assembly mechanism and functions can be exploited to design vaccine components and drugs that bind the pili and disarm them or, alternatively, use them to gain entry into the bacterial cells.

High-resolution structure determination of T4P is a daunting task. The filaments themselves are too large to be studied by nuclear magnetic resonance spectroscopy (NMR) and too flexible and heterogeneous in length to grow in regular crystal arrays for X-ray crystallographic structure determination. A common approach to structural analysis of large macromolecular complexes is to examine the smaller components or subunits of the complex, whose sizes are more amenable to X-ray crystallography and NMR methods. However, as is the case with many such complexes, the pilin subunits are not soluble as monomers. In fact, the Type IV pilins appear to have two stable states: one being anchored in the bacterial inner membrane, which acts as a molecular chaperone for the pilin subunits during the transition from newly translated protein to a polymeric filament; and the other being the polymeric state of the pilus filament. Thus, Type IV pilins can be considered to be both membrane proteins and fibre-forming proteins, both of which present problems for structure determination. Two strategies have been used successfully to obtain monomeric pilin: (i) dissociation of pilus filaments and solubilization of the pilin subunits in detergent; and (ii) recombinant expression of pilin subunits lacking the polymerization domain. The first Type IV pilin structure, GC pilin from *Neisseria gonorrhoeae*, was solved by X-ray crystallography by dissociating pilus filaments in detergent and crystallizing the subunits in the presence of detergent (Parge *et al.*, 1995). In this structure the N-terminal ~53 residue segment, which is conserved in all Type IV pilins (Fig. 1.2), forms an extended α-helix that looks like a handle for the less-conserved C-terminal globular domain (Fig. 1.3A). It was reasoned that the protruding half of the N-terminal α-helix could be deleted without affecting the structure of the globular domain, and would result in a soluble protein that would pack more readily into ordered crystals for X-ray crystallography and also be more amenable to NMR spectroscopy. Indeed this was the case, and high resolution crystal structures have been determined by X-ray crystallography for N-terminally truncated PilA from *Pseudomonas aeruginosa* strains K (PAK pilin) and K122–4 (Hazes *et al.*, 2000; Audette *et al.*, 2004) and TcpA from *Vibrio cholerae* (Craig *et al.*, 2003), and by NMR for K122–4 pilin, PilS from *Salmonella typhi* and BfpA from enteropathogenic *E. coli* (EPEC bundlin) (Keizer *et al.*, 2001; Xu *et al.*, 2004; Ramboarina *et al.*, 2005). In addition, a second full-length crystal structure has

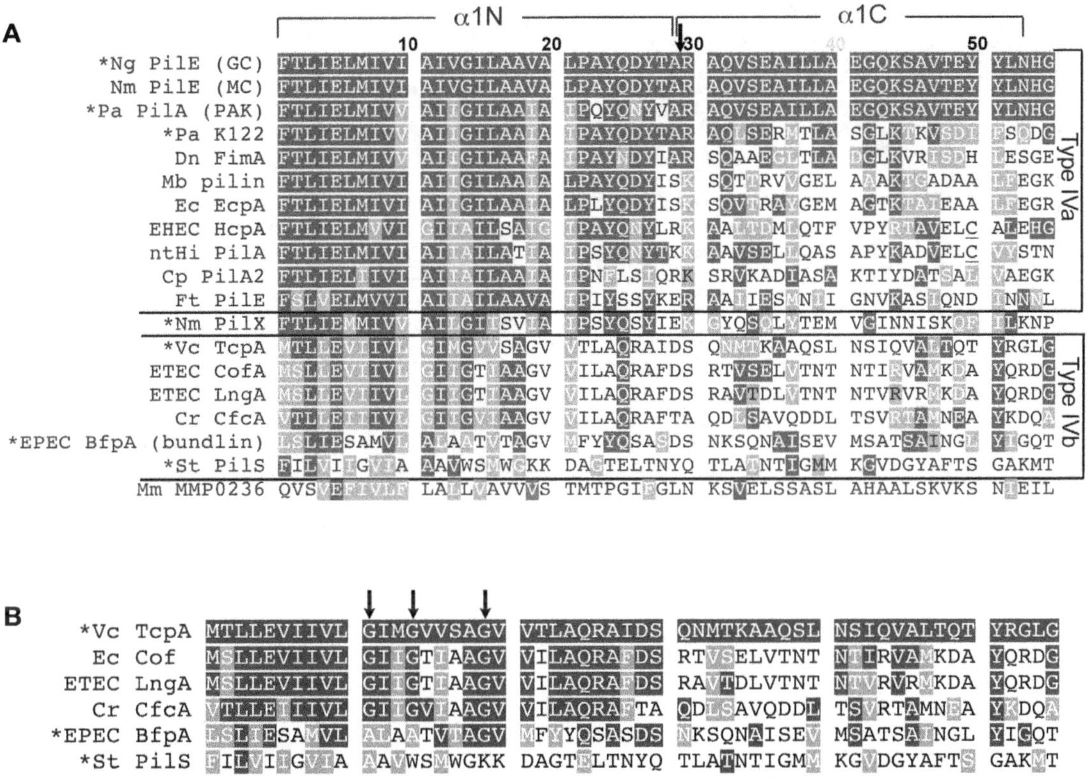

Figure 1.2 Amino acid sequence alignment of the conserved N-terminal 55 residues of Type IV pilins from bacterial pathogens. **A** Alignment of amino acid residues 1–55 of the mature Type IV pilin sequences to GC pilin from *N. gonorrhoeae*. Asterisk (*) indicates that the atomic structure is available. Dark shading indicates residues that are identical to GC pilin and light shading indicates residues that are homologous. Cysteines are underlined. Arrow indicates the approximate site of N-terminal truncation. This site varies for the different pilin structures. Ng, *N. gonorrhoeae* strain MS11 (Accession number P02974); Nm, *N. meningitidis* (CAA30557); Pa, *Pseudomonas aeruginosa* (PAK, P02973, K122–4, P17838); Dn, *Dichelobacter nodosus* (X52403); Mb, *Moraxella bovis* (AAA53087); Ec, *Eikenella corrodens* (P35645); EHEC, enterohaemorrhagic *E. coli* O157-H7 (AAG54412); ntHi, non-typeable *Haemophilus influenzae*; Cp, *Clostridium perfringens* (Gram-positive, NP_563200); Ft, *Francisella tularensis* Schu S4 FTT0889c, CAG45522), Vc, *Vibrio cholerae* (ABQ19609); ETEC, enterotoxigenic *E. coli* (CofA, BAA07174; LngA, AAC33154); Cr, *Citrobacter rodentium* (AAO17805); EPEC, enteropathogenic *E. coli* (Z68186); St, *Salmonella typhi* (AAF81215); Mm, *Methanococcus maripaludis* (45357799). MMP0236 is a putative Type IV pilin from the archaeal species *M. maripaludis*. Nm PilX is a minor pilin subunit. **B** Alignment of N-terminal mature sequence of the Type IVb pilins to *V. cholerae* TcpA. Arrows indicate conserved glycines.

been determined for PAK pilin using detergent solubilization (Craig *et al.*, 2003).

Cryo-electron microscopy (cryo-EM) coupled with three-dimensional image processing is rapidly becoming the method of choice for determining intermediate resolution structures of helical filaments such as T4P. Cryo-EM structures determined using classical Fourier-Bessel methods of helical reconstruction in some cases approach atomic resolution, seen for microtubules (8 Å) and bacterial flagella (4 Å) (Li *et al.*, 2002; Yonekura *et al.*, 2003). However, T4P structures have been intractable by these methods because of their narrow dimensions and featurelessness, and possibly because of heterogeneity in their helical symmetry. A new method of image processing, the Iterative Helical Real Space Reconstruction method has circumvented such problems in other helical filaments (Egelman, 2000), and produced the first intermediate resolution (12.5 Å) structure of a T4P, for *N. gonorrhoeae* (Craig *et al.*, 2006), as well as an archaeal pilus that has T4P-like properties (Wang *et al.*, 2008). This method promises to provide

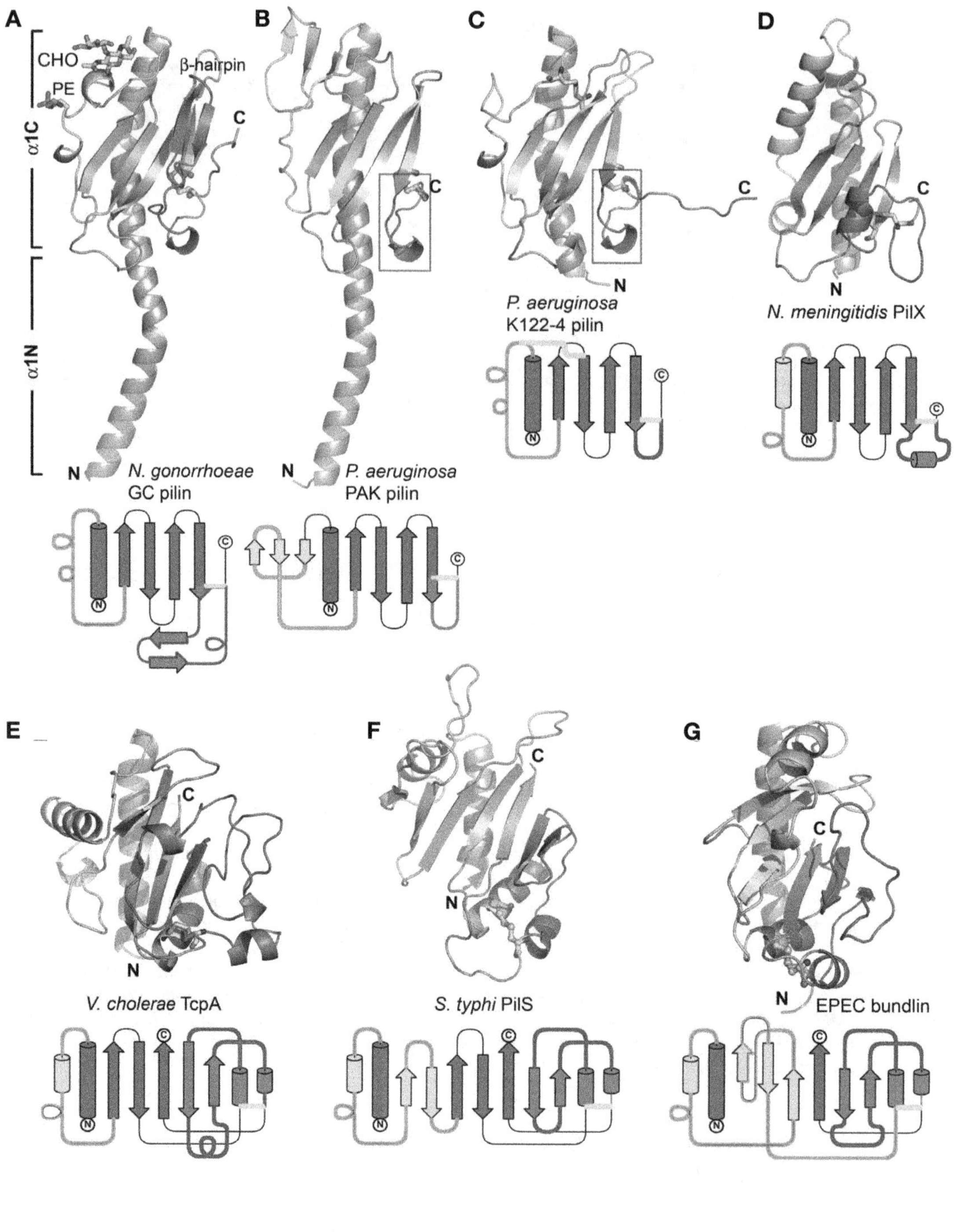

Figure 1.3 Atomic structures and secondary structure arrangement of Type IV pilins and the minor pilin PilX. Coloured figures are available online at http://www.sfu.ca/mbb/People/Craig/horizon-t4p-figs.html. The conserved α-helix/β-sheet core is coloured grey, and structurally variable αβ-loops and D-regions are coloured green and magenta, respectively. Cysteines are shown in ball-and stick representation with carbons in cyan and sulphurs in yellow. The posttranslational modifications are shown as sticks. **A** *N. gonorrhoeae* GC pilin (Craig *et al.*, 2006; Parge *et al.*, 1995). CHO is the *O*-linked disaccharide at Ser63 and PE is phosphoethanolamine at Ser68. **B** *P. aeruginosa* PAK pilin (Craig *et al.*, 2003). **C** *P. aeruginosa* K122–4 pilin (Audette *et al.*, 2004). **D** *N. meningitidis* minor pilin PilX (Helaine *et al.*, 2007). **E** *V. cholerae* TcpA (classical biotype) (Craig *et al.*, 2003). **F** *S. typhi* PilS (Xu *et al.*, 2004). **G** EPEC bundlin (Ramboarina *et al.*, 2005). Boxed regions of **B** and **C** are the *P. aeruginosa* receptor binding regions shown in close-up view in Fig. 1.6A and B, respectively. The structures shown in **C–G** lack the N-terminal 25–28 residues.

new and higher resolution T4P structures. Cryo-EM reconstructions can be used as molecular envelopes in which to fit high-resolution pilin subunit structures determined by X-ray crystallography and NMR to obtain 'pseudo-atomic' resolution structures of the pili. In addition to cryo-EM, mass spectrometry has emerged as a powerful method of probing pilus structure and identifying post-translational modifications, and epitope mapping continues to provide valuable information regarding surface exposure. Pilus structures and models determined by these integrated methods, taken together with a wealth of genetic, immunological, spectroscopic and other data, are revealing atomic-level details that can help to explain how pilin subunits assemble into thin and flexible but very strong filaments, how they perform their diverse functions, and how they might be targeted for antibacterial therapies.

The pilin subunit

The T4P were identified as a class long before their molecular structures were obtained, based on amino acid sequence homology in the N-terminal segment, particularly in the first ~25 residues (Fig. 1.2A) (Strom and Lory, 1993). This region comprises mostly hydrophobic residues, with the notable exception of a glutamic acid at residue 5, which is present in all but one Type IV pilin sequenced to date. The N-terminal residue is methylated at the terminal nitrogen (Pasloske and Paranchych, 1988; Lory and Strom, 1997). The C-terminal segments are less similar, but all of the Type IV pilins from Gram-negative bacteria have a pair of disulphide-bridged cysteines in this region. Several Type IV pilins also have a second pair of cysteines (Hobbs *et al.*, 1991; Gil *et al.*, 2004), yet the newly identified Type IV pilin, PilA2, of the Gram-positive *Clostridium perfringens* has none (Varga *et al.*, 2006). The Type IV pilins are further grouped into two sub-classes, Type IVa and IVb, based on the length of their leader sequence, the length of the mature sequence, and the identity of the N-terminal residue. The N-terminal segments of the Type IVb pilins are more similar to each other than to the Type IVa pilins (Fig. 1.2B) (Strom and Lory, 1993). The Type IVa pilins have a 6- to 8-residue leader peptide, a ~144- to 160-residue mature sequence, and an N-terminal phenylalanine. These pilins are found in a broad range of Gram-negative bacteria with disparate host and tissue specificity. The Type IVb pilins have an unusually long (10 to 30 residues) leader sequence, a mature sequence of 170–208 amino acids and the N-terminal residue varies but typically contains a bulky hydrophobic side chain. The Type IVb pilins are thus far found only in Gram-negative bacteria that infect the gut. *C. perfringens* PilA2 has an intermediate leader sequence length (12 residues) and mature sequence (170 residues) and thus does not fall clearly into either Type IV pilin sub-class.

The N-terminal ~53 residues assume an extended α-helix, α1, in the full-length GC and PAK pilin structures (Fig. 1.3A and B) (Parge *et al.*, 1995; Craig *et al.*, 2003; Craig *et al.*, 2006). The N-terminal half of α1, α1N, protrudes from the C-terminal globular domain. α1N is very hydrophobic and is likely to be flexible when not packed into a pilus filament, which explains in part why full-length Type IV pilins are so challenging to crystallize. In the N-terminally truncated pilin structures the C-terminal half of α1, α1C, is also α-helical. In fact, both a full-length and a truncated structure of PAK pilin are essentially identical in the globular domain, indicating that α1N does not influence the fold in this region (Hazes *et al.*, 2000; Craig *et al.*, 2003). α1N is predicted to be helical in all Type IV pilins, based on amino acid sequence homology with this region in GC and PAK pilin (Fig. 1.2A). α1 of GC and PAK pilin has a gently curving S-shape due to kinks induced by a proline at position 22 and a glycine or a proline at position 42. Pro22 is conserved in all of the Type IVa pilins, and most have a glycine or proline at position 42. Both proline and glycine can disrupt α-helical secondary structure. Residues at these positions are not conserved in the Type IVb pilins, but several have glycines at positions 11, 14 and 19 in the central segment of α1N, or elsewhere in α1 (Fig. 1.2B), which are expected to perturb the α-helical structure. The possible significance of this will be discussed below in the context of the filament structure.

Surprisingly, even though the amino acid sequence similarity is limited in many cases to the first ~50 residues, the overall structures of all Type IV pilins solved to date are remarkably similar. α1C forms the spine of the globular domain and interacts via hydrophobic side chains with a four- or five-stranded anti-parallel β-sheet. These segments are together referred to as the conserved structural core (Fig. 1.3). This conserved core is flanked by structurally variable regions: the αβ-loop between α1 and the β-sheet; and the D-region, which is delineated by the disulphide-bridged cysteines. The conserved structural core is present in both Type IVa and IVb pilins, yet these two sub-classes are topologically distinct. In the Type IVa pilins, the four-stranded β-sheet has continuous nearest-neighbour connectivity, with each strand folding back and hydrogen-bonding with the previous one in an anti-parallel arrangement. The D-region is formed between a cysteine at the C-terminus of the final β-strand, and a cysteine near the C-terminus of the protein. This topology is observed in all of the Type IVa pilin structures solved to date: GC pilin, PAK pilin and K122–4 pilin (Parge *et al.*, 1995; Hazes *et al.*, 2000; Craig *et al.*, 2003; Audette *et al.*, 2004), as well as the *N. meningitidis* minor pilin, PilX (Fig. 1.3A–D) (Helaine *et al.*, 2007).

The β-sheet connectivity is very different for *V. cholerae* TcpA, *S. typhi* PilS and EPEC bundlin (Craig *et al.*, 2003; Xu *et al.*, 2004; Ramboarina *et al.*, 2005), which are all Type IVb pilins. The topology of the first few strands of β-sheet differs among these pilins, two of which have additional strands inserted before the core β-sheet, but the C-terminal half of the core β-sheet and the D-regions are all very similar (Fig. 1.3E–G). The polypeptide chain exits the β-sheet after the first few strands and forms an α-helix that runs under the β-sheet and roughly parallel to it. The chain then returns to form the terminal two strands of the β-sheet, followed by a short α-helix that lies on the periphery of the globular domain, and finally terminates as the central strand of the β-sheet. Most of the polypeptide chain between the first few strands and the central strand of the β-sheet comprises the D-region, and the D-region includes the two terminal strands of the β-sheet. The two D-region α-helices, which are unique to the Type IVb pilins, are covalently linked by the disulphide bond, and associate with the D-region β-strands via hydrophobic side chain interactions. Thus, for the Type IVb pilins, the D-region is complex and integral to the β-sheet, whereas in the Type IVa pilins the D-region is adjacent to the β-sheet and connects the C-terminal loop to the β-sheet.

Considerable length and structural variations are observed among the αβ-loops of the different Type IV pilins. Of the three Type IVa pilin structures available, plus the minor pilin PilX, there are three distinct αβ-loop structures, although they are all comparable in their bulk and their orientation to the conserved α-helix/β-sheet structural core (Fig. 1.3A–D). GC and K122–4 pilin αβ-loops lack regular secondary structure, while PAK pilin has three short β-strands and PilX has an α-helix. The αβ-loop of GC pilin contains two post-translational modifications, a disaccharide at Ser63 and a phosphate moiety at Ser68 (Parge *et al.*, 1995; Hegge *et al.*, 2004; Craig *et al.*, 2006; Aas *et al.*, 2007), and in K122–4 pilin, the αβ-loop is disulphide-bonded to strand 3 of the β-sheet. The αβ-loops of the Type IVb pilins also differ substantially from one another and from the Type IVa pilins. All start with an irregular stretch of polypeptide followed by an α-helix with a varying number of turns and then another segment that varies in length and structure prior to joining the core β-sheet (Fig. 1.3E–G). In TcpA, the second irregular segment of the αβ-loop feeds directly into the first strand of the core β-sheet, whereas in PilS this segment forms two short antiparallel β-strands that precede but are continuous with the core β-sheet. In bundlin, which is the most atypical of the Type IV pilins, this segment forms three β-strands with non-nearest-neighbour connectivity. These strands replace the first two continuous β-strands seen in the core β-sheet of the other Type IV pilins. The αβ-loops of all the Type IV pilins, regardless of their structure, contact α1C via side chain interactions, and provide each pilin with a unique surface at one edge of the globular domain.

As discussed above, the D-regions of the Type IVa pilins are distinct from those of the Type IVb pilins, and within each sub-class structural variation occurs in the form of 'structural insertions'. In the simplest case of the *Pseudomonas* pilins, the D-region forms a 12-residue double turn structure (Fig. 1.3B and C). GC pilin has a similar structure but with a β-hairpin inserted prior to the turns, and PilX has a short α-helix followed by a short β-hairpin turn (Fig. 1.3A and D). These structural insertions lie on top of the core β-sheet and protrude from the surface of the globular domain. The D-regions of the Type IVb pilins are structurally similar to each other, differing mainly in the size of the irregular loop between the fourth and fifth β-strands (Fig. 1.3E and F). Thus, larger loops seen in TcpA and bundlin could be considered structural insertions for the Type IVb D-regions. These loops also lie across the D-region β-strands and protrude from the pilin surface, whereas the D-region α-helices lie beneath these strands. So the D-region of both Type IVa and IVb pilins forms a second structurally variable edge of the globular domain, which, together with the αβ-loop, flanks the conserved α-helix/β-sheet core. Structural analyses of intact pilus filaments in combination with functional studies suggest that the conserved structural core directs pilus assembly, mainly via α1N, whereas the flanking αβ-loop and D-region are exposed on the pilus surface and define the diverse pilus functions.

The pilus filament

The N-terminal segment of the pilin subunit was predicted to be the polymerization domain based on its amino acid sequence conservation and hydrophobicity (Watts *et al.*, 1982; Schoolnik *et al.*, 1984). This hypothesis was supported by the first Type IV pilin structure for *N. gonorrhoeae*, which showed that the N-terminus is an extended α-helix (Parge *et al.*, 1995). Based on the GC pilin structure and fibre diffraction data for *P. aeruginosa* PAK pilin (Watts *et al.*, 1982), an attractive computational model was proposed for pilus assembly, whereby the N-terminal α-helices formed a spiralling bundle in the filament core and the β-sheets formed a continuous β-ribbon that wound up the filament axis. Further support for this model came with the crystal structure of N-terminally truncated PAK pilin, which shared the α-helix/β-sheet core and could potentially form a filament like the GC pilus model (Hazes *et al.*, 2000). Next came the *V. cholerae* TcpA structure, which also had the conserved core α-helix/β-sheet core, albeit with a different topology (Craig *et al.*, 2003). However, the bulky αβ-loop and D-region meant that TcpA could not assemble according to the GC pilus model without significant steric clashes. A model for the *V. cholerae* toxin-coregulated pilus (TCP) was proposed based on the filament dimensions and EM-derived symmetry data, and on the helical arrangement of N-terminally truncated TcpA subunits in the crystal lattice. Subsequently, new pilin structures were determined and new filament models were proposed, but the most definitive model was obtained for GC pili by cryo-EM and three-dimensional image reconstruction (Craig *et al.*, 2006). This structure confirmed that the N-terminal α-helices do indeed cement the subunits together, but that the globular domain interactions are minimal. These findings were supported by a hydrogen/deuterium exchange mass spectrometry (DXMS) study of the *V. cholerae* TCP, which showed that the N-terminal α-helices are buried in the TCP filament by subunit–subunit contacts, but that the globular domains are in fact very accessible to solvent and thus do not contribute substantially to the structural stability of the pili (Li *et al.*, 2008).

Given the homology in the N-terminus, the conserved nature of the pilin subunit structure, and the morphological similarity of the T4P, one might expect them all to share the same architecture and assembly mechanism. However, the available structures and models of T4P indicate differing helical symmetries, which may suggest different assembly mechanisms. To navigate through these seemingly contradictory data, it is useful to arm oneself with a basic understanding of helical symmetry. In a helically symmetrical filament, subunits are typically arranged along a one-start or primitive helix, with each subunit positioned equidistant from the filament axis, and related to its neighbouring subunits by two 'symmetry operations': an axial translation (rise) along the filament axis and an azimuthal rotation about the filament axis (Fig. 1.4A). This means that all subunits in the filament are connected by a single path that twists along the filament axis. In addition to the one-start helix, other paths can be followed, defining, for example, a three-start helix, a four-start helix, etc. (Fig. 1.4B). The handedness of the starts is defined in the same way that the handedness of an α-helix is defined: hold your right hand with the thumb pointing up along the filament axis and your fingers curved to represent the helical start(s). If the path of the subunits in a given start twists upward in the direction your fingers point, this is a right-handed helix; if they

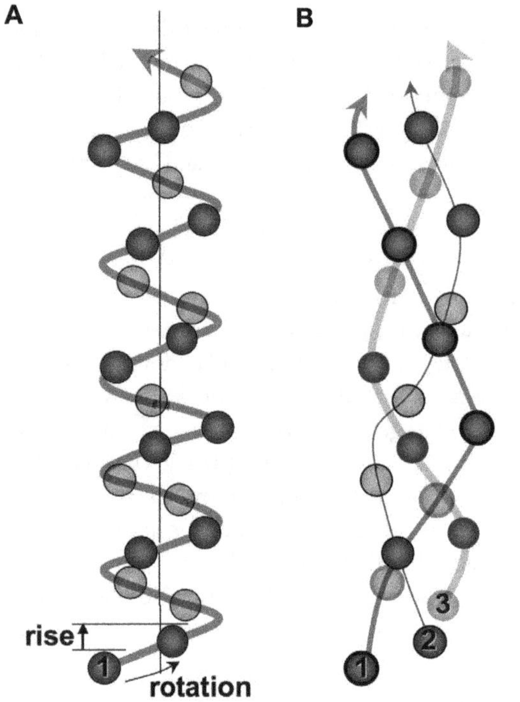

Figure 1.4 Relationship between subunits in filaments with helical symmetry. **A** Subunits are represented as spheres connected by a right-handed one-start helix. Each subunit is related to its neighbouring subunits by an axial rise and an azimuthal rotation. **B** The same subunits are also connected by a left-handed three-start helix. Other paths or helical starts can also be envisioned.

twist in the opposite direction it is left-handed. Fig. 1.4A shows a right-handed one-start helix and 1.4B shows a left-handed three-start helix for the same helical filament. In fact, there are an infinite number of helical starts, but only a few will be evident (or meaningful) for a given filament. Information on the helical symmetry of a filament can be obtained, in some cases, by calculating a Fourier transform of a digitized EM image of the filament, or from X-ray fibre diffraction data.

The GC pilus cryo-EM reconstruction provided sufficient detail to unambiguously dock in the X-ray crystal structure of the pilin subunit. A 'pseudo-atomic' resolution structure of the filament was then generated by applying the experimentally derived symmetry operators for the one-start helix, a rise of 10.5 Å and a rotation of 100.8°, to the docked pilin subunit (Fig. 1.5A and B). This docking identified the one-start helix as right-handed, but the dominant helix is a left-handed three-start helix. The designation of helical starts is essential in determining a filament structure by image processing methods, but it is also useful for describing filament architecture and contacts among subunits. Subunits in the GC pilus filament are much more loosely packed than was first predicted. They are held together primarily by hydrophobic interactions among the N-terminal α-helices, which form a hydrophobic core in the centre of the filament (Fig. 1.5C). The resolution of the reconstruction (12.5 Å) is not high enough to precisely position the N-terminal α-helices, or to visualize any conformational changes that might occur to this flexible region, so the subunits were fit into the reconstruction as rigid bodies, and only minor refinements of the positions of the N-terminal four residues were performed to eliminate steric clashes. The single charged residue in α1N, Glu5, is oriented in a position to potentially neutralize the positively charged N-terminal phenylalanine of the adjacent subunit in the one-start helix (Fig. 1.5D). Glu5 is conserved among all Type IV pilins, with the exception of PilS, and mutating this residue results in disruption of pilus expression (Pasloske *et al.*, 1989), suggesting that this charge neutralization in the otherwise hydrophobic filament core is critical for pilus assembly. In contrast to the tight packing of the N-terminal α-helices, the globular domains are only loosely packed on the filament surface, leaving grooves between the subunits. This arrangement gives the GC pilus filament a highly corrugated appearance.

Since α1N is the GC pilus polymerization domain and this segment is conserved among the Type IV pilins, it might be expected that all T4P would thus share the GC pilus architecture. The conserved Glu5 would be precisely positioned to neutralize the positively charged N-terminus of an adjacent subunit in the right-handed one-start helix, thereby locking in the register of the subunits and generating filaments with identical symmetry. Although there are no cryo-EM reconstructions available for the other T4P, structure-based pilus models display helical symmetries that differ from that of GC pili. The most well-studied in this respect is *V. cholerae* TCP, which is modeled based on the TcpA crystal structure and crystal packing, EM-derived symmetry data and solvent accessibility as determined by DXMS (Li *et al.*, 2008). Like GC pili, the TCP model has

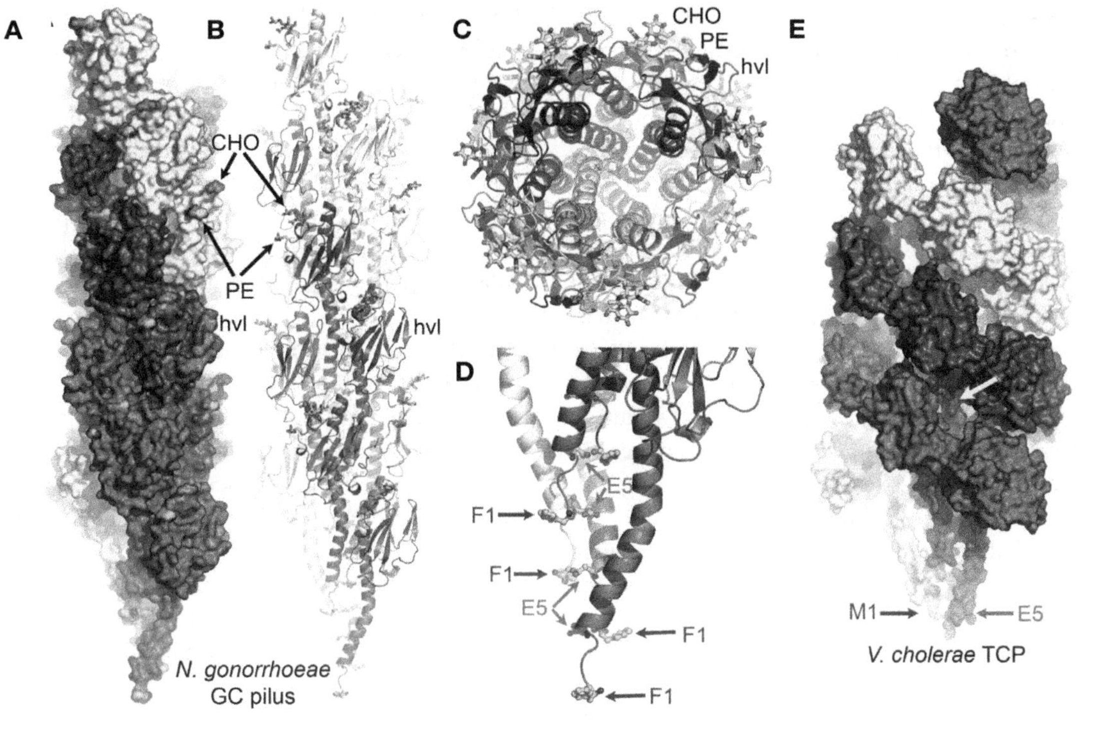

Figure 1.5 Cryo-EM reconstruction of *N. gonorrhoeae* GC pilus and structure-based model of the *V. cholerae* TCP. Coloured figures are available online at http://www.sfu.ca/mbb/People/Craig/horizon-t4p-figs.html. Subunits are coloured corresponding to a left-handed three-start helix. GC pilus reconstruction at 12.5 Å resolution shown in side view as **A** a surface representation and **B** a cartoon, and **C** as an end view. **D** Close-up view of the N-termini. The disaccharide (CHO) and phosphoethanolamine (PE) are indicated. The hypervariable loop (hvl) is coloured magenta for one central subunit in **A** and **B**. The positively charged N-terminal phenylalanine (F1) and negatively charged glutamate (E5) are coloured green at the tips of the N-terminal segments. These residues are predicted to neutralize each other in the otherwise hydrophobic core of the filament. **E** Surface representation of TCP showing the putative interaction between the N-terminal methionine (M1) and Glu5 (E5). The white arrow indicates the central segment of α1N that is exposed in the gaps or cavities between the loosely packed globular domains.

a dominant left-handed 3-start helix (Fig. 1.5E), and Glu5 is positioned to neutralize the Met1 N-terminus in the one-start helix. However, an important difference between the TCP model and the GC reconstruction is the hand of the one-start helix: TCP has a left-handed one-start helix whereas GC has a right-handed one. The symmetry operation for the TCP one-start helix is also substantially different, having a rise of 7.5 Å and a rotation of 140°, compared with a 10.5 Å rise and a 100.8° rotation for GC pili. Since the T4P most likely share a common assembly mechanism, with subunits added to the growing filament one at a time, it is surprising that they would be added in one direction (i.e. right-handed) for GC pili and in the opposite direction (left-handed) for TCP. However, the TCP model would require only very small shifts in subunit positions to attain the same symmetry as that of GC pili.

On the other hand, the new 16-Å cryo-EM reconstruction of a putative T4P from the archaeal species *Methanococcus maripaludis* looks dramatically different from the GC pilus reconstruction and other T4P models (Wang *et al.*, 2008), having a hollow lumen much like that seen for the *Salmonella* flagella (Yonekura *et al.*, 2003). Furthermore, these pili are heterogeneous with respect to their helical symmetry. Two packing arrangements were observed, one with one-start helical symmetry with a subunit rise of ~12 Å and a rotation of ~221°, and the other with fourfold rotational symmetry, having a rise of ~43.8 Å and a variable rotation. The putative *M. maripaludis* pilin subunit (MMP0236) has a hydrophobic N-terminus and a glutamic acid at residue 5 (Fig. 1.2A), which are typical for Type IV pilins, but the amino acid sequence similarity is not extensive and the C-terminus does not have the conserved cysteines seen in most Type IV pilins. These differences may mean that the subunit fold of the archaeal pilins differs from that of the bacterial Type IV pilins, which would help to explain apparent differences in pilus architecture. Clearly, new and higher-resolution structures for other T4P filaments, as well as high-resolution structures of full-length pilin subunits, are needed to determine the extent of variability among T4P architectures and assembly mechanisms.

While questions remain regarding the precise symmetry of the *V. cholerae* TCP filament, the current model is supported by biological data and enables us to visualize important findings from the DXMS analysis (Li *et al.*, 2008). Most of the N-terminal α-helix is poorly labelled by deuterium, consistent with this region being buried in the filament core. In contrast, much of the globular domain is well labelled and thus solvent-accessible, including a portion of the αβ-loop and much of the D-region. This means that these two regions do not participate substantially in subunit–subunit interactions and are instead exposed on the filament surface. Surprisingly, a segment of α1N, from residue ~13–23, is readily labelled by deuterium and is thus solvent-exposed. In the TCP model this segment is exposed between globular domains of neighbouring subunits (Fig. 1.5E). This was unexpected given the degree of hydrophobicity in the N-terminus. However, the central segment of α1N for TcpA and the other Type IVb pilins is actually amphipathic, and has three conserved glycines that might perturb the α-helix. The glycines, which lack side chains, may also allow better access of the deuterium label to the amide nitrogens in this segment. Since the amino acid sequence of this segment is conserved in other Type IVb pilins (Fig. 1.2B) it may be exposed in these pili as well. The DXMS data, together with the GC pilus reconstruction support a generalized T4P model where subunits are held together primarily by hydrophobic interactions among their N-terminal α-helices, leaving the globular domains only loosely connected on the filament surface with gaps and grooves running between them. The extensive contacts among the N-terminal segments would provide mechanical strength for the filaments, and the loose packing of the globular domains would allow for flexibility. This architecture would also maximize the exposure of the structurally variable αβ-loops and D-regions to impart distinct functional characteristics on pili from different bacterial species.

Functional implications of pilus structure

As mentioned in the introduction, the T4P perform a remarkable array of functions, and these functions, which are particularly important in bacterial pathogenesis, vary depending on the organism on which the pili are present (Craig *et al.*, 2004). In *N. gonorrhoeae*, the T4P mediate twitching motility, adhesion to host cells, immune escape, microcolony formation and DNA uptake, among other tasks. In *P. aeruginosa*, the pili also induce twitching motility and host cell adhesion, and have recently been shown to bind DNA (van Schaik *et al.*, 2005), although the biological significance of this observation is not clear. The *V. cholerae* TCP are required for microcolony formation and serve as the high-affinity receptor for the cholera toxin phage, which is responsible for imparting toxicity on this *Vibrio* species. In addition, the TCP apparatus serves as a secretion system for an essential colonization factor, TcpF (Kirn *et al.*, 2003). Many of these pilus-related functions have been localized to specific sites on the pilin subunits within the D-region and, to a lesser extent, the αβ-loop.

Neisseria gonorrhoeae Type IV pili

N. gonorrhoeae causes the sexually transmitted disease gonorrhoea. A fascinating feature of *N. gonorrhoeae* is its ability to alter the amino acid sequence of the pilin subunit from one cell generation to the next. This means that antibodies raised to GC pili of one bacterial cell generation may not recognize pili from subsequent generations, thereby allowing *N. gonorrhoeae* to escape immune surveillance. Many sites on GC pilin undergo antigenic variation but the segment spanning residues 128–141, which is located on the β-hairpin of the D-region, is hypervariable (hvl, Figs. 1.3A and 1.5A–C). Forest and colleagues replaced this β-hairpin loop with a short linker sequence, and expressed this construct in *E. coli* as an N-terminal truncation (Hansen *et al.*, 2007). They showed that this soluble pilin, which also lacks the disaccharide and phosphoryl post-translational modifications in the αβ-loop, did not elicit a strong T-cell-dependent B-cell response in mice, whereas an analogous construct containing the β-hairpin did. This suggests that the hypervariable loop is immunodominant, even in the absence of the quaternary pilus structure. The β-hairpin and loop that house the hypervariable loop lie on a ridge in the D-region of the globular domain that protrudes from the pilus surface and would be more accessible to antibodies than more conserved regions of the subunit (Fig. 1.5A–C). Thus, antigenic variation of this protrusion would be an effective means for the bacteria to escape a protective host immune response.

The αβ-loop of the pilin subunit represents a second dominant feature on the pilus filament, in part due to the presence of two post-translational modifications: an *O*-linked disaccharide at Ser63 and a phosphate moiety at Ser68. The 2.3-Å electron density map of GC pilin from *N. gonorrhoeae* strain MS11 showed electron density at Ser63 (Craig *et al.*, 2006) consistent with a disaccharide identified by mass spectrometry (MS) for the same strain: 2,4-diacetamido-2,4,6-trideoxyhexose sugar linked to a distal hexose (Hegge *et al.*, 2004; Aas *et al.*, 2007). This disaccharide was identified as α-D-galactopyranosyl-(1→3)-2,4-diacetamido-2,4-dideoxy-β-D-glucopyranoside in the crystal structure. Electron density at the glucose C6 position suggested a 2,4-dideoxyglucose rather than the 2,4,6-trideoxy species, but this extra hydroxyl group was not observed by MS analysis. The trideoxyhexose was also demonstrated for a trisaccharide bound in this same αβ-loop region of *N. meningitidis* pilin (Stimson *et al.*, 1995). Recently, *N. meningitidis* was shown to possess an alternative, previously unrecognized trideoxyhexose, glyceramidoacetamidotrideoxyhexose (Chamot-Rooke *et al.*, 2007). This new sugar is the product of an altered allele of the *pglB* gene, *pglB2*, which results in a glyceramido group replacing one acetamido group on the proximal sugar. The *pglB2* allele is present in approximately half of the *N. meningitidis* clinical isolates tested (Power *et al.*, 2003). Thus, while it seems that the nature of the *O*-linked sugar in the D-region can vary for the *Neisseria* pilins, the presence of the extra electron density seen at the glucose C6 position of in the GC pilin crystal structure remains a mystery. In addition to the sugar moiety at residue 63, GC pilin possesses a covalently attached phosphoethanolamine or phosphocholine at Ser68 (Hegge *et al.*, 2004; Craig *et al.*, 2006). Both the glycan and phosphate modifications are analogous to modifications of bacterial LPS, but their physiological significance is not understood. Like the hypervariable loop in the D-region, these modifications are prominently displayed on the pilus surface (Fig. 1.5A–C). The exposed disaccharide might mimic host glycans and thus not be recognized as a foreign antigen to the immune system. GC pili are displayed peritrichously on the neisserial cell surface and should be readily recognized by host defenses. However, by exposing epitopes on the pili that mimic host molecules together with those that are hypervariable, the neisserial cells would escape detection by the immune system.

An important step in host colonization is pilus-mediated aggregation of the neisserial cells, also called microcolony formation. For *N. meningitidis* this process requires the presence of a pilin-like protein, PilX, a minor component of the T4P apparatus that shares N-terminal sequence homology with the Type IV pilins (Helaine *et al.*, 2005). Immuno-gold labelling studies showed that PilX is distributed sparsely throughout the pilus filaments. The X-ray crystal structure of N-terminally truncated PilX revealed a Type IVa pilin topology (Fig. 1.3D), which suggested that PilX may be incorporated into the pilus filament as if it were a pilin subunit (Helaine *et al.*, 2007). The D-region of PilX contains a short α-helix 'pigtail' followed by a hairpin turn or 'hook' that sits on the core β-sheet and would protrude from the pilus surface, as does the hypervariable loop/β-hairpin of GC pilin. The pigtail was found to be critical for microcolony formation, and hence for adhesion to host cells. The authors proposed that PilX subunits are incorporated into the pilus filaments and form interaction sites to bring pili together, and that these interactions stabilize the pili and counteract the effects of the PilT retraction motor. The T4P of *N. meningitidis*, like *N. gonorrhoeae*, are hypervariable, which may explain why these pili employ a minor pilin with a non-variable sequence to facilitate pilus–pilus interactions. It is notable, however, that an analogous minor pilin has not been identified for *N. gonorrhoeae*, which has hypervariable pili and would presumably also require non-variable proteins to form stable aggregates.

Pseudomonas aeruginosa Type IV pili

Type IV pili are important in establishing lung infections for the opportunistic pathogen *P. aeruginosa*. Laboratory strains of *P. aeruginosa* pili bind to the glycosphingolipids asialo-GM1 and asialo-GM2 on host epithelial surfaces. The receptor binding region was localized to the C-terminal loop of the pilin D-region (Fig. 1.6A and B) (Sheth *et al.*, 1994). Surprisingly this region is not well conserved among *P. aeruginosa* strains, even though they bind the same receptor. Furthermore, antibodies raised against the C-terminus of PAK pilin cross-react with pili from related *P. aeruginosa* strains (Doig *et al.*, 1990). The 13-residue D-regions of both PAK and K122–4 pilins adopt tight loops containing two consecutive turns (Figs 1.3 and 1.6) (Hazes *et al.*, 2000; Audette *et al.*, 2004). Many of the side chains are oriented away from the loop, exposing mostly main chain atoms at the globular domain surface. Thus, it was proposed that both antibodies and host cell receptors bind primarily to backbone residues of the constrained D-region epitope rather than to side chains. The most C-terminal segment, residues 143–144 for PAK pilin and 143–150 for K122–3, appear to obscure the D-region in the crystal structures, but the NMR structure of K122–4 pilin indicates that this segment is highly flexible and may not interfere with the receptor interaction (Keizer *et al.*, 2001). Antibodies to the D-region of PAK pili, as well as gold-conjugated asialo-GM1, bind to the tips of the pili but not along the pilus length, suggesting that the D-region is only exposed at the filament tip (Lee *et al.*, 1994). *P. aeruginosa* pilus models have been proposed based on the atomic structures of the pilins and fibre diffraction data, but none of these appear to expose the D-region sufficiently to allow antibody to bind (Keizer *et al.*, 2001; Craig *et al.*, 2004). A cryo-EM structure of a *P. aeruginosa* pilus would help us to better understand the interaction between the D-region and host cells receptors and antibodies.

Vibrio cholerae TCP

The *V. cholerae* TCP are essential for colonization of the human small intestine, which can lead to the severe diarrhoeal disease cholera.

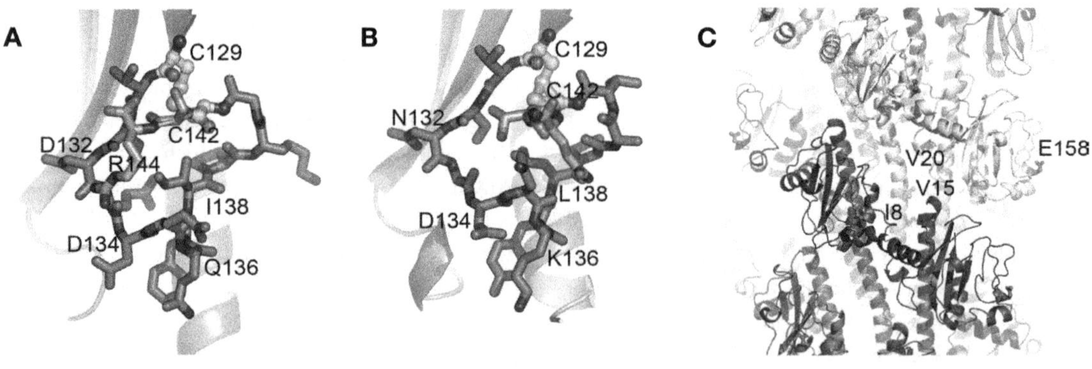

Figure 1.6 Functional regions of the Type IV pili. Coloured figures are available online at http://www.sfu.ca/mbb/People/Craig/horizon-t4p-figs.html. Close-ups of the receptor-binding regions of *P. aeruginosa* **A** PAK pili and **B** K122–4 showing the double-turn structure that exposes main chain atoms but orients the side chains away from the antibody or receptor binding sites. Pilin subunits are coloured as in Fig. 1.3. The C-terminal two residues of PAK pilin are shown in grey coming out of the page in **A**, whereas only the first two of the eight residues of the C-terminus (beyond the D-region) of K122–4 are shown in **B**. **C** Close-up view of the TCP model showing the gap between globular domains that exposes N-terminal residues ~13–23. Residue Glu158 is shown in magenta as a ball-and-stick representation. Glu158 is important for pilus–pilus interaction. Individual mutations at residues I8, V15 and V20 (green) on the exposed α1N segment can rescue a colonization defect in an E158L mutant.

TCP do not appear to interact directly with host cells, but instead bind to each other to hold the cells together in microcolonies. Microcolonies protect the bacteria from host defenses and help to concentrate the secreted cholera toxin, which is responsible for the diarrhoeal symptoms. Unlike *Neisseria* pili, TCP do not undergo antigenic variation. The D-region of the TCP subunit TcpA has been implicated in microcolony formation: mutational analysis showed that a number of charged residues in the D-region are involved in pilus–pilus interactions (Kirn *et al.*, 2000), and passive immunization studies in mice showed that antibodies directed to D-region peptides are protective (Sun *et al.*, 1997). The TcpA D-region, like that of GC pilin, protrudes from the pilus surface, as shown by DXMS analysis (Li *et al.*, 2008), and could potentially intercalate into cavities between globular domains in adjacent TCP filaments. This inter-digitation between D-region bulges and cavities along the filament length would provide an extensive interaction surface to hold pili and hence *Vibrio* cells together. In support of this model, a D-region mutation, E158L, which abrogates pilus bundling, microcolony formation and colonization of the infant mouse, was rescued by three individual secondary mutations (Kirn *et al.*, 2000). Surprisingly, all of these suppressor mutations were located in or near the exposed region of α1N: I8V, V15G and V20G (Fig. 1.6C). This suggests that the D-region bulge of one filament may penetrate the cavity between globular domains of an adjacent filament deep enough to come in contact with the exposed N-terminal side chains. Mutations of large bulky valines to glycine, which lacks a side chain, may allow the bulky E158L mutation on the D-region to fit better into the cavity between the globular domains. Although leucine is no bulkier than glutamic acid, the loss of a negative charge at this site might decrease complementarity between the D-region bulge of one filament and the cavity of an adjacent one, and secondary mutation of a valine to a glycine in α1N would make more room in the cavity. Thus, the conserved D-region of TcpA seems to act like the corresponding region in PilX of *N. meningitidis*, forming a key interface for pilus–pilus interactions. TCP are not known to be retractile, so extensive interactions all along the filament surface may be beneficial in establishing tightly associated *V. cholerae* microcolonies. In contrast, the irregular distribution of PilX molecules on the *N. meningitidis* pilus filament would provide a looser interaction among these pili, sufficient to allow bacterial adhesion to host cells, but not be so tight that the pili cannot eventually be retracted, perhaps as a mechanism to disperse the bacteria and spread infection.

Structural analyses of both *N. gonorrhoeae* GC pili and *V. cholerae* TCP reveal highly corrugated surfaces with grooves and cavities, and protruding functional regions. This was somewhat of a surprise, as these filaments look relatively featureless by negative-stain EM. However, the loose packing of the globular domains on the filament surface is consistent with the flexibility observed for these and other T4P, as it would provide gaps for compression on one side of the filament as it bends. The grooves that run along the three-start helices of GC pili are positively charged and may provide non-specific binding sites for DNA during natural transformation. The extensive overlapping interactions among the hydrophobic N-terminal α-helices in the filament core would provide the tensile strength needed by the pili to overcome high stress forces encountered in the intestines, genital tract and airways. The C-terminal disulphide bonds present on the Type IV pilins would provide further filament stabilization by anchoring the protruding and functionally important D-regions to the conserved structural core and preventing stress-induced unfolding of the pilin subunits. A disulphide bond is also conserved in the P pilus subunits of uropathogenic *E. coli*. P pili have dramatically different subunit structures and assembly mechanisms, but are also thin and flexible and have to withstand high flow forces in the bladder and urethra (Sauer *et al.*, 1999). Furthermore, the sortase-assembled pili of Gram-positive bacteria have two intramolecular lysine-asparagine isopeptide bonds in their pilin subunit, and mutating a residue involved in bond formation results in pili that are much more susceptible to proteolysis (Kang *et al.*, 2007). Thus, the presence of intermolecular cross-links appears to be a common feature to stabilize these long and extremely thin adhesive filaments.

Type IV pilus assembly mechanism

What do these structures of Type IV pilin subunits and pilus filaments tell us about the pilus assembly mechanism? Pilus assembly involves a number of different proteins, including an inner membrane-bound pilin-specific signal peptidase, a cystoplasmic inner-membrane-associated nucleotide binding protein, an integral inner membrane binding protein of unknown function, an outer membrane secretin, and of course, the pilin subunit itself. Homologues to these proteins are present in the Type II secretion (T2S) machinery, which assembles trans-periplasmic 'pseudopili' to secrete toxins and other proteins across the Gram-negative bacterial outer membrane (Johnson *et al.*, 2006). The N-terminal segment of the T2S pseudopilins is homologous to α1N of the Type IV pilins, including the conserved glutamic acid at position 5. These pseudopilins also share the α-helix/β-sheet core topology of the Type IVa pilins, and have a structurally variable αβ-loop, but lack the conserved disulphide bond and hence the D-region (Kohler *et al.*, 2004; Korotkov and Hol, 2008; Yanez *et al.*, 2008a; Yanez *et al.*, 2008b). Thus, T4P and pseudopilus assembly may proceed via the same mechanism. Our working model for the T4P assembly mechanism is that the pilin subunits diffuse in the plane of the inner membrane, anchored by α1N, and are drawn to a growing pilus filament by electrostatic attractions between their negatively charged Glu5 and the positively charged N-terminal residue in the terminal pilin of the filament (Craig *et al.*, 2006; Li *et al.*, 2008). ATP hydrolysis by the hexameric nucleotide binding protein on the cytoplasmic side of the inned membrane results in a conformational change that may be transmitted to the pilus filament via the integral membrane protein, thereby extruding the filament outward a short distance (~10 Å) (Yamagata and Tainer, 2007). This motion would position the newly added terminal residue such that it can now attract a new pilin subunit via complementarity between its positively charged N-terminus and the negative Glu5 of the incoming subunit, both of which are embedded in the hydrophobic lipid bilayer. A 25 Å resolution negative stain EM reconstruction for the *Klebsiella oxytoca* pseudopilus supports this pilus assembly model, having a one-start helix with a shift of ~10.3 Å/subunit (Kohler *et al.*, 2004), similar to that of the GC pilus cryo-EM reconstruction. Our assembly model would require that each subunit be added sequentially along the one-start helix, and it is expected that this helical start would have the same hand in all T4P and pseudopili. However, both the *K. oxytoca* pseudopilus model and the current *V. cholerae* TCP model have left-handed one-start helices, whereas the GC pilus one-start helix is right-handed. New higher resolution cryo-EM reconstructions of these and other pili and pseudopili should help to determine whether or not they indeed share a similar symmetry, architecture and assembly mechanism. The commonalities between the T4P and the T2S systems extend to their functions as well, since the T4P apparatus acts as a secretion system in some bacteria. *V. cholerae* TCP secrete a soluble protein, TcpF, which is essential for efficient colonization of the infant mouse, but its function is not known (Kirn and Taylor, 2005). In addition, *Francisella tularensis* and *Dichelobacter nodosus* T4P systems secrete proteases and other hydrolytic enzymes (Hager *et al.*, 2006; Han *et al.*, 2007). Thus, structural information obtained for components of both the T4P system and the T2S system are likely to be directly relevant to each other.

Conclusions

The structural information that is accruing for the T4P has provided insights into their complex assembly mechanism, explained their impressive mechanical properties, and suggested how they might perform some of their diverse functions in bacterial survival and pathogenesis. The conserved α-helical N-terminus mediates filament formation by forming helical arrays in the pilus core, and these extended, mostly hydrophobic, interactions impart tensile strength on the filament. The N-terminal α-helix, together with the β-sheet of the globular domain, acts as a structural scaffold on which to display the structurally and functionally variable αβ-loop and D-region on the pilus surface. These regions define the shape and chemistry of the filament surface and interact with other extracellular factors, such as pili, cell surface receptors and DNA, to carry out key pilus functions. The loose packing of the globular domains on the filament surface optimizes the exposure of these regions and allows for filament flexibility. Commonalities in the amino acid sequences and structures of the pilin subunits, and in pilus morphology and molecular models, suggest that all T4P possess the same architecture and assembly mechanism, but further study is required to determine the level to which these similarities extend.

Future trends

While there are a number of high-resolution structures of N-terminally truncated Type IV pilins and pseudopilins, there are only two full-length pilin structures, and only one cryo-EM reconstruction of a bona fide T4P filament. Full-length pilin structures are difficult to obtain because of the presence of the flexible and hydrophobic α1N, but these should be pursued as they will be important in fitting pilin structures into cryo-EM reconstructions. These reconstructions are also challenging but new image processing software has already produced high quality structures of the GC pilus and an archaeal pilus, and provides hope that newer, higher-resolution T4P structures will be obtained in the near future. These structures will help us to understand the precise interactions that occur among the N-terminal α-helices in the filament core, and will provide symmetry information that will establish whether or not all T4P have the same helical symmetry and assembly mechanism. The GC pilus reconstruction and TCP filament model both show a highly corrugated surface exposing structurally and functionally variable regions. Given the variation in amino acid sequence and structure of the αβ-loops and D-regions, it is likely that pilus surfaces from different species will differ substantially. Thus, pilin structures from other bacterial species should be sought, in combination with EM reconstructions of these pili. An atomic-level understanding of these molecular surfaces will allow us to clarify specific functions, and to design highly specific binding agents that are effective antibacterial therapeutics. The deep grooves and cavities identified for both GC pili and TCP provide ideal drug binding targets. Such molecules would bind with high specificity and affinity, and would either disrupt key pilus functions or act within the periplasm upon pilus retraction. Furthermore, integrated structural methods should be applied to determining the structures of the many components of the pilus biogenesis apparatus to aid in understanding pilus assembly. While many of these components are membrane bound proteins, and thus challenging to crystallize, the structures of the soluble domains are often attainable and can be highly informative. This approach has been very successful for the T2S apparatus (Johnson *et al.*, 2006). Close attention should be paid to new developments in T2S systems as they will likely be highly relevant to the T4P system. Our understanding of the pilus biogenesis process would greatly benefit from being studied as an intact machine, as has been done for the Type III secretion system, the bacterial flagella motor, and, most recently, the Type IV secretion system (Blocker *et al.*, 2003; Minamino *et al.*, 2008; Fronzes *et al.*, 2009). The Gram-positive *Clostridium* T4P apparatus may offer a simplified system to study as it lacks the outer membrane secretin and associated proteins. Finally, new genomic sequences make it easy to readily identify T4P systems in newly sequenced bacteria. In summary, the pilus surface has been 'scratched', and is poised for ambitious investigations by many structural biology groups in collaboration with biochemists, geneticists and bioinformaticians. This continues to be a challenging system to study, but one that promises to provide great returns.

Acknowledgments

L.C. is supported by salary grants from the Michael Smith Foundation for Health Research and the Canadian Institutes for Health Research (CIHR). Work in the Craig lab is supported by grants from CIHR, the National Institute of Health, the Natural Sciences and Engineering Research Council and the Canadian Foundation for Innovation. Figures 1.3, 1.5 and 1.6 were prepared using PyMol (http://www.pymol.org).

References

Aas, F.E., Vik, A., Vedde, J., Koomey, M., and Egge-Jacobsen, W. (2007). *Neisseria gonorrhoeae O*-linked pilin glycosylation: functional analyses define both the biosynthetic pathway and glycan structure. Mol. Microbiol. 65, 607–624.

Audette, G.F., Irvin, R.T., and Hazes, B. (2004). Crystallographic analysis of the *Pseudomonas aeruginosa* strain K122-4 monomeric pilin reveals a conserved receptor-binding architecture. Biochemistry *43,* 11427–11435.

Blocker, A., Komoriya, K., and Aizawa, S. (2003). Type III secretion systems and bacterial flagella: insights into their function from structural similarities. Proc. Natl. Acad. Sci. U S A *100,* 3027–3030.

Chamot-Rooke, J., Rousseau, B., Lanternier, F., Mikaty, G., Mairey, E., Malosse, C., Bouchoux, G., Pelicic, V., Camoin, L., Nassif, X., and Dumenil, G. (2007). Alternative *Neisseria* spp. Type IV pilin glycosylation with a glyceramido acetamido trideoxyhexose residue. Proc. Natl. Acad. Sci. U.S.A. *104,* 14783–14788.

Craig, L., Pique, M.E., and Tainer, J.A. (2004). Type IV pilus structure and bacterial pathogenicity. Nat. Rev. Microbiol. *2,* 363–378.

Craig, L., Taylor, R.K., Pique, M.E., Adair, B.D., Arvai, A.S., Singh, M., Lloyd, S.J., Shin, D.S., Getzoff, E.D., Yeager, M., Forest, K.T., and Tainer, J.A. (2003). Type IV pilin structure and assembly: X-ray and EM analyses of *Vibrio cholerae* toxin-coregulated pilus and *Pseudomonas aeruginosa* PAK pilin. Mol. Cell *11,* 1139–1150.

Craig, L., Volkmann, N., Arvai, A.S., Pique, M.E., Yeager, M., Egelman, E.H., and Tainer, J.A. (2006). Type IV pilus structure by cryo-electron microscopy and crystallography: implications for pilus assembly and functions. Mol. Cell *23,* 651–662.

Doig, P., Sastry, P.A., Hodges, R.S., Lee, K.K., Paranchych, W., and Irvin, R.T. (1990). Inhibition of pilus-mediated adhesion of *Pseudomonas aeruginosa* to human buccal epithelial cells by monoclonal antibodies directed against pili. Infect. Immun. *58,* 124–130.

Egelman, E.H. (2000). A robust algorithm for the reconstruction of helical filaments using single-particle methods. Ultramicroscopy *85,* 225–234.

Fronzes, R., Schafer, E., Wang, L., Saibil, H.R., Orlova, E.V., and Waksman, G. (2009). Structure of a type IV secretion system core complex. Science *323,* 266–268.

Gil, H., Benach, J.L., and Thanassi, D.G. (2004). Presence of pili on the surface of *Francisella tularensis.* Infect. Immun. *72,* 3042–3047.

Hager, A.J., Bolton, D.L., Pelletier, M.R., Brittnacher, M.J., Gallagher, L.A., Kaul, R., Skerrett, S.J., Miller, S.I., and Guina, T. (2006). Type IV pili-mediated secretion modulates *Francisella virulence.* Mol. Microbiol. *62,* 227–237.

Han, X., Kennan, R.M., Parker, D., Davies, J.K., and Rood, J.I. (2007). Type IV fimbrial biogenesis is required for protease secretion and natural transformation in *Dichelobacter nodosus.* J. Bacteriol. *189,* 5022–5033.

Hansen, J.K., Demick, K.P., Mansfield, J.M., and Forest, K.T. (2007). Conserved regions from *Neisseria gonorrhoeae* pilin are immunosilent and not immunosuppressive. Infect. Immun. *75,* 4138–4147.

Hazes, B., Sastry, P.A., Hayakawa, K., Read, R.J., and Irvin, R.T. (2000). Crystal structure of *Pseudomonas aeruginosa* PAK pilin suggests a main-chain-dominated mode of receptor binding. J. Mol. Biol. *299,* 1005–1017.

Hegge, F.T., Hitchen, P.G., Aas, F.E., Kristiansen, H., Lovold, C., Egge-Jacobsen, W., Panico, M., Leong, W.Y., Bull, V., Virji, M., Morris, H.R., Dell, A., and Koomey, M.(2004). Unique modifications with phosphocholine and phosphoethanolamine define alternate antigenic forms of *Neisseria gonorrhoeae* Type IV pili. Proc. Natl. Acad. Sci. U.S.A. *101,* 10798–10803.

Helaine, S., Carbonnelle, E., Prouvensier, L., Beretti, J.L., Nassif, X., and Pelicic, V. (2005). PilX, a pilus-associated protein essential for bacterial aggregation, is a key to pilus-facilitated attachment of *Neisseria meningitidis* to human cells. Mol. Microbiol. *55,* 65–77.

Helaine, S., Dyer, D.H., Nassif, X., Pelicic, V., and Forest, K.T. (2007). 3D structure/function analysis of PilX reveals how minor pilins can modulate the virulence properties of Type IV pili. Proc. Natl. Acad. Sci. U.S.A. *104,* 15888–15893.

Hobbs, M., Dalrymple, B.P., Cox, P.T., Livingstone, S.P., Delaney, S.F., and Mattick, J.S. (1991). Organization of the fimbrial gene region of *Bacteroides nodosus*: class I and class II strains. Mol. Microbiol. *5,* 543–560.

Johnson, T.L., Abendroth, J., Hol, W.G., and Sandkvist, M. (2006). Type II secretion: from structure to function. FEMS Microbiol. Lett. *255,* 175–186.

Kang, H.J., Coulibaly, F., Clow, F., Proft, T., and Baker, E.N. (2007). Stabilizing isopeptide bonds revealed in Gram-positive bacterial pilus structure. Science *318,* 1625–1628.

Keizer, D.W., Slupsky, C.M., Kalisiak, M., Campbell, A.P., Crump, M.P., Sastry, P.A., Hazes, B., Irvin, R.T., and Sykes, B.D. (2001). Structure of a pilin monomer from *Pseudomonas aeruginosa*: implications for the assembly of pili. J. Biol. Chem. *276,* 24186–24193.

Kirn, T.J., Bose, N., and Taylor, R.K. (2003). Secretion of a soluble colonization factor by the TCP type 4 pilus biogenesis pathway in *Vibrio cholerae.* Mol. Microbiol. *49,* 81–92.

Kirn, T.J., Lafferty, M.J., Sandoe, C.M., and Taylor, R.K. (2000). Delineation of pilin domains required for bacterial association into microcolonies and intestinal colonization by *Vibrio cholerae.* Mol. Microbiol. *35,* 896–910.

Kirn, T.J., and Taylor, R.K. (2005). TcpF is a soluble colonization factor and protective antigen secreted by El Tor and classical O1 and O139 *Vibrio cholerae* serogroups. Infect. Immun. *73,* 4461–4470.

Kohler, R., Schafer, K., Muller, S., Vignon, G., Deiderichs, K., Philippsen, A., Ringler, P., Pugsley, A.P., Engel, A., and Welte, W. (2004). Structure and assembly of the pseudopilin PulG. Mol. Microbiol. *54,* 647–664.

Korotkov, K.V., and Hol, W.G. (2008). Structure of the GspK-GspI-GspJ complex from the enterotoxigenic *Escherichia coli* type 2 secretion system. Nat. Struct. Mol. Biol. *15,* 462–468.

Lee, K.K., Sheth, H.B., Wong, W.Y., Sherburne, R., Paranchych, W., Hodges, R.S., Lingwood, C.A., Krivan, H., and Irvin, R.T. (1994). The binding of *Pseudomonas aeruginosa* pili to glycosphingolipids is a tip-associated event involving the C-terminal region of the structural pilin subunit. Mol. Microbiol. *11,* 705–713.

Li, H., DeRosier, D., Nicholson, W., Nogales, E., and Downing, K. (2002). Microtubule structure at 8 Å resolution. Structure (Cambridge) *10,* 1317–1328.

Li, J., Lim, M.S., Li, S., Brock, M., Pique, M.E., Woods, V.L., Jr., and Craig, L. (2008). *Vibrio cholerae* toxin-coregulated pilus structure analysed by hydrogen/

deuterium exchange mass spectrometry. Structure *16*, 137–148.

Lory, S., and Strom, M.S. (1997). Structure–function relationship of type-IV prepilin peptidase of *Pseudomonas aeruginosa* – a review. Gene *192*, 117–121.

Maier, B., Potter, L., So, M., Seifert, H.S., and Sheetz, M.P. (2002). Single pilus motor forces exceed 100 pN. Proc. Natl. Acad. Sci. U.S.A. *99*, 16012–16017.

Minamino, T., Imada, K., and Namba, K. (2008). Mechanisms of type III protein export for bacterial flagellar assembly. Mol. Biosyst. *4*, 1105–1115.

Parge, H.E., Forest, K.T., Hickey, M.J., Christensen, D.A., Getzoff, E.D., and Tainer, J.A. (1995). Structure of the fibre-forming protein pilin at 2.6 A resolution. Nature *378*, 32–38.

Pasloske, B.L., and Paranchych, W. (1988). The expression of mutant pilins in *Pseudomonas aeruginosa*: fifth position glutamate affects pilin methylation. Mol. Microbiol. *2*, 489–495.

Pasloske, B.L., Scraba, D.G., and Paranchych, W. (1989). Assembly of mutant pilins in *Pseudomonas aeruginosa*: formation of pili composed of heterologous subunits. J. Bacteriol. *171*, 2142–2147.

Power, P.M., Roddam, L.F., Rutter, K., Fitzpatrick, S.Z., Srikhanta, Y.N., and Jennings, M.P. (2003). Genetic characterization of pilin glycosylation and phase variation in *Neisseria meningitidis*. Mol. Microbiol. *49*, 833–847.

Ramboarina, S., Fernandes, P.J., Daniell, S., Islam, S., Simpson, P., Frankel, G., Booy, F., Donnenberg, M.S., and Matthews, S. (2005). Structure of the bundle-forming pilus from enteropathogenic *Escherichia coli*. J. Biol. Chem. *48*, 40252–40260.

Sauer, F.G., Futterer, K., Pinkner, J.S., Dodson, K.W., Hultgren, S.J., and Waksman, G. (1999). Structural basis of chaperone function and pilus biogenesis. Science *285*, 1058–1061.

Schoolnik, G.K., Fernandez, R., Tai, J.Y., Rothbard, J., and Gotschlich, E.C. (1984). Gonococcal pili. Primary structure and receptor binding domain. J. Exp. Med. *159*, 1351–1370.

Sheth, H.B., Lee, K.K., Wong, W.Y., Srivastava, G., Hindsgaul, O., Hodges, R.S., Paranchych, W., and Irvin, R.T. d. (1994). The pili of *Pseudomonas aeruginosa* strains PAK and PAO bind specifically to the carbohydrate sequence beta GalNAc(1–4)β-Gal found in glycosphingolipids asialo-GM1 and asialo-GM2. Mol. Microbiol. *11*, 715–723.

Stimson, E., Virji, M., Makepeace, K., Dell, A., Morris, H.R., Payne, G., Saunders, J.R., Jennings, M.P., Barker, S., Panico, M., Blench, I., and Moxon, E.R. (1995). Meningococcal pilin: a glycoprotein substituted with digalactosyl 2,4-diacetamido-2,4,6-trideoxyhexose. Mol. Microbiol. *17*, 1201–1214.

Strom, M.S., and Lory, S. (1993). Structure-function and biogenesis of the Type IV pili. Annu. Rev. Microbiol. *47*, 565–596.

Sun, D., Lafferty, M.J., Peek, J.A., and Taylor, R.K. (1997). Domains within the *Vibrio cholerae* toxin coregulated pilin subunit that mediate bacterial colonization. Gene *192*, 79–85.

van Schaik, E.J., Giltner, C.L., Audette, G.F., Keizer, D.W., Bautista, D.L., Slupsky, C.M., Sykes, B.D., and Irvin, R.T. (2005). DNA binding: a novel function of *Pseudomonas aeruginosa* Type IV pili. J. Bacteriol. *187*, 1455–1464.

Varga, J.J., Nguyen, V., O'Brien, D.K., Rodgers, K., Walker, R.A., and Melville, S.B. (2006). Type IV pili-dependent gliding motility in the Gram-positive pathogen *Clostridium perfringens* and other Clostridia. Mol. Microbiol. *62*, 680–694.

Wang, Y.A., Yu, X., Ng, S.Y. M., Jarrell, K.F., and Egelman, E.H. (2008). The structure of an archaeal pilus. J. Mol. Biol. *381*, 456–466.

Watts, T.H., Kay, C.M., and Paranchych, W. (1982). Dissociation and characterization of pilin isolated from *Pseudomonas aeruginosa* strains PAK and PAO. Can. J. Biochem. *60*, 867–872.

Xu, X.F., Tan, Y.W., Lam, L., Hackett, J., Zhang, M., and Mok, Y.K. (2004). NMR structure of a Type IVb pilin from *Salmonella typhi* and its assembly into pilus. J. Biol. Chem. *279*, 31599–31605.

Yamagata, A., and Tainer, J.A. (2007). Hexameric structures of the archaeal secretion ATPase GspE and implications for a universal secretion mechanism. EMBO J. *26*, 878–890.

Yanez, M.E., Korotkov, K.V., Abendroth, J., and Hol, W.G. (2008a). The crystal structure of a binary complex of two pseudopilins: EpsI and EpsJ from the type 2 secretion system of *Vibrio vulnificus*. J. Mol. Biol. 375, 471–486.

Yanez, M.E., Korotkov, K.V., Abendroth, J., and Hol, W.G. (2008b). Structure of the minor pseudopilin EpsH from the Type 2 secretion system of *Vibrio cholerae*. J. Mol. Biol. *377*, 91–103.

Yonekura, K., Maki-Yonekura, S., and Namba, K. (2003). Complete atomic model of the bacterial flagellar filament by electron cryomicroscopy. Nature *424*, 643–650.

Type IV Pilus Biogenesis, Structure and Function: Lessons from Type IVa Pilin Systems

2

Michael Koomey

Abstract

Type IV pili are unabashed celebrities of the world of prokaryotic appendages and their fame and attraction are well deserved. Not only are they expressed by a vast number of phlyogenetically distinct species, they are associated with a dazzling array of diverse processes and phenotypes that impact heavily on microbial biology. Here, I give an overview of where we are today with regard to understanding the biogenesis, dynamics, structure and function of the Type IVa class of Type IV pili. In particular, I highlight studies of organelle biology in two systems that have played prominent roles in the development of the field: *Pseudomonas aeruginosa* and *Neisseria gonorrhoeae*. New emerging systems and outstanding issues are also addressed.

Historical perspective

Organelles now recognized as Type IV pili (T4P) first gained notoriety due to their association with a form of flagellar-independent surface movement displayed by *Acinetobacter*, *Moraxella* and *Pseudomonas* species, termed twitching motility (Henrichsen, 1975, 1983). Contemporaneous studies correlated both phenotypes with competence for natural genetic transformation in *Moraxella* and pathogenic *Neisseria* species (Bovre and Froholm, 1972; Froholm *et al.*, 1973) and with sensitivity to some bacteriophages in *Pseudomonas aeruginosa* (Bradley, 1972a, b, 1974). Based on morphological similarities and these shared phenotypes, Ottow defined these as Group 4 pili and this nomenclature survived as the term Type 4 or Type IV pili (Ottow, 1975). This classification scheme has held up remarkably well with a slight modification to recognize two classes of pilin subunits (IVa and IVb) which differ in the average lengths of their leader peptide, which residues reside at the N-terminal position of the processed protein, and the size of the domain flanked by intramolecular disulphide bonds.

Looking through the retroscope, *P. aeruginosa* has been at the forefront of T4P biology. This situation reflected the fact that biochemical studies of *P. aeruginosa* pili pioneered by Frost, Paranchych and colleagues were well advanced as well as the robustness of this model system for the selection of mutants based on pilus-specific phage resistance and screening of altered colony morphology (Frost and Paranchych, 1977; Frost *et al.*, 1978; Watts *et al.*, 1982a–c, 1983a,b). In a series of papers starting in the early seventies, Bradley examined the associations between T4P expression, twitching motility, and phage susceptibility and established the pilus retraction model (Bradley, 1972a,b, 1974). He declared in the definitive 1980 work that 'There is no doubt that fully functional retractile pili are the mechanical basis for twitching motility' (Bradley, 1980). The next advance in the field came through the observation that heterologous pilin subunits could be assembled in *P. aeruginosa* but not in *E. coli* (Finlay *et al.*, 1986; Elleman and Peterson, 1987). These results suggested a conserved biogenesis machinery that was distinct from that documented for the chaperone/usher assembly pathway being characterized predominantly in *E. coli* at that time. Subsequent reverse and forward genetic studies carried out primarily by the

Mattick (Whitchurch *et al.*, 1991; Hobbs *et al.*, 1993; Whitchurch and Mattick, 1994; Alm and Mattick, 1995, 1996; Martin *et al.*, 1995; Alm *et al.*, 1996a,b; Watson *et al.*, 1996a; Whitchurch *et al.*, 1996) and Lory (Johnson *et al.*, 1986; Strom and Lory, 1986; Johnson and Lory, 1987; Ishimoto and Lory, 1989; Nunn *et al.*, 1990; Totten *et al.*, 1990; Nunn and Lory, 1991; Strom and Lory, 1991; Ishimoto and Lory, 1992; Koga *et al.*, 1993) groups identified what appears to be the entire repertoire of components required for T4P expression (Table 2.1). Meanwhile, Pugsley and colleagues reported the characterization of a gene cluster in *K. pneumonia* required for pullulanase secretion and noted not only the similarities of the components to the T4P assembly proteins but also of the outer membrane protein PulD to the filamentous phage-encoded protein, pIV (Pugsley, 1993). Likewise, T4P-related assembly components were discovered to be involved in competence for natural genetic transformation in both *H. influenzae* (Tomb *et al.*, 1991) and *Bacillus subtilis* (Albano and Dubnau, 1989). Thus, it became absolutely clear that a conserved machinery was involved in superficially diverse processes associated with the movement of macromolecules across membranes. Subsequently, the family grew to include the so-called Tad (tight adherence macromolecular transport) systems (Kachlany *et al.*, 2000), as well as archaeal flagellar biogenesis (Bayley and Jarrell, 1998) and sugar uptake systems (Elferink *et al.*, 2001). Thus, the basic components arose early in prokaryotic evolution and the machinery was recruited to carry out a variety of tasks.

By correcting the motility defect in one of Bradley′s mutants expressing non-retractile pili by complementation, Mattick and colleagues identified the PilT protein, now recognized as the retraction ATPase (Whitchurch *et al.*, 1991). The structural similarities and conserved sequences between the $PilB_{Pa}$ assembly ATPase and PilT immediately suggested that they functioned similarly but antagonistically and, taken together with Bradley's retraction model, that the former promoted the incorporation of pilins into pili and the latter acted to disassemble them. Further understanding of T4P dynamics came through

Table 2.1 Defined components of T4P biogenesis, structure and function (Type IVa)

	P. aeruginosa	*N. gonorrhoeae*	Other relevant species
Core assembly elements			
Assembly ATPase	PilB	PilF	$PilB_{Mx}$
Prepilin peptidase	PilD	PilD	$PilD_{Mx}$
Others	PilM, N, O, P	PilM,N,O,P	PilM, N, O, P_{Tth}
Pilus homeostasis effectors			
Pilin-like proteins	FimT,U,PilV,W,X	PilH-L	PilH-K,X_{Nm}
	PilE?	PilV	–
	PilE?	ComP	–
High mol. wt T4P-associated protein	PilY1	PilC	$PilC_{Nm}$
Others	T4PY/T4PZ	?	?
	PilF	PilW	$PilW_{Nm}$ Tgl_{Mx}
Surface localization			
Secretin	PilQ	PilQ	$PilQ_{Nm}$, $PilQ_{Mx}$
Pilus retraction			
Retraction ATPase	PilT	PilT	$PilT_{Nm}$, $PilT_{Mx}$

Tth, *Thermus thermophilus*;Nm, *Neisseria meningiditis*; Mx, *Myxococcus xanthus*.

the observation that dramatic defects in pilus expression in 'biogenesis' mutants could be suppressed by *pilT* null mutations in *N. gonorrhoeae* (Wolfgang *et al.*, 1998; Wolfgang *et al.*, 2000). Thus, proteins characterized as assembly factors were in fact components that control T4P dynamics by preventing or inhibiting PilT-mediated retraction. These results demonstrated that PilT could conditionally antagonize stable pilus expression and showed that T4P biogenesis entailed genetically dissociable steps of fibre formation and surface localization of the intact organelle and an intermediate step at which T4P dynamics were modulated. These findings thus made it possible to assign empirically the step at which various components function and, based on tests of epistasis, to establish the order in which they must act. Finally, two studies directly established that T4P in *P. aeruginosa* and *N. gonorrhoeae* actually undergo retraction and that force associated with retraction could account for twitching motility (Merz *et al.*, 2000; Skerker and Berg, 2001).

Genetic analyses of T4P expression

The assembly of surface filamentous appendages exemplified by pili and flagella requires the temporal and spatial coordination of two critical processes: translocation of the subunits across membranes (subunit secretion) and subunit polymerization. In chaperone/usher systems, release of subunits from the cytoplasmic membrane is mediated by binding to periplasmic chaperones (Hung and Hultgren, 1998). These complexes are then targeted to the outer membrane where their interaction with the usher outer membrane protein appears to lead to chaperone dissociation and exposure of subunit-interactive surfaces, which in turn drives assembly into pilus fibres (Sauer *et al.*, 1999). Thus, premature pilin oligomerization is prevented by interaction with the chaperone and assembly and outer membrane translocation are intrinsically coupled events. Flagellar biogenesis is a more complex process that involves >30 protein components with flagellin being extruded through the filament channel and hook and incorporated into the distal end of the nascent structure (Macnab, 2003). In this case then, polymerization is delayed until the subunit is external to the cell (Fig. 2.1).

Although components required for T4P expression had been defined by mutant analyses, it remained unclear as to how they functioned or interacted (save for the PilD prepilin peptidase that processes the pilin subunit) (Nunn and Lory, 1991). Attempts to rationalize T4P biogenesis in the context of either the chaperone/usher or flagellar systems were untenable since none of the T4P biogenesis components were structurally related to those found there. This situation changed when it was found in *N. gonorrhoeae* that loss-of-function mutations in *pilT* alleviated the biogenesis requirement for the PilC adhesin, establishing for the first time that PilC was not a true assembly component and that PilT could function as an antagonist of organelle expression (Wolfgang *et al.*, 1998). This result led to similar studies of what effect PilT might have in other *N. gonorrhoeae* biogenesis mutants and in particular those lacking the PilQ secretin which is found in the outer membrane (Tonjum *et al.*, 1998). Remarkably, the absence of PilT in this background led to the expression of T4P fibres that failed to reach the cell surface and remain localized in membrane protrusions (Wolfgang *et al.*, 2000). Therefore, PilQ was not a true assembly component and T4P assembled inside the periplasm.

Core assembly elements

Using these criteria and approaches, the core assembly elements required for pilus formation could be defined on the basis that they are essential regardless of the status of the retraction ATPase. These include the membrane embedded prepilin peptidases that cleave off an N-terminal extension (PilD) and cytoplasmic soluble ATPases ($PilB_{Pa}$, $PilF_{Ng}$). Prepilin peptidase pilin processing is essential to pilins becoming polymerization competent either because it is required for pilins to interact with one another or with the other core assembly elements. There are no structural data on the T4P assembly ATPases although they are clearly members of the bacterial secretion ATPase superfamily which includes the Type IV secretion system traffic ATPase HP0525 (*Helicobacter pylori*), the type II secretion ATPase EpsE (*Vibrio cholerae*), and the archaeal type II secretion ATPase GspE (*Archaeoglobus fulgidus*) proteins (Robien *et al.*, 2003; Savvides *et al.*,

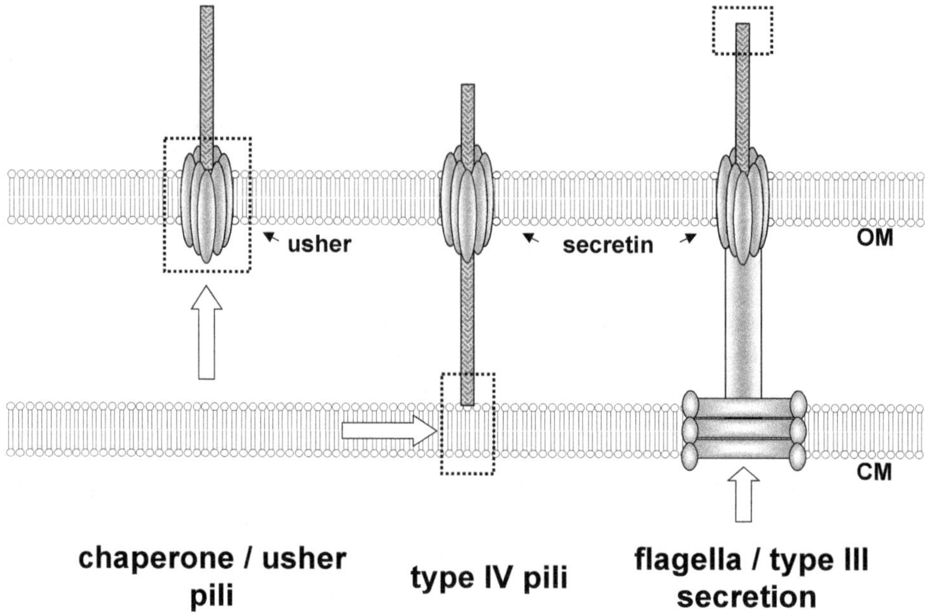

Figure 2.1 Three solutions to the tempero-spatial coordination of subunit membrane translocation and polymerization evident in bacterial surface filament systems. In the chaperone/usher systems, pilin subunits assemble at the level of the outer membrane via interaction with the usher while in flagella/type III secretion systems, polymerization occurs at the distal tip of the growing filament (dotted boxes denote assembly sites, large arrows show traffic flow of subunits to assembly sites). Although not shown, premature subunit interactions are prevented in both systems by dedicated chaperones. In T4P, subunits are assembled at or on the cytoplasmic membrane and flow to the sites of assembly by lateral diffusion in the membrane. This configuration underlies the reversibility of the system (the ability of T4P to extend and retract) and the ability to harness the chemical energy generated via nucleotide hydrolysis by assembly and retraction ATPases localized to the cytoplasmic leaflet of the membrane. Consistent with the simple bidirectional remodelling of pilins between their integral membrane and polymerized states, no classical subunit-dedicated chaperones have been found in T4P systems. See text for further details.

2003; Yamagata and Tainer, 2007). Missense substitutions in the conserved domains implicated in ATP hydrolysis have been shown to disrupt T4P expression (Turner *et al.*, 1993). The mechanisms by which assembly ATPases function are probably best inferred from data obtained from studies of the PilT retraction ATPase (see below). The integral inner membrane proteins of the $PilC_{Pa}$/$PilG_{Ng}$ family have long been considered to be core elements of assembly, primarily due to their very high degree of conservation in both T4P and type II secretion systems and tight genetic linkage to the genes for prepilin peptidase and the assembly ATPase (Nunn *et al.*, 1990; Tonjum *et al.*, 1995). However, a study in *N. meningitidis* concluded that this protein was not an assembly component as the expression defect seen in a $PilG_{Nm}$ null mutant was suppressed by a mutation in the PilT retraction ATPase (Carbonnelle *et al.*, 2006). This astonishing finding has been uniformly greeted with scepticism as it seems counterintuitive to all other information available. It will thus be important to independently confirm this result in *N. meningitidis* and other systems. Further studies in *N. meningitidis* have extended the core biogenesis repertoire to include the conserved PilM, PilN, PilO and perhaps PilP proteins whose genes are linked to one another and highly conserved in T4P-expressing species (Carbonnelle *et al.*, 2006). Studies in the thermophilic bacterium *Thermus thermophilus* have shown that PilM, N and O all fractionate with the inner membrane (Rumszauer *et al.*, 2006). However, the roles of PilM, N and O in biogenesis remain to be determined. In summary, most if not all core biogenesis components identified to date are associated with the inner membrane, consistent with the notion that subunit polymerization takes place at this site.

Surface localization of intact T4P

As defined in *pilT* mutants of *Neisseria* species, translocation of T4P across the membrane appears to require only the secretin PilQ (Wolfgang *et al.*, 2000; Carbonnelle *et al.*, 2006). Like its homologues in the type II secretion and filamentous phage systems, PilQ forms large multimers that are believed to form gated channels through which the pilus (or other folded substrates) pass (Bayan *et al.*, 2006). It should be noted here that there is, in fact, no direct evidence that discriminates between whether T4P and other substrates actually go through the PilQ multimer channel or that the PilQ multimers form a higher ordered structure that facilitate T4P membrane translocation. In some T4P systems, members of the PilP lipoprotein family (whose gene is invariably linked to *pilQ*) are thought to be required for PilQ function and, a priori, PilP should also be essential to surface translocation (Martin *et al.*, 1995). However, this does not appear to be the case in *N. meningitidis,* as it was reported that *pilP* mutations were devoid of pili regardless of whether PilT was present or not (Carbonnelle *et al.*, 2006). This finding may not be broadly applicable, as our own studies in *N. gonorrhocac* arc cquivocal; somc *pilP* null alleles in a *pilT* background lead to periplasmic T4P while others do not (unpublished data, H. Winther-Larsen). Structural studies have also provided evidence for specific interactions between PilP and PilQ that are likely relevant to functionality (Balasingham *et al.*, 2007). Thus, the bulk of evidence suggests that PilP and PilQ act in concert but more work is required to resolve how PilP contributes to the system.

Effectors of pilus homeostasis

The remaining components dramatically influencing T4P expression at the cell surface fall into the category of what have been defined as effectors of pilus homeostasis. In otherwise wild-type backgrounds, null mutations at loci encoding these components result in the absence or dramatically reduced levels of T4P but when examined in a *pilT* background, these same mutations have no quantitative defect in T4P expression. These molecules are thus not canonical assembly factors but rather act by promoting T4P expression in the presence of the PilT retraction ATPase. Since the documentation of these phenotypes in *pilC*$_{Ng}$ and *pilQ*$_{Ng}$ mutants, other *N. gonorrhoeae* molecules implicated in acting in this fashion have been defined and characterized (Winther-Larsen *et al.*, 2005). In particular, a set of five pilin-like molecules, named PilH-L, was identified and the absence of any one of them leads to greatly reduced T4P levels. The *pilH-L* genes are arranged in tandem suggesting that they are part of an operon. As seen by immunoblotting, PilH-L are expressed at levels well below that of the T4P subunit. In wild-type and *pilT* cells, all five proteins co-purify with T4P while in a *pilT* background, the absence of any one (by nonpolar mutation) leads to dramatically decreased levels of the others in this fraction. In addition, levels of the T4P co-purifying PilC$_{Ng}$ adhesin are dramatically reduced in these latter backgrounds and, conversely, the levels of PilH-L are reduced in T4P from a *pilC, pilT* background. These observations suggest PilH-L and the PilC adhesin act at a common step and may act as a multimeric complex. Moreover like *pilC, pilT* mutants, piliated *pilH–L, pilT* mutants are incapable of adhering to human epithelial cells and defective in natural genetic transformation at the level of DNA binding/uptake transformation (Winther-Larsen *et al.*, 2005). Subsequent work in *N. meningitidis* was largely confirmatory with regard to PilH-L save for differences in the phenotype of *N. meningitidis* mutants lacking PilX (the PilL orthologue – see below) (Helaine *et al.*, 2007).

In an effort to better understand relationships and interactions among PilH–L, an analysis of related genes and proteins was performed and showed that the genomes of many T4P expressing species contain at least one group of linked ORFs whose products have COG designations identical to those for PilH–K and that are arranged in the same gene order as seen in the *pilH–L locus* (Russell and Darzins, 1994; Alm and Mattick, 1995; Alm *et al.*, 1996b; Alm and Mattick, 1996; Winther-Larsen *et al.*, 2005). Of particular note was the finding that ORFs sharing the same COG designation as the neisserial PilC adhesin protein lie just downstream of each cluster. Conserved gene organization and structure across phylogenetic lineages and linkage to *pilC* orthologous genes documents a common ancestry and predicts conserved, interactive functions for each set of proteins. Mutagenesis and complementation studies in *P. aeruginosa* have been interpreted as indicating that the products of most of these genes (including the PilY1 homologue of $PilC_{Ng}$) are essential for T4P expression (Russel and Darzins, 1994; Alm and Mattick, 1995, 1996; Alm *et al.*, 1996). It should be interesting to determine if the *Pseudomonas* gene products are canonical biogenesis factors or if they influence T4P dynamics in a fashion seen in neisserial studies.

The PilW outer membrane lipoprotein of *N. meningitidis* also falls into the category of being an effector of pilus homeostasis as defined by the conditional effects of PilT on T4P expression in *pilW* null mutants (Carbonnelle *et al.*, 2005). PilW mutants are associated with a failure of the PilQ secretin to form a high-molecular weight multimer, a property which is believed to be essential to PilQ function. Although piliation was restored in the *pilW*, *pilT* background, PilQ multimerization was not. Thus, either multimerization formation is not a sine qua non of PilQ function or there is a PilQ-independent mechanism for T4P surface expression in this background. Molecules orthologous to PilW are found in many T4P expressing species and two have been characterized in more detail. *P. aeruginosa* expresses a PilW orthologue named PilF that, like its counterpart, is an outer membrane lipoprotein required for T4P expression (Koo *et al.*, 2008). Likewise, *pilF* null mutations result in the failure to detect the PilQ multimer. The equivalent Tgl lipoprotein of *Myxococcus xanthus* is also required for T4P expression and the assembly of PilQ multimers (Nudleman *et al.*, 2006). Remarkably, the piliation and PilQ multimerization defects in *M. xanthus* can be corrected by a process termed contact stimulation. In a truly impressive series of works, Kaiser and colleagues have shown that piliation and associated phenotypes (including PilQ multimerization) can be restored in nonpiliated *tgl* mutants by mixing them with Tgl expressing cells via the physical transfer of the protein between cells (Rodriguez-Soto and Kaiser, 1997; Wall *et al.*, 1998; Nudleman *et al.*, 2005). The hypothesis is that the role of contact stimulation in wild-type cells is to coordinate polar T4P expression in cells aligned in an end to end fashion during social motility. Taken together, the data suggest that PilW, PilF and Tgl carry out analogous function in their endogenous species although it remains to be seen if piliation is restored by *pilT* mutations in the *pilF* and *tgl* situations. Crystal structures for both PilW and PilF have also been determined allowing speculation as to how these proteins might impact on PilQ function (Koo *et al.*, 2008; Trindade *et al.*, 2008).

Another class of pilus homeostasis factors are the T4PY and T4PZ proteins found in some strains of *P. aeruginosa*. These proteins attracted attention as their genes mapped adjacent to the *pilA* pilin subunit gene and appeared to be limited in their distribution to specific *pilA* alleles (Kus *et al.*, 2004). T4PY and Z are predicted to encode inner membrane proteins with approximately 50% amino acid sequence similarity to one another within three N-terminal transmembrane domains. When introduced alone into a standard *P. aeruginosa* strain (lacking endogenous PilA), the *pilA* genes associated with *tfpY* and *tfpZ* led to low amounts of T4P and defects in twitching motility while when introduced into an equivalent *pilA*, *pilT* background, high levels of T4P expression were seen (Asikyan *et al.*, 2008). When the *pilA* alleles were introduced together with their cognate T4PY/T4PZ protein into the *pilA* background, piliation and twitching motility phenotypes were restored to wild-type levels. Intriguingly, the *pilA*, *tfpY*/ *tfpZ* interactions were allele-specific suggesting that corresponding specific interactions between PilA and T4PZ/Y were involved in promoting T4P expression in the presence of PilT retraction activity.

A diagram encompassing the findings discussed in this section is included here as Fig. 2.2.

Mechanisms of T4P dynamics

As dynamic filamentous polymers, T4P undergo rounds of extension and retraction that, by all available data, correspond to pilin subunit polymerization and depolymerization events. In formulating a model to understand organelle dynamics, let us make two reasonable assumptions. First, steady-state levels of T4P are dictated then by the relative frequencies of extension/polymerization event initiation versus the frequencies with which

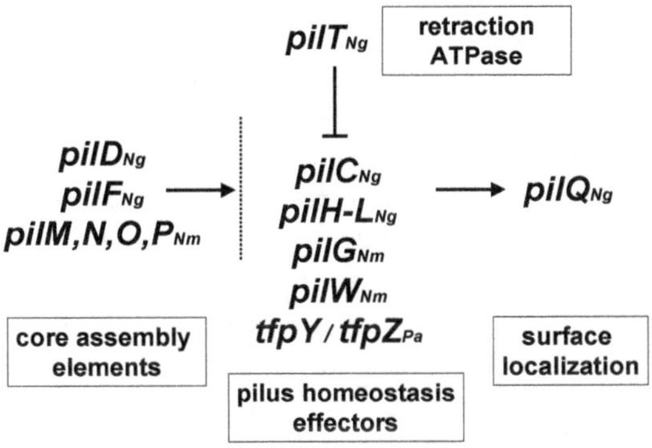

Figure 2.2 A genetically inferred pathway for T4P assembly and dynamics. The diagram incorporates data from studies in *N. gonorrhoeae*, *N. meningitidis* and *P. aeruginosa* and summarizes the ordered activities of components responsible for fibre formation, surface localization and organelle dynamics. Only core assembly elements are essential for pilus formation (pilin subunit polymerization). In the absence of any of the components to the right of the dotted partition, PilT-mediated retraction of the organelle predominates. See text for details.

extension/polymerization events are terminated by retraction/disassembly pilin polymerization and depolymerization processes are driven to non-equilibrium by the assembly ($PilF_{Ng}$) and retraction (PilT) ATPases. Second, hexameric forms of PilF and PilT localizing to the pilus base (or associated components at that site) act sequentially and in a mutually exclusive fashion. The state of growth or retraction for an individual pilus therefore becomes a matter of occupancy and whether the $PilF_{Ng}$ or PilT is operating at the pilus assembly site. But what factors dictate whether the assembly or retraction ATPase is resident and active at that site? Pilus homeostasis effectors are obviously excellent candidates to play a role here and one can easily envision that the dramatic reductions in steady-state T4P levels in mutants lacking these factors are gross distortions or amplifications of normal T4P dynamics. For example, reduction in T4P levels seen during prolonged contact of *N. meningitidis* with human cell lines is associated with downregulation of $pilC_{Nm}$ transcription (Morand *et al.*, 2004). While downregulation of pilus homeostasis effectors might be responsible for synchronous T4P retraction at a global level within a cell or populations of cells, it cannot account for the behaviour seen during twitching motility where individual pili are extending and retracting independently of one another on a single cell (Skerker and Berg, 2001). Therefore, there must be other events and processes that operate locally (at the base of individual pilus filaments) to satisfy the temporal and spatial requirements needed for this behaviour. My group proposed earlier that pilus homeostasis effectors are parts of stimulus responsive elements that control local T4P dynamics (Winther-Larsen *et al.*, 2005). Direct visualization of dynamics in wild-type *P. aeruginosa* suggests that retraction events are preceded by binding of the pilus tip to the surface substratum (Skerker and Berg, 2001) while in *M. xanthus* it has been proposed that retraction requires pilus binding to extracellular fibrils (Li *et al.*, 2003). Assuming that the cells themselves are firmly anchored, external force associated with restriction of pilus extension or imposition of a physical load on the growing pilus might perturb the activity, stability or localization of the pilus homeostasis effectors. This idea is perhaps best exemplified by the phenotype seen in *N. gonorrhoeae pilQ* mutants where the load imposed on the growing pilus by the barrier of the outer membrane shifts dynamics in favor of retraction (Wolfgang *et al.*, 2000).

The simplest model is that the pilus homeostasis effectors impact on T4P dynamics by inhibiting the recruitment of PilT. To account for this activity, one model would be that these molecules might form a cap structure at the base of the pilus that precludes PilT recognition or activation. In the specific case of the minor pilins that act in this manner, it is easy to envision how they might interact with the system as they are structurally related to the pilin subunit. Thus, they fulfill the structural requirements to interact with pilin and intersect other membrane-associated core elements of the biogenesis pathway. Alternatively, they might influence the structure and conformation of pilin itself or another component of the inner membrane complex with which PilT engages. There is also good evidence that the status of the pilin subunit itself influences dynamics. When expressed together with a wild-type pilin subunit, mutant pilins carrying substitutions at the fifth position glutamate (that are incapable of being assembled on their own) exert a dominant-negative effect resulting in very low levels of piliation (Aas *et al.*, 2007). However in a *pilT* background, piliation is restored and the mutant pilin is now incorporated into heteropolymeric T4P. When taken together with the examples of the PilH–L_{Ng}, $PilC_{Ng}$ adhesin, the $PilW_{Nm}$/$PilF_{Ng}$/Tgl as well as the *P. aeruginosa* T4PZ/Y mutants, the only feature common to all the situations that might signal the recruitment or activation of PilT is a perturbation in the kinetics of pilin assembly or the integrity of the assembly complex.

Another factor influencing dynamics includes the relative levels of PilT protein itself. Reduction in the levels of PilT (below that seen in wild-type cells) resulted in a reduction of the frequency of retraction events while an increase led to an increase in retraction events (Clausen *et al.*, 2008). Currently, there are no reports on the effects of altered levels of the assembly ATPases on retraction dynamics. It nonetheless seems that relative levels of the assembly and retraction ATPases could play a role and that this system could perhaps be modeled by simple chemical kinetics and the law of mass action. In wild-type *N. gonorrhoeae*, retraction is a highly irreversible process as seen using an optical tweezers system (Merz *et al.*, 2000). However, when levels of PilT are reduced, the application of high external forces mediated by tethering can induce a switch to pilus elongation (Maier *et al.*, 2004). This implies that a force-dependent mechanism can alter ATPase occupancy presumably by inducing a conformational or compositional change in the pilus or related machinery. I would refer the reader here to the original works for further details on the use of optical tweezers to measure T4P dynamics and the elementary step lengths occurring during T4P extension and retraction (Maier, 2005; Clausen *et al.*, 2008).

T4P as protein polymerization motors

I feel a useful viewpoint for thinking about T4P biology and associated phenotypes begins with the recognition that these organelles are polymerization motors. Specifically, T4P fulfil virtually all the requirements to be defined as protein-based biological machines that convert chemical energy into mechanical energy and force via subunit multimerization. As exemplified by eukaryotic actin and microtubule based systems, polymerization and depolymerization of protein subunits generate force for cell motility and organelle trafficking (Theriot, 2000). In the absence of external energy however, polymerization/depolymerization runs to equilibrium and stops. External energy required to drive these systems to chemical nonequilibrium is generated by nucleotide binding and hydrolysis associated with the subunits themselves. Although reversible, T4P pilin polymerization and depolymerization events differ from the actin and microtubule systems in a number of ways. Actin and microtubule monomers are freely soluble in the cytosol (together with readily available ATP and GTP) and polymerization/depolymerization events can occur at both ends of the polar fibres. Unassembled pilin monomers are embedded in the inner membrane and thus, polymerization/depolymerization can only occur at one end of the polar filament. T4P also have an exploratory character similar to that of the microtubule cytoskeleton (Mitchison and Kirschner, 1984) and clathrin-based endocytic machinery (Ehrlich *et al.*, 2004). After a nucleating step, all three complexes grow until they undergo a ‘catastrophic event’ and disassemble, an abortive event attributed to the failure to mature efficiently or recruit the appropriate partner. Thus, PilT driven retraction/disassembly has aspects of a quality control mechanism as it appears to monitor the functionality and efficacy of the assembly complex. This is especially interesting as, in contrast to the other pilus and flagella-related systems, no dedicated chaperones have been identified in T4P systems. Rather, defects or perturbations in assembly are sensed and remedied by virtue of the PilT ATPase attacking and disassembling the pilus growth complex, which in turn allows the process to start over again.

Understanding how the energy generated by $PilF_{Pa}$ and PilT – mediated nucleotide hydrolysis is transduced to drive bidirectional remodelling of pilin between its integral membrane and polymer forms remains a formidable task. A major breakthrough relevant to this topic came from the crystallographic snapshots of *Aquifex aeolicus* PilT caught in a variety of states (Satyshur *et al.*, 2007). As seen in studies of other secretion ATPases, PilT forms dynamic, hexameric complexes that undergo large nucleotide-dependent domain rearrangements. Although the various forms captured may not represent the truly functional form of the PilT multimer, they suggest that major domain movements could be harnessed to generate the large forces associated with pilus retraction and disassembly of pilin subunit from the pilus. It seems intuitive that analogous structural changes in hexameric assembly ATPases might drive polymerization of pilin and pilus extrusion but this remains to be confirmed. A complicating factor here is the presence in some species of a third secretion ATPase termed PilU whose gene maps downstream of *pilT*. In *N. gonorrhoeae*, *pilU* null mutations reduce T4P-associated autoaggregation and lead to hyperadherence for human epithelial cells (Park *et al.*, 2002). Mutations in *pilT* are epistatic to those in *pilU* and thus PilU exerts its effects in conjunction with, or downstream of, PilT. PilU mutants in *P. aeruginosa* and *D. nodosus* are defective in twitching motility but show evidence for residual pilus retraction activity (Whitchurch and Mattick, 1994; Chiang *et al.*, 2008; Han *et al.*, 2008).

Another conundrum here stems from the fact that structures and models of PilF and PilT family members offer few if any insights as to how these proteins might interact with membrane bound or assembled pilin. Nonetheless, these proteins do fractionate with membrane fractions suggesting that they might dock with the membrane, presumably through heterotypic protein–protein interactions (Brossay *et al.*, 1994). Likely candidates for such components are the $PilC_{Pa}$ family members, other defined assembly proteins (PilM, N and O family members for example) and perhaps pilin/pili itself. Genetic evidence from the Xcp type II secretion system of *P. aeruginosa* suggests direct interaction between a T4P pilin and the assembly ATPase (Kagami *et al.*, 1998). Specifically, a temperature-sensitive mutation in the *xcpT* gene (encoding the major subunit of the T4P-related pseudopilus) was suppressed by a mutation in the *xcpR* gene (which encodes the functional equivalent of the T4P assembly ATPase). If the assembly and retraction ATPase does contact pili/pilin directly, how does it gain access to the membrane embedded subunits? Evidence for indirect interaction comes from studies of the *E. coli* bundle forming pilus (bfp) system employing a variety of approaches which strongly suggest that BfpE (the $PilC_{Pa}$ homologue) interacts directly with the resident assembly and retraction ATPases (BfpD and F respectively) and acts perhaps as a stimulant and conduit for energy transduction (Crowther *et al.*, 2004; Crowther *et al.*, 2005).

Evidence for force generation imparted by pilin polymerization comes from the formation of membrane protrusions seen in the absence of the *N. gonorrhoeae* PilQ secretin (and the PilT retraction ATPase) (Wolfgang *et al.*, 2000). The levels of force required for outer membrane deformation are unknown but are likely to be considerable. Because of the severe growth defects owing to membrane disruption, suppressor mutants in which normal growth was restored were recovered and shown to carry missense mutations in the pilin subunit and when these mutants were reintroduced into an otherwise wild-type background, T4P expression was abolished or dramatically reduced. Moreover, epistasis experiments confirmed that the formation of membrane protrusions in the PilQ-, PilT- background was dependent on the PilD prepilin peptidase as well as the $PilF_{Ng}$ assembly ATPase. Taken together, these findings demonstrated that pilin polymerization generates significant protrusive force. It is worth mentioning that although they are orders of magnitude smaller in size, the membrane protrusions seen in this system were strikingly similar to the eukaryotic membrane extensions generated by some intracellular pathogens in conjunction with actin filament formation (Tilney and Portnoy, 1989). T4P retraction also generates significant force as optical tweezers measurements have shown that retraction of a single filament can generate forces in the range of 50–100 piconewtons (Merz *et al.*, 2000). This force is clearly sufficient to power twitching motility using a grappling hook-type mechanism in which the pilus is extended, binds to substratum via a tip-associated event and retracts.

Analyses of T4P associated functions

T4P play central roles in the expression of many phenotypes including adherence, motility, multicellularity, sensitivity to bacteriophages and DNA binding and uptake during natural genetic transformation. While studies of non-piliated mutants have been crucial to establishing these relationships, they have not provided much insight into the nature of the associations. With the exception of *pilT* mutants however (and self-evident defects in motility and phage sensitivity due to the absence of pilus retraction), surprisingly few functionally defective mutants retaining uncompromised T4P expression have been identified. This situation may be due to the facts that certain functions are intrinsically linked to expression of the pilus filament itself and that many proteins dispensable for assembly but essential to function impact on T4P dynamics by acting as effectors of pilus homeostasis. Nonetheless, in pathogenic *Neisseria* species, mutations at single loci resulting in loss of function (but not T4P expression defects) have been reported.

DNA binding and uptake in natural genetic transformation

In *N. gonorrhoeae*, sequence-specific DNA uptake requires three molecules that are dispensable for T4P expression: ComP (a pilin subunit-like molecule), PilT (the retraction ATPase) and ComE (a periplasmic protein with intrinsic DNA -binding activity). By conditionally altering the levels and timing of ComP and PilT expression, it was shown that PilT is dispensable for functional DNA binding and therefore contributes specifically to uptake while ComP functions at the level of sequence-specific DNA binding (Aas *et al.*, 2002b). Given that PilT functions specifically in DNA uptake, it seems reasonable to surmise that DNA translocation is physically linked to T4P retraction into the periplasm, either because the DNA is bound to the pilus itself (or a T4P-like structure) or because of transient outer membrane pores that are created in the wake of the retracting structure (Graupner *et al.*, 2001). As it seems unlikely that retraction alone accounts for the complete uptake of large DNA molecules (based on direct observation of T4P retraction events (Merz *et al.*, 2000; Skerker and Berg, 2001)), retractile pili and/or PilT likely facilitate transfer of DNA from a primary surface receptor to a periplasmic receptor and associated uptake machinery. ComE is an abundant periplasmic protein with non-specific DNA-binding activity and Chen and Gotschlich hypothesized that it may act as just such a secondary receptor that engages DNA which has been delivered by the T4P-related machinery (Chen and Gotschlich, 2001). Based on the structural similarities between ComP and the T4P pilin subunit, one can envision that a pilus-like heteropolymer is formed and ComP does co-purify with T4P but whether it is an integral part of the pilus fibre or whether it indirectly requires T4P for surface exposure remains unclear. Together with models from the Gram-positive transformation systems, T4P and related structures function in transformation at two levels: one in which they promote the surface display of a DNA binding apparatus and another in which they facilitate transfer of the DNA across the barriers of the outer membrane and peptidoglycan to the machinery involved in uptake across the cytoplasmic membrane (Chen and Dubnau, 2004).

T4P-mediated adherence

The property of adherence for human cells has also been dissociated from T4P expression in pathogenic *Neisseria* species. Piliated, *N. gonorrhoeae* mutants lacking the PilV pilin-like protein fail to adhere to primary human epithelial cells and this phenotype is most readily attributable to a defect in the display of the PilC_{Ng} adhesin in the context of the pilus fibre (Winther-Larsen *et al.*, 2001). As in the case of ComP, structural similarities between PilV and the pilin subunit suggest heteropolymer formation and PilV does co-purify with T4P. Interestingly, PilV and ComP share extended sequence conservation and *pilV* null mutants are hypertransformable in conjunction with enhanced DNA binding and uptake (Aas *et al.*, 2002a). Moreover, the absence of PilV resulted in the presence of enhanced levels of T4P co-purifying ComP while PilV overexpression reduces transformability at the level of DNA uptake (Aas *et al.*, 2002a). Both effects of altered *pilV* expression occurred independently of changes in T4P expression levels. Taken together with other observations of genetic interaction between *comP* and *pilV*, these findings suggest that ComP and PilV might compete to control the expression of two distinct organelles: one of which is dedicated to DNA binding and uptake and another that promotes human cell adherence.

An analogous dissociation of cellular adherence from T4P expression has been reported for *N. meningitidis* mutants lacking the PilX prepilin-like protein (equivalent to $PilL_{Ng}$) (Helaine *et al.*, 2007). These findings are, however, inconsistent with those made by another group, who found that *N. meningitidis pilX* mutants were reduced in T4P expression (Lappann *et al.*, 2006). Likewise, *N. gonorrhoeae* mutants lacking the orthologous PilL protein were also reduced in T4P expression and, in contrast to the situation in *N. meningitidis*, the adherence defect was not suppressed in a *pilT* background (Winther-Larsen *et al.*, 2005). It is also curious that the *N. meningitidis* mutants were reported to have unchanged levels of T4P-associated $PilC_{Nm}$ adhesin while the gonococcal mutants had clearly reduced levels of it (Winther-Larsen *et al.*, 2005). Owing to the high frequency of neisserial pilin antigenic variation (few if any strains express the same pilin allele), some of the discrepancies here may be a result of subunit differences in the backgrounds. Nonetheless, it is surprising that such major discrepancies could arise in what otherwise appear to be highly related systems.

Studies in *P. aeruginosa* present another interesting facet of T4P mediated adherence. As exhibited by *P. aeruginosa* PAK, adherence is thought to require a receptor-binding domain located within the C-terminal region of the PilA pilin subunit itself (Irvin *et al.*, 1989). When expressed as T4P in *N. gonorrhoeae,* however, adherence for human cells mediated by $PilA_{PAK}$ requires the resident accessory proteins PilV and PilC (Winther-Larsen *et al.*, 2007). It seems then that adherence potentially mediated by a $PilA_{PAK}$ intrinsic domain is not manifested in the gonococcal background. It is also noteworthy that *P. aeruginosa* also expresses homologues of the gonococcal PilH-L proteins and $PilC_{Ng}$ implicated in neisserial adherence and that the organization of these genes is conserved in these as well in other T4P expressing species. As *P. aeruginosa* mutants lacking any one of these proteins fail to express T4P, it would be very interesting to see if the expression defects could be suppressed by the absence of PilT. If so, it would then be possible to assess the relative contributions these ancillary proteins to adherence function.

T4P-associated multicellularity

T4P expression has been linked to a form of cellular interaction termed autoagglutination in a number of species and early on, this property in *N. gonorrhoeae* was proposed to result from the capacity of T4P to form thick, branched structures creating a fibrous network tethering cells to one another (Todd *et al.*, 1984). Non-autoagglutinating gonococcal mutants retaining uncompromised T4P expression were first isolated using a unique genetic selection and all three resulted from unique, single residue substitutions in the highly conserved N-terminus of pilin (Chiang *et al.*, 1995; Park *et al.*, 2001). The autoagglutination defects were suppressed in a *pilT* background suggesting a role for pilus retraction in the phenotype. Remarkably, analogous altering of the equivalent residues into the pilin subunit in the Type IVb Tcp pilus of *V. cholerae* also specifically disrupted autoagglutination, as well as associated microcolony formation and the ability to intestinally colonize infant mice (Chiang *et al.*, 1995). A more thorough, reverse genetic approach in this system identified a number of other pilin residues important to aggregation and led to the isolation of intragenic suppressors (Kirn *et al.*, 2000). This work established a connection between lateral bundling of T4P filaments and autoagglutination and current models invoke that ensuing changes in antiparallel pilus intercalation reflect altered subunit registration and pilus surface chemistry (Craig *et al.*, 2004). Unfortunately, this hypothesis has yet to be confirmed by high-resolution structural studies. Caution here is also warranted as not all autoagglutination mutants are altered in pilus bundling or fibre morphology (Kirn *et al.*, 2000; Park *et al.*, 2001). More work is therefore needed to identify the structural basis for autoagglutination and whether alterations seen in the pilin mutants are a simple consequence of altered pilin–pilin interactions or defects in the ability of the mutant subunit to interact with other proteins or cell surface constituents.

T4P as primary bacteriophage receptors

Many bacteriophages utilize pili as their primary receptor to infect cells. Productive infection here involves adsorption of phage particles to the sides or tip of the pilus and retraction of the pilus, which brings the phage into contact with a second, cell surface-associated receptor. Phage components involved in molecular recognition of pili have been well defined in a number of systems. For example, minor coat proteins of the g3p family serve as pilus receptors in many filamentous phages. In stark contrast, the structural features and components of pili that participate in these interactions have yet to be defined in any system. With regard to T4P-bacteriophage systems, strain *P. aeruginosa* PAK and the pilus-dependent bacteriophage PO4 have been utilized extensively. Here, the expression of a diverse group of exogenous T4P pilin subunits in *P. aeruginosa* PAK backgrounds lacking the endogenous pilin gene has been reported to restore twitching motility (indicative of pilus retraction) and productive PO4 infection (Watson *et al.*, 1996b; Roine *et al.*, 1998). Similarly, three distinct pilin genes were reported to restore PO4 sensitivity in a *pilA* strain of *P. stutzeri* (Graupner *et al.*, 2000). Based on the observed relaxed specificity of PO4 phage vis-á-vis subunit composition, it was proposed that a 'minor' pilus-associated protein might actually be serving as the primary receptor (Watson *et al.*, 1996b). Alternatively, PO4 phage might recognize a common motif distributed in T4P polymers or pilin subunits. To further cloud this issue, one study reported that *P. aeruginosa* strain PAO is susceptible to PO4 (Bradley, 1973), while another concluded that strain PAO became PO4 sensitive only when complemented by the PAK pilin gene (Watson *et al.*, 1996b). The molecular components of T4P required for phage binding and recognition in *P. aeruginosa*, or for that matter in any T4P-expressing species remain poorly characterized. Work from our group recently showed that expression of the *P. aeruginosa* PAK PilA pilin in *N. gonorrhoeae* was sufficient and necessary for binding of PO4 phage (although the strain remained phage resistant, presumably due to the absence of a functional second receptor) (Winther-Larsen *et al.*, 2007). Additionally, EM micrographs clearly showed that, like what was seen in *P. aeruginosa*, this interaction involved phage tail fibres that are likely to be composed of a single protein. In another work, Perham and colleagues have documented the ability of the purified filamentous phage Pf1 g3p minor coat protein to bind to *P. aeruginosa* PAK pili (Holland *et al.*, 2006).

The molecular characterization of phage protein–T4P interactions represents an underexploited yet potentially powerful means to probe the structure and surface architecture of T4P. There are many models of T4P structure derived from various physical data and modelling, but very few of them have been rigorously tested. For instance, although a large number of monoclonal antibodies reactive with purified *N. gonorrhoeae* T4P have been described, the epitopes recognized have only been identified in a few, rare cases. Additionally, I know of only one monoclonal antibody (20D9) that reacts well with native, neisserial T4P (as seen using immunolabelling) and to my knowledge its epitope has yet to be defined (Carbonnelle *et al.*, 2005). The situation for *Neisseria* T4P may be unusual due to the high degree of antigenic variation such there are no conserved surface-exposed epitopes. In contrast, the situation for *V. cholerae* Tcp (a Type IVb system) is considerably better, as well defined batteries of pilin missense mutants, antisera and monoclonal antibodies have been used to delineate parts of the surface architecture. In any event, the phage protein approach offers advantages over the use of antibodies. First, purified phage proteins are likely to be monovalent and less bulky than antibodies giving a higher degree of resolution in imaging and detection. Second, it should be straightforward to use a phage display-type approach to pan for proteins with higher avidity or altered patterns of reactivity. The power of this approach has probably already been demonstrated in works that may have fallen off the radar screen as they were focused on more ecological and evolutionary aspects of phage biology. The RNA phage phi-6 initiates infection of susceptible *Pseudomonas* species hosts via the interaction between its P3 protein and T4P (Mindich *et al.*, 1976). Burch and colleagues have isolated phage mutants with expanded host range and identified a number of adaptive missense mutations in the p3 gene (Duffy *et al.*, 2006; Duffy *et al.*, 2007; Ferris *et al.*, 2007; Knies *et al.*, 2008). Although not yet specifically shown, it seems very likely that these changes in P3 impart on the phage the ability to bind to T4P built up of different pilin serotypes. It should be possible then exploit this system to characterize the T4P-P3 interactions at the atomic level and ultimately map the surface architecture of the pilus. Such approaches might also be amenable to addressing the nature of the pilus tip itself (a topic which has yet to be directly addressed in any study) since some T4P specific phages are known to bind to the distal end of the pilus. In these instances, it remains possible that the pilus ligand may be a minor component protein although it could also be encompassed in the pilin itself but exposed only at the end of the polymer. T4P-binding, phage-based receptors might also have pragmatic applications as therapeutic anti-infectives or colonization blocking compounds.

Twitching motility

Based on the direct observation and biophysical studies of T4P retraction, it seems straightforward to understand how T4P and extension and retraction promote bacterial motility across solid surfaces. Certainly, extension, binding and retraction are consistent with the jerky movement originally described as having a 'twitching' character. However, motility is only useful and beneficial if it is regulated. It was thus extremely interesting when genes whose products proteins were related to classical flagellar chemotaxis (Che) turned up repeatedly in studies of T4P biology.

A chemosensory system that influences twitching motility in *P. aeruginosa* has been mapped to two linked loci: the *pilGHIJK* (Darzins, 1993, 1994, 1995) and *chpABC* (Whitchurch *et al.*, 2004) gene clusters. A null mutation in *pilG* encoding a protein related to the response regulator CheY disrupted T4P expression while introduction of null mutations into any one of the other genes at this locus led to either nonpiliation or defective patterns of motility. Although *pilG*, *pilI* and *pilJ* mutants were reported to be highly deficient in T4P expression, they remained susceptible to some bacteriophages and, therefore, retained some level of retractile T4P. The central component encoded within the *chpABC* locus appears to be ChpA that contains multiple target sites for phosphorylation as well as a CheY-like receiver domain at its C-terminus (Leech and Mattick, 2006).

Synechocystis phototaxis motility mutants isolated from forward genetic screens identified genes encoding chemosensory-related proteins and, in all cases, T4P expression levels were aberrant (Bhaya *et al.*, 2001). Mutations within *taxP1–taxD1–taxAY1* locus led to either hyperpiliated or non-motile phenotypes while lesions within the *taxD3* and *taxAY3* loci resulted in either non-piliation or non-motility. The TaxAY proteins are mosaic structures consisting of both CheA- and CheY-related domains, whereas the TaxD1 protein which is a methyl-accepting chemotaxis protein homologue has two segments related to chromophore-binding domains found in phytochromes. The Ctr1 protein (which controls *pilA* transcription in this background) is most related to *P. aeruginosa* PilJ and maps in a gene cluster whose organization and products most resemble those associated with the *P. aeruginosa pilGHIJK* locus (Chung *et al.*, 2001).

As part of their social gliding behaviour, *M. xanthus* cells undergo periodic reversals of cell movement (Blackhart and Zusman, 1985). During these reversals, T4P are believed to be disassembled (retracted?) at one pole and reassembled at the other. This process is disrupted by mutations in the *frz* gene cluster whose products include counterparts of most of the canonical Che components (McBride *et al.*, 1989). Null mutations in *frz* genes decrease reversal frequencies (McBride *et al.*, 1989), whereas a gain-of-function mutation in *frzCD* increases reversal frequencies presumably by constitutively active signalling (Blackhart and Zusman, 1985). An important breakthrough in understanding this system came when it was found that the *frz* regulatory pathway acts to control the pole to pole, oscillatory relocalization of FrzS, which in turn correlates with switching the sites of T4P extrusion from one cell end to the other (Mignot *et al.*, 2005). It is attractive to think that PilT mediated retraction might play a role in disassembly of pili during polar switching although this idea has yet to be tested. *M. xanthus* also carries a second system termed *dif* that encodes a second set of chemotaxis-related homologues required for social motility (Yang *et al.*, 2000). *dif* mutants express retractile T4P but lack glycoprotein surface fibrils that promote cell–cell aggregation and which are thought to anchor and trigger T4P retraction (Li *et al.*, 2003). Despite the significant advances in knowledge of Che-related systems impacting on T4P biology, the key question as to how tactic signals are transduced and ultimately feed into the systems remains unanswered. Nonetheless, it is encouraging that there appears to be a fairly high degree of conservation in the components in distantly related T4P expressing species suggesting common mechanisms. It will also be interesting to see what role Che-like proteins might play in the archaeal flagella systems that share more features with T4P systems than they do with classical flagellar systems.

New, emerging systems

T4P are being increasingly uncovered in unexpected places. Perhaps the most eventful discovery of late was the finding that T4P are expressed in clostridial species where they are associated with twitching motility (Varga *et al.*, 2006). This finding was not entirely unexpected as associations between polar pili and twitching motility in Gram-positive organisms previously existed in the literature (Henriksen and Henrichsen, 1975; Rakotoarivonina *et al.*, 2002). Nonetheless, the work of Varga and colleagues formally confirmed this and demonstrated that the outer membrane was dispensable for T4P biogenesis, as was hypothesized from the detection of intact T4P in the periplasm of *pilQ* mutants in *N. gonorrhoeae* (Wolfgang *et al.*, 2000). Moreover, amongst all the T4P components essential for biogenesis and dynamics seen in Gram-negatives, all are represented in the clostridial genomes, save for those encoding the PilQ secretin and the associated PilP lipoprotein. This, it seems that the basic elements and features of T4P systems established in Gram-negatives will apply to Gram-positives as well. One critical difference in *C. perfringens* was the observation that a *pilT* null mutation abolished detectable T4P expression. This was remarkable since in all other instances, the loss or absence of PilT leads to hyperpiliation. Counter to the paradigms established in *N. gonorrhoeae, P. aeruginosa* and *M. xanthus*, it appears then that PilT impacts on pilus biosynthesis in this case. Assuming that this result is real and not attributable to some trivial explanation (incorrect annotation, etc), what might be the mechanism involved?

An even more puzzling set of findings on T4P-related biology comes from recent efforts made in studies of *Francisella tularensis*, the causative agent of tularemia, and its close relatives. Two main subspecies are associated with human disease: *Francisella tularensis* subsp. *tularensis* is highly virulent while *Francisella tularensis* subsp. *holarctica* produces a milder form of the disease in humans but retains full virulence in mice. A live, attenuated vaccine strain (LVS) was developed from a *F. tularensis* subsp. *holarctica* strain. A third *F. tularensis* subspecies, subsp. *novicida* is pathogenic only for immunocompromised individuals but can cause disease in mice. The genomes of both *Francisella tularensis* subsp. *tularensis* and *F. tularensis* subsp. *novicida* carry a nearly full repertoire of genes required for Type IVa T4P expression and function. A variety of observations suggest that these genes impact on virulence but in surprising ways. First, a spontaneous mutant of *F. tularensis* subsp. *holarctica* in which a deletion has occurred in a pilin-like protein encoding locus shows reduced virulence and an analogous rearrangement is present in the LVS derivative (Forslund *et al.*, 2006). In *F. tularensis* subsp. *novicida,* a number of proteins were identified whose secretion was abolished in mutants carrying mutations in genes predicted to encode T4P and ancillary factors (Hager *et al.*, 2006). This suggested that T4P in this species might also be associated with a type II secretion system. At first blush, this situation thus appears be very similar to that operating in *Dichelobacter nodosus* and *V. cholerae* where the integrity of the T4P system is required for secretion of virulence and colonization factors (Kirn *et al.*, 2003; Han *et al.*, 2008). However, *F. tularensis* subsp. *novicida* mutants lacking the secretion substrates or carrying mutations in the T4P associated loci had increased virulence in the mouse model (Hager *et al.*, 2006). This observation was reconciled by the observation that PepO, one of the secreted substrates, is a zinc protease that promotes increased vasoconstriction and thus inhibits tissue spread and dissemination. Moreover, PepO is absent from the human virulent *Francisella* lineages.

While these findings are fascinating, the key questions as to how and if these phenotypes are related to the expression of bona fide T4P remain unanswered. This is because despite many efforts, T4P have yet to be identified in *Francisella tularensis* subsp. *tularensis* while reports of T4P in *F. tularensis* subsp. *holarctica* are poorly substantiated.

In 2004, Gil *et al* reported that *F. tularensis* subsp. *holarctica* LVS expressed pilus-like appendages and carried genes related to T4P biogenesis that were transcribed (Gil *et al.*, 2004). This work is often cited as showing that this species expresses T4P but this was clearly not the case as the organelles were not characterized either biochemically or immunologically. In fact, it has never been shown that they are even proteinaceous in nature. In addition, efforts of others have failed to identify pilus-like organelles in these strains. Recently, the same group published a report claiming that a null mutation in the gene for the apparent T4P assembly ATPase resulted in the absence of the organelles in LVS (Chakraborty *et al.*, 2008). Furthermore, mutations in the gene encoding a homologue of the PilT retraction ATPase also abolished organelle expression, as reported for T4P in *C. perfringens*. What makes this story even more amazing is that *pilT* in *F. tularensis* subsp. *holarctica* LVS appears to be a pseudogene whose truncated polypeptide would be considered by all available data on this class of ATPases to be non-functional. Again, it is remarkable that, as yet, no effort has been made to confirm the identity of the organelles affected. This situation becomes even more bizarre since as noted earlier, *F. tularensis* subsp. *holarctica* LVS carries a deletion and associated nonsense mutations that inactivate the genes encoding the most likely candidates for T4P subunits (based on structural similarities and synteny) (Forslund *et al.*, 2006). Thus, despite reductions in epithelial cell adherence and virulence seen in mutants failing to express the putative assembly and retraction ATPases, it remains unclear if these phenotypes are associated either directly or indirectly with changes in T4P expression.

The story doesn't get any clearer in *F. tularensis* subsp. *novicida*. Here, a recent paper by Zogaj and colleagues reported that null mutations in the putative assembly and retraction ATPase genes abolished expression of what are claimed to be T4P (Zogaj *et al.*, 2008). However, null mutations in the genes encoding the homologues of the $PilC_{Pa}$ and PilQ assembly proteins were unperturbed in organelle expression but defective in protein secretion and virulence in the mouse model. Moreover, a mutant lacking the three genes encoding pilin-like proteins (those lacking in *F. tularensis* subsp. *holarctica* LVS) likewise retained organelle expression but was defective in protein secretion and reduced in virulence. Conversely, a mutant defective in expressing a pilin-like protein termed PilE4 retained protein secretion and virulence but lacked the organelles. The authors conclude that T4P expression and protein secretion can be dissociated in *F. tularensis* subsp. *novicida* and that PilE4 is the likely major subunit of T4P. Again however, conclusive evidence that the organelles seen are in fact T4P is conspicuously lacking.

Unfortunately, one cannot address all the interesting facets found in the works cited above (for example, the fact that *F. tularensis* species lack a homologue of the canonical PilM T4P assembly factor (Gil *et al.*, 2004) is particularly intriguing). In summary, it appears assured that homologues of T4P structural and biogenesis components impact on the pathogenesis of *F. tularensis* subsp. *holarctica* and *F. tularensis* subsp. *novicida* infections and a similar story is likely to be seen for *Francisella tularensis* subsp. *tularensis*. The potential role of T4P per se in these phenotypes however remains to be established.

Another interesting case involves the bacterial symbiont *Bdellovibrio bacteriovorus*, which carries a complete repertoire of genes required for Type IVa T4P expression and function (Evans *et al.*, 2007). In one case, a knockout mutant defective in the expression of the likely T4P pilin subunit was found to be defective in predation of its *E. coli* host while using a forward genetic approach, a number of predation defective mutants were found to carry inactivating mutations in genes encoding homologues of consensus T4P biogenesis components and PilT (Medina *et al.*, 2008). T4P-like organelles at the anterior pole of cells have also been documented but it has not yet been demonstrated that these structures are, in fact, T4P nor has it been shown that these structures are absent or altered in the mutants mentioned earlier. Nonetheless, it has been hypothesized that *B. bacteriovorus* T4P might promote attachment to host cells and/or a subsequent step of invasion into the host periplasm. If so, T4P would also be implicated in bacterial–bacterial pathogenic interactions along with the previously well established roles in bacterial diseases in mammals and plants.

A role for the PilT ATPase in T4P assembly?

In contrast to the well documented expression of pili (and hyperpiliation) seen in *pilT* null mutants of all other species examined, two studies mentioned previously have reported that analogous mutants are non-piliated suggestion a role for PilT in assembly. While both works offer a variety of potential explanations for these findings, none of them are very satisfying. Additionally, the level of resolution of the data presented fails to resolve if the mutants truly make absolutely no pili, as opposed to dramatically reduced amounts of pili or short pili incapable of detection under the conditions used. These are important distinctions that need to be resolved before one can begin rewriting the story of PilT biology. I would like to offer a potential explanation for these findings based on a model that predicts that T4P assembly is kinetically different in the presence versus the absence of PilT in all species. Based on data from type II secretion systems showing that pseudopilin subunits are translocated by the Sec/SRP pathway (Francetic *et al.*, 2007), T4P subunits are likely to enter the assembly machinery laterally from a membrane associated pool. Based on estimated extrusion rates of 1000–1500 subunits per pilus per second, pilin would have to diffuse in the two-dimensional bilayer at fairly high rates to feed assembly. However, rates of protein diffusion in membranes are thought to be relatively low and it seems reasonable that localized pilin concentration might be a rate limiting step for assembly. However, there is evidence that pilins re-entering the membrane following pilus retraction are reused in subsequent rounds of assembly (Skerker and Berg, 2001). As such, pilus assembly in the absence of retraction (in a *pilT* background) would simply be limited by the rates of steady-state diffusion of pilin to the assembly site. If, however, multiple rounds of pilus extrusion and retraction occur at single assembly sites, the relative concentration of pilin subunits at that site over time is determined primarily by the amounts of pilin delivered directly back to that site following retraction. It remains unknown what the relative ratio of extrusion to retraction events is in a single wild-type cell, but in a *P. aeruginosa pilT* background both the number of pilus filaments and the average length of pili increases dramatically. Thus, one can infer that a significant number of rounds of extrusion and retraction are probably taking place (or at least a significant number of extrusion events are terminated by *pilT* – mediated retraction). If we make a fairly conservative estimate that for every instance in which an extruded pilus is detected there have been five rounds of extrusion/retraction, localized pools of pilin would be substantially greater in a wild-type background versus a *pilT* mutant (Fig. 2.3).

This 'pump priming' model might account for the discrepancies between *pilT* mutants in *C. perfringens* and *Francisella novicida* versus all the other backgrounds in a number of ways. First, a simple basis for the differences could be that the relative levels of pilins subunits per cell is substantially lower in *C. perfringens* and *Francisella novicida* versus the other cases. There is no good data on this although it is generally agreed that *P. aeruginosa* and *N. gonorrhoeae* synthesize

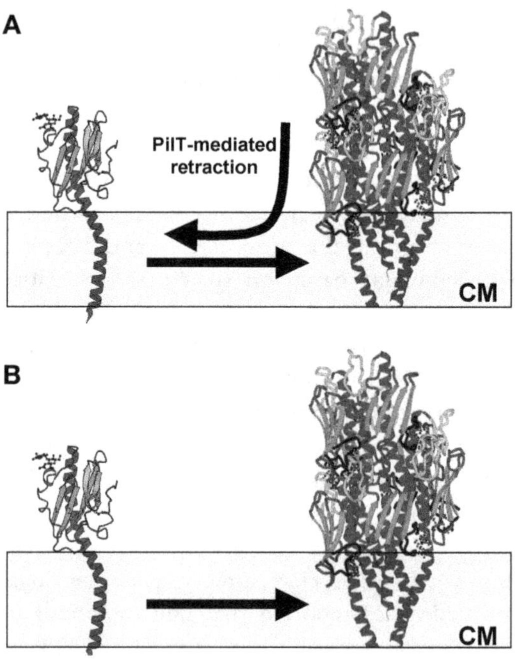

Figure 2.3 A model for how PilT-mediated retraction might influence local membrane pools of pilin and thus T4P assembly. **(A)** In wild-type cells, levels of pilin pools localized to assembly sites are dictated by both lateral diffusion and pilin re-entering the membrane following T4P retraction. **(B)** In the absence of functional PilT, pilin localizes solely to these sites via lateral diffusion. See text for details.

relatively high levels of subunits. Additionally or alternatively, levels of pilin degradation/turnover might be substantially higher in *C. perfringens* and *Francisella novicida pilT* mutants than they are in the other instances. Phospholipid membranes are also highly dynamic systems, with the rate of lipid diffusion being orders of magnitude times higher than that of membrane proteins. As such, differences in phospholipid constitution might also impact here. In fact, the potential role of lipid membrane biochemistry in T4P assembly and dynamics has been largely ignored.

Another confounding factor here is that little is known about what effects relative pilin subunit abundance and component stoichiometry have on dictating the number of pilus assembly/nucleation sites per cell versus influencing the average length of the pili themselves. For example in the *C. perfringens* and *Francisella novicida* mutants, reduced availability of subunits or an increase in the number of assembly sites might lead to a situation where there are many assembly – competent sites but not enough pilin at any one to lead to the synthesis of a recognizable pilus. Suffice to say that there are many possible explanations for the observations made for the *C. perfringens* and *Francisella novicida pilT* mutants to be considered before we need to envision means by which PilT might act as a true assembly component. Admittedly, devising experiments to address these possibilities in a straightforward way remains a tough task.

Future trends

Over the last 40 years, a significant transition from mystery to molecular mechanism has taken place in the biology of T4P. But at this point, most if not all of the low-hanging fruit has been plucked. T4P and related systems will continue to be uncovered, particularly through genomic and metagenomic sequencing endeavours and each system will undoubtedly offers its unique facets of how a highly conserved set of elements can be remodelled in different ways. As such, comparative analyses have significant potential for enriching the knowledge base of Type IV pilus systems. However, certain systems will continue to be the premier models in which to unravel the intricacies of organelle biogenesis, structure and function. For example, well defined phenotypes relating to adherence and competence for natural genetic transformation and the lack of impinging regulatory elements continue to make *N. gonorrhoeae* an attractive model system. Nonetheless, *N. gonorrhoeae* doesn't use T4P for vectorial motility and accordingly, *P. aeruginosa* and *M. xanthus* will be likely to rule in this arena. Here, the findings that Che-related components influence T4P expression raise a whole new set of questions to be addressed. How does one extrapolate between the gross alterations in Type IV pilus expression levels seen in the Che knockout mutants and the functions normally served by Che-related factors? And at what levels of the system does the signal transduction pathway exert its impact?

Another area of fundamental research begging attention involves the molecular mechanisms by which effectors of pilus homeostasis inhibit, preclude or neutralize PilT. In particular, it will be interesting and important to see if the findings made in *N. gonorrhoeae* and *N. meningitidis* are applicable to other Type IV pilus expressing species. And finally, it will be crucial to determine how pilin-like proteins interact with Type IV pilin subunits, T4P themselves or the Type IV pilus associated machinery to exert their control on Type IV pilus dynamics and associated function. Of course, newer technologies in structural characterization, biophysics and the like will undoubtedly play a greater role in all of these efforts. By way of example, the structure of an archaeal T4P was recently solved before virtually any of the genetic or biochemical details were known (Wang *et al.*, 2008). But barring a major biochemical breakthrough in which the whole system (or significant parts thereof) can be reconstituted *in vitro*, genetic approaches and analyses will continue to be forces to be reckoned with.

Acknowledgements

I gratefully acknowledge the contributions of past and current T4P aficionados (thanks for the information!) and apologize to those whose works could not be included here. Work in my research group is supported by funding from the Norwegian Research Council.

References

Aas, F.E., Lovold, C., and Koomey, M. (2002a). An inhibitor of DNA binding and uptake events dictates the proficiency of genetic transformation in *Neisseria gonorrhoeae*: mechanism of action and links to Type IV pilus expression. Mol. Microbiol. *46*, 1441–1450.

Aas, F.E., Winther-Larsen, H.C., Wolfgang, M., Frye, S., Lovold, C., Roos, N., van Putten, J.P., and Koomey, M. (2007). Substitutions in the N-terminal alpha helical spine of *Neisseria gonorrhoeae* pilin affect Type IV pilus assembly, dynamics and associated functions. Mol. Microbiol. *63*, 69–85.

Aas, F.E., Wolfgang, M., Frye, S., Dunham, S., Lovold, C., and Koomey, M. (2002b). Competence for natural transformation in *Neisseria gonorrhoeae*: components of DNA binding and uptake linked to Type IV pilus expression. Mol. Microbiol. *46*, 749–760.

Albano, M., and Dubnau, D.A. (1989). Cloning and characterization of a cluster of linked *Bacillus subtilis* late competence mutations. J. Bacteriol. *171*, 5376–5385.

Alm, R.A., Bodero, A.J., Free, P.D., and Mattick, J.S. (1996a). Identification of a novel gene, pilZ, essential for type 4 fimbrial biogenesis in *Pseudomonas aeruginosa*. J. Bacteriol. *178*, 46–53.

Alm, R.A., Hallinan, J.P., Watson, A.A., and Mattick, J.S. (1996b). Fimbrial biogenesis genes of *Pseudomonas aeruginosa*: *pilW* and *pilX* increase the similarity of type 4 fimbriae to the GSP protein-secretion systems and *pilY1* encodes a gonococcal PilC homologue. Mol. Microbiol. *22*, 161–173.

Alm, R.A., and Mattick, J.S. (1995). Identification of a gene, *pilV*, required for type 4 fimbrial biogenesis in *Pseudomonas aeruginosa*, whose product possesses a pre-pilin-like leader sequence. Mol. Microbiol. *16*, 485–496.

Alm, R.A., and Mattick, J.S. (1996). Identification of two genes with prepilin-like leader sequences involved in type 4 fimbrial biogenesis in *Pseudomonas aeruginosa*. J. Bacteriol. *178*, 3809–3817.

Asikyan, M.L., Kus, J.V., and Burrows, L.L. (2008). Novel proteins that modulate Type IV pilus retraction dynamics in *Pseudomonas aeruginosa*. J. Bacteriol. *190*, 7022–7034.

Balasingham, S.V., Collins, R.F., Assalkhou, R., Homberset, H., Frye, S.A., Derrick, J.P., and Tonjum, T. (2007). Interactions between the lipoprotein PilP and the secretin PilQ in *Neisseria meningitidis*. J. Bacteriol. *189*, 5716–5727.

Bayan, N., Guilvout, I., and Pugsley, A.P. (2006). Secretins take shape. Mol. Microbiol. *60*, 1–4.

Bayley, D.P., and Jarrell, K.F. (1998). Further evidence to suggest that archaeal flagella are related to bacterial Type IV pili. J. Mol. Evol. *46*, 370–373.

Bhaya, D., Takahashi, A., and Grossman, A.R. (2001). Light regulation of Type IV pilus-dependent motility by chemosensor-like elements in *Synechocystis* PCC6803. Proc. Natl. Acad. Sci. U.S.A. *98*, 7540–7545.

Blackhart, B.D., and Zusman, D.R. (1985). 'Frizzy' genes of *Myxococcus xanthus* are involved in control of frequency of reversal of gliding motility. Proc. Natl. Acad. Sci. U.S.A. *82*, 8767–8770.

Bovre, K., and Froholm, L.O. (1972). Competence in genetic transformation related to colony type and fimbriation in three species of *Moraxella*. Acta Pathol. Microbiol. Scand. B Microbiol. Immunol. *80*, 649–659.

Bradley, D.E. (1972a). Evidence for the retraction of *Pseudomonas aeruginosa* RNA phage pili. Biochem. Biophys. Res. Commun. *47*, 142–149.

Bradley, D.E. (1972b). Shortening of *Pseudomonas aeruginosa* pili after RNA-phage adsorption. J. Gen. Microbiol. *72*, 303–319.

Bradley, D.E. (1973). Basic characterization of a *Pseudomonas aeruginosa* pilus-dependent bacteriophage with a long noncontractile tail. J. Virol. *12*, 1139–1148.

Bradley, D.E. (1974). The adsorption of *Pseudomonas aeruginosa* pilus-dependent bacteriophages to a host mutant with nonretractile pili. Virology *58*, 149–163.

Bradley, D.E. (1980). A function of *Pseudomonas aeruginosa* PAO polar pili: twitching motility. Can. J. Microbiol. *26*, 146–154.

Brossay, L., Paradis, G., Fox, R., Koomey, M., and Hebert, J. (1994). Identification, localization, and distribution of the PilT protein in *Neisseria gonorrhoeae*. Infect. Immun. *62*, 2302–2308.

Carbonnelle, E., Helaine, S., Nassif, X., and Pelicic, V. (2006). A systematic genetic analysis in *Neisseria meningitidis* defines the Pil proteins required for assembly, functionality, stabilization and export of Type IV pili. Mol. Microbiol. *61*, 1510–1522.

Carbonnelle, E., Helaine, S., Prouvensier, L., Nassif, X., and Pelicic, V. (2005). Type IV pilus biogenesis in *Neisseria meningitidis*: PilW is involved in a step occurring after pilus assembly, essential for fibre stability and function. Mol. Microbiol. *55*, 54–64.

Chakraborty, S., Monfett, M., Maier, T.M., Benach, J.L., Frank, D.W., and Thanassi, D.G. (2008). Type IV pili in *Francisella tularensis*: roles of *pilF* and *pilT* in fiber assembly, host cell adherence, and virulence. Infect. Immun. *76*, 2852–2861.

Chen, I., and Dubnau, D. (2004). DNA uptake during bacterial transformation. Nat. Rev. Microbiol. *2*, 241–249.

Chen, I., and Gotschlich, E.C. (2001). ComE, a competence protein from *Neisseria gonorrhoeae* with DNA-binding activity. J. Bacteriol. *183*, 3160–3168.

Chiang, P., Sampaleanu, L.M., Ayers, M., Pahuta, M., Howell, P.L., and Burrows, L.L. (2008). Functional role of conserved residues in the characteristic secretion NTPase motifs of the *Pseudomonas aeruginosa* Type IV pilus motor proteins PilB, PilT and PilU. Microbiology *154*, 114–126.

Chiang, S.L., Taylor, R.K., Koomey, M., and Mekalanos, J.J. (1995). Single amino acid substitutions in the N-terminus of *Vibrio cholerae* TcpA affect colonization, autoagglutination, and serum resistance. Mol. Microbiol. *17*, 1133–1142.

Chung, Y.H., Cho, M.S., Moon, Y.J., Choi, J.S., Yoo, Y.C., Park, Y.I., Lee, K.M., Kang, K.W., and Park, Y.M. (2001). *ctr1*, a gene involved in a signal transduction pathway of the gliding motility in the cyanobacterium *Synechocystis* sp. PCC 6803. FEBS Lett. *492*, 33–38.

Clausen, M., Koomey, M., and Maier, B. (2008). Dynamics of Type IV pili are controlled by switching between multiple states. Biophys J. *96*, 1169–1177.

Craig, L., Pique, M.E., and Tainer, J.A. (2004). Type IV pilus structure and bacterial pathogenicity. Nat. Rev. Microbiol. *2*, 363–378.

Crowther, L.J., Anantha, R.P., and Donnenberg, M.S. (2004). The inner membrane subassembly of the enteropathogenic *Escherichia coli* bundle-forming pilus machine. Mol. Microbiol. *52*, 67–79.

Crowther, L.J., Yamagata, A., Craig, L., Tainer, J.A., and Donnenberg, M.S. (2005). The ATPase activity of BfpD is greatly enhanced by zinc and allosteric interactions with other Bfp proteins. J. Biol. Chem. *280*, 24839–24848.

Darzins, A. (1993). The *pilG* gene product, required for *Pseudomonas aeruginosa* pilus production and twitching motility, is homologous to the enteric, single-domain response regulator CheY. J. Bacteriol. *175*, 5934–5944.

Darzins, A. (1994). Characterization of a *Pseudomonas aeruginosa* gene cluster involved in pilus biosynthesis and twitching motility: sequence similarity to the chemotaxis proteins of enterics and the gliding bacterium *Myxococcus xanthus*. Mol. Microbiol. *11*, 137–153.

Darzins, A. (1995). The *Pseudomonas aeruginosa pilK* gene encodes a chemotactic methyltransferase (CheR) homologue that is translationally regulated. Mol. Microbiol. *15*, 703–717.

Duffy, S., Burch, C.L., and Turner, P.E. (2007). Evolution of host specificity drives reproductive isolation among RNA viruses. Evolution *61*, 2614–2622.

Duffy, S., Turner, P.E., and Burch, C.L. (2006). Pleiotropic costs of niche expansion in the RNA bacteriophage phi 6. Genetics *172*, 751–757.

Ehrlich, M., Boll, W., Van Oijen, A., Hariharan, R., Chandran, K., Nibert, M.L., and Kirchhausen, T. (2004). Endocytosis by random initiation and stabilization of clathrin-coated pits. Cell *118*, 591–605.

Elferink, M.G., Albers, S.V., Konings, W.N., and Driessen, A.J. (2001). Sugar transport in *Sulfolobus solfataricus* is mediated by two families of binding protein-dependent ABC transporters. Mol. Microbiol. *39*, 1494–1503.

Elleman, T.C., and Peterson, J.E. (1987). Expression of multiple types of N-methyl Phe pili in *Pseudomonas aeruginosa*. Mol. Microbiol. *1*, 377–380.

Evans, K.J., Lambert, C., and Sockett, R.E. (2007). Predation by *Bdellovibrio bacteriovorus* HD100 requires Type IV pili. J. Bacteriol. *189*, 4850–4859.

Ferris, M.T., Joyce, P., and Burch, C.L. (2007). High frequency of mutations that expand the host range of an RNA virus. Genetics *176*, 1013–1022.

Finlay, B.B., Pasloske, B.L., and Paranchych, W. (1986). Expression of the *Pseudomonas aeruginosa* PAK pilin gene in *Escherichia coli*. J. Bacteriol. *165*, 625–630.

Forslund, A.L., Kuoppa, K., Svensson, K., Salomonsson, E., Johansson, A., Bystrom, M., Oyston, P.C., Michell, S.L., Titball, R.W., Noppa, L., Frithz-Lindsten, E., Forsman, M., and Forsberg, A. (2006). Direct repeat-mediated deletion of a Type IV pilin gene results in major virulence attenuation of *Francisella tularensis*. Mol. Microbiol. *59*, 1818–1830.

Francetic, O., Buddelmeijer, N., Lewenza, S., Kumamoto, C.A., and Pugsley, A.P. (2007). Signal recognition particle-dependent inner membrane targeting of the PulG pseudopilin component of a type II secretion system. J. Bacteriol. *189*, 1783–1793.

Froholm, L.O., Jyssum, K., and Bovre, K. (1973). Electron microscopical and cultural features of *Neisseria meningitidis* competence variants. Acta Pathol. Microbiol. Scand. [B] Microbiol. Immunol. *81*, 525–537.

Frost, L.S., Carpenter, M., and Paranchych, W. (1978). N-methylphenylalanine at the N-terminus of pilin isolated from *Pseudomonas aeruginosa* K. Nature *271*, 87–89.

Frost, L.S., and Paranchych, W. (1977). Composition and molecular weight of pili purified from *Pseudomonas aeruginosa* K.J. Bacteriol. *131*, 259–269.

Gil, H., Benach, J.L., and Thanassi, D.G. (2004). Presence of pili on the surface of *Francisella tularensis*. Infect. Immun. 72, 3042–3047.

Graupner, S., Frey, V., Hashemi, R., Lorenz, M.G., Brandes, G., and Wackernagel, W. (2000). Type IV pilus genes *pilA* and *pilC* of *Pseudomonas stutzeri* are required for natural genetic transformation, and *pilA* can be replaced by corresponding genes from non-transformable species. J. Bacteriol. *182*, 2184–2190.

Graupner, S., Weger, N., Sohni, M., and Wackernagel, W. (2001). Requirement of novel competence genes *pilT* and *pilU* of *Pseudomonas stutzeri* for natural transformation and suppression of *pilT* deficiency by a hexahistidine tag on the Type IV pilus protein PilAI. J. Bacteriol. *183*, 4694–4701.

Hager, A.J., Bolton, D.L., Pelletier, M.R., Brittnacher, M.J., Gallagher, L.A., Kaul, R., Skerrett, S.J., Miller, S.I., and Guina, T. (2006). Type IV pili-mediated secretion modulates *Francisella virulence*. Mol. Microbiol. *62*, 227–237.

Han, X., Kennan, R.M., Davies, J.K., Reddacliff, L.A., Dhungyel, O.P., Whittington, R.J., Turnbull, L., Whitchurch, C.B., and Rood, J.I. (2008). Twitching motility is essential for virulence in *Dichelobacter nodosus*. J. Bacteriol. *190*, 3323–3335.

Helaine, S., Dyer, D.H., Nassif, X., Pelicic, V., and Forest, K.T. (2007). 3D structure/function analysis of PilX reveals how minor pilins can modulate the virulence properties of Type IV pili. Proc. Natl. Acad. Sci. U.S.A. *104*, 15888–15893.

Henrichsen, J. (1975). The occurrence of twitching motility among Gram-negative bacteria. Acta. Pathol. Microbiol. Scand. B *83*, 171–178.

Henrichsen, J. (1983). Twitching motility. Annu. Rev. Microbiol. 37, 81–93.

Henriksen, S.D., and Henrichsen, J. (1975). Twitching motility and possession of polar fimbriae in spreading *Streptococcus sanguis* isolates from the human throat. Acta Pathol. Microbiol. Scand. B 83, 133–140.

Hobbs, M., Collie, E.S., Free, P.D., Livingston, S.P., and Mattick, J.S. (1993). PilS and PilR, a two-component transcriptional regulatory system controlling expression of type 4 fimbriae in *Pseudomonas aeruginosa*. Mol. Microbiol. 7, 669–682.

Holland, S.J., Sanz, C., and Perham, R.N. (2006). Identification and specificity of pilus adsorption proteins of filamentous bacteriophages infecting *Pseudomonas aeruginosa*. Virology *345*, 540–548.

Hung, D.L., and Hultgren, S.J. (1998). Pilus biogenesis via the chaperone/usher pathway: an integration of structure and function. J. Struct. Biol. *124*, 201–220.

Irvin, R.T., Doig, P., Lee, K.K., Sastry, P.A., Paranchych, W., Todd, T., and Hodges, R.S. (1989). Characterization of the *Pseudomonas aeruginosa* pilus adhesin: confirmation that the pilin structural protein subunit contains a human epithelial cell-binding domain. Infect. Immun. *57*, 3720–3726.

Ishimoto, K.S., and Lory, S. (1989). Formation of pilin in *Pseudomonas aeruginosa* requires the alternative sigma factor (RpoN) of RNA polymerase. Proc. Natl. Acad. Sci. U.S.A. *86*, 1954–1957.

Ishimoto, K.S., and Lory, S. (1992). Identification of *pilR*, which encodes a transcriptional activator of the *Pseudomonas aeruginosa* pilin gene. J. Bacteriol. *174*, 3514–3521.

Johnson, K., and Lory, S. (1987). Characterization of *Pseudomonas aeruginosa* mutants with altered piliation. J. Bacteriol. *169*, 5663–5667.

Johnson, K., Parker, M.L., and Lory, S. (1986). Nucleotide sequence and transcriptional initiation site of two *Pseudomonas aeruginosa* pilin genes. J. Biol. Chem. *261*, 15703–15708.

Kachlany, S.C., Planet, P.J., Bhattacharjee, M.K., Kollia, E., DeSalle, R., Fine, D.H., and Figurski, D.H. (2000). Nonspecific adherence by *Actinobacillus actinomycetemcomitans* requires genes widespread in bacteria and archaea. J. Bacteriol. *182*, 6169–6176.

Kagami, Y., Ratliff, M., Surber, M., Martinez, A., and Nunn, D.N. (1998). Type II protein secretion by *Pseudomonas aeruginosa*: genetic suppression of a conditional mutation in the pilin-like component XcpT by the cytoplasmic component XcpR. Mol. Microbiol. *27*, 221–233.

Kirn, T.J., Bose, N., and Taylor, R.K. (2003). Secretion of a soluble colonization factor by the TCP type 4 pilus biogenesis pathway in *Vibrio cholerae*. Mol. Microbiol. *49*, 81–92.

Kirn, T.J., Lafferty, M.J., Sandoe, C.M., and Taylor, R.K. (2000). Delineation of pilin domains required for bacterial association into microcolonies and intestinal colonization by *Vibrio cholerae*. Mol. Microbiol. *35*, 896–910.

Knies, J.L., Dang, K.K., Vision, T.J., Hoffman, N.G., Swanstrom, R., and Burch, C.L. (2008). Compensatory evolution in RNA secondary structures increases substitution rate variation among sites. Mol. Biol. Evol. *25*, 1778–1787.

Koga, T., Ishimoto, K., and Lory, S. (1993). Genetic and functional characterization of the gene cluster specifying expression of *Pseudomonas aeruginosa* pili. Infect. Immun. *61*, 1371–1377.

Koo, J., Tammam, S., Ku, S.Y., Sampaleanu, L., Burrows, L.L., and Howell, P.L. (2008). PilF is an outer membrane lipoprotein required for multimerization and localization of the *Pseudomonas aeruginosa* Type IV pilus secretin. J. Bacteriol. *190*, 6961–6969.

Kus, J.V., Tullis, E., Cvitkovitch, D.G., and Burrows, L.L. (2004). Significant differences in Type IV pilin allele distribution among *Pseudomonas aeruginosa* isolates from cystic fibrosis (CF) versus non-CF patients. Microbiology *150*, 1315–1326.

Lappann, M., Haagensen, J.A., Claus, H., Vogel, U., and Molin, S. (2006). Meningococcal biofilm formation: structure, development and phenotypes in a standardized continuous flow system. Mol. Microbiol. *62*, 1292–1309.

Leech, A.J., and Mattick, J.S. (2006). Effect of site-specific mutations in different phosphotransfer domains of the chemosensory protein ChpA on *Pseudomonas aeruginosa* motility. J. Bacteriol. *188*, 8479–8486.

Li, Y., Sun, H., Ma, X., Lu, A., Lux, R., Zusman, D., and Shi, W. (2003). Extracellular polysaccharides mediate pilus retraction during social motility of *Myxococcus xanthus*. Proc. Natl. Acad. Sci. U.S.A. *100*, 5443–5448.

Macnab, R.M. (2003). How bacteria assemble flagella. Annu. Rev. Microbiol. *57*, 77–100.

Maier, B. (2005). Using laser tweezers to measure twitching motility in *Neisseria*. Curr. Opin. Microbiol. *8*, 344–349.

Maier, B., Koomey, M., and Sheetz, M.P. (2004). A force-dependent switch reverses Type IV pilus retraction. Proc. Natl. Acad. Sci. U.S.A. *101*, 10961–10966.

Martin, P.R., Watson, A.A., McCaul, T.F., and Mattick, J.S. (1995). Characterization of a five-gene cluster required for the biogenesis of type 4 fimbriae in *Pseudomonas aeruginosa*. Mol. Microbiol. *16*, 497–508.

McBride, M.J., Weinberg, R.A., and Zusman, D.R. (1989). 'Frizzy' aggregation genes of the gliding bacterium *Myxococcus xanthus* show sequence similarities to the chemotaxis genes of enteric bacteria. Proc. Natl. Acad. Sci. U.S.A. *86*, 424–428.

Medina, A.A., Shanks, R.M., and Kadouri, D.E. (2008). Development of a novel system for isolating genes involved in predator–prey interactions using host independent derivatives of *Bdellovibrio bacteriovorus* 109J. BMC Microbiol. *8*, 33.

Merz, A.J., So, M., and Sheetz, M.P. (2000). Pilus retraction powers bacterial twitching motility. Nature *407*, 98–102.

Mignot, T., Merlie, J.P., Jr., and Zusman, D.R. (2005). Regulated pole-to-pole oscillations of a bacterial gliding motility protein. Science *310*, 855–857.

Mindich, L., Sinclair, J.F., and Cohen, J. (1976). The morphogenesis of bacteriophage phi6: particles formed by nonsense mutants. Virology *75*, 224–231.

Mitchison, T., and Kirschner, M. (1984). Dynamic instability of microtubule growth. Nature *312*, 237–242.

Morand, P.C., Bille, E., Morelle, S., Eugene, E., Beretti, J.L., Wolfgang, M., Meyer, T.F., Koomey, M., and Nassif, X. (2004). Type IV pilus retraction in pathogenic *Neisseria* is regulated by the PilC proteins. EMBO J. *23*, 2009–2017.

Nudleman, E., Wall, D., and Kaiser, D. (2005). Cell-to-cell transfer of bacterial outer membrane lipoproteins. Science *309*, 125–127.

Nudleman, E., Wall, D., and Kaiser, D. (2006). Polar assembly of the Type IV pilus secretin in *Myxococcus xanthus*. Mol. Microbiol. *60*, 16–29.

Nunn, D., Bergman, S., and Lory, S. (1990). Products of three accessory genes, *pilB*, *pilC*, and *pilD*, are required for biogenesis of *Pseudomonas aeruginosa* pili. J. Bacteriol. *172*, 2911–2919.

Nunn, D.N., and Lory, S. (1991). Product of the *Pseudomonas aeruginosa* gene *pilD* is a prepilin leader peptidase. Proc. Natl. Acad. Sci. U.S.A. *88*, 3281–3285.

Ottow, J.C. (1975). Ecology, physiology, and genetics of fimbriae and pili. Annu. Rev. Microbiol. *29*, 79–108.

Park, H.S., Wolfgang, M., and Koomey, M. (2002). Modification of Type IV pilus-associated epithelial cell adherence and multicellular behavior by the PilU protein of *Neisseria gonorrhoeae*. Infect. Immun. *70*, 3891–3903.

Park, H.S., Wolfgang, M., van Putten, J.P., Dorward, D., Hayes, S.F., and Koomey, M. (2001). Structural alterations in a Type IV pilus subunit protein result in concurrent defects in multicellular behaviour and adherence to host tissue. Mol. Microbiol. *42*, 293–307.

Pugsley, A.P. (1993). The complete general secretory pathway in Gram-negative bacteria. Microbiol. Rev. *57*, 50–108.

Rakotoarivonina, H., Jubelin, G., Hebraud, M., Gaillard-Martinie, B., Forano, E., and Mosoni, P. (2002). Adhesion to cellulose of the Gram-positive bacterium *Ruminococcus albus* involves Type IV pili. Microbiology *148*, 1871–1880.

Robien, M.A., Krumm, B.E., Sandkvist, M., and Hol, W.G. (2003). Crystal structure of the extracellular protein secretion NTPase EpsE of *Vibrio cholerae*. J Mol Biol *333*, 657–674.

Rodriguez-Soto, J.P., and Kaiser, D. (1997). The *tgl* gene: social motility and stimulation in *Myxococcus xanthus*. J. Bacteriol. *179*, 4361–4371.

Roine, E., Raineri, D.M., Romantschuk, M., Wilson, M., and Nunn, D.N. (1998). Characterization of Type IV pilus genes in *Pseudomonas syringae* pv. tomato DC3000. Mol. Plant Microbe Interact. *11*, 1048–1056.

Rumszauer, J., Schwarzenlander, C., and Averhoff, B. (2006). Identification, subcellular localization and functional interactions of PilMNOWQ and PilA4 involved in transformation competency and pilus biogenesis in the thermophilic bacterium *Thermus thermophilus* HB27. FEBS J. 273, 3261–3272.

Russell, M.A., and Darzins, A. (1994). The pilE gene product of *Pseudomonas aeruginosa*, required for pilus biogenesis, shares amino acid sequence identity with the N-termini of type 4 prepilin proteins. Mol. Microbiol. *13*, 973–985.

Satyshur, K.A., Worzalla, G.A., Meyer, L.S., Heiniger, E.K., Aukema, K.G., Misic, A.M., and Forest, K.T. (2007). Crystal structures of the pilus retraction motor PilT suggest large domain movements and subunit cooperation drive motility. Structure *15*, 363–376.

Sauer, F.G., Futterer, K., Pinkner, J.S., Dodson, K.W., Hultgren, S.J., and Waksman, G. (1999). Structural basis of chaperone function and pilus biogenesis. Science *285*, 1058–1061.

Savvides, S.N., Yeo, H.J., Beck, M.R., Blaesing, F., Lurz, R., Lanka, E., Buhrdorf, R., Fischer, W., Haas, R., and Waksman, G. (2003). VirB11 ATPases are dynamic hexameric assemblies: new insights into bacterial Type IV secretion. EMBO J. *22*, 1969–1980.

Skerker, J.M., and Berg, H.C. (2001). Direct observation of extension and retraction of Type IV pili. Proc. Natl. Acad. Sci. U.S.A. *98*, 6901–6904.

Strom, M.S., and Lory, S. (1986). Cloning and expression of the pilin gene of *Pseudomonas aeruginosa* PAK in *Escherichia coli*. J. Bacteriol. *165*, 367–372.

Strom, M.S., and Lory, S. (1991). Amino acid substitutions in pilin of *Pseudomonas aeruginosa*. Effect on leader peptide cleavage, amino-terminal methylation, and pilus assembly. J. Biol. Chem. *266*, 1656–1664.

Theriot, J.A. (2000). The polymerization motor. Traffic *1*, 19–28.

Tilney, L.G., and Portnoy, D.A. (1989). Actin filaments and the growth, movement, and spread of the intracellular bacterial parasite, *Listeria monocytogenes*. J. Cell Biol. *109*, 1597–1608.

Todd, W.J., Wray, G.P., and Hitchcock, P.J. (1984). Arrangement of pili in colonies of *Neisseria gonorrhoeae*. J. Bacteriol. *159*, 312–320.

Tomb, J.F., el-Hajj, H., and Smith, H.O. (1991). Nucleotide sequence of a cluster of genes involved in the transformation of *Haemophilus influenzae* Rd. Gene *104*, 1–10.

Tonjum, T., Caugant, D.A., Dunham, S.A., and Koomey, M. (1998). Structure and function of repetitive sequence elements associated with a highly polymorphic domain of the *Neisseria meningitidis* PilQ protein. Mol. Microbiol. *29*, 111–124.

Tonjum, T., Freitag, N.E., Namork, E., and Koomey, M. (1995). Identification and characterization of *pilG*, a highly conserved pilus-assembly gene in pathogenic *Neisseria*. Mol. Microbiol. *16*, 451–464.

Totten, P.A., Lara, J.C., and Lory, S. (1990). The *rpoN* gene product of *Pseudomonas aeruginosa* is required for expression of diverse genes, including the flagellin gene. J. Bacteriol. *172*, 389–396.

Trindade, M.B., Job, V., Contreras-Martel, C., Pelicic, V., and Dessen, A. (2008). Structure of a widely conserved Type IV pilus biogenesis factor that affects the stability of secretin multimers. J. Mol. Biol. *378*, 1031–1039.

Turner, L.R., Lara, J.C., Nunn, D.N., and Lory, S. (1993). Mutations in the consensus ATP-binding sites of XcpR and PilB eliminate extracellular protein secretion and pilus biogenesis in *Pseudomonas aeruginosa*. J. Bacteriol. *175*, 4962–4969.

Varga, J.J., Nguyen, V., O'Brien, D.K., Rodgers, K., Walker, R.A., and Melville, S.B. (2006). Type IV pili-dependent gliding motility in the Gram-positive pathogen *Clostridium perfringens* and other Clostridia. Mol. Microbiol. *62*, 680–694.

Wall, D., Wu, S.S., and Kaiser, D. (1998). Contact stimulation of Tgl and Type IV pili in *Myxococcus xanthus*. J. Bacteriol. *180*, 759–761.

Wang, Y.A., Yu, X., Ng, S.Y., Jarrell, K.F., and Egelman, E.H. (2008). The structure of an archaeal pilus. J. Mol. Biol. *381*, 456–466.

Watson, A.A., Alm, R.A., and Mattick, J.S. (1996a). Identification of a gene, *pilF*, required for type 4 fimbrial biogenesis and twitching motility in *Pseudomonas aeruginosa*. Gene *180*, 49–56.

Watson, A.A., Mattick, J.S., and Alm, R.A. (1996b). Functional expression of heterologous type 4 fimbriae in *Pseudomonas aeruginosa*. Gene *175*, 143–150.

Watts, T.H., Kay, C.M., and Paranchych, W. (1982). Dissociation and characterization of pilin isolated from *Pseudomonas aeruginosa* strains PAK and PAO. Can. J. Biochem. *60*, 867–872.

Watts, T.H., Kay, C.M., and Paranchych, W. (1983). Spectral properties of three quaternary arrangements of *Pseudomonas* pilin. Biochemistry 22, 3640–3646.

Watts, T.H., Sastry, P.A., Hodges, R.S., and Paranchych, W. (1983). Mapping of the antigenic determinants of *Pseudomonas aeruginosa* PAK polar pili. Infect. Immun. *42*, 113–121.

Watts, T.H., Scraba, D.G., and Paranchych, W. (1982). Formation of 9-nm filaments from pilin monomers obtained by octyl-glucoside dissociation of *Pseudomonas aeruginosa* pili. J. Bacteriol. *151*, 1508–1513.

Watts, T.H., Worobec, E.A., and Paranchych, W. (1982). Identification of pilin pools in the membranes of *Pseudomonas aeruginosa*. J. Bacteriol. *152*, 687–691.

Whitchurch, C.B., Alm, R.A., and Mattick, J.S. (1996). The alginate regulator AlgR and an associated sensor FimS are required for twitching motility in *Pseudomonas aeruginosa*. Proc. Natl. Acad. Sci. U.S.A. *93*, 9839–9843.

Whitchurch, C.B., Hobbs, M., Livingston, S.P., Krishnapillai, V., and Mattick, J.S. (1991). Characterisation of a *Pseudomonas aeruginosa* twitching motility gene and evidence for a specialised protein export system widespread in eubacteria. Gene *101*, 33–44.

Whitchurch, C.B., Leech, A.J., Young, M.D., Kennedy, D., Sargent, J.L., Bertrand, J.J., Semmler, A.B., Mellick, A.S., Martin, P.R., Alm, R.A., Hobbs, M., Beatson, S.A., Huang, B., Nguyen, L., Commolli, J.C, Engel, J.N., Darzins, A., Mattick, J.S. (2004). Characterization of a complex chemosensory signal transduction system which controls twitching motility in *Pseudomonas aeruginosa*. Mol. Microbiol. *52*, 873–893.

Whitchurch, C.B., and Mattick, J.S. (1994). Characterization of a gene, *pilU*, required for twitching motility but not phage sensitivity in *Pseudomonas aeruginosa*. Mol. Microbiol. *13*, 1079–1091.

Winther-Larsen, H.C., Hegge, F.T., Wolfgang, M., Hayes, S.F., van Putten, J.P., and Koomey, M. (2001). *Neisseria gonorrhoeae* PilV, a Type IV pilus-associated protein essential to human epithelial cell adherence. Proc. Natl. Acad. Sci. U.S.A. *98*, 15276–15281.

Winther-Larsen, H.C., Wolfgang, M., Dunham, S., van Putten, J.P., Dorward, D., Lovold, C., Aas, F.E., and Koomey, M. (2005). A conserved set of pilin-like molecules controls Type IV pilus dynamics and organelle-associated functions in *Neisseria gonorrhoeae*. Mol. Microbiol. *56*, 903–917.

Winther-Larsen, H.C., Wolfgang, M.C., van Putten, J.P., Roos, N., Aas, F.E., Egge-Jacobsen, W.M., Maier, B., and Koomey, M. (2007). *Pseudomonas aeruginosa* Type IV pilus expression in *Neisseria gonorrhoeae*: effects of pilin subunit composition on function and organelle dynamics. J. Bacteriol. *189*, 6676–6685.

Wolfgang, M., Park, H.S., Hayes, S.F., van Putten, J.P., and Koomey, M. (1998). Suppression of an absolute defect in Type IV pilus biogenesis by loss-of-function mutations in *pilT*, a twitching motility gene in *Neisseria gonorrhoeae*. Proc. Natl. Acad. Sci. U.S.A. *95*, 14973–14978.

Wolfgang, M., van Putten, J.P., Hayes, S.F., Dorward, D., and Koomey, M. (2000). Components and dynamics of fiber formation define a ubiquitous biogenesis pathway for bacterial pili. EMBO J. *19*, 6408–6418.

Yamagata, A., and Tainer, J.A. (2007). Hexameric structures of the archaeal secretion ATPase GspE and implications for a universal secretion mechanism. EMBO J. *26*, 878–890.

Yang, Z., Ma, X., Tong, L., Kaplan, H.B., Shimkets, L.J., and Shi, W. (2000). *Myxococcus xanthus* dif genes are required for biogenesis of cell surface fibrils essential for social gliding motility. J. Bacteriol. *182*, 5793–5798.

Zogaj, X., Chakraborty, S., Liu, J., Thanassi, D.G., and Klose, K.E. (2008). Characterization of the *Francisella tularensis subsp. novicida* Type IV pilus. Microbiology *154*, 2139–2150.

The Bundle-forming Pilus and Other Type IVb Pili

3

Ekaterina Milgotina and Michael S. Donnenberg

Abstract
Type IV pili are remarkable multifunctional organelles expressed by diverse pathogenic bacteria. Type IVb pili have proven highly amenable to study, in part because of the arrangement of the genes encoding their synthesis in contiguous clusters. In this chapter we review the biogenesis and function of Type IVb pili, with an emphasis on the bundle-forming pilus produced by enteropathogenic *Escherichia coli*. In particular, we discuss the structure and function of the adhesive pilin protein and the cellular localization and known interactions among components of the biogenesis machinery.

Introduction

Type IV pili (T4P) or fimbriae are long flexible surface appendages produced by many Gram-negative and some Gram-positive bacteria. The functional roles of T4P vary depending on the microorganism, but in many cases pili are vital instruments for bacteria to become established and survive in an environmental niche. In the case of pathogenic microorganisms, T4P are often critical to pathogenesis. T4P may play roles in bacterial adherence to host cells, colonization, biofilm formation, motility, autoaggregation, and horizontal gene transfer (Table 3.1).

Despite a similar morphology to pili of the chaperone-usher family, T4P are unrelated to these better understood organelles. Constituents of T4P do not share homology with the components of the chaperone-usher system and T4P have a distinct, but poorly understood mechanism of biogenesis. According to current models of T4P biogenesis, T4P are composed of multiple copies of a major pilin subunit assembled on the inner membrane and spanning the periplasmic space to cross the outer membrane. T4P are also endowed with the remarkable ability to retract from the cell surface.

It is apparent from the presence of conserved functional homologues that T4P share a common ancestry with the Type II secretion system (T2SS), filamentous bacteriophages, the Gram-positive competence system, and archaeal flagella (Fla), suggesting that they are derived from a primordial machinery for translocating macromolecules across membranes (Peabody *et al.*, 2003). Proteins common among T4P, T2SS, and Fla systems include the pre-pilin peptidase, the ATPase, and a polytopic transmembrane protein. In addition, T4P and T2SS share pre-pilin peptidase substrates and, along with Type 3 secretion systems, the outer membrane secretin.

The main structural subunit of T4P is the 15 to 20 kDa pilin protein, which is synthesized as a pre-pilin. T4P can be subdivided into two groups – Type IVa and Type IVb – based on several criteria including the length of the leader peptide, the N-terminal amino acid of the mature protein, and whether the biogenesis machinery is encoded within a linked gene cluster or scattered around the genome. Type IVa pre-pilins have a short leader peptide (usually 5–7 amino acids), the mature amino terminus is usually *N*-methyl phenylalanine and the genes required for biogenesis are encoded in several unlinked operons. In contrast, Type IVb pre-pilins possess longer leader sequences, may have a variety

Table 3.1 Type IVb pili and associated phenotypes. The name of the pilus, if there is one, the pathotype, serovar or other characteristics of the bacterium in which it is expressed, and the phenotypes shown to be associated with expression are listed

Name	Species or type	Associated phenotypes	References
Bundle-forming pilus	Enteropathogenic *Escherichia coli*	Cell adherence, autoaggregation, biofilm formation, cell invasion, virulence	Girón *et al.*, 1991; Donnenberg *et al.*, 1992; Bieber *et al.*, 1998; Moreira *et al.*, 2006
Toxin co-regulated pilus	*Vibrio cholerae*	Cell adherence, autoaggregation, colonization, virulence	Taylor *et al.*, 1987; Herrington *et al.*, 1988
Longus pilus	Enterotoxigenic *E. coli*	Autoaggregation,	Girón *et al.*, 1997
CFA/III pilus	Enterotoxigenic *E. coli*	Cell adherence	Taniguchi *et al.*, 2001
Thin pilus	*E. coli* R64 plasmid	Autoaggregation, DNA transfer	Yoshida *et al.*, 1998
	Salmonella enterica serovar Typhi	Cell adherence, autoaggregation, cell invasion	Zhang *et al.*, 2000; Morris *et al.*, 2003; Tsui *et al.*, 2003; Xu *et al.*, 2004
Fimbrial low-molecular- weight protein (Flp) pilus	*Aggregatibacter (Actinobacillus) actinomycetemcomitans*	Cell adherence, autoaggregation, colonization, biofilm formation	Kachlany *et al.*, 2001; Tomich *et al.*, 2007

of *N*-methylated hydrophobic amino acids (L, M, V) at the mature N-terminus, and the genes encoding the biogenesis machinery for Type IVb pili are typically contiguous. In addition, the genes encoding the Type IVa pre-pilins are often smaller than those encoding the Type IVb pre-pilins and the solved structures of the Type IVb pilin proteins show a more complex fold (Craig *et al.*, 2004; Ramboarina *et al.*, 2005). The ability of the pili to aggregate longitudinally to form bundles (Fig. 3.1A), an attribute associated with bacterial auto-aggregation is common among IVb pili, but is also displayed by some IVa pili, such as those expressed by *Neisseria gonorrhoeae* (Craig *et al.*, 2004; Craig *et al.*, 2006). Among the Type IVb systems, the bundle forming pilus (BFP) of enteropathogenic *Escherichia coli* (EPEC), the toxin-co-regulated pilus (TCP) of *Vibrio cholerae,* and thin pilus of self-transmissible plasmid R64 have been studied intensively.

The BFP of EPEC is a very attractive model system for studying T4P as all components specifically comprising the BFP machinery have been identified. The genes coding for Bfp proteins are organized into one cluster located within the EAF (EPEC adherence factor) plasmid (Stone *et al.*, 1996), which greatly facilitates genetic manipulations. In addition, the BFP may be expressed in a laboratory strain of *E. coli* and functionality of the BFP apparatus can be evaluated in assays of epithelial cell adherence and autoaggregation–disaggregation phenotypes. In this chapter, we discuss the functions and biogenesis of Type IVb pili, primarily using BFP as the prototype. We describe the genetic organization and regulation of the *bfp* operon as well as possible functional roles of Bfp proteins and molecular assembly of the BFP apparatus.

Organization of the *bfp* cluster

The genes required for BFP biogenesis are located on a ~90-kb EAF plasmid that defines typical EPEC strains (Stone *et al.*, 1996; Sohel *et al.*, 1996). To form functional BFP filaments, EPEC utilizes the products of thirteen genes, many of which have orthologues in other Type IVa and IVb systems (Fig. 3.2). A fourteenth open reading frame, *bfpH,* appears to be a pseudogene. Twelve genes are required for BFP biogenesis. The only identified structural component of BFP, bundlin, is synthesized as a pre-pilin from the mRNA-transcript of the *bfpA* gene. Pre-bundlin is processed

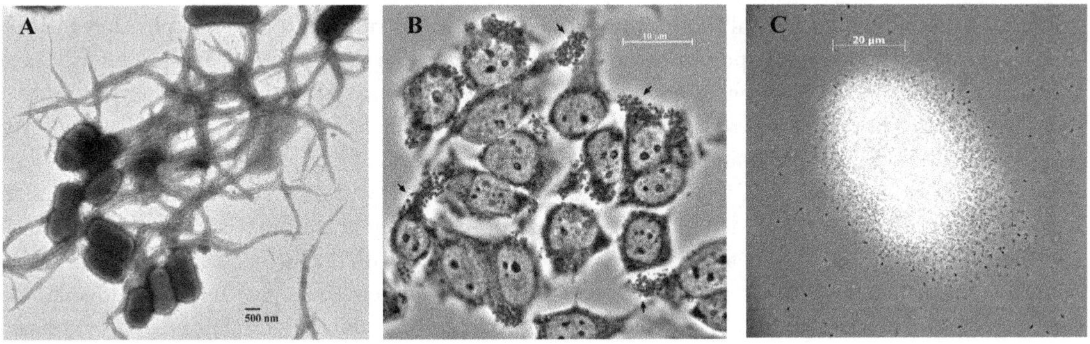

Figure 3.1 The bundle-forming pilus and associated phenotypes of EPEC. **A** EPEC bundle-forming pili were visualized by transmission electron microscopy. Wild type bacteria were grown under BFP-inducing conditions for 4.5 h, fixed with glutaraldehyde and paraformaldehyde, and stained with 1% phosphotungstic acid for 30 s. **B** Phase-contrast image stained with Giemsa of wild type EPEC exhibiting localized adherence to HeLa cells. Bacterial colonies are indicated by arrows. **C** Phase contrast micrograph of wild type bacterial auto-aggregation in liquid culture upon incubation under BFP-inducing conditions. Scale bars are indicated.

into its mature form by the pre-pilin peptidase BfpP encoded by the corresponding gene *bfpP*. BfpP also utilizes as substrates products of the *bfpI*, *bfpJ*, and *bfpK* genes coding for so-called pre-pilin-like proteins. The *bfpB* gene encodes an outer membrane lipoprotein of the secretin family. An accessory protein of BfpB, BfpG, is encoded by the *bfpG* gene. The BFP machinery also includes two ATPases, BfpD and BfpF, that act in opposing processes: BfpD is required for pilus assembly, while BfpF mediates BFP retraction. The roles of cytoplasmic membrane proteins BfpC and BfpE are not clear yet. However, there is evidence that they might serve as connecting elements between cytoplasmic and periplasmic components of the BFP apparatus. Disruption of any of the genes, with the exception of *bfpH* and *bfpF*, abolishes BFP biogenesis, while mutation in *bfpF* leads to abnormalities in BFP function.

The genes constituting the *bfp* cluster following *bfpA* either overlap or are separated by sequences of no more than 22 nucleotides. In contrast, a 51 nucleotide gap between the *bfpA* and *bfpG* genes is predicted to form a stem-loop

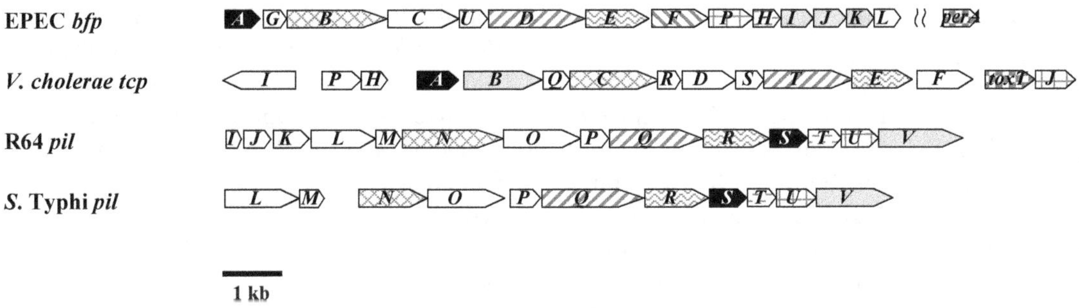

Figure 3.2 Organization of T4bP gene clusters. Genes are depicted as arrows showing the direction of transcription and filled according to the proteins that they encode as follows. Those encoding the pre-pilin proteins are shown in black, the secretins have cross hatches, the ATPases have diagonal lines (the lines of the retraction ATPase run from top left to bottom right), the polytopic transmembrane proteins have wavy lines, the pre-pilin peptidases have horizontal and vertical lines, the lytic transglycosylases have dashed lines, the pre-pilin-like proteins are grey, and regulators have a checkerboard pattern. Genes with no sequence homologues are shown in white, although they may share functional roles with proteins in other systems. Non-contiguous genes are separated by two wavy lines. A scale bar shows the distance of 1000 bases.

structure, representing a potential transcriptional regulation site that may be responsible for the difference in levels of mRNA transcripts of *bfpA* and downstream genes (Puente *et al.*, 1996). The genes coding for proteins BfpG and BfpU begin with uncommon start codons GUG and UUG. The only genes that are not preceded by a consensus ribosome binding site, *bfpP* and *bfpH*, were also the only two genes whose expression was not detected *in vitro* (Stone *et al.*, 1996). However, as we discuss later, the functionality of *bfpP* coding for the pre-pilin peptidase was confirmed *in vivo* (Zhang *et al.*, 1994). The sequence of *bfpH* places its product in a family of lytic transglycosylases catalysing the hydrolysis of bacterial peptidoglycans. However, its predicted product is missing several key residues within the catalytic site (Lehnherr *et al.*, 1998). Furthermore, the absence of a detectable phenotype of a *bfpH* mutant suggests that this ORF is a pseudo-gene, which lost its functionality during evolution, perhaps facilitated by the presence of functional orthologues elsewhere in the EPEC genome (Anantha *et al.*, 2000). In contrast, it appears that a BfpH orthologue, PilT, from the T4P encoded on the self-transmissible plasmid R64, is a functional protein. Mutation in this *pilT* gene had a moderate effect on pilus formation and dramatically decreased the efficiency of horizontal gene transfer, but did not influence phage sensitivity (Yoshida *et al.*, 1999). The effect on gene transfer is in agreement with prior observations regarding conjugative plasmids, lytic transglycosylase proteins, and plasmid transfer (Bayer *et al.*, 1995).

Bundlin

The first gene of the *bfp* operon, *bfpA*, encodes pre-bundlin, the precursor of the only known structural element of BFP. A typical T4b pilin, bundlin is synthesized as a pre-protein with a long (13 amino acids) leader sequence followed by a hydrophobic trans-membrane motif of 22 amino acids and a globular periplasmic domain. The hydrophilic positively charged leader peptide is cleaved by the pre-pilin peptidase BfpP at a canonical site **G↓LSLIE** (Zhang *et al.*, 1994). Within this sequence, Gly (−1) is invariant and Glu (+5) is highly conserved among known Type IV pilins. However, the Type IVb pilin PilS from *Salmonella enterica* serovar Typhi (*S.* Typhi) is an exception, with an Asp residue in the (+4) position and Ala residue in the (+5) position. The presence of Leu, not Phe in the +1 position is another feature that characterizes bundlin as a Type IVb pilin. We have confirmed that the mature amino-terminus begins either with *N*-methyl leucine or leucine in approximately equal proportions (unpublished). Bundlin, as with the majority of known pilins, also has two cysteine residues forming a disulphide bond in the globular domain. To be exported to the periplasm, pre-bundlin does not require accessory proteins encoded by the *bfp* operon. Instead, it is presumed that bundlin is targeted to the inner membrane via the signal recognition particle (SRP)-Sec pathway as was shown for the pseudopilin XcpT of the *Pseudomonas aeruginosa* T2SS (Arts *et al.*, 2007b).

Structures of three T4b pilins lacking the hydrophobic amino terminal residues that interfere with solubility have been solved to date by x-ray diffraction (TcpA) or NMR (PilS and bundlin) (Craig *et al.*, 2003; Xu *et al.*, 2004; Ramboarina *et al.*, 2005) (see Fig. 1.3). Despite the overall resemblance of the ladle-like Type IV pilins, each of the T4b pilins exhibits a larger globular domain than those of the T4a counterparts with additional beta sheets forming a layer enveloping the domains common to T4a pilins. Additionally, bundlin has a unique arrangement of beta sheets in comparison to those of the other T4b structures. The most pronounced dissimilarities among the T4b pilins are found in the αβ-loop region, which is shorter in bundlin and inserts a small beta strand into the overall beta sheet prior to adopting the fold common to the other T4b pilins. In contrast, the D-regions of all three proteins share significant structural and sequence homology. A comprehensive structural analysis of bundlin and its homologues is provided in Chapter 1.

Variations in bundlin sequence

An analysis of *bfpA* sequences from diverse EPEC strains has revealed ten alleles (Blank *et al.*, 2000; Blank *et al.*, 2003; Lacher *et al.*, 2007). These genetic variants fall into two groups, α and β, according to the degree of sequence similarity. Pre-bundlins encoded by the α1, α2, and α3 alleles are 97% identical, while β1 through β7 alleles are more divergent. Proteins encoded by the β alleles have only 89% identical amino acid residues and are 79.7% identical to the α group. As α and β alleles were found in both of the two major EPEC lineages, EPEC1 and EPEC2, it appears that *bfpA* alleles were spread between EPEC strains by means of horizontal gene transfer.

Sequence analysis reveals that amino acid variation is greater near the C-terminus, with excess non-synonymous substitutions located between cysteine residues in a region proposed to be exposed on the surface of the BFP filament (Ramboarina *et al.*, 2005). These substitutions do not seem to influence BFP biogenesis or function, as all *bfpA* alleles were able to complement a *bfpA* null mutant for autoaggregation and strains expressing different bundlin alleles are able to form mixed aggregates (Fernandes *et al.*, 2007; Hyland *et al.*, 2008). Rather it appears that the human immune response may provide the selective pressure for diversification in this region. Volunteers challenged for the first time with an EPEC strain that expressed α1 bundlin develop significant increases in antibody titers against the purified homologous bundlin molecule, but not against purified β6 bundlin (Fernandes *et al.*, 2007). Similarly, monoclonal antibodies that recognize an epitope composed of residues belonging to the D-region of TcpA and localized on the surface of TCP are protective against *V. cholerae* in a murine model (Sun *et al.*, 1991; Sun *et al.*, 1997; Craig *et al.*, 2003).

Bundlin as an adhesin

Enteropathogenic *E. coli* form compact colonies on the surface of epithelial cells *in vitro* and *in vivo*, a property termed localized adherence (LA) (Fig. 3.1B) (Scaletsky *et al.*, 1984). In addition, EPEC strains autoaggregate *in vitro* under conditions that induce BFP expression (Fig. 3.1C) (Vuopio-Varkila and Schoolnik, 1991). Strains cured of the EAF plasmid are deficient in LA and do not produce BFP and mutants that cannot make BFP fail to perform LA or autoaggregation (Knutton *et al.*, 1987; Girón *et al.*, 1991; Donnenberg *et al.*, 1992; Sohel *et al.*, 1996; Anantha *et al.*, 2000). A BFP antiserum also reduces EPEC adherence to host cells (Girón *et al.*, 1991). Differential BFP-dependent adherence to human versus murine cell lines and a dose-dependent binding of purified pre-bundlin to human cells provide further evidence that BFP serves as an adhesin (Tobe and Sasakawa, 2002). Bona fide adhesins display specific binding to a cognate receptor on the eukaryotic cell surface. Although several candidate receptors have been proposed, N-acetyllactosamine (LacNAc) has emerged as the probable receptor for BFP composed of α bundlin variants (Vanmaele *et al.*, 1999). LacNAc conjugated to BSA or gold nanoparticles competitively inhibits adhesion of EPEC to human epithelial and intestinal cell lines and human intestinal explants (Hyland *et al.*, 2006a; Hyland *et al.*, 2006b). Furthermore, LacNAc–gold particles were found by transmission electron microscopy to bind specifically to BFP (Hyland *et al.*, 2006b). Further evaluation of the bundlin-LacNAc interaction by ultra-sensitive nanoelectrospray mass spectrometry (nanoES-MS) revealed a very low binding affinity ($K_a \sim 10^3 M^{-1}$), although the presence of at least two binding sites on each bundlin monomer and the multimeric structure of BFP might compensate for this low-efficiency monovalent interaction (Hyland *et al.*, 2008). In addition, upon incubation with LacNAc, EPEC retract the BFP leading to bundlin degradation. Surprisingly, binding to LacNAc could not be demonstrated for EPEC strains carrying β-alleles of *bfpA*, suggesting that different bundlin types may recognize different receptors on the host surface. Since EPEC strains expressing divergent bundlin proteins can form mixed auto-aggregates, it appears that autoaggregation and LA functions of BFP are distinct.

Precedent for the existence of multiple receptors for a Type IVb pilus may be found in the R64 thin pili system, which consist of PilS pilin monomers with variant PilV pilin adhesins located on the tip of the pili. The presence of four shufflon determinants downstream of the *pilV* gene provides seven different C-terminal variants of PilV that can combine with the constant N-terminus. The C-terminal segments appear to recognize different lipopolysaccharides on the surface of recipient cells during liquid mating (Yoshida *et al.*, 1998; Ishiwa and Komano, 2004).

The first extracellular domain of the cystic fibrosis transmembrane conductance regulator (CFTR) was reported to be a potential receptor of the Type IVb pili of *S. enterica* serovar Typhi (Pier *et al.*, 1998; Zhang *et al.*, 2000; Tsui *et al.*,

2003; Xu *et al.*, 2004). A peptide representing residues 103–117 of CFTR inhibited *S.* Typhi entry to INT407 cells by 70% at ca. 2 μM, while addition of full-length PilS at equimolar concentration overcame the peptide. The latter observation was validated in the reciprocal experiment (Zhang *et al.*, 2000; Tsui *et al.*, 2003). However, only full-length PilS, but not its truncated version t-PilS, was able to inhibit bacterial entry to the host cell (Xu *et al.*, 2004). In addition, the shorter version of the CFTR peptide 108–117 was unable to bind to the t-PilS protein upon titration in an NMR experiment (Xu *et al.*, 2004). Thus, the evidence that PilS–CFTR represent a bona fide adhesin–receptor pair is not yet definitive.

Pre-pilin peptidase

Type IV pre-pilins are processed to the mature protein by pre-pilin peptidases. In the case of BFP, pre-bundlin is processed by BfpP (Zhang *et al.*, 1994). Type IV pre-pilin peptidases (T4PP) are polytopic membrane proteins (Akahane *et al.*, 2005) that not only cleave the leader peptide, but also catalyse *N*-methylation (Strom *et al.*, 1993b) or some other unidentified modifications (Yoshida *et al.*, 1998) of the newly formed N-terminus of mature pilin. These two modifications are performed by T4PP at distinct catalytic sites on the cytoplasmic side of the inner membrane in accord with the membrane topology of pre-pilins (Strom and Lory, 1987). The role of methylation remains unclear as the absence of this modification does not affect pilus formation and functioning. In contrast, piliation is abolished when pre-pilins remain unprocessed (Strom and Lory, 1991).

Originally, based on studies of *P. aeruginosa* PilD, pre-pilin peptidases were considered to belong to a novel class of thiol proteases lacking a consensus catalytic sequence. In a postulated model, methyltransferase and proteolytic functions were attributed to adjacent regions sharing four cysteine residues within the large cytoplasmic domain 1 (Strom *et al.*, 1993a). However, comprehensive alignment analysis of known T4PP revealed that some of them lack domain 1 including the putative catalytic tetracysteine motif (LaPointe and Taylor, 2000; Akahane *et al.*, 2005). LaPointe and Taylor found that two totally conserved aspartate residues within TcpJ, the pre-pilin peptidase of TCP from *V. cholerae*, were required for the endopeptidase activity of T4PP. Furthermore, mutations of analogous Asp residues in VcpD, a second *V. cholerae* T4PP processing a Type IVa pre-pilin, confirmed the universal mode of action of T4PP. Among the novel properties of T4PP are a non-canonical catalytic motif, neutral optimum pH, and resistance to inhibition by pepstatin. These properties indicate that T4PP can be categorized as non-pepsin-like aspartic acid proteases.

Despite a limited degree of homology between T4PP, some of them are interchangeable between species (Strom *et al.*, 1993b; De Groot *et al.*, 1994; Köhler *et al.*, 2004). Thus, bundlin can be fully processed by the PilD T4PP when expressed in a *P. aeruginosa* background, and EPEC *bfpP* was found to be able to complement a *pilD* mutant in *P. aeruginosa* (Zhang *et al.*, 1994). In addition, some T4PP are required both for biogenesis of T4P and for T2SS (Pepe *et al.*, 1996; Marsh and Taylor, 1998).

DsbA

Most Type IV pilins have two cysteine residues near the C-terminus that form a disulphide bond. This reaction is catalysed by DsbA, a periplasmic thiol-disulphide isomerase (Bardwell *et al.*, 1991). EPEC transposon mutants with disruptions in *dsbA* fail to perform LA due to dramatically reduced levels of bundlin (Zhang and Donnenberg, 1996). Moreover, mutation of either Cys129 or Cys179 to Ser also led to undetectable level of bundlin, suggesting that disulphide bond formation is necessary for stability of the protein. This hypothesis was confirmed by demonstrating a drastically reduced half-life for newly synthesized bundlin in the *dsbA* background. Interestingly, the effect of DsbA on stability of pre-bundlin in the absence of BfpP was similar to the effect on bundlin. Moreover, oxidized and reduced pre-bundlin and oxidized and reduced bundlin could all be detected at early time points after bundlin synthesis in the wild type strain. These observations indicate that the pre-pilin peptidase and DsbA act simultaneously and independently and confirm the topology of nascent pre-bundlin in the inner membrane with its amino terminus accessible to BfpP while its carboxyl terminus is accessible to DsbA (Zhang and Donnenberg, 1996; Donnenberg *et al.*, 1997) (Fig. 3.3).

Interestingly, mutation in the gene coding for TcpG, a DsbA homologue from *V. cholerae* (Yu *et al.*, 1992), only slightly affected TCP formation and the pilin retained its disulphide bond. However, pili synthesized by the *tcpG* mutant were less hydrophobic; and the bacteria did not show autoagglutination or fucose-resistant haemagglutination, and failed to colonize mice (Peek and Taylor, 1992). The presence of disulphide-bonded pilin in the *tcpG* mutant might be expected to occur as a result of atmospheric oxidation, though pili malfunction may be attributed to the protein misfolding in the absence of catalyst (Peek and Taylor, 1992).

Assembly of BFP and other Type IVb pili

The BFP biogenesis machinery spans the inner membrane, periplasmic space, and outer membrane with part of the complex situated on the cytoplasmic side of the inner membrane (Fig. 3.4). Based on the cellular localization of BFP proteins (Table 3.2) and emerging information on protein interactions, the BFP machinery can be conceptually divided into a cytoplasmic membrane (CM) subassembly, an outer membrane (OM) subassembly, and the pilus itself. However, it is expected that further experimentation will reveal a single connected functional machine.

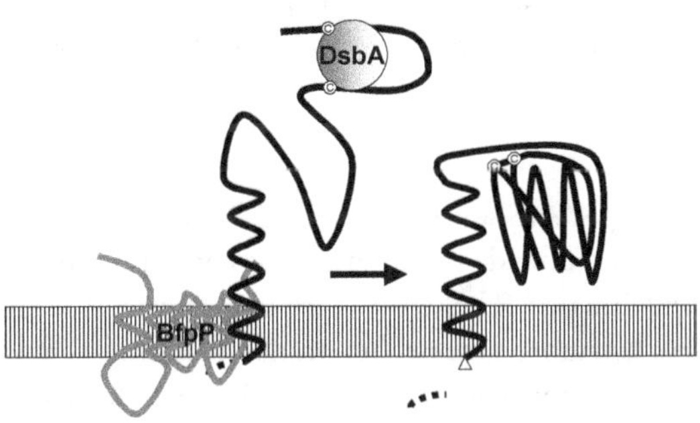

Figure 3.3 A model of post-translational processing of pre-bundlin. The pre-pilin is directed to the cytoplasmic membrane by the general secretory pathway, where the hydrophobic helical domain of the mature amino-terminus acts as a transmembrane domain. In this orientation, the pre-pilin is accessible for simultaneous and independent modification by the pre-pilin peptidase BfpP and the disulphide isomerase DsbA. BfpP with its active sites on the cytoplasmic face of the membrane, removes the signal sequence (dotted line) and *N*-methylates the new amino terminus (triangle). DsbA, localized to the periplasm, catalyses the formation of a disulphide bond between the two cysteine residues (circles labelled with the letter C). This reaction is critical to proper folding and stability.

Outer membrane subassembly

At least three proteins, BfpB, BfpG, and BfpU, are components of the OM subassembly of the BFP apparatus. Mutant strains of *bfpG*, *bfpB*, and *bfpU* are deficient in AA and LA. However, they have normal levels and localization of bundlin in the inner and outer membranes (Ramer *et al.*, 1996; Schreiber *et al.*, 2002; Ramer *et al.*, 2002; Daniel *et al.*, 2006). The central element of the OM subassembly, BfpB, is a lipoprotein with homology to the pIV-PulD secretin protein superfamily, which includes, for example, PulD of the *Klebsiella oxytoca* T2SS, PilQ of the *Neisseria meningitidis* T4aP, InvG of the T3SS in *Salmonella enterica* serovar Typhimurium, and filamentous phage secretin pIV (Russel and Kazmierczak, 1993; Crago and Koronakis, 1998; Nouwen *et al.*, 2000; Collins *et al.*, 2003). Secretins form pores in the OM for translocation of macromolecules. Similar to the other members of the family, BfpB forms highly stable ring-shaped dodecamer complexes with outer and inner diameters of about 20 nm and 7 nm, respectively. However, BfpB is somewhat unusual as it is a lipoprotein (Ramer *et al.*, 1996). It appears that formation of the stable BfpB multimer complexes might be affected in the absence of BfpG. The high molecular weight BfpB complex is normally resistant to 4% SDS upon incubation at temperatures up to 65°C. However BfpB from a *bfpG* mutant strain migrates as a monomer on SDS-PAGE after pre-incubation in 4% SDS buffer at 22°C (Schmidt *et al.*, 2001). Nevertheless, BfpB localizes to the OM in the absence of BfpG or any other Bfp proteins (Daniel *et al.*, 2006). In agreement with its putative role as an OM pore through which the pilus fibre passes, there is evidence that BfpB forms an incompletely gated transmembrane channel in the OM (Schmidt *et al.*, 2001). In two studies, affinity-tagged BfpB was recovered from EPEC after cross-linking. In one study, ten of the Bfp proteins (including BfpB) were identified in the eluate by immunoblotting. Only BfpP, BfpH, BfpI and BfpK could not be detected (Hwang *et al.*,

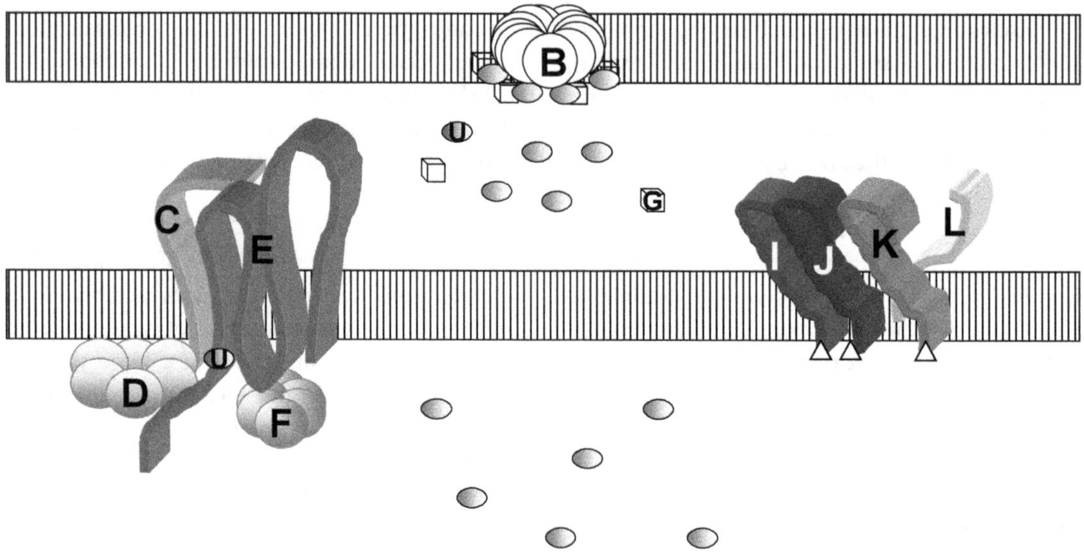

Figure 3.4 A model of components of the BFP biogenesis machine. An inner membrane subassembly is composed of the hexameric ATPase BfpD, recruited to the membrane by the bitopic and polytopic transmembrane proteins BfpC and BfpE. BfpF, the putative retraction ATPase interacts with the small cytoplasmic loop of BfpE. The polymeric outer membrane secretin BfpB interacts with BfpU and BfpG, which are both targeted to the periplasm by the general secretory pathway. Surprisingly BfpU is also found in the cytoplasm and associated with the inner membrane, perhaps in association with BfpE. The pilin-like proteins BfpI, BfpJ, and BfpK are processed to their mature form by the pre-pilin peptidase BfpP (not shown), which presumably also *N*-methylates the mature amino termini (triangles). These proteins may interact with one another, but their placement in the machine has yet to be determined. BfpL probably adopts a topology similar to those of the pilin-like proteins, although it lacks a pre-pilin peptidase cleavage site. It is not yet clear how the inner and outer membrane subassemblies come together.

2003). In another study in which antibodies only to BfpU, BfpG and bundlin were used, BfpU and BfpG, but not bundlin were detected in the eluate (Daniel *et al.*, 2006).

BfpG, encoded by the second gene of the *bfp* cluster, does not have recognizable sequence homologues. The BfpG protein, despite having a putative signal peptidase II cleavage site, is actually processed by signal peptidase I at two distinct sites (Daniel *et al.*, 2006). The stability of BfpG is influenced by BfpB, as BfpG cannot be detected in a *bfpB* deletion mutant, but it is present in a non-polar *bfpB* insertion mutant, in which the N-terminus of BfpB is still expressed (Schmidt *et al.*, 2001; Ramer *et al.*, 2002; Daniel *et al.*, 2006). Furthermore, BfpG localizes to the OM of wild-type EPEC, but to the periplasmic space of a *bfpB* mutant (Daniel *et al.*, 2006). These results are consistent with the observation that the amino termini of secretins protrude into the periplasmic space (Shevchik *et al.*, 1997; Opalka *et al.*, 2003). Indeed, BfpG interacts with the amino-terminus of BfpB in a yeast two-hybrid system. Furthermore, when functional histidine-tagged BfpG is recovered after cross-linking, BfpB and BfpU, but not bundlin are present in the eluate. These data suggest a direct interaction between BfpG and the amino-terminus of BfpB, but this has not been demonstrated.

The fifth gene of the operon, *bfpU*, encodes a 17-kDa protein, BfpU with no homology to known T4P proteins. BfpU is synthesized as a pre-protein and processed by signal peptidase I. Surprisingly, given its overall hydrophilicity, mature BfpU can be detected not only in the periplasm, but in the cytoplasm, CM and OM (Schreiber *et al.*, 2002; Daniel *et al.*, 2006). However, in the absence of BfpB, BfpU can no longer be detected in the OM fraction. In contrast to BfpG, when functional histidine-tagged BfpU

Table 3.2 Known and predicted locations and topology of Bfp proteins

Protein	Cellular localization and topology	Reference
Bundlin (BfpA)	Inner membrane, pilus	Girón *et al.*, 1991; Anantha *et al.*, 2000
BfpB	Outer membrane	Ramer *et al.*, 1996
BfpC	Inner membrane, with single transmembrane domain and amino terminus in cytoplasm	Crowther *et al.*, 2004
BfpD	Cytoplasm, inner face of inner membrane	Crowther *et al.*, 2004
BfpE	Inner membrane with four transmembrane domains and amino terminus in cytoplasm	Blank and Donnenberg, 2001
BfpF	Cytoplasm?	Predicted
BfpG	Periplasm, outer membrane	Schmidt *et al.*, 2001; Daniel *et al.*, 2006
BfpI	Inner membrane	Ramer *et al.*, 2002
BfpJ	Inner membrane	Ramer *et al.*, 2002
BfpK	Inner membrane	Ramer *et al.*, 2002
BfpL	Inner membrane, outer membrane	Ramer *et al.*, 2002
BfpP	Inner membrane	Predicted
BfpU	Cytoplasm, periplasm, outer membrane, inner membrane	Schreiber *et al.*, 2002; Ramer *et al.*, 2002; Daniel *et al.*, 2006

is recovered after cross-linking, bundlin as well as BfpB and BfpG can be detected in the eluate, suggesting that BfpU may be part of more than one type of complex (Daniel *et al.*, 2006). However, its localization to multiple compartments suggests that BfpU may interact with other components of the machine as well.

The *V. cholerae* outer membrane secretin TcpC is also a lipoprotein that forms a multimeric OM complex (Bose and Taylor, 2005). However, TcpC is undetectable in the absence of the periplasmic protein TcpQ and *vice versa*. In addition, the normal OM localization of TcpC is altered upon complementation *in trans* of the double *tcpC, tcpQ* mutant with *tcpC*. In this case, TcpC is targeted to the IM.

Most secretins of the pIV-PulD family are not lipoproteins but instead require small auxiliary lipoproteins called pilotins for the stability, OM targeting, and assembly into multimer complexes (Bayan *et al.*, 2006). In contrast, BfpB and TcpC, along with the PilN secretin of R64, are lipoproteins themselves. Ancillary proteins like BfpG and TcpQ are not lipoproteins and their functions have yet to be defined. Thus, BfpB, TcpC and PilN may represent a distinct functional subclass within the pIV-PulD secretin superfamily.

Inner membrane subassembly

The inner membrane subassembly is composed of four proteins, BfpC, BfpD, BfpE and BfpF. BfpC is a bitopic IM protein with a single transmembrane domain, the N-terminus in the cytoplasm, and the C-terminus in the periplasmic space (Crowther *et al.*, 2004). While there are no proteins with significant sequence homology to BfpC identified so far, it is believed to play a role equivalent to that of transmembrane proteins in other T4P and T2S systems.

BfpD belongs to the large and highly conserved GspE superfamily of ATPases. ATPases are indispensable components of T4P, T2S, DNA transfer, archaeal flagella and other systems. They are believed to provide energy for the assembly of functional molecular complexes. BfpD shares all the common features of GspE ATPases, namely, Walker boxes A and B, an Asp box, and a His box (Rivas *et al.*, 1997; Peabody *et al.*, 2003).

Mutation of key residues within the BfpD ATP-binding Walker A motif prevents BFP biogenesis (Bieber *et al.*, 1998).

As with many ATPases, BfpD forms multimers upon ATP binding (Shiue *et al.*, 2006; Yamagata and Tainer, 2007). By electron microscopy these multimers take the form of ring-shaped complexes with six-fold symmetry. Surprisingly, however, the molecular radius of R64 ATPase PilQ as determined by gel filtration chromatography was more consistent with an octomer (Sakai *et al.*, 2001). However, structural studies have not been reported. TcpT, the BfpD homologue in the *V. cholerae* TCP system, can form dimers and its C-terminus is implicated in promoting TcpT self-interactions (Tripathi and Taylor, 2007). Similar observations were made for OutE, a T2SS ATPase in *Erwinia chrysanthemi* (Py *et al.*, 1999).

BfpE might be considered a central element of the IM complex and the entire biogenesis machine, as the abundance of a number of Bfp proteins was decreased in the *bfpE* mutant background (Ramer *et al.*, 2002). BfpE is a member of the GspF family of polytopic inner membrane proteins highly conserved among bacterial T4P, T2SS and archaeal flagella (Peabody *et al.*, 2003). The topology of these proteins has been the subject of some debate (Blank and Donnenberg, 2001; Peabody *et al.*, 2003). Based on computer algorithms, most members of the GspF family are predicted to contain three transmembrane (TM) domains with the N-terminal domain and a large loop between TM2 and TM3 located on the cytoplasmic side of the IM and the small loop between TM1 and TM2 in the periplasm. Indeed, there is experimental evidence using C-terminal truncations fused to periplasmic reporter proteins to support this prediction in the case of OutF from *Erwinia carotovora* T2SS and XcpS from *P. aeruginosa* T2SS (Thomas *et al.*, 1997; Arts *et al.*, 2007a). However, computer predictions for BfpE reveal either three or four transmembrane domains, with a short cytoplasmic loop between TM2 and TM3 and a large periplasmic loop between TM3 and TM4. Indeed, an extensive analysis, using C-terminal deletions fused to both cytoplasmic and periplasmic reporter proteins, 'sandwich' fusions in which reporter proteins were inserted into the molecule without deletions, and sucrose flotation gradient analysis confirmed the four-transmembrane topology (Blank and Donnenberg, 2001). Furthermore, an interaction between the small cytoplasmic loop of BfpE and the cytoplasmic protein BfpF (see below) is consistent with the four TM domain model. A bioinformatics study concluded that in many GspF homologues, the putative second transmembrane domain is long enough to span the membrane twice (Peabody *et al.*, 2003). Recently, the three-dimensional electron microscopy structure of the *N. meningitidis* PilG GspF protein was published (Collins *et al.*, 2007). A topological model of PilG based on the structure and on computer prediction of the secondary structure has four TM domains similar to BfpE. It remains to be resolved whether all GspF homologues have the same topology.

The cytoplasmic domain of BfpC, BfpD, and the cytoplasmic N-terminus of BfpE not only interact with each other, but form a stable complex. BfpD is a soluble cytoplasmic protein when expressed alone, but is recruited to the membrane when BfpC and BfpE are also present. Formation of an analogous three-component complex was reported in *E. chrysanthemi* between OutL, OutE, and OutF and in *P. aeruginosa* between XcpY, XcpR, and XcpS (Py *et al.*, 2001; Arts *et al.*, 2007a). OutE and XcpR are homologues of BfpD, and OutF and XcpS are homologues of BfpE, while OutL and XcpY thus appear to be functional homologues of BfpC.

TcpT of *V. cholerae* T4P, a homologue of BfpD, is tethered to the IM by the cytoplasmic domain of a bitopic IM protein TcpR and localized in the cytoplasm in the *tcpR* mutant (Tripathi and Taylor, 2007). However, the stability of both proteins is altered in the absence of the other TCP components. Interaction of a stable TcpR-PhoA fusion with TcpT causes conformational changes in TcpT and results in its degradation. The N-terminal domain of TcpT composed of the first 100 amino acid residues forms a stable complex with TcpR, while the region responsible for destabilizing interactions is probably located within residues 101–503. Analogous data were provided for T2SS proteins OutE and OutL, and EpsE and EpsL in *E. chrysanthemi* and *V. cholerae*, respectively (Sandkvist *et al.*, 1995; Py *et al.*, 1999; Sandkvist *et al.*, 2000; Py *et al.*, 2001;

Abendroth *et al.*, 2005). In addition, a complex formed between copurified *V. cholerae* EpsE and EpsL was reported to express ATPase activity, which was further greatly enhanced by addition of the acidic phospholipid cardiolipin (Camberg *et al.*, 2007). The formation of these analogous IM complexes in different systems in spite of limited or absent amino acid sequence similarities may suggest a unifying mechanism required for the interaction among other components of the machines, and transfer energy for either assembly of the filament structure or export of the secretion substrates.

The 'pilin-like' proteins

All T4P and T2S systems include several essential proteins that share characteristics of Type IV pilins including a pre-pilin peptidase cleavage site, a hydrophobic domain at the amino-terminus of the mature protein, and usually two cysteine residues predicted to form a disulphide bond. These proteins have variously been referred to as pilin-like proteins, pseudopilins and pre-pilin peptidase substrates. Indeed, the structures of such proteins from T2SSs solved to date clearly demonstrate a Type IV pilin fold (Köhler *et al.*, 2004; Yanez *et al.*, 2008) and several can polymerize into T4P when overexpressed (Sauvonnet *et al.*, 2000a; Sauvonnet *et al.*, 2000b). Whether these proteins are incorporated into the pilus has been a subject of some controversy. There is some evidence that they co-purify with the *Neisseria gonorrhoeae* T4a pilus (Winther-Larsen *et al.*, 2005). Thus far, very little has been published regarding the function of the BfpI, BfpJ, and BfpK pilin-like proteins of the BFP system or of any similar T4b proteins. BfpI, BfpJ, and BfpK all migrate more slowly in a *bfpP* mutant, suggesting that they are indeed processed by the pre-pilin peptidase in EPEC (Ramer *et al.*, 2002). Furthermore, the steady state levels of BfpI and BfpK are reduced in the absence of each other or in the absence of BfpJ. As previously mentioned, BfpI and BfpK were the only proteins that could not be detected after cross-linking and elution of tagged BfpB from a *bfpB* mutant (Hwang *et al.*, 2003).

The final gene in the *bfp* operon encodes BfpL, which resembles the pilin-like proteins in predicted topology, but lacks a pre-pilin peptidase cleavage motif and has no sequence homology with other proteins. Mutation of *bfpL* abolishes pilus biogenesis and dramatically affects the stability of the BfpI, BfpJ, and BfpK pilin-like proteins (Ramer *et al.*, 2002). Despite the obvious functional importance of BfpL, its precise role within the BFP machinery is unknown. In the *V. cholerae* T2SS, the IM complex between EpsL and EpsE, homologues of BfpC and BfpE, was further stabilized by the bitopic IM protein EpsM, which has a single transmembrane domain and one periplasmic domain. Because of its similar topology, EpsM might be a functional analog of BfpL. Alternatively, EpsL and EspM might both correspond to BfpC, which has a large periplasmic domain lacking in EspL.

BFP retraction

A fascinating property of T4P is their ability to retract from the surface. Retraction is associated with a form of surface translocation known as twitching motility, in which the bacterium uses the pilus, with its distal end attached to a surface, as a rope that when retracted moves the cell forward. In this process, pilin subunits most likely disassemble from the filament and return to the inner membrane. Pilus retraction is a very rapid process (about 1200 nm/s) driven by the chemical energy of ATP hydrolysis. *N. gonorrhoeae* ATPase PilT was shown to be the most powerful molecular motor known, generating forces exceeding 100 pN (Maier *et al.*, 2002; Maier *et al.*, 2004; Satyshur *et al.*, 2007).

BfpF is implicated in the retraction of the BFP filament. Like BfpD, BfpF is predicted to be a hexameric ATPase belonging to the GspE superfamily. A *bfpF* mutant is hyperpiliated and forms aggregates larger then those formed by the wild type, which fail to disperse over time. The *bfpF* mutant also adheres to the epithelial surface in greater numbers then the wild type bacteria. However, the mutant appears to be less virulent in humans than the wild type strain, as higher doses are required to cause diarrhoea (Anantha *et al.*, 1998; Bieber *et al.*, 1998).

One function of pilus retraction could be to bring the bacteria closer to the epithelial cell. However, there is no evidence to support this hypothesis (Anantha *et al.*, 1998). Furthermore, *bfpF* plays no detectable role in pedestal formation and actin polymerization phenotypes, the signatures of the prolonged infection (Anantha *et al.*, 1998; Bieber *et al.*, 1998).

BfpF is linked to the IM protein complex by binding to the cytoplasmic loop of BfpE. Overexpression of the loop in the wild type strain under BFP-inducing conditions caused the bacteria to lose the ability to disaggregate, a phenotype indistinguishable from that of a *bfpF* mutant (Crowther *et al.*, 2004). This result substantiates the membrane topology of BfpE and identifies it as the first protein identified to be involved in T4P retraction process other than the putative ATPase.

Interestingly, although common features of T4a systems, homologues of BfpF are not present in the gene clusters of the other T4b pili. Whether these pili retract remains to be determined.

Regulation of the *bfp* cluster

Expression of T4b pilus production is coordinately regulated as expected of a process that plays an essential role in the virulence of many pathogens, but one that is required only at certain phases of the infectious cycle. The details differ substantially among the various species that produce them. BFP are expressed poorly in bacterial laboratory media, but highly in tissue culture media, and temperature and pH have a dramatic effect on bundlin synthesis (Vuopio-Varkila and Schoolnik, 1991; Puente *et al.*, 1996). Expression of the *bfp* cluster is regulated by PerA, the product of the *perA* gene located within the *perABC* operon (*per* is for plasmid-encoded regulator) 6.4kb downstream of the *bfpL* gene (Gómez-Duarte and Kaper, 1995; Tobe *et al.*, 1996). PerA belongs to the AraC/XylS family of transcriptional regulators and shares compelling homology with several transcriptional activators of virulence genes, including ToxT, a transcriptional regulator of cholera toxin and TCP in *V. cholerae* (DiRita *et al.*, 1991). According to sequence alignment analysis, PerA contains two conserved C-terminal helix–turn–helix motifs required for DNA binding, but, in contrast to many members of AraC-family, including ToxT, PerA lacks a coiled-coil motif at the amino-terminus domain, which mediates homo- and hetero-oligomerization. PerA binds to a promoter region of the *bfp* cluster, and to a sequence within the *bfpA* gene (Ibarra *et al.*, 2003). It also auto-activates the transcription of the *perA* gene (Martínez-Laguna *et al.*, 1999).

Expression of *bfpA* and other EPEC virulence genes is negatively regulated *in vitro* under acidic conditions by GadX, another member of the AraC/XylS family of transcriptional regulators that affects the expression of the acid resistance mechanism genes *gadA* and *gadB* through binding to the promoter regions of *perA*, *gadA* and *gadB*. However, participation of additional factors mediating the GadX regulatory effect was not excluded (Shin *et al.*, 2001). In the absence of PerA, expression of *bfp* is slightly repressed by nucleoid-associated protein (H-NS) and by integration host factor (IHF). PerA greatly enhances transcription of *bfp* independently of H-NS and IHF. An unusual weak positive effect of H-NS and weak negative effect of IHF on the *per* promoter may be indirectly mediated by GadX, as H-NS was shown to function as a transcriptional repressor of *gadX* (Shin *et al.*, 2001; Porter *et al.*, 2004; Giangrossi *et al.*, 2005).

In contrast to PerA, *V. cholerae* ToxT self-activates by binding to the *tcpA* promoter and activating expression of *tcp* operon (Hulbert and Taylor, 2002). In this case RNA polymerase reads through *toxT* gene located immediately upstream of the last gene of *tcp* operon, *tcpJ*. Expression of *toxT* is also regulated by a tandem of transcription activators TcpPH/ToxRS, which are, in turn, regulated by multiple quorum-sensing systems through a complex cascade of enzymatic and DNA- and RNA-binding reactions (Kaper and Sperandio, 2005; Matson *et al.*, 2007; Stonehouse *et al.*, 2008). However, quorum-sensing is not a universal mechanism, as some *V. cholerae* strains (for example El Tor N16961 and O395) have a mutation in a central element, negative regulator *hapR*, and are not sensitive to fluctuations in bacterial density (Zhu *et al.*, 2002).

Future trends

Despite significant advances in our knowledge of T4P, we have far to go before we understand how they are assembled and retracted. T4b pilus systems and the BFP system in particular offer excellent models for further advancing our understanding. Probing the interactions between different constituents of BFP and characterization of the IM and OM protein subassemblies have shed light on connections between some components of the machinery, but more pieces have to come together before we can visualize the puzzle. One crucial missing element is the link between the IM and OM subassemblies of BFP, as it is clear that the BFP apparatus spans the periplasmic space. One of the most fascinating questions is how bundlin, which begins as an integral membrane protein (Fig. 3.3), gets extracted from the membrane and engaged into the machinery. Another is how the chemical energy of ATP hydrolysis on the cytoplasmic side of the IM is transformed to the mechanical energy required for this process. A better understanding of the function of the biogenesis machinery will require the solution of the structures of all the individual components and of the machine itself. Crystal structures and solution structures at the atomic level and cryoelectron structures of larger complexes will be critical to this advance. Thus far, only structures of related T2SS components have been published. To facilitate our understanding, a complete interaction map among the components will also be required. These efforts are under way, but remain incomplete. Biophysical approaches, which have already added tremendous insight by visualizing and measuring pilus retraction in real time, will continue to advance our understanding as they are applied to the machine itself. It will be necessary to explore conformational changes that occur with ATP hydrolysis using structural, genetic, biochemical and biophysical methods to gain insight into the dynamics of the process. As more details are added, the differences between T4P and T2S systems as well as between T4a and T4b systems will require attention.

Conclusions

Studies of BFP and other T4b systems have advanced considerably over recent years. The structures of three T4b pilin proteins have been solved and a better understanding of the post-translational processing of the pre-pilins has emerged (Fig. 3.3). Strong data have been reported that support the roles of some pilin proteins as adhesins critical for infection. Once mature pilin is synthesized, it must be recruited to the biogenesis machine, which must then extract it from the cytoplasmic membrane so that it can be polymerized into pili. This process probably involves the concerted efforts of an ATPase and two integral cytoplasmic membrane proteins, all which interact in a complex (Fig. 3.4). Polymerized pili must penetrate the outer membrane through the multimeric secretin protein. Two essential proteins, BfpU and BfpG, are directed to the periplasm by the general secretory pathway and interact with the amino terminus of the secretin in the outer membrane. Pilin-like proteins are also essential to pilin biogenesis, but their precise cellular location and interaction partners have yet to be resolved. Many T4P are retractable, a process that requires a second protein with ATPase motifs that interacts at a distinct domain of the conserved polytopic transmembrane protein. Presumably this ATPase supplies energy required to reverse the biogenesis process and return pilin subunits from the pilus to the cytoplasmic membrane.

Acknowledgements

This work was supported by Public Health Service award R01 AI-37606 from the National Institutes of Health. The authors thank Ru-ching Hsia and Wensheng Luo for technical assistance in preparation of Fig. 3.1.

References

Abendroth, J., Murphy, P., Sandkvist, M., Bagdasarian, M., and Hol, W.G. (2005). The X-ray structure of the type II secretion system complex formed by the N-terminal domain of EpsE and the cytoplasmic domain of EpsL of *Vibrio cholerae*. J. Mol. Biol. *348*, 845–855.

Akahane, K., Sakai, D., Furuya, N., and Komano, T. (2005). Analysis of the pilU gene for the prepilin peptidase involved in the biogenesis of Type IV pili encoded by plasmid R64. Mol. Genet. Genomics 273, 350–359.

Anantha, R.P., Stone, K.D., and Donnenberg, M.S. (1998). The role of BfpF, a member of the PilT family of putative nucleotide-binding proteins, in Type IV pilus biogenesis and in interactions between enteropathogenic *Escherichia coli* and host cells. Infect. Immun. *66*, 122–131.

Anantha, R.P., Stone, K.D., and Donnenberg, M.S. (2000). Effects of *bfp* mutations on biogenesis of functional enteropathogenic *Escherichia coli* Type IV pili. J. Bacteriol. *182*, 2498–2506.

Arts, J., De Groot, A., Ball, G., Durand, E., El, K.M., Filloux, A., Tommassen, J., and Koster, M. (2007a). Interaction domains in the *Pseudomonas aeruginosa* type II secretory apparatus component XcpS (GspF). Microbiology *153*, 1582–1592.

Arts, J., van Boxtel, R., Filloux, A., Tommassen, J., and Koster, M. (2007b). Export of the pseudopilin XcpT of the *Pseudomonas aeruginosa* type II secretion system via the signal recognition particle-Sec pathway. J. Bacteriol. *189*, 2069–2076.

Bardwell, J.C., McGovern, K., and Beckwith, J. (1991). Identification of a protein required for disulfide bond formation *in vivo*. Cell *67*, 581–589.

Bayan, N., Guilvout, I., and Pugsley, A.P. (2006). Secretins take shape. Mol. Microbiol. *60*, 1–4.

Bayer, M., Eferl, R., Zellnig, G., Teferle, K., Dijkstra, A., Koraimann, G., and Hogenauer, G. (1995). Gene 19 of plasmid R1 is required for both efficient conjugative DNA transfer and bacteriophage R17 infection. J. Bacteriol. *177*, 4279–4288.

Bieber, D., Ramer, S.W., Wu, C.Y., Murray, W.J., Tobe, T., Fernandez, R., and Schoolnik, G.K. (1998). Type IV pili, transient bacterial aggregates, and virulence of enteropathogenic *Escherichia coli*. Science *280*, 2114–2118.

Blank, T.E. and Donnenberg, M.S. (2001). Novel topology of BfpE, a cytoplasmic membrane protein required for Type IV fimbrial biogenesis in enteropathogenic *Escherichia coli*. J. Bacteriol. *183*, 4435–4450.

Blank, T.E., Lacher, D.W., Scaletsky, I.C.A., Zhong, H.L., Whittam, T.S., and Donnenberg, M.S. (2003). Enteropathogenic *Escherichia coli* O157 strains from Brazil. Emerg. Infect. Dis. *9*, 113–115.

Blank, T.E., Zhong, H., Bell, A.L., Whittam, T.S., and Donnenberg, M.S. (2000). Molecular variation among Type IV pilin (*bfpA*) genes from diverse enteropathogenic *Escherichia coli* strains. Infect. Immun. *68*, 7028–7038.

Bose, N. and Taylor, R.K. (2005). Identification of a TcpC-TcpQ outer membrane complex involved in the biogenesis of the toxin-coregulated pilus of *Vibrio cholerae*. J. Bacteriol. *187*, 2225–2232.

Camberg, J.L., Johnson, T.L., Patrick, M., Abendroth, J., Hol, W.G., and Sandkvist, M. (2007). Synergistic stimulation of EpsE ATP hydrolysis by EpsL and acidic phospholipids. EMBO J. *26*, 19–27.

Collins, R.F., Ford, R.C., Kitmitto, A., Olsen, R.O., Tonjum, T., and Derrick, J.P. (2003). Three-dimensional structure of the *Neisseria meningitidis* secretin PilQ determined from negative-stain transmission electron microscopy. J. Bacteriol. *185*, 2611–2617.

Collins, R.F., Saleem, M., and Derrick, J.P. (2007). Purification and three-dimensional electron microscopy structure of the *Neisseria meningitidis* Type IV pilus biogenesis protein PilG. J. Bacteriol. *189*, 6389–6396.

Crago, A.M. and Koronakis, V. (1998). *Salmonella* InvG forms a ring-like multimer that requires the InvH lipoprotein for outer membrane localization. Mol. Microbiol. *30*, 47–56.

Craig, L., Pique, M.E., and Tainer, J.A. (2004). Type IV pilus structure and bacterial pathogenicity. Nat. Rev. Microbiol. *2*, 363–378.

Craig, L., Taylor, R.K., Pique, M.E., Adair, B.D., Arvai, A.S., Singh, M., Lloyd, S.J., Shin, D.S., Getzoff, E.D., Yeager, M., Forest, K.T., and Tainer, J.A. (2003). Type IV pilin structure and assembly: X-ray and EM analyses of *Vibrio cholerae* toxin-coregulated pilus and *Pseudomonas aeruginosa* PAK pilin. Mol. Cell *11*, 1139–1150.

Craig, L., Volkmann, N., Arvai, A.S., Pique, M.E., Yeager, M., Egelman, E.H., and Tainer, J.A. (2006). Type IV pilus structure by cryo-electron microscopy and crystallography: implications for pilus assembly and functions. Mol. Cell *23*, 651–662.

Crowther, L.J., Anantha, R.P., and Donnenberg, M.S. (2004). The inner membrane subassembly of the enteropathogenic *Escherichia coli* bundle-forming pilus machine. Mol. Microbiol. *52*, 67–79.

Daniel, A., Singh, A., Crowther, L.J., Fernandes, P.J., Schreiber, W., and Donnenberg, M.S. (2006). Interaction and localization studies of enteropathogenic *Escherichia coli* Type IV bundle-forming pilus outer membrane components. Microbiology *152*, 2405–2420.

De Groot, A., Heijnen, I., De Cock, H., Filloux, A., and Tommassen, J. (1994). Characterization of Type IV pilus genes in plant growth- promoting *Pseudomonas putida* WCS358. J. Bacteriol. *176*, 642–650.

DiRita, V.J., Parsot, C., Jander, G., and Mekalanos, J.J. (1991). Regulatory cascade controls virulence in *Vibrio cholerae*. Proc. Natl. Acad. Sci. U.S.A. *88*, 5403–5407.

Donnenberg, M.S., Girón, J.A., Nataro, J.P., and Kaper, J.B. (1992). A plasmid-encoded Type IV fimbrial gene of enteropathogenic *Escherichia coli* associated with localized adherence. Mol. Microbiol. *6*, 3427–3437.

Donnenberg, M.S., Zhang, H.-Z., and Stone, K.D. (1997). Biogenesis of the bundle-forming pilus of enteropathogenic *Escherichia coli*: reconstitution of fimbriae in recombinant *E- coli* and role of DsbA in pilin stability – a review. Gene *192*, 33–38.

Fernandes, P.J., Guo, Q., and Donnenberg, M.S. (2007). Functional consequences of sequence variation in bundlin, the enteropathogenic *Escherichia coli* Type IV pilin protein. Infect. Immun. *75*, 4687–4696.

Giangrossi, M., Zattoni, S., Tramonti, A., De Biase, D., and Falconi, M. (2005). Antagonistic role of H-NS and GadX in the regulation of the glutamate decarboxylase-dependent acid resistance system in *Escherichia coli*. J. Biol. Chem. *280*, 21498–21505.

Girón, J.A., Gómez-Duarte, O.G., Jarvis, K.G., and Kaper, J.B. (1997). Longus pilus of enterotoxigenic *Escherichia coli* and its relatedness to other type-4 pili – a minireview. Gene *192*, 39–43.

Girón, J.A., Ho, A.S.Y., and Schoolnik, G.K. (1991). An inducible bundle-forming pilus of enteropathogenic *Escherichia coli*. Science *254*, 710–713.

Gómez-Duarte, O.G. and Kaper, J.B. (1995). A plasmid-encoded regulatory region activates chromosomal *eaeA* expression in enteropathogenic *Escherichia coli*. Infect. Immun. *63*, 1767–1776.

Herrington, D.A., Hall, R.H., Losonsky, G., Mekalanos, J.J., Taylor, R.K., and Levine, M.M. (1988). Toxin, toxin-coregulated pili, and the *toxR* regulon are essential for *Vibrio cholerae* pathogenesis in humans. J. Exp. Med. *168*, 1487–1492.

Hulbert, R.R. and Taylor, R.K. (2002). Mechanism of ToxT-dependent transcriptional activation at the *Vibrio cholerae tcpA* promoter. J. Bacteriol. *184*, 5533–5544.

Hwang, J., Bieber, D., Ramer, S.W., Wu, C.Y., and Schoolnik, G.K. (2003). Structural and topographical studies of the Type IV bundle-forming pilus assembly complex of enteropathogenic *Escherichia coli*. J. Bacteriol. *185*, 6695–6701.

Hyland, R.M., Beck, P., Mulvey, G.L., Kitov, P.I., and Armstrong, G.D. (2006a). N-acetyllactosamine conjugated to gold nanoparticles inhibits enteropathogenic *Escherichia coli* colonization of the epithelium in human intestinal biopsy specimens. Infect. Immun. *74*, 5419–5421.

Hyland, R.M., Griener, T.P., Mulvey, G.L., Kitov, P.I., Srivastava, O.P., Marcato, P., and Armstrong, G.D. (2006b). Basis for N-acetyllactosamine-mediated inhibition of enteropathogenic *Escherichia coli* localized adherence. J. Med. Microbiol. *55*, 669–675.

Hyland, R.M., Sun, J., Griener, T.P., Mulvey, G.L., Klassen, J.S., Donnenberg, M.S., and Armstrong, G.D. (2008). The bundlin pilin protein of enteropathogenic *Escherichia coli* is an N-acetyllactosamine-specific lectin. Cell Microbiol. *10*, 177–187.

Ibarra, J.A., Villalba, M.I., and Puente, J.L. (2003). Identification of the DNA binding sites of PerA, the transcriptional activator of the *bfp* and *per* operons in enteropathogenic *Escherichia coli*. J. Bacteriol. *185*, 2835–2847.

Ishiwa, A. and Komano, T. (2004). PilV adhesins of plasmid R64 thin pili specifically bind to the lipopolysaccharides of recipient cells. J. Mol. Biol. *343*, 615–625.

Kachlany, S.C., Planet, P.J., DeSalle, R., Fine, D.H., Figurski, D.H., and Kaplan, J.B. (2001). *flp-1*, the first representative of a new pilin gene subfamily, is required for non-specific adherence of *Actinobacillus actinomycetemcomitans*. Mol. Microbiol. *40*, 542–554.

Kaper, J.B. and Sperandio, V. (2005). Bacterial cell-to-cell signaling in the gastrointestinal tract. Infect. Immun. *73*, 3197–3209.

Knutton, S., Baldini, M.M., Kaper, J.B., and McNeish, A.S. (1987). Role of plasmid-encoded adherence factors in adhesion of enteropathogenic *Escherichia coli* to HEp-2 cells. Infect. Immun. *55*, 78–85.

Köhler, R., Schafer, K., Muller, S., Vignon, G., Diederichs, K., Philippsen, A., Ringler, P., Pugsley, A.P., Engel, A., and Welte, W. (2004). Structure and assembly of the pseudopilin PulG. Mol. Microbiol. *54*, 647–664.

Lacher, D.W., Steinsland, H., Blank, T.E., Donnenberg, M.S., and Whittam, T.S. (2007). Molecular evolution of typical enteropathogenic *Escherichia coli*: clonal analysis by multilocus sequence typing and virulence gene allelic profiling. J. Bacteriol. *189*, 342–350.

LaPointe, C.F. and Taylor, R.K. (2000). The type 4 prepilin peptidases comprise a novel family of aspartic acid proteases. J. Biol. Chem. *275*, 1502–1510.

Lehnherr, H., Hansen, A.M., and Ilyina, T. (1998). Penetration of the bacterial cell wall: a family of lytic transglycosylases in bacteriophages and conjugative plasmids. Mol. Microbiol. *30*, 454–457.

Maier, B., Koomey, M., and Sheetz, M.P. (2004). A force-dependent switch reverses Type IV pilus retraction. Proc. Natl. Acad. Sci. U.S.A. *101*, 10961–10966.

Maier, B., Potter, L., So, M., Seifert, H.S., and Sheetz, M.P. (2002). Single pilus motor forces exceed 100 pN. Proc Natl Acad Sci U.A *99*, 16012–16017.

Marsh, J.W. and Taylor, R.K. (1998). Identification of the *Vibrio cholerae* type 4 prepilin peptidase required for cholera toxin secretion and pilus formation. Mol. Microbiol. *29*, 1481–1492.

Martínez-Laguna, Y., Calva, E., and Puente, J.L. (1999). Autoactivation and environmental regulation of *bfpT* expression, the gene coding for the transcriptional activator of *bfpA* in enteropathogenic *Escherichia coli*. Mol. Microbiol. *33*, 153–166.

Matson, J.S., Withey, J.H., and DiRita, V.J. (2007). Regulatory networks controlling *Vibrio cholerae* virulence gene expression. Infect. Immun. *75*, 5542–5549.

Moreira, C.G., Palmer, K., Whiteley, M., Sircili, M.P., Trabulsi, L.R., Castro, A.F., and Sperandio, V. (2006). Bundle-forming pili and EspA are involved in biofilm formation by enteropathogenic *Escherichia coli*. J. Bacteriol. *188*, 3952–3961.

Morris, C., Yip, C.M., Tsui, I.S., Wong, D.K., and Hackett, J. (2003). The shufflon of *Salmonella enterica* serovar Typhi regulates Type IVB pilus-mediated bacterial self-association. Infect. Immun. *71*, 1141–1146.

Nouwen, N., Stahlberg, H., Pugsley, A.P., and Engel, A. (2000). Domain structure of secretin PulD revealed by limited proteolysis and electron microscopy. EMBO J. *19*, 2229–2236.

Opalka, N., Beckmann, R., Boisset, N., Simon, M.N., Russel, M., and Darst, S.A. (2003). Structure of the filamentous phage pIV multimer by cryo-electron microscopy. J. Mol. Biol. *325*, 461–470.

Peabody, C.R., Chung, Y.J., Yen, M.R., Vidal-Ingigliardi, D., Pugsley, A.P., and Saier, M.H., Jr. (2003). Type II protein secretion and its relationship to bacterial Type IV pili and archaeal flagella. Microbiology *149*, 3051–3072.

Peek, J.A. and Taylor, R.K. (1992). Characterization of a periplasmic thiol:disulfide interchange protein required for the functional maturation of secreted virulence factors of *Vibrio cholerae*. Proc. Natl. Acad. Sci. U.S.A. *89*, 6210–6214.

Pepe, C.M., Eklund, M.W., and Strom, M.S. (1996). Cloning of an *Aeromonas hydrophila* Type IV pilus biogenesis gene cluster: Complementation of pilus assembly functions and characterization of a Type IV leader peptidase *N*- methyltransferase required for extracellular protein secretion. Mol. Microbiol. *19*, 857–869.

Pier, G.B., Grout, M., Zaidi, T., Meluleni, G., Mueschenborn, S.S., Banting, G., Ratcliff, R., Evans, M.J., and Colledge, W.H. (1998). *Salmonella typhi* uses CFTR to enter intestinal epithelial cells. Nature *393*, 79–82.

Porter, M.E., Mitchell, P., Roe, A.J., Free, A., Smith, D.G., and Gally, D.L. (2004). Direct and indirect transcriptional activation of virulence genes by an AraC-like protein, PerA from enteropathogenic *Escherichia coli*. Mol. Microbiol. *54*, 1117–1133.

Puente, J.L., Bieber, D., Ramer, S.W., Murray, W., and Schoolnik, G.K. (1996). The bundle-forming pili of enteropathogenic *Escherichia coli*: Transcriptional regulation by environmental signals. Mol. Microbiol. *20*, 87–100.

Py, B., Loiseau, L., and Barras, F. (1999). Assembly of the type II secretion machinery of *Erwinia chrysanthemi*: direct interaction and associated conformational change between OutE, the putative ATP-binding component and the membrane protein OutL. J Mol. Biol 289, 659–670.

Py, B., Loiseau, L., and Barras, F. (2001). An inner membrane platform in the type II secretion machinery of Gram-negative bacteria. EMBO Rep. 2, 244–248.

Ramboarina, S., Fernandes, P.J., Daniell, S., Islam, S., Simpson, P., Frankel, G., Booy, F., Donnenberg, M.S., and Matthews, S. (2005). Structure of the bundle-forming pilus from enteropathogenic *Escherichia coli*. J. Biol. Chem. *280*, 40252–40260.

Ramer, S.W., Bieber, D., and Schoolnik, G.K. (1996). BfpB, an outer membrane lipoprotein required for the biogenesis of bundle-forming pili in enteropathogenic *Escherichia coli*. J. Bacteriol. *178*, 6555–6563.

Ramer, S.W., Schoolnik, G.K., Wu, C.Y., Hwang, J., Schmidt, S.A., and Bieber, D. (2002). The Type IV pilus assembly complex: biogenic interactions among the bundle-forming pilus proteins of enteropathogenic *Escherichia coli*. J. Bacteriol. *184*, 3457–3465.

Rivas, S., Bolland, S., Cabezon, E., Goni, F.M., and de la Cruz, F. (1997). TrwD, a protein encoded by the IncW plasmid R388, displays an ATP hydrolase activity essential for bacterial conjugation. J. Biol. Chem. *272*, 25583–25590.

Russel, M. and Kazmierczak, B. (1993). Analysis of the structure and subcellular location of filamentous phage pIV. J. Bacteriol. *175*, 3998–4007.

Sakai, D., Horiuchi, T., and Komano, T. (2001). ATPase activity and multimer formation of PilQ protein are required for thin pilus biogenesis in plasmid R64. J. Biol. Chem. *276*, 17968–17975.

Sandkvist, M., Bagdasarian, M., Howard, S.P., and DiRita, V.J. (1995). Interaction between the autokinase EpsE and EpsL in the cytoplasmic membrane is required for extracellular secretion in *Vibrio cholerae*. EMBO J. *14*, 1664–1673.

Sandkvist, M., Keith, J.M., Bagdasarian, M., and Howard, S.P. (2000). Two regions of EpsL involved in species-specific protein–protein interactions with EpsE and EpsM of the general secretion pathway in *Vibrio cholerae*. J. Bacteriol. *182*, 742–748.

Satyshur, K.A., Worzalla, G.A., Meyer, L.S., Heiniger, E.K., Aukema, K.G., Misic, A.M., and Forest, K.T. (2007). Crystal structures of the pilus retraction motor PilT suggest large domain movements and subunit cooperation drive motility. Structure *15*, 363–376.

Sauvonnet, N., Gounon, P., and Pugsley, A.P. (2000a). PpdD Type IV pilin of *Escherichia coli* K-12 can be assembled into pili in *Pseudomonas aeruginosa*. J. Bacteriol. *182*, 848–854.

Sauvonnet, N., Vignon, G., Pugsley, A.P., and Gounon, P. (2000b). Pilus formation and protein secretion by the same machinery in *Escherichia coli*. EMBO J. *19*, 2221–2228.

Scaletsky, I.C.A., Silva, M.L.M., and Trabulsi, L.R. (1984). Distinctive patterns of adherence of enteropathogenic *Escherichia coli* to HeLa cells. Infect. Immun. *45*, 534–536.

Schmidt, S.A., Bieber, D., Ramer, S.W., Hwang, J., Wu, C.Y., and Schoolnik, G. (2001). Structure-function analysis of BfpB, a secretin-like protein encoded by the bundle-forming-pilus operon of enteropathogenic *Escherichia coli*. J. Bacteriol. *183*, 4848–4859.

Schreiber, W., Stone, K.D., Strong, M.A., DeTolla, L.J., Jr., Hoppert, M., and Donnenberg, M.S. (2002). BfpU, a soluble protein essential for Type IV pilus biogenesis in enteropathogenic *Escherichia coli*. Microbiology *148*, 2507–2518.

Shevchik, V.E., Robert-Baudouy, J., and Condemine, G. (1997). Specific interaction between OutD, an *Erwinia chrysanthemi* outer membrane protein of the general secretory pathway, and secreted proteins. EMBO J *16*, 3007–3016.

Shin, S., Castanie-Cornet, M.P., Foster, J.W., Crawford, J.A., Brinkley, C., and Kaper, J.B. (2001). An activator of glutamate decarboxylase genes regulates the expression of enteropathogenic *Escherichia coli* virulence genes through control of the plasmid-encoded regulator, Per. Mol. Microbiol. *41*, 1133–1150.

Shiue, S.J., Kao, K.M., Leu, W.M., Chen, L.Y., Chan, N.L., and Hu, N.T. (2006). XpsE oligomerization triggered by ATP binding, not hydrolysis, leads to its association with XpsL. EMBO J. *25*, 1426–1435.

Sohel, I., Puente, J.L., Ramer, S.W., Bieber, D., Wu, C.-Y., and Schoolnik, G.K. (1996). Enteropathogenic *Escherichia coli*: identification of a gene cluster coding for bundle-forming pilus morphogenesis. J. Bacteriol. *178*, 2613–2628.

Stone, K.D., Zhang, H.-Z., Carlson, L.K., and Donnenberg, M.S. (1996). A cluster of fourteen genes from enteropathogenic *Escherichia coli* is sufficient for biogenesis of a Type IV pilus. Mol. Microbiol. *20*, 325–337.

Stonehouse, E., Kovacikova, G., Taylor, R.K., and Skorupski, K. (2008). Integration host factor positively regulates virulence gene expression in *Vibrio cholerae*. J. Bacteriol. *190*, 4736–4748.

Strom, M.S., Bergman, P., and Lory, S. (1993a). Identification of active-site cysteines in the conserved domain of PilD, the bifunctional Type IV pilin leader peptidase/*N*- methyltransferase of *Pseudomonas aeruginosa*. J. Biol. Chem. *268*, 15788–15794.

Strom, M.S. and Lory, S. (1987). Mapping of export signals of *Pseudomonas aeruginosa* pilin with alkaline phosphatase fusions. J. Bacteriol. *169*, 3181–3188.

Strom, M.S. and Lory, S. (1991). Amino acid substitutions in pilin of *Pseudomona aeruginosa*. Effect on leader peptide cleavage, amino-terminal methylation, and pilus assembly. J. Biol. Chem. *266*, 1656–1664.

Strom, M.S., Nunn, D.N., and Lory, S. (1993b). A single bifunctional enzyme, PilD, catalyzes cleavage and N-methylation of proteins belonging to the Type IV pilin family. Proc. Natl. Acad. Sci. U.S.A. *90*, 2404–2408.

Sun, D., Lafferty, M.J., Peek, J.A., and Taylor, R.K. (1997). Domains within the *Vibrio cholerae* toxin coregulated pilin subunit that mediate bacterial colonization. Gene *192*, 79–85.

Sun, D., Seyer, J.M., Kovari, I., Sumrada, R.A., and Taylor, R.K. (1991). Localization of protective epitopes within the pilin subunit of the *Vibrio cholerae* toxin-coregulated pilus. Infect. Immun. *59*, 114–118.

Taniguchi, T., Akeda, Y., Haba, A., Yasuda, Y., Yamamoto, K., Honda, T., and Tochikubo, K. (2001). Gene cluster for assembly of pilus colonization factor antigen III of enterotoxigenic *Escherichia coli*. Infect. Immun. *69*, 5864–5873.

Taylor, R.K., Miller, V.L., Furlong, D.B., and Mekalanos, J.J. (1987). Use of *phoA* gene fusions to identify a pilus colonization factor coordinately regulated with cholera toxin. Proc. Natl. Acad. Sci. U.S.A. *84*, 2833–2837.

Thomas, J.D., Reeves, P.J., and Salmond, G.P.C. (1997). The general secretion pathway of *Erwinia carotovora* subsp. *carotovora*: analysis of the membrane topology of OutC and OutF. Microbiology *143*, 713–720.

Tobe, T. and Sasakawa, C. (2002). Species-specific cell adhesion of enteropathogenic *Escherichia coli* is mediated by Type IV bundle-forming pili. Cell Microbiol. *4*, 29–42.

Tobe, T., Schoolnik, G.K., Sohel, I., Bustamante, V.H., and Puente, J.L. (1996). Cloning and characterization of *bfpTVW*, genes required for the transcriptional activation of *bfpA* in enteropathogenic *Escherichia coli*. Mol. Microbiol. *21*, 963–975.

Tomich, M., Planet, P.J., and Figurski, D.H. (2007). The *tad* locus: postcards from the widespread colonization island. Nat. Rev. Microbiol. *5*, 363–375.

Tripathi, S.A. and Taylor, R.K. (2007). Membrane association and multimerization of TcpT, the cognate ATPase ortholog of the *Vibrio cholerae* toxin-coregulated-pilus biogenesis apparatus. J. Bacteriol. *189*, 4401–4409.

Tsui, I.S., Yip, C.M., Hackett, J., and Morris, C. (2003). The Type IVB pili of *Salmonella enterica* serovar Typhi bind to the cystic fibrosis transmembrane conductance regulator. Infect. Immun. *71*, 6049–6050.

Vanmaele, R.P., Heerze, L.D., and Armstrong, G.D. (1999). Role of lactosyl glycan sequences in inhibiting enteropathogenic *Escherichia coli* attachment. Infect. Immun. *67*, 3302–3307.

Vuopio-Varkila, J. and Schoolnik, G.K. (1991). Localized adherence by enteropathogenic *Escherichia coli* is an inducible phenotype associated with the expression of new outer membrane proteins. J. Exp. Med. *174*, 1167–1177.

Winther-Larsen, H.C., Wolfgang, M., Dunham, S., van Putten, J.P., Dorward, D., Løvold, C., Aas, F.E., and Koomey, M. (2005). A conserved set of pilin-like molecules controls Type IV pilus dynamics and organelle-associated functions in *Neisseria gonorrhoeae*. Mol. Microbiol. *56*, 903–917.

Xu, X.F., Tan, Y.W., Lam, L., Hackett, J., Zhang, M., and Mok, Y.K. (2004). NMR structure of a Type IVb pilin from *Salmonella typhi* and its assembly into pilus. J. Biol. Chem. *279*, 31599–31605.

Yamagata, A. and Tainer, J.A. (2007). Hexameric structures of the archaeal secretion ATPase GspE and implications for a universal secretion mechanism. EMBO J. *26*, 878–890.

Yanez, M.E., Korotkov, K.V., Abendroth, J., and Hol, W.G. (2008). The crystal structure of a binary complex of two pseudopilins: EpsI and EpsJ from the type 2 secretion system of *Vibrio vulnificus*. J. Mol. Biol. *375*, 471–486.

Yoshida, T., Furuya, N., Ishikura, M., Isobe, T., Haino-Fukushima, K., Ogawa, T., and Komano, T. (1998). Purification and characterization of thin pili of IncI1 plasmids ColIb-P9 and R64: Formation of PilV-specific cell aggregates by Type IV pili. J. Bacteriol. *180*, 2842–2848.

Yoshida, T., Kim, S.R., and Komano, T. (1999). Twelve *pil* genes are required for biogenesis of the R64 thin pilus. J. Bacteriol. *181*, 2038–2043.

Yu, J., Webb, H., and Hirst, T.R. (1992). A homologue of the *Escherichia coli* DsbA protein involved in disulphide bond formation is required for enterotoxin biogenesis in *Vibrio cholerae*. Mol. Microbiol. *6*, 1949–1958.

Zhang, H.-Z. and Donnenberg, M.S. (1996). DsbA is required for stability of the Type IV pilin of enteropathogenic *Escherichia coli*. Mol. Microbiol. *21*, 787–797.

Zhang, H.-Z., Lory, S., and Donnenberg, M.S. (1994). A plasmid-encoded prepilin peptidase gene from enteropathogenic *Escherichia coli*. J. Bacteriol. *176*, 6885–6891.

Zhang, X.L., Tsui, I.S., Yip, C.M., Fung, A.W., Wong, D.K., Dai, X., Yang, Y., Hackett, J., and Morris, C. (2000). *Salmonella enterica* serovar Typhi uses Type IVB pili To enter human intestinal epithelial cells. Infect. Immun. *68*, 3067–3073.

Zhu, J., Miller, M.B., Vance, R.E., Dziejman, M., Bassler, B.L., and Mekalanos, J.J. (2002). Quorum-sensing regulators control virulence gene expression in *Vibrio cholerae*. Proc. Natl. Acad. Sci. U.S.A. *99*, 3129–3134.

Structure, Function and Biogenesis of Pili Formed by the Chaperone/Usher Pathway

4

Han Remaut and Gabriel Waksman

Abstract

The chaperone/usher pathway is used by a wide range of Gram-negative bacteria to expose adhesive filaments, called pili or fimbriae, on their outer surface. In pathogenic strains, pili constitute important virulence factors that allow host recognition and immune evasion via specific attachment, cell invasion and/or biofilm formation. Pilus biogenesis involves a periplasmic chaperone and an outer membrane protein called the usher. The periplasmic chaperone aids pilus subunit folding in the periplasm and maintains subunits in a polymerization-prone folding state. The usher recruits chaperone–subunit complexes to the outer membrane, facilitates their ordered polymerization and is responsible for pilus translocation to the outer surface. The tremendous advancements in the structural molecular characterization of pilus components and the chaperone/usher assembly machinery over the last two decades has brought the mechanistic understanding of this biosynthetic pathway to a point where strategies can now be developed for the targeted disarmament of bacteria of these virulence organelles.

Introduction

The chaperone/usher (CU) pathway is responsible for the synthesis of adhesive, linear multisubunit fibres, called pili or fimbriae, first observed in *Escherichia coli* more then half a century ago (Duguid *et al.*, 1955). The availability of multiple bacterial genome sequences now shows CU pili form by far the most abundant group of bacterial cell surface appendages, mainly present in Enterobacteriaceae such as *Escherichia, Shigella, Klebsiella, Proteus, Yersinia* and *Salmonella*, but also found in several *Pseudomonas, Haemophilus, Bordetella, Xyllella, Burkholderia, Acinetobacter* and *Ralstonia* species. These fibrous organelles assembled by the CU pathway are typically encoded in individual gene clusters, located on the chromosome or plasmids. Pathogenic strains often carry a multitude of different CU pathways, the distribution and exact combination of which determine strain virulence and host and tissue tropism (Townsend *et al.*, 2001; Kaper *et al.*, 2004; Fitzhenry *et al.*, 2006). Enteroheamorrhagic and uropathogenic *E. coli* (EHEC and UPEC, respectively) strains, for example, carry up to 10 different CU fimbrial clusters, several of which are strain specific and associated with the ability to cause enteric disease or colonize the urinary tract (Perna *et al.*, 2001; Welch *et al.*, 2002; Kaper *et al.*, 2004; Chen *et al.*, 2006; Wright and Hultgren, 2006). In pathogenic strains, CU pili are major contributors to virulence, responsible for host cell recognition and attachment and/or the evasion of host responses by the induction of cell invasion or through capsule and biofilm formation (Kuehn *et al.*, 1992; Mulvey *et al.*, 1998; Du *et al.*, 2002; Ulett *et al.*, 2007; Wright *et al.*, 2007).

Pilus morphology

CU-assembled organelles are formed by linear, unbranched polymers of several hundreds to thousands of pilus subunits. The pilus subunits, also called pilins, range in size from ~12 to ~20 kDa and comprise a single immunoglobulin-like domain, preceded by a short stretch (10–20 residues) of unstructured sequence, called the

N-terminal extension (Nte). Subunits polymerize through non-covalent head-to-tail interactions mediated by their N-terminal extension peptides (see below). The resulting fibres form single-start filaments with flexible, extended conformations (fibrillae) or more rigid, helically wound conformations (rod-like pili), the only exception reported thus far being *H. influenzae* Hib pili, alleged to form a left-handed three-start helical structure (Mu *et al.*, 2002). The organelles are usually peritrichously distributed on the bacterial cell surface, though exceptions occur as is seen at least in the case of lpf pili (long polar fimbriae) found in *Salmonella* and *Escherichia* spp. (Baumler and Heffron, 1995; Torres *et al.*, 2002).

CU organelles differ widely in complexity, ranging from homopolymers to highly choreographed heteropolymers composed of up to six or seven different subunit types (Fig. 4.1). Filament morphology ranges from thin fibrillar material, ~2–3 nm in diameter, to ~7–8 nm thick rigid fimbriae composed of a helically wound rod and a distinct fibrillar tip structure. Different CU organelle complexities and morphologies appear to cluster along a number of evolutionary lines as judged from the phylogenetic analysis of their corresponding usher sequences (CU organelles are encoded in individual operons that contain a single and unique usher gene). Comparing sequences in all 189 known CU operons, Nuccio and Baümler now suggest a classification system that divides CU systems into six major clades: α, β, γ (subdivided into γ_1, γ_2, γ_3 and γ_4), κ, π and σ-fimbriae, based on common ancestry of their ushers (Nuccio and Baümler, 2007).

Fibrillar organelles are formed by CU systems belonging to the κ- (*E. coli* K88- and K99-related fimbriae) and γ_3-fimbrial clades. The latter is also termed the 'FGL-chaperone' clade because the chaperone involved in assembling these organelles contains a long structural loop located between the F_1 and G_1 strands (see below for more details) (Hung *et al.*, 1996; Nuccio and Baumler, 2007). FGL-chaperone assembled organelles (γ_3 fimbriae) are simple in architecture, containing just a single polymerizing subunit, possibly accompanied by a non-polymerizing subunit that caps the fibres at their distal end (Fig. 4.1) (Zavialov *et al.*, 2007). At least in the case of *E. coli* Afa/Dr adhesins, this additional, non-polymerizing subunit serves as an invasin, inducing uptake of the bacteria by the host cell (Garcia *et al.*, 2000). FGL systems assemble into thin flexible fibres of ~2–3 nm thickness (often referred to as fibrillae) that frequently curl up into a capsule-like amorphous mass on the bacterial surface (Keller *et al.*, 2002; Anderson *et al.*, 2004; Salih *et al.*, 2008). The elusive nature of these organelles in electron micrographs (FGL-chaperone assembled fibrillae are not usually visible unless immuno-gold labelled) has led some of these structures to be referred to as 'afimbrial' or 'afimbrial sheats' in the past (Garcia *et al.*, 1996). However, subsequent structural and biochemical characterization of the *Y. pestis* F1 capsular antigen, *E. coli* Afa/Dr adhesins and *S. enterica* Saf fimbriae has shown that these sheets or masses are in fact formed by aggregated single start, highly flexible fibrillae (Zavialov *et al.*, 2002; Zavialov *et al.*, 2003; Anderson *et al.*, 2004; Remaut *et al.*, 2006; Salih *et al.*, 2008). κ fimbrial systems contain three to five pilus subunit types, which assemble into thin flexible fibrillae of 2–5 nm thickness that do not usually aggregate but extend individually from the bacterial surface (Hahn *et al.*, 2000; Nuccio and Baumler, 2007). The role and localization of the multiple subunits types in these organelles are poorly understood.

Rod-like, or 'typical', fimbrial organelles are found in the α-, γ- and π- fimbrial clades, with UPEC P and Type 1 pili (π- and γ_1-fimbrial clades, respectively) and CS1 fimbriae in enterotoxigenic *E. coli* (ETEC; α-fimbrial clade) as prototypical examples. These CU systems typically assemble 7–10-nm thick rigid pili with an often remarkably complex quaternary organization made up of multiple subunit types (Fig. 4.1). P pili, for example, comprise no fewer than six different subunit types, arranged into two distinct subassemblies, a flexible distal tip fibrillum of ~2 nm diameter, displayed onto a long, rigid 6.8-nm wide pilus rod (Kuehn *et al.*, 1992). The P pilus rod is formed by a homopolymer of >1000 copies of the PapA subunit. The PapA homopolymer organizes into a right-handed, one-start helical structure with a 2.5 nm pitch and 3.3 subunits per turn, containing a central cavity of ~2.5 × ~1.5 nm, connected to the exterior by a set of radial channels. The P pilus tip fibrillum contains a specialized adhesive subunit, PapG, present in a single copy at the

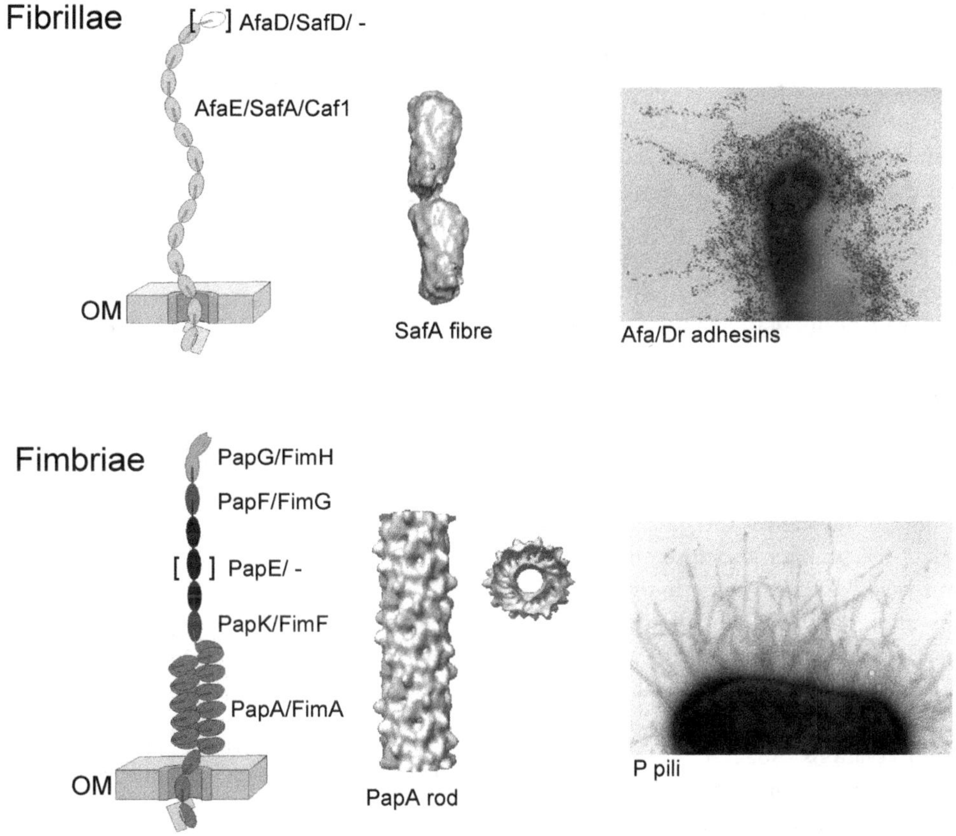

Figure 4.1 Fibrillar and fimbrial morphologies observed in CU organelles. **Top left**: schematic diagram of fibrillar organelles of the γ_3 fimbrial clade, labelled according to subunit type as found in EPEC Afa/Dr adhesins, *S. enterica* Saf pili and the *Y. pestis* F1 capsular antigen, respectively (OM: outer membrane). **Top middle and right**: 3D EM reconstitution of a 2-subunit segment of Saf pili and immuno-gold-stained EM imaging of Afa/Dr adhesins at the EPEC cell surface, respectively (Keller *et al.*, 2002; Salih *et al.*, 2008). **Bottom left**: schematic diagram of rod-like fimbriae, labelled according to subunit type as observed in UPEC P and Type 1 pili, respectively. **Bottom middle and right**: 3D EM reconstitution of the P pilus rod and negative stain EM imaging of P-piliated UPEC, respectively.

distal end, displayed onto a short extended and flexible polymer of PapE subunits (5–10). Two additional subunit types, PapF and PapK, serve as single copy linker subunits between the tip adhesin and the PapE polymer, and between the PapE polymer and the PapA rod, respectively (Fig. 4.1). Finally, the sixth pilus subunit, PapH, serves as terminator subunit, halting pilus growth at the basal end by arresting PapA polymerization (Baga *et al.*, 1987; Verger *et al.*, 2006). Type 1 pili (γ_1-fimbrial clade) display a very similar architecture, but form shorter tip fibrillae due to the lack of a polymerizing tip fibrillum subunit (PapE in P pili) (Hahn *et al.*, 2002). α-fimbrial organelles (alternate chaperone/usher pathway) like ETEC CS1 and CFA/I fimbriae lack a visible tip fibrillum all together and here the adhesive subunit is connected directly to the polymerizing major pilus subunit (Evans *et al.*, 1979; Levine *et al.*, 1984). No functional or structural information is available for σ-fimbriae: this family was defined entirely based on usher sequence and the fimbriae associated to this system have not yet been observed.

From a functional perspective, perhaps the most significant difference between fibrillar and rod-like organelles is the structure of the adhesive unit. In γ_3-fimbriae and at least in the *E. coli* K88 and K99 κ-fimbriae, adhesion is mediated by the polymerizing subunit forming the fibre (De Greve *et al.*, 2007; Zavialov *et al.*, 2007). This is in contrast to the 'typical' rod-like pili assembled by the α-, γ-, π- fimbrial clades, where adhesion is mediated by a specialized two-domain subunit (or 'adhesin') present in a single copy at the tip of the fibre (Dodson *et al.*, 2001). Likely, the flexible nature of the γ_3 and κ-fimbrial organelles provides the conformational freedom to enable multivalent adhesion. In rigid rod-like pili, the adhesive subunit is present in a single copy at the distal end of long, protruding and rigid fibres, thereby maximizing exposure of the adhesin. An example of the importance of topological considerations for receptor binding can be seen from the Type 1 pilus tissue tropism in UPEC, *K. pneumoniae* and *S.* Typhimurium, where differences in the receptor specificity of the tip adhesion subunit (FimH) were found to depend on the pilus architecture, rather than the primary structure of the adhesin (Duncan *et al.*, 2005). Whilst individual FimH adhesins show very similar affinities for high-mannose glycans, their receptor specificity is altered when grafted onto different pilus shafts (Duncan *et al.*, 2005; Bouckaert *et al.*, 2006). These observations suggest pathogens have diversified tissue and host tropism by recycling the same high-affinity mannose-binding adhesin, FimH, into structurally different organelles.

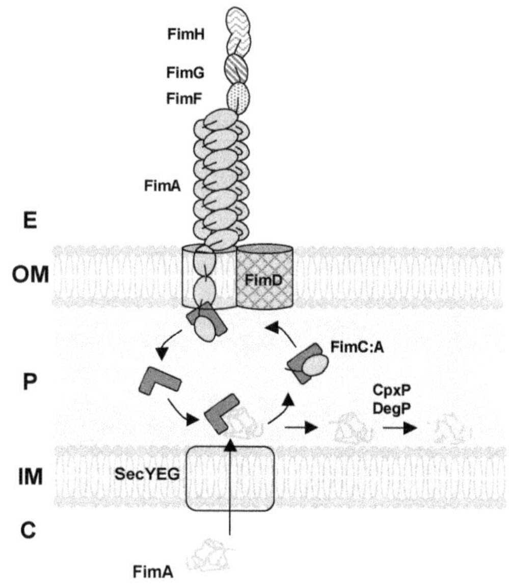

Figure 4.2 Schematic diagram of chaperone/usher pilus assembly pathway. Inner and outer membrane and extracellular, periplasmic and cytoplasmic space are labelled IM, OM, E, P and C, respectively. Chaperone (FimC), usher (FimD), tip adhesin (FimH) and pilus subunits (FimG, FimH and FimA) are labelled according the UPEC Type 1 pilus nomenclature.

Pilus assembly

CU usher organelles are encoded in individual gene clusters that in addition to sequences encoding pilus subunits and possible transcriptional regulators, also contain genes encoding two proprietary assembly proteins: a specialized periplasmic chaperone and an outer membrane assembly platform, called the usher (Sauer *et al.*, 2004). Inner membrane passage of pilus subunits occurs via the general secretory pathway, after which the chaperone and usher mediate the periplasmic and outer membrane stages of pilus assembly (Fig. 4.2).

Chaperone–subunit interactions

After release from the Sec translocase, nascent pilus subunits travel to the outer membrane usher as binary chaperone–subunit complexes. The chaperone aids subunit folding in the periplasm and maintains subunits in a polymerization-prone folding state (Barnhart *et al.*, 2000; Sauer *et al.*, 2002; Zavialov *et al.*, 2003; Vetsch *et al.*, 2004). The periplasmic chaperones are ~25 kDa proteins formed of two immunoglobulin-like (Ig-like) domains that meet at a right angle (Holmgren and Bränden, 1989; Kuehn *et al.*, 1993). The interactive surfaces that contact the pilus subunit are formed by two conserved basic residues in the linker sequence between the domains (R8 and K112, following Pap numbering) and by the N-terminal and seventh β-strands (strands A_1 and G_1, respectively) of the N-terminal Ig-domain of the chaperone (Fig. 4.3). Pilus subunits are characterized by an incomplete, C-terminally truncated, immunoglobulin-like fold, lacking the C-terminal β-strand (Figs. 4.3 and 4.4) (Choudhury *et al.*, 1999; Sauer *et al.*, 1999). As a result of this missing structural element, pilus subunits are unstable on their own, aggregate, and are degraded by the periplasmic protease DegP after activation of the Cpx stress pathway (Jones *et al.*, 1997; Isaac *et al.*, 2005). Early studies on the chaperone–subunit interactions revealed that the C-terminal strand of pilus subunits (strand F) is specifically recognized by the chaperone (Kuehn *et al.*, 1993; Soto *et al.*, 1998). The conserved basic pair in the linker between the N- and C-

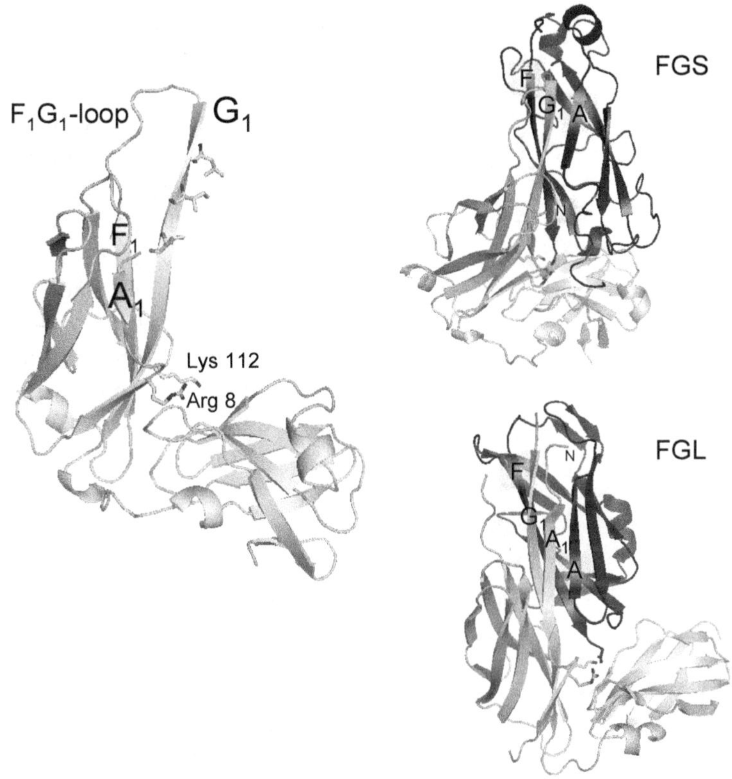

Figure 4.3 Structure of the periplasmic chaperone. **Left**: ribbon diagram of the P pilus chaperone PapD (1PDK), showing the conserved basic pair (Arg 8 and Lys 112) and the alternating motive of hydrophobic residues on the G_1 donor strand in stick representation. Labels indicate the chaperone A_1, F_1, G_1 strand and F_1G_1-loop. **Top and bottom right**: ribbon diagram of chaperone–subunit complexes (coloured grey–black) belonging to the FGS (UPEC PapD–PapK; PDB entry 1PDK) and FGL (*S. enterica* SafB–SafA; PDB entry 2CO6) class periplasmic chaperones, respectively. The chaperone G_1 donor strand and A_1 strand are represented labelled accordingly, the chaperone N-terminus is labelled 'N'.

terminal domains of the chaperone binds the subunit carboxy-terminus, whilst the subunit F strand aligns with the chaperone G_1 strand (Fig. 4.3). These interactions are essential for chaperone–subunit complex formation and nucleate subunit folding (Soto *et al.*, 1998; Barnhart *et al.*, 2000). In the mature chaperone–subunit complexes, the chaperone G_1 strand occupies the hydrophobic groove on the subunit surface, left by the absence of the C-terminal strand in the pilin. Thus, the chaperone complements the incomplete fold of the pilus subunit by donating part of its G_1 strand (donor strand), a process termed 'donor-strand complementation (DSC)' (Figs 4.3 and 4.4) (Choudhury *et al.*, 1999; Sauer *et al.*, 1999). By providing, *in trans*, the missing structural information in the pilus Ig-like fold, the chaperone functions as a structural platform catalysing folding of pilus subunits (Barnhart *et al.*, 2000; Bann *et al.*, 2004; Vetsch *et al.*, 2004).

Hung *et al.* (1996) described the presence of two chaperone subfamilies based on conserved structural differences in the flexible loop connecting the F_1 and G_1 strands (F_1–G_1 loop; Fig. 4.3) and the chaperone N-terminus. Chaperones in the CU systems of the γ_3 fimbrial clade are characterized by a longer F_1–G_1 loop and an extended N-terminus compared to chaperones in rod-forming CU systems. As a result, these two classes of chaperones were called FGL- and FGS-chaperones, 'FGL' and 'FGS' standing for '$\underline{F}_1\underline{G}_1$-$\underline{l}$ong' and '$\underline{F}_1\underline{G}_1$-$\underline{s}$hort', respectively (Hung *et al.*, 1996; Zavialov *et al.*, 2007)}. In chaperone–subunit interactions involving FGS chaperones, the chaperone's G_1 donor strand inserts in between

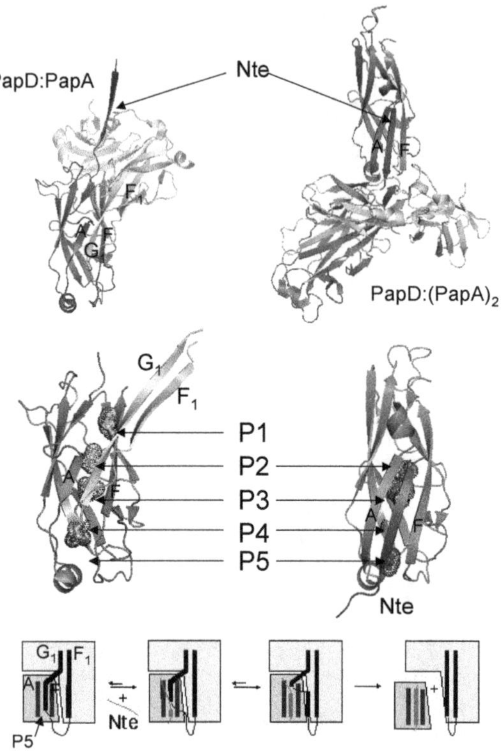

Figure 4.4 Donor-strand exchange and subunit–subunit interactions. **Top left and right**: ribbon diagram of the P pilis PapD–PapA chaperone–subunit (coloured yellow–magenta; PDB entry 2UY6) and PapD–PapA–PapA chaperone–subunit–subunit complexes (coloured yellow–magenta–blue), respectively. **Middle left and right**: ribbon diagram of the PapA subunit in donor-strand complementation and donor-strand exchange with the chaperone G_1 strand and PapA N-terminal extension (Nte), respectively. Labels indicate the P1, P2, P3, P4 and P5 sites and the subunit A and F strands. **Bottom**: schematic representation of the concerted donor-strand exchange mechanism. The diagrams show the F_1-G_1 β-hairpin in the chaperone (coloured yellow) and A and F strands in the subunit (coloured blue). The attacking Nte is coloured red. A colour version of this figure is located in the plate section at the back of the book.

the subunit's F and A strands, which are part of the same β-sheet in the Ig-like fold (Choudhury *et al.*, 1999; Sauer *et al.*, 1999; Verger *et al.*, 2006). In FGL systems, the A and F strands form the edge strands of the two opposing sheets in the Ig-like β-sandwich (Zavialov *et al.*, 2003; Anderson *et al.*, 2004; Remaut *et al.*, 2006): the chaperone G_1 donor strand pairs with the subunit's F strand, while the extended N-terminal strand A_1 aligns with the subunit's A strand (Fig. 4.3).

Subunit–subunit interactions

CU pilus subunits polymerize through stable non-covalent interactions between the pilin's 10–20-residues-long N-terminal extension peptide (Nte) and the Ig-domain of a preceding subunit in a fold complementation mechanism similar to that seen in the chaperone–subunit interaction. During subunit polymerization the complementing β-strand (G_1) donated by the chaperone is replaced by the Nte peptide of a newly incoming subunit, a process termed 'donor-strand exchange (DSE)' (Fig. 4.4) (Sauer *et al.*, 2002; Zavialov *et al.*, 2003; Remaut *et al.*, 2006). Donor-strand exchange is a spontaneous process, driven by folding energy preserved in the pilus subunits when bound to the chaperone (Sauer *et al.*, 2002; Zavialov *et al.*, 2003; Zavialov *et al.*, 2005). During donor-strand complementation, the chaperone donor strand (G_1) inserts a motif of four or five non-polar residues into the hydrophobic groove formed by the missing C-terminal strand in the pilus subunit. For clarity, these alternating hydrophobic residues are called P1 through P5 residues, while their corresponding acceptor sites in the subunit groove are called P1 through P5 pockets (Fig. 4.4). After donor-strand exchange, these same hydrophobic sites in the subunit's groove are occupied by non-polar residues (also termed 'P1–P5 residues') in the complementing Nte peptide of the incoming subunit. In FGS systems, the chaperone G_1 strand and subunit Nte bind sites P1–P4 and P2–P5, respectively, resulting in a shift in register upon donor-strand exchange (Fig. 4.4). In FGL systems, the P5 residue in the chaperone' G_1 strand occupies the P5 site intermittently, in an equilibrium between a bound and unbound state (see below).

When bound to the chaperone, the subunit's hydrophobic core is in a more open, less tightly packed conformation, compared to that seen in the subunit–Nte complex (Fig. 4.4) (Sauer *et al.*, 2002; Zavialov *et al.*, 2003). This more open conformation is the result of the presence of more bulky hydrophobic side chains in the P1–P4 positions of the chaperone G_1 strand as compared to the equivalent positions in the subunit's Nte. In addition, in the chaperone–subunit complex, the G_1 donor strand runs parallel to the subunit F strand, resulting in the subunit adopting a non-canonical Ig-fold. Instead, in the subunit–Nte interaction, the subunit's Nte runs antiparallel to strand F. These two factors contribute to the chaperone donor strand forcing a more open conformation onto the subunit's hydrophobic core, stabilizing the subunit in a higher energy folding intermediate, thereby priming the subunit for subsequent DSE. This stored folding energy drives donor-strand exchange as it is released upon transition of the subunit to its final, compact folding state when bound to a neighbouring subunit's Nte (Zavialov *et al.*, 2005).

Zavialov and coworkers postulated this transition could occur in either one of two ways (Zavialov *et al.*, 2003). In its simplest form, when bound to the outer membrane usher, the chaperone would dissociate from the subunit and thereby liberate the hydrophobic cleft on the subunit for binding with a new subunit's Nte. Alternatively, donor-strand exchange could occur via a zippering process in which the chaperone's donor strand is gradually displaced by the Nte of the incoming subunit. In support of a concerted process, Vetsch *et al.* showed that an excess of free chaperone, effectively reducing the concentration of free subunit, does not alter the rate of DSE in UPEC FimA polymerization (Vetsch *et al.*, 2006). Independently, we directly identified the existence of ternary chaperone–subunit–Nte complexes during *in vitro* DSE reactions of *Salmonella* SafA subunits using non-denaturing mass spectrometry (Remaut *et al.*, 2006). Crucially, the formation of this ternary complex forms the rate-limiting step in DSE and was shown to depend on the accessibility of the P5 site in the accepting pilus subunit's hydrophobic groove (Fig. 4.4).

As mentioned above, in FGS chaperone-assembled pilus subunits, the P5 site is not occupied by the chaperone G_1 strand and is therefore freely accessible to the P5 residue of the Nte of an incoming pilus subunit (Fig. 4.4). In FGL chaperone-assembled systems, like Saf, the chaperone occupies the P5 site, but appears to do so in an equilibrium with an unbound, accessible state (Remaut *et al.*, 2006). In the Saf system, mutating the chaperone's G_1 residue occupying the P5 site to a larger hydrophobic side chain shifts the equilibrium towards the closed state, suppresses ternary complex formation and drastically reduces the DSE rate. Similarly, decreasing the hydrophobic bulk of the P5 residue on the incoming subunit's Nte negatively affects ternary complex formation and DSE rate. Thus, the P5 pocket on a chaperone-bound subunit forms the primary acceptor site for an incoming subunit's Nte, and interaction of the subunit's Nte (through its P5 residue) with this site initiates the process of DSE. Single-site mutagenesis of the P5, P4 and P3 residues in the subunits' Nte to alanine revealed a gradient of decreasing DSE efficiency moving away from the P5 initiation site, suggesting a zippering strand exchange, with the new Nte gradually displacing the chaperone's G_1 donor strand in a stepwise process from P5 to P1 (shown schematically in the lower part of Fig. 4.4 for FGS systems). In Saf, this process is concluded by a capping residue N-terminal of P1 (termed 'P*'; an equivalent bulky hydrophobic is also present in the N-termini of other FGL-chaperone assembled subunits), effectively preventing reversal of the donor strand exchange reaction (Remaut *et al.*, 2006).

Pilus assembly at the usher

In vivo, subunit polymerization occurs at the outer membrane (OM) usher. The usher recruits binary chaperone–subunit complexes to the OM, mediates their ordered polymerization and is responsible for translocation of the growing pilus to the outer surface (Fig. 4.2) (Dodson *et al.*, 1993; Saulino *et al.*, 1998; Nishiyama *et al.*, 2005; Nishiyama *et al.*, 2008; Remaut *et al.*, 2008). OM ushers are ~800-residue-long proteins comprising four functional domains. The N-terminal ~125 residues form a distinct soluble periplasmic domain that forms the initial targeting site for chaperone–subunit complexes (Ng *et al.*, 2004; Nishiyama *et al.*, 2005). This domain discriminately recognizes different chaperone–subunit complexes by directly binding to the N-terminal domain of the periplasmic chaperone and part of the chaperone-bound subunit (Nishiyama *et al.*, 2005; Eidam *et al.*, 2008). In CU pili with complex quaternary organizations such as UPEC P and Type 1 pili, the usher shows decreasing affinities roughly according to subunit 'rank' in the pilus assembly chain, a process believed to be at least in part responsible for the tightly orchestrated subunit ordering seen in these fibres (Saulino *et al.*, 1998). In addition to this usher-driven control mechanism, donor-strand exchange rates in absence of the usher have also been found to vary amongst the different subunit–Nte interactions and show a positive correlation with the final subunit order in the pilus fibre (Lee *et al.*, 2007; Rose *et al.*, 2008).

A second periplasmic domain is formed by the C-terminal ~170 residues. Though essential, the exact role of this domain is unclear. The central part of the protein gives rise to the usher translocation channel. This consists of a 24-stranded β-pore with an approximate inner diameter of 45 × 25 Å, compatible with the passage of folded pilus subunits (Fig. 4.5A) (Remaut *et al.*, 2008). This β-barrel domain is interrupted at strands 6 and 7 by a soluble β-sandwich 'plug' domain (~110 residues; corresponding to the fourth predicted usher domain) that blocks the channel in its non-activated form. Differences in protease susceptibility in the Type 1 pilus usher FimD show the usher undergoes a conformational change upon binding of the N-terminal domain of the tip adhesin FimH (Saulino *et al.*, 1998; Munera *et al.*, 2007). The adhesin-induced conformational change constitutes a crucial activation step required for subunit polymerization and translocation by the usher. Given its position in the non-activated PapC translocation pore, it is clear that such an activation step would minimally involve a reordering of the plug domain, either by an upward rotation inside the translocation pore, or by its complete exclusion from the pore into the periplasm (Remaut *et al.*, 2008). Complementation studies on non-functional PapC mutants suggest that C-terminal truncates fail to activate the translocation pore, possibly implicating the C-terminal domain in the usher activation events (So and Thanassi, 2006).

Cryo-electron microscopy (EM) images of 2D crystals of the UPEC P pilus usher PapC reconstituted in *E. coli* lipids reveal it forms homodimeric channels in the OM. Complementation studies of PapC loss-of-function mutants subsequently established that these usher dimers represent the functional unit (Li *et al.*, 2004; So and Thanassi, 2006). The X-ray structure of the PapC translocation domain confirms the twinned pore complex seen in the 2D crystals and reveals a dimerization interface that is likely mediated by lipid rather than direct protein–protein contacts (Fig. 4.5B)(Remaut *et al.*, 2008).

As part of the same study, Remaut *et al.* (2008) successfully isolated the UPEC Type 1 pilus tip fibrillum bound to its OM usher, FimD. By deletion of the Type 1 pilus rod subunit, FimA, the pilus assembly process stalls in a three-subunit intermediate comprising the FimD dimer bound to the tip adhesin, FimH, and a single copy each of the linker subunits FimG and FimF. FimH, FimG and FimF are in donor-strand exchange with one another, with the last incorporated subunit, FimF, still bound to its periplasmic chaperone. The cryo-EM reconstruction of this three-subunit assembly intermediate shows that, remarkably, just a single usher pore in the dimeric complex is used for pilus translocation (Fig. 4.5) (Remaut *et al.*, 2008). Whilst the tip adhesin is already found on the outer surface side of the complex, the last incorporated subunit, FimF, is located on its periplasmic face, still in donor-strand complementation with the FimC chaperone, and bound by the N-terminal domain of one of the two ushers. The FimF subunit is in donor-strand exchange with the preceding subunit, FimG, which resides in the translocating usher pore and is itself in donor-strand exchange with the surface-exposed FimH adhesin. In the isolated complex, the second usher pore lies idle, in its non-activated, closed form. The requirement for this second usher copy to make a functional assembly complex is believed to stem from the need for two chaperone–subunit recruitment sites to allow polymerization at the periplasmic face of the assembly platform. During pilus assembly, the most recently incorporated subunit remains bound to its periplasmic chaperone at the N-terminal recruitment domain of either usher pore (in the isolated three-subunit intermediate, this most likely is the translocating usher copy), until it undergoes donor-strand exchange with a newly recruited subunit. Based on this new structural information, Remaut *et al.* propose a model in which both usher protomers alternate in the recruitment of new chaperone–subunit complexes to the assembly platform through their N-terminal domain, with the penultimate chaperone–subunit complex at the base of the fibre

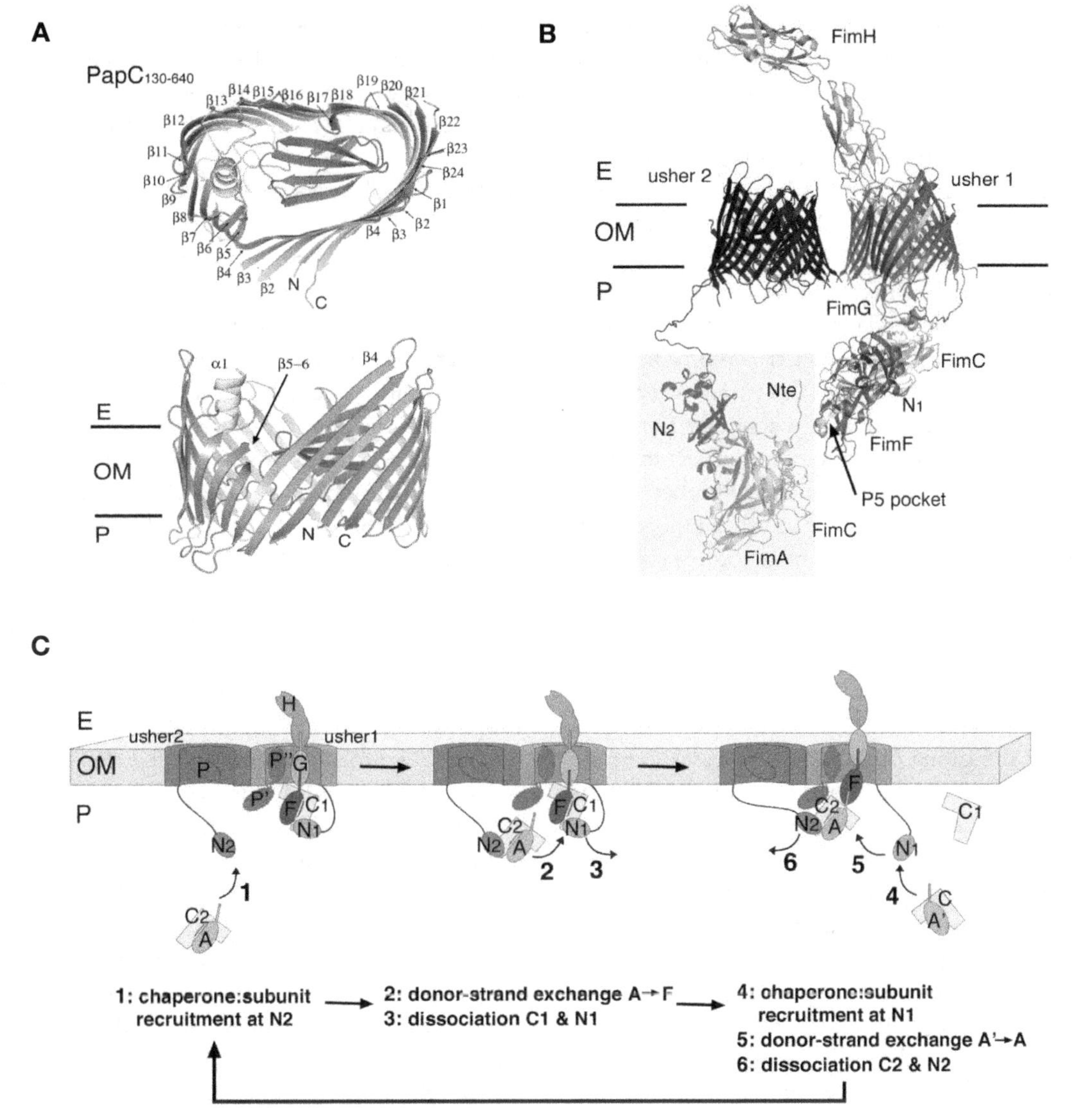

Figure 4.5 Pilus assembly at the usher. **A** Ribbon diagram of the P pilus usher (PapC) translocation pore (PDB entry 2VQI), view from the extracellular side (upper panel) or within the membrane (lower panel) (extracellular space, the outer membrane and periplasm are labelled E, OM, P, respectively). The 24-stranded β-barrel, β5–6 hairpin and central plug domain are coloured blue, orange and magenta, respectively. **B** Atomic model of the usher-bound 3-subunit intermediate of the UPEC Type 1 pilus (FimD usher 1 and 2, the FimH adhesin, FimG and FimF subunits and FimC chaperone are coloured blue, dark blue, green, orange, red and yellow, respectively) (Remaut *et al.*, 2008). Shaded in grey, a new chaperone–subunit complex (FimC–FimA, coloured yellow–cyan) bound to the N-terminal domain of usher pore 2 is modelled in. Labels indicate the P5 initiation site for donor-strand exchange on the penultimate, chaperone-bound subunit. **C** Schematic diagram of fibre assembly at the usher. A colour version of this figure is located in the plate section at the back of the book.

bound at the N-terminal domain of the opposing usher protomer (Fig. 4.5B and C) (Remaut *et al.*, 2008).

Subunit polymerization and fibre formation at the usher is a process that occurs independently of cellular energy (Jacob-Dubuisson *et al.*, 1994). *In vitro* (in the absence of the usher), subunit polymerization is a spontaneous process that is driven by folding energy stored in the chaperone-bound subunits (see above) (Sauer *et al.*, 2002; Zavialov *et al.*, 2005). Chaperone-bound subunits are maintained in a high-energy folding intermediate that relaxes upon donor-strand exchange with the Nte of a new incoming subunit. However, this process is slow and occurs on the scale of hours to days. In contrast, pilus assembly *in vivo* occurs in a matter of minutes (Jacob-Dubuisson *et al.*, 1994) and thus it has long been speculated that the usher acts as a polymerization catalyst. In a recent study of UPEC Type 1 pilus formation, Nishiyama *et al.* directly show, for the first time, that the usher indeed increases the rate of subunit polymerization (Nishiyama *et al.*, 2008). In the presence of the FimD usher, *in vitro* FimA polymerization is two orders of magnitude faster and occurs in a processive manner, resulting in fewer, longer pili compared to a high number of low molecular weight polymers in usher-independent DSE. The observed catalytic function of the usher required pre-incubation with the FimC–FimH chaperone–adhesin complex, corroborating the reported FimH-dependent activation step of FimD (Saulino *et al.*, 1998; So and Thanassi, 2006; Munera *et al.*, 2007). This step likely involves the gating of the usher translocation pore (see above). The absence of FimF or FimG results in a lag phase in usher-mediated catalysis of FimA polymerization, corresponding to the low reactivity between the Nte of FimA and the groove of the FimH pilin domain, and in agreement with the adaptor function of the FimG and FimF subunits.

As described above, *in vitro* DSE is a concerted process that starts with the binding of the incoming subunit's Nte in the P5 pocket of the subunit undergoing strand exchange. Interestingly, the orientation of the last incorporated, chaperone-bound subunit in the cryo-EM reconstruction of the Type 1 pilus tip complex is such that the P5 pocket is exposed and points towards the chaperone–subunit recruitment site on the second usher copy (Fig. 4.5B). When the N-terminal domain of this second usher copy is modelled in, it can be oriented such that the new chaperone–subunit complex recruited at this site is positioned such that its N-terminal extension comes within reach of the P5 pocket of the penultimate chaperone–subunit complex. Based on this model, Remaut *et al.* proposed a mechanism of subunit polymerization at the usher in which the spontaneous donor strand exchange process at the base of the fibre would be catalysed by the effect of increased local concentration and the favourable orientation of the Nte and P5 pocket in the attacking and accepting subunits, respectively (Fig. 4.5B and C) (Remaut *et al.*, 2008).

CU organelles as antibacterial target

Because of their important role in bacterial virulence, CU pili have received considerable attention in vaccine development programs and in the search for novel antibacterials. For example, both the pilus assembly process of Type 1 and P pili and the ligand recognition by their tip adhesins have been targeted in the search for chemical antagonists of UPEC- urinary tract infections (Ohlsson *et al.*, 2002; Pinkner *et al.*, 2006; Wellens *et al.*, 2008). UPEC employ Type 1 pili to gain access to, invade and persist inside the human bladder epithelium. The FimH adhesin targets the bacteria to the bladder by binding high-mannose-containing glycoproteins (mostly uroplakin Ia) on the superficial umbrella cells that line the uroepithelium (Krogfelt *et al.*, 1990; Mulvey *et al.*, 1998). During cystitis, the Type 1 pilus-dependent attachment to the uroepithelium induces internalization of a number of bacteria, which then multiply and develop into biofilm-like intracellular bacterial communities (IBCs) that serve as reservoirs for the secondary infection of neighbouring tissues and further colonization of the bladder (Martinez *et al.*, 2000; Justice *et al.*, 2004; Wright *et al.*, 2007). In pyelonephretic strains, the PapG adhesin in P pili binds Gal-α(1–4)-Gal-containing glycolipid receptors on the kidney epithelium (Roberts *et al.*, 1994; Dodson *et al.*, 2001). For both adhesins, carbohydrate-derived molecules have been shown to act as competitive inhibitors for receptor binding. Alkyl- and aryl-α-D-mannosides bind FimH with nano- to micromolar affinities, block bacterial adhesion on uroepithelial cells and counteract internalization and biofilm formation (area 1 in Fig. 4.6) (Bouckaert *et al.*, 2005; Wellens *et al.*, 2008) whilst *p*-methoxy-phenyl derivatives of galabiose (galabiosides) form potent inhibitors of PapG, with low micromolar IC_{50} values (Ohlsson *et al.*, 2002).

In addition to this ligand-based approach, compounds have been identified that block the pilus assembly process. A family of bicyclic 2-pyridones, termed pilicides, were rationally designed to bind the conserved basic pair in the joint between the two domains in the chaperone. This molecular surface provides the first interaction with unfolded pilus subunits and is essential for subunit folding (area 2 in Fig. 4.6) (Soto *et al.*, 1998). The designed compounds inhibit mannose-sensitive haemagglutination and biofilm formation in UPEC by suppressing Type 1 and P pilus formation and were shown by NMR titration experiments to bind the PapD chaperone with millimolar affinity (Hedenstrom *et al.*, 2005; Pinkner *et al.*, 2006). Surprisingly, the X-ray structure of a pilicide bound to the P pilus chaperone, PapD, revealed that the compound bound a conserved hydrophobic patch on the strands F_1, C_1 and D_1 on the back of the N-terminal domain of the chaperone. This surface is known to interact with the usher (area 3 in Fig. 4.6) and surface plasmon resonance experiments indeed confirmed that pilicides interfered with the binding of chaperone–subunit complexes to the usher (Nishiyama *et al.*, 2005; Pinkner *et al.*, 2006; Eidam *et al.*, 2008).

Discussion and future trends

Over the past decade, our structural molecular understanding of pilus biogenesis by the chaperone/usher pathway has matured into a near complete atomic picture of the assembly process. CU pilus subunits interact non-covalently through a fold-complementation mechanism, a novel protein–protein interaction mechanism that has since been recognized in a number of biological processes such as in SFC ubiquitin ligases and viral proteases (Remaut and Waksman, 2006). It has also become clear that periplasmic chaperones, in addition to their role in preventing premature subunit interactions, catalyse subunit folding and maintain subunits in an 'activated' folding intermediate in order to drive their polymerization in the absence of a hydrolysable energy source. The actual polymerization step involves a concerted mechanism in which a new subunit's N-terminal extension peptide competes off the chaperone donor strand that complements the accepting subunit's Ig-fold. In order to promote strand exchange, the attacking N-terminal extension binds in an accessible site in the hydrophobic groove of the chaperone-

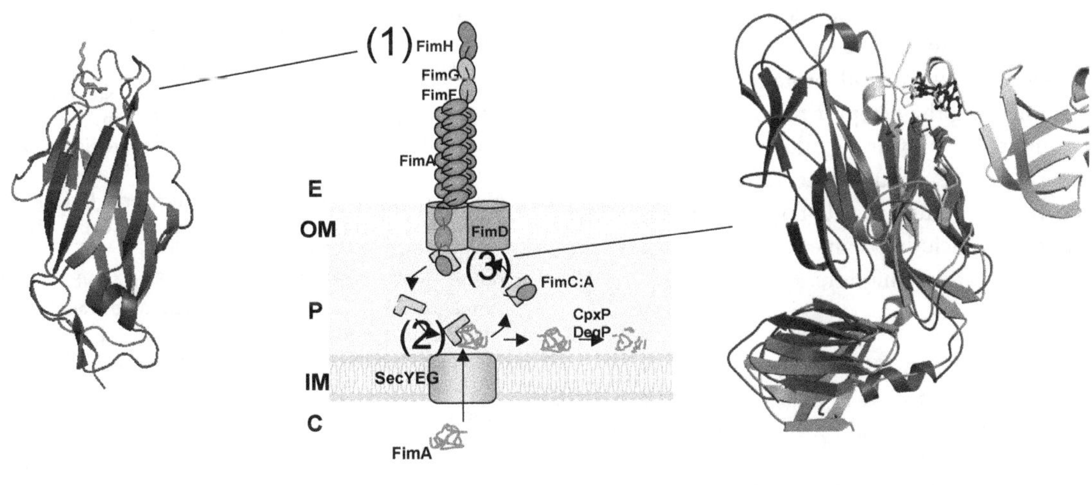

Figure 4.6 Chemical antagonist of CU assembly and receptor recognition. **Left**: ribbon diagram of the FimH adhesin domain bound to the synthetic ligand butyl α-D-mannose (in stick representation). **Middle**: schematic diagram of chaperone–usher pilus assembly pathway. The diagram indicates three areas of interest in CU antagonist development (1) receptor binding, (2) chaperone–subunit interaction and subunit folding and (3) chaperone–subunit recruitment to the usher. **Right**: ribbon diagram of the P pilus chaperone PapD (coloured blue) bound to a pilicide (in stick representation) superimposed onto the Type 1 pilus FimC–FimH chaperone–adhesin complex (coloured magenta–green) bound to the N-terminal domain of the FimD usher (coloured yellow). A patch of conserved hydrophobic residues on the usher/pilicide binding surface are shown in stick representation. A colour version of this figure is located in the plate section at the back of the book.

bound subunit, the P5 site. The strand exchange equilibrium is shifted towards product formation by the 'collapse' of the accepting subunit's hydrophobic core to its lower energy conformation when bound to the N-terminal extension. These processes occur at the bacterial outer membrane, and are catalysed by the usher, which is further responsible for the translocation of the growing fibre across the membrane. In comparison with our understanding of chaperone–subunit and subunit–subunit interactions, the molecular details of the integrated pilus formation at the usher are only starting to emerge. The recent structural information on this translocator reveals it forms a twinned-pore dimeric complex in which just one pore is gated for fibre translocation, whilst both pores are likely involved in alternating chaperone–subunit recruitment. With this structural model of pilus biogenesis, the stage is now set for future research into the mechanistic details of the usher. Important questions remain on how the protein catalyses subunit polymerization, how it drives subunit translocation and how it gets activated upon binding of the chaperone–adhesin complex.

The better understanding of these processes not only addresses fundamental questions in protein translocation but is also likely to open new avenues to the production of pilus assembly antagonists. Current attempts have concentrated on inhibiting the receptor-recognition process of mature pili or preventing pilus assembly by blocking chaperone–subunit recruitment to the usher. Alternative routes to CU-targeted antivirulence drugs may include polymerization inhibitors and usher activation inhibitors.

References

Anderson, K.L., Billington, J., Pettigrew, D., Cota, E., Simpson, P., Roversi, P., Chen, H.A. Urvil, P., du Merle, L., Barlow, P.N., Medof, M.E., Smith, R.A., Nowicki, B., Le, B.ouguénec C., Lea, S.M., and Matthews, S. (2004). An atomic resolution model for assembly, architecture, and function of the Dr adhesins. Mol. Cell *15*, 647–657.

Baga, M., Norgren, M., and Normark, S. (1987). Biogenesis of *E. coli* Pap pili: papH, a minor pilin subunit involved in cell anchoring and length modulation. Cell *49*, 241–251.

Bann, J.G., Pinkner, J.S., Frieden, C., and Hultgren, S.J. (2004). Catalysis of protein folding by chaperones in pathogenic bacteria. Proc. Natl. Acad. Sci. U.S.A. *101*, 17389–17393.

Barnhart, M.M., Pinkner, J.S., Soto, G.E., Sauer, F.G., Langermann, S., Waksman, G., Frieden, C., and Hultgren, S.J. (2000). PapD-like chaperones provide the missing information for folding of pilin proteins. Proc. Natl. Acad. Sci. U.S.A. *97*, 7709–7714.

Baümler, A.J., and Heffron, F. (1995). Identification and sequence analysis of lpfABCDE, a putative fimbrial operon of *Salmonella typhimurium*. J. Bacteriol. *177*, 2087–2097.

Baümler, A.J., and Heffron, F. (1995) Identification and sequence analysis of lpfABCDE, a putative fimbrial operon of *Salmonella typhimurium*. J. bacteriol. *177*, 2087–2097.

Bouckaert, J., Berglund, J., Schembri, M., De Genst, E., Cools, L., Wuhrer, M., Hung, C.S., Pinkner, J., Slattegard, R., Zavialov, A., Choudhury, D., Langermann, S., Hultgren, S.J., Wyns, L., Klemm, P., Oscarson, S., Knight, S.D., and De Greve, H. (2005). Receptor binding studies disclose a novel class of high-affinity inhibitors of the *Escherichia coli* FimH adhesin. Mol. Microbiol. *55*, 441–455.

Bouckaert, J., Mackenzie, J., de Paz, J.L., Chipwaza, B., Choudhury, D., Zavialov, A., Mannerstedt, K., Anderson, J., Piérard, D., Wyns, L., Seeberger, P.H., Oscarson, S., De Greve, H., Knight, S.D. (2006). The affinity of the FimH fimbrial adhesin is receptor-driven and quasi-independent of *Escherichia coli* pathotypes. Mol. Microbiol. *61*, 1556–1568.

Chen, S.L., Hung, C.S., Xu, J., Reigstad, C.S., Magrini, V., Sabo, A., Blasiar, D., Bieri, T., Meyer, R.R., Ozersky, P., Armstrong, J.R., Fulton, R.S., Latreille, J.P., Spieth, J., Hooton, T.M., Mardis, E.R., Hultgren, S.J., and Gordon, J.I.(2006). Identification of genes subject to positive selection in uropathogenic strains of *Escherichia coli*: A comparative genomics approach. Proc. Natl. Acad. Sci. U.S.A. *103*, 5977–5982.

Choudhury, D., Thompson, A., Stojanoff, V., Langermann, S., Pinkner, J.S., Hultgren, S.J., and Knight, S.D. (1999). X-ray structure of the FimC–FimH chaperone–adhesin complex from uropathogenic *Escherichia coli*. Science *285*, 1061–1066.

De Greve, H., Wyns, L., and Bouckaert, J. (2007). Combining sites of bacterial fimbriae. Curr. Opin. Struct. Biol. *17*, 506–512.

Dodson, K.W., Jacob-Dubuisson, F., Striker, R.T., and Hultgren, S.J. (1993). Outer-membrane PapC molecular usher discriminately recognizes periplasmic chaperone-pilus subunit complexes. Proc. Natl. Acad. Sci. U.S.A. *90*, 3670–3674.

Dodson, K.W., Pinkner, J.S., Rose, T., Magnusson, G., Hultgren, S.J., and Waksman, G. (2001). Structural basis of tropism of pyelonephritic *E. coli* for the human kidney. Cell *105*, 733–743.

Du, Y., Rosqvist, R., and Forsberg, A. (2002). Role of fraction 1 antigen of *Yersinia pestis* in inhibition of phagocytosis. Infect. Immun. *70*, 1453–1460.

Duguid, J.P., Smith, I.W., Dempster, G., and Edmunds, P.N. (1955). Non-flagellar filamentous appendages (fimbriae) and haemagglutinating activity in *Bacterium coli*. J. Pathol. Bacteriol. *70*, 335–348.

Duncan, M.J., Mann, E.L., Cohen, M.S., Ofek, I., Sharon, N., and Abraham, S.N. (2005). The distinct binding specificities exhibited by enterobacterial type 1 fimbriae are determined by their fimbrial shafts. J. Biol. Chem. *280*, 37707–37716.

Eidam, O., Dworkowski, F.S., Glockshuber, R., Grütter, M.G., and Capitani, G. (2008). Crystal structure of the ternary FimC–FimF(t)–FimD(N) complex indicates conserved pilus chaperone–subunit complex recognition by the usher FimD. FEBS Lett. *582*, 651–655.

Evans, D.G., Evans, D.J., Clegg, S., and Pauley, J.A. (1979). Purification and characterization of the CFA/I antigen of enterotoxigenic *Escherichia coli*. Infect. Immun. *25*, 738–748.

Fitzhenry, R., Dahan, S., Torres, A., Chong, Y., Heuschkel, R., Murch, S., Thomson, M., Kaper, J., Frankel, G., and Phillips, A. (2006). Long polar fimbriae and tissue tropism in *Escherichia coli* O157:H7. Microbes Infect. *8*, 1741–1749.

Garcia, M.I., Gounon, P., Courcoux, P., Labigne, A., and Le Bouguenec, C. (1996). The afimbrial adhesive sheath encoded by the afa-3 gene cluster of pathogenic *Escherichia coli* is composed of two adhesins. Mol. Microbiol. *19*, 683–693.

Garcia, M.I., Jouve, M., Nataro, J.P., Gounon, P., and Le Bouguénec, C. (2000). Characterization of the AfaD-like family of invasins encoded by pathogenic *Escherichia coli* associated with intestinal and extra-intestinal infections. FEBS Lett. *479*, 111–117.

Hahn, E., Wild, P., Hermanns, U., Sebbel, P., Glockshuber, R., Häner, M., Taschner, N., Burkhard, P., Aebi, U., and Müller, S.A. (2002). Exploring the 3D molecular architecture of *Escherichia coli* Type 1 Pili. J. Mol. Biol. *323*, 845–857.

Hahn, E., Wild, P., Schraner, E.M., Bertschinger, H.U., Häner, M., Müller, S.A., and Aebi, U. (2000). Structural analysis of F18 fimbriae expressed by porcine toxigenic *Escherichia coli*. J. Struct. Biol. *132*, 241–250.

Hedenstrom, M., Emtenas, H., Pemberton, N., Aberg, V., Hultgren, S.J., Pinkner, J.S., Tegman, V., Almqvist, F., Sethson, I., and Kihlberg, J. (2005). NMR studies of interactions between periplasmic chaperones from uropathogenic *E. coli* and pilicides that interfere with chaperone function and pilus assembly. Org. Biomol. Chem. *3*, 4193–4200.

Holmgren, A., and Brändén, C.I. (1989). Crystal structure of chaperone protein PapD reveals an immunoglobulin fold. Nature *342*, 248–251.

Hung, D.L., Knight, S.D., Woods, R.M., Pinkner, J.S., and Hultgren, S.J. (1996). Molecular basis of two subfamilies of immunoglobulin-like chaperones. EMBO J. *15*, 3792–3805.

Isaac, D.D., Pinkner, J.S., Hultgren, S.J., and Silhavy, T.J. (2005). The extracytoplasmic adaptor protein CpxP is degraded with substrate by DegP. Proc. Natl. Acad. Sci. U.S.A. *102*, 17775–17779.

Jacob-Dubuisson, F., Striker, R., and Hultgren, S.J. (1994). Chaperone-assisted self-assembly of pili independent of cellular energy. J. Biol. Chem. *269*, 12447–12455.

Jones, C.H., Danese, P.N., Pinkner, J.S., Silhavy, T.J., and Hultgren, S.J. (1997). The chaperone-assisted membrane release and folding pathway is sensed by two signal transduction systems. EMBO J. *16*, 6394–6406.

Justice, S.S., Hung, C., Theriot, J.A., Fletcher, D.A., Anderson, G.G., Footer, M.J., and Hultgren, S.J. (2004). Differentiation and developmental pathways of uropathogenic *Escherichia coli* in urinary tract pathogenesis. Proc. Natl. Acad. Sci. U.S.A. *101*, 1333–1338.

Kaper, J., Nataro, J., and Mobley, H. (2004). Pathogenic *Escherichia coli*. Nat. Rev. Microbiol. *2*, 123–140.

Keller, R., Ordonez, J.G., de Oliveira, R.R., Trabulsi, L.R., Baldwin, T.J., and Knutton, S. (2002). Afa, a diffuse adherence fibrillar adhesin associated with enteropathogenic *Escherichia coli*. Infect. Immun. *70*, 2681–2689.

Krogfelt, K.A., Bergmans, H., and Klemm, P. (1990). Direct evidence that the FimH protein is the mannose-specific adhesin of *Escherichia coli* type 1 fimbriae. Infect. Immun. *58*, 1995–1998.

Kuehn, M.J., Heuser, J., Normark, S., and Hultgren, S.J. (1992). P pili in uropathogenic *E. coli* are composite fibres with distinct fibrillar adhesive tips. Nature *356*, 252–255.

Kuehn, M.J., Ogg, D.J., Kihlberg, J., Slonim, L.N., Flemmer, K., Bergfors, T., and Hultgren, S.J. (1993). Structural basis of pilus subunit recognition by the PapD chaperone. Science *262*, 1234–1241.

Lee, Y.M., Dodson, K.W., and Hultgren, S.J. (2007). Adaptor function of PapF depends on donor strand exchange in P-pilus biogenesis of *Escherichia coli*. J. Bacteriol. *189*, 5276–5283.

Levine, M.M., Ristaino, P., Marley, G., Smyth, C., Knutton, S., Boedeker, E., Black, R., Young, C., Clements, M.L., and Cheney, C. (1984). Coli surface antigens 1 and 3 of colonization factor antigen II-positive enterotoxigenic *Escherichia coli*: morphology, purification, and immune responses in humans. Infect. Immun. *44*, 409–420.

Li, H., Qian, L., Chen, Z., Thibault, D., Liu, G., Liu, T., and Thanassi, D.G. (2004). The outer membrane usher forms a twin-pore secretion complex. J. Mol. Biol. *344*, 1397–1407.

Martinez, J.J., Mulvey, M.A., Schilling, J.D., Pinkner, J.S., and Hultgren, S.J. (2000). Type 1 pilus-mediated bacterial invasion of bladder epithelial cells. EMBO J. *19*, 2803–2812.

Mu, X.Q., Egelman, E.H., and Bullitt, E. (2002). Structure and function of Hib pili from *Haemophilus influenzae* Type b. J. Bacteriol. *184*, 4868–4874.

Mulvey, M.A., Lopez-Boado, Y.S., Wilson, C.L., Roth, R., Parks, W.C., Heuser, J. and Hultgren, S.J. (1998). Induction and evasion of host defenses by Type 1-piliated uropathogenic *Escherichia coli*. Science *282*, 1494–1497.

Munera, D., Hultgren, S.J., and Fernandez, L. (2007). Recognition of the N-terminal lectin domain of FimH adhesin by the usher FimD is required for type 1 pilus biogenesis. Mol. Microbiol. *64*, 333–346.

Ng, T.W., Akman, L., Osisami, M., and Thanassi, D.G. (2004). The usher N terminus is the initial targeting site for chaperone–subunit complexes and participates in subsequent pilus biogenesis events. J. Bacteriol. *186*, 5321–5331.

Nishiyama, M., Horst, R., Eidam, O., Herrmann, T., Ignatov, O., Vetsch, M., Bettendorff, P., Jelesarov, I., Grutter, M.G., Wüthrich, K., Glockshuber, R., and Capitani, G. (2005). Structural basis of chaperone–

subunit complex recognition by the type 1 pilus assembly platform FimD. EMBO J. *24*, 2075–2086.

Nishiyama, M., Ishikawa, T., Rechsteiner, H., and Glockshuber, R. (2008). Reconstitution of pilus assembly reveals a bacterial outer membrane catalyst. Science *320*, 376–379.

Nuccio, S., and Baumler, A. (2007). Evolution of the chaperone/usher assembly pathway: fimbrial classification goes greek. Microbiol. Mol. Biol. Rev. *71*, 551–575.

Ohlsson, J., Jass, J., Uhlin, B., Kihlberg, J., and Nilsson, U. (2002). Discovery of potent inhibitors of PapG adhesins from uropathogenic *Escherichia coli* through synthesis and evaluation of galabiose derivatives. ChemBioChem. *3*, 772–779.

Perna, N.T., Plunkett 3rd, G., Burland, V., and Mau, B. (2001). Genome sequence of enterohaemorrhagic *Escherichia coli* O157: H7. Nature *409*, 529–533.

Pinkner, J.S., Remaut, H., Buelens, F., Miller, E., Aberg, V., Pemberton, N., Hedenström, M., Larsson, A., Seed, P., Waksman, G., Hultgren, S.J, and Almqvist, F.(2006). Rationally designed small compounds inhibit pilus biogenesis in uropathogenic bacteria. Proc. Natl. Acad. Sci. U.S.A. *103*, 17897–17902.

Remaut, H., Rose, R.J., Hannan, T.J., Hultgren, S.J., Radford, S.E., Ashcroft, A.E., and Waksman, G. (2006). Donor-strand exchange in chaperone-assisted pilus assembly proceeds through a concerted β strand displacement mechanism. Mol. Cell *22*, 831–842.

Remaut, H., Tang, C., Henderson, N.S., Pinkner, J., Wang, T., Hultgren, S.J., Thanassi, D.G., Waksman, G., and Li, H. (2008). Fibre formation across the bacterial outer membrane by the chaperone/usher pathway. Cell *133*, 640–652.

Remaut, H., and Waksman, G. (2006). Protein–protein interaction through beta-strand addition. Trends Biochem. Sci. *31*, 436–444.

Roberts, J.A., Marklund, B.I., Ilver, D., Haslam, D., Kaack, M.B., Baskin, G., Louis, M., Möllby, R., Winberg, J., and Normark, S. (1994). The Gal(alpha 1–4) Gal-specific tip adhesin of *Escherichia coli* P-fimbriae is needed for pyelonephritis to occur in the normal urinary tract. Proc. Natl. Acad. Sci. U.S.A. *91*, 11889–11893.

Rose, R.J., Verger, D., Daviter, T., Remaut, H., Waksman, G., Ashcroft, A.E., and Radford, S.E. (2008). Unravelling the molecular basis of subunit specificity in P pilus assembly by mass spectrometry. Proc. Natl. Acad. Sci. U.S.A. *105*, 12873–12878

Salih, O., Remaut, H., Waksman, G., and Orlova, E. (2008). Structural analysis of the Saf pilus by electron microscopy and image processing. J. Mol. Biol. *379*, 174–187.

Sauer, F.G., Fütterer, K., Pinkner, J.S., Dodson, K.W., Hultgren, S.J., and Waksman, G. (1999). Structural basis of chaperone function and pilus biogenesis. Science *285*, 1058–1061.

Sauer, F.G., Pinkner, J.S., Waksman, G., and Hultgren, S.J. (2002). Chaperone priming of pilus subunits facilitates a topological transition that drives fiber formation. Cell *111*, 543–551.

Sauer, F.G., Remaut, H., Hultgren, S.J., and Waksman, G. (2004). Fiber assembly by the chaperone-usher pathway. Biochim. Biophys. Acta *1694*, 259–267.

Saulino, E., Thanassi, D., Pinkner, J., and Hultgren, S. (1998). Ramifications of kinetic partitioning on usher-mediated pilus biogenesis. EMBO J. *17*, 2177–2185.

So, S.S., and Thanassi, D.G. (2006). Analysis of the requirements for pilus biogenesis at the outer membrane usher and the function of the usher C-terminus. Mol. Microbiol. *60*, 364–375.

Soto, G.E., Dodson, K.W., Ogg, D., Liu, C., Heuser, J., Knight, S.D., Kihlberg, J., Jones, C.H., and Hultgren, S.J. (1998). Periplasmic chaperone recognition motif of subunits mediates quaternary interactions in the pilus. EMBO J. *17*, 6155–6167.

Torres, A.G., Giron, J.A., Perna, N.T., Burland, V., Blattner, F.R., Avelino-Flores, F., and Kaper, J.B. (2002). Identification and characterization of lpfABCC'DE, a fimbrial operon of enterohemorrhagic *Escherichia coli* O157:H7. Infect. Immun. *70*, 5416–5427.

Townsend, S.M., Kramer, N.E., Edwards, R., Baker, S., Hamlin, N., Simmonds, M., Stevens, K., Maloy, S., Parkhill, J., Dougan, G., and Bäumler, A.J. (2001). *Salmonella enterica* serovar typhi possesses a unique repertoire of fimbrial gene sequences. Infect. Immun. *69*, 2894–2901.

Ulett, G.C., Mabbett, A.N., Fung, K.C., Webb, R.I., and Schembri, M.A. (2007). The role of F9 fimbriae of uropathogenic *Escherichia coli* in biofilm formation. Microbiology *153*, 2321–2331.

Verger, D., Miller, E., Remaut, H., Waksman, G., and Hultgren, S. (2006). Molecular mechanism of P pilus termination in uropathogenic *Escherichia coli*. EMBO Rep. *7*, 1228–1232.

Vetsch, M., Erilov, D., Molière, N., Nishiyama, M., Ignatov, O., and Glockshuber, R. (2006). Mechanism of fibre assembly through the chaperone-usher pathway. EMBO Rep. *7*, 734–738.

Vetsch, M., Puorger, C., Spirig, T., Grauschopf, U., Weber-Ban, E.U., and Glockshuber, R. (2004). Pilus chaperones represent a new type of protein-folding catalyst. Nature *431*, 329–333.

Welch, R.A., Burland, V., Plunkett III, G., Redford, P., Roesch, P., Rasko, D., Buckles, E.L., Liou, S.-R., Boutin, A., Hackett, J., Stroud, D., Mayhew, G.F., Rose, D.J., Zhou, S., Schwartz, D.C., Perna, N.T., Mobley, H.L., Donnenberg, M.S., and Blattner, F.R. (2002). Extensive mosaic structure revealed by the complete genome sequence of uropathogenic *Escherichia coli*. Proc. Natl. Acad. Sci. U.S.A. *99*, 17020–17024.

Wellens, A., Garofalo, C., Nguyen, H., Van Gerven, N., Slättegård, R., Hernalsteens, J., Wyns, L., Oscarson, S., De Greve, H., Hultgren, S., and Bouckaert, J. (2008). Intervening with urinary tract infections using anti-adhesives based on the crystal structure of the FimH–oligomannose-3 Complex. PLoS ONE 3, e2040.

Wright, K.J., and Hultgren, S.J. (2006). Sticky fibers and uropathogenesis: bacterial adhesins in the urinary tract. Future Microbiol. *1*, 75–87.

Wright, K.J., Seed, P.C., and Hultgren, S.J. (2007). Development of intracellular bacterial communities of uropathogenic *Escherichia coli* depends on type 1 pili. Cell Microbiol. *9*, 2230–2241.

Zavialov, A., Zav'yalova, G., Korpela, T., and Zav'yalov, V. (2007). FGL chaperone-assembled fimbrial polyadhesins: anti-immune armament of Gram-negative bacterial pathogens. FEMS Microbiol. Rev. *31*, 478–514.

Zavialov, A.V., Berglund, J., Pudney, A.F., Fooks, L.J., Ibrahim, T.M., MacIntyre, S., and Knight, S.D. (2003). Structure and biogenesis of the capsular F1 antigen from *Yersinia pestis* preserved folding energy drives fiber formation. Cell *113*, 587–596.

Zavialov, A.V., Kersley, J., Korpela, T., Zav'yalov, V.P., MacIntyre, S., and Knight, S.D. (2002). Donor strand complementation mechanism in the biogenesis of non-pilus systems. Mol. Microbiol. *45*, 983–995.

Zavialov, A.V., Tischenko, V.M., Fooks, L.J., Brandsdal, B.O., Aqvist, J., Zav'yalov, V.P., Macintyre, S., and Knight, S.D. (2005). Resolving the energy paradox of chaperone/usher-mediated fibre assembly. Biochem. J. *389*, 685–694.

Gram-positive Bacterial Pili and the Host–Pathogen Interface

5

Anjali Mandlik, Andrew H. Gaspar, Anu Swaminathan, Arunima Mishra, Asis Das and Hung Ton-That

Abstract
The cell walls of many Gram-positive bacteria harbour covalently linked protein polymers, known as pili or fimbriae, which enable these bacteria to adhere to specific host tissues and initiate a pathogenic program. A typical pilus contains a major pilin forming the shaft and one or more minor pilin subunits. The heteromeric pilus is assembled by specific transpeptidase enzymes called sortases. While the joining of individual pilins into a polymer is catalysed by a pilus-specific sortase, anchoring of the pilus to the cell wall peptidoglycan is carried out by the housekeeping sortase. This chapter reviews our current understanding of the mechanism of pilus assembly and the roles of pili in bacterial pathogenesis.

Introduction

Gram-positive bacteria utilize their thick cell wall as a scaffold to harbour a variety of surface proteins that are crucial for bacterial virulence and fitness. A striking feature of the cell envelope of Gram-positive bacteria is that many of their surface proteins are anchored to the cell wall covalently. The archetype of these cell wall-anchored proteins is the protein A of *Staphylococcus aureus,* which can be released from the bacterial surface only upon treatment with a cell wall hydrolase (Sjoquist *et al.*, 1972). Protein A contains a specific C-terminal domain, named the cell wall sorting signal (CWSS), which allows its anchoring to the cell wall (Schneewind *et al.*, 1992). This CWSS consists of three modules: the LPXTG motif, a hydrophobic domain and a positively charged tail (Schneewind *et al.*, 1992). Homologues of the CWSS of protein A are found in most Gram-positive bacteria, and the CWSS is necessary and sufficient to anchor a protein on the surface of a Gram-positive microbe (Schneewind *et al.*, 1993).

How does the CWSS mediate cell wall anchoring of a surface protein? It turns out that the LPXTG motif of a surface protein is cleaved between threonine and glycine (Navarre and Schneewind, 1999), and the cleaved polypeptide is linked to the amino group of the bacterial cell wall crossbridge (Schneewind *et al.*, 1995). The enzyme that catalyses both of these reactions has been identified; it is a transpeptidase, named sortase SrtA (surface protein sorting A) (Mazmanian *et al.*, 1999). We now know that Gram-positive bacteria contain at least one sortase homologue (Comfort and Clubb, 2004; Dramsi *et al.*, 2005). Sortase-less bacterial mutants fail to anchor surface proteins to the cell wall peptidoglycan (Mazmanian *et al.*, 2001), and they display an attenuated phenotype in animal models of infections (Paterson and Mitchell, 2004).

Sortase also catalyses the covalent cross-linking of proteins required for pilus assembly. That pilus assembly may involve sortase was first suggested by the finding that the major subunit of a pilus polymer found in *Actinomyces* spp. contains the CWSS (Yeung and Ragsdale, 1997; Yeung *et al.*, 1998). Subsequently, molecular genetic and biochemical investigation of several pili in *Corynebacterium diphtheriae* has shown that pilus polymerization is in fact catalysed by distinct pilus-specific sortases (Ton-That and Schneewind, 2004). Since then, pilus assembly has been studied in many other Gram-positive bacteria, which revealed that the sortase-catalysed pilus assembly is a general feature of these organisms and that the pilus is an important determinant of bacterial virulence. Sortase is therefore an important target of anti-infective therapy. This chapter will focus on two fundamental questions: how pili are assembled and what function they play in bacterial pathogenesis.

Cell wall anchoring of surface proteins in Gram-positive bacteria

Staphylococcus aureus protein A and sortase SrtA

Our basic understanding of cell wall anchoring of proteins is derived from molecular dissection of the protein A of *S. aureus* as an experimental model (Mazmanian *et al.*, 2001). Like typical secreted proteins, protein A contains in its N-terminus a signal peptide essential for the export of the protein by the general secretion machinery. After its synthesis in the cytoplasm, the signal peptide of protein A allows its translocation through the SecYEG apparatus and is ultimately cleaved off the precursor polypeptide. As mentioned above, the C-terminus of protein A contains the CWSS. The hydrophobic domain of the CWSS and the charged tail that follows allow the protein to be embedded in the membrane and retained within the exoplasm for its subsequent processing by the sortase SrtA. SrtA cleaves the LPXTG motif between threonine and glycine, forming an acyl enzyme intermediate in which the Thr residue is linked to the catalytic Cys residue. A nucleophilic attack by the amino group of the pentaglycine crossbridge within the lipid II precursor *then* resolves this intermediate. This transpeptidation reaction thus joins protein A to the peptidoglycan (Fig. 5.1).

Sortase is an integral membrane protein that captures its membrane bound substrates and catalyses the transpeptidation reaction with the lipid II precursor also embedded in the membrane (Ton-That *et al.*, 2004a). Although lateral diffusion could bring the enzyme and its substrates together, sortase might be localized in close proximity to the Sec machinery, essentially forming an assembly line or 'portal' for coordinate delivery of surface proteins and anchoring of the proteins to the cell wall (Rosch and Caparon, 2005). Indeed, some support for this conjecture has recently come from the finding that sortase SrtA is colocalized with the SecA translocase in the membrane of *Streptococcus mutans* (Hu *et al.*, 2008). As we discuss later, this coupling of secretion and sortase reactions is likely to be a key feature of pilus biogenesis in Gram-positive bacteria.

Architecture of pili and assembly mechanism

Actinomyces fimbriae represent the first nonflagellar filament from a Gram-positive bacterial surface to be characterized by biochemistry and

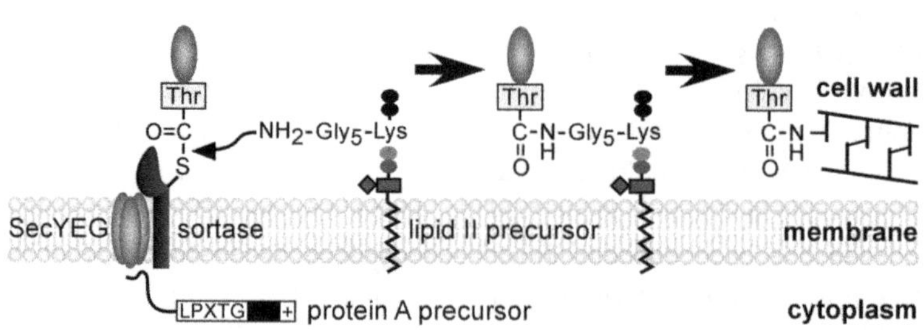

Figure 5.1 Sortase-catalysed anchoring of *Staphylococcus aureus* protein A to the bacterial cell wall. After being synthesized in the cytoplasm as a precursor harbouring an N-terminal signal sequence and a C-terminal sorting signal, the protein is translocated across the membrane via the Sec machinery. The membrane-anchored sortase enzyme (black) cleaves the sorting signal between threonine and glycine to form an acyl-enzyme intermediate. This thioester linkage intermediate is resolved by the nucleophilic attack of the amino group of the pentaglycine crossbridge within the lipid II precursor. The lipid II linked protein is then incorporated into the cell wall by a series of transglycosylation reactions.

molecular genetics (Girard and Jacius, 1974; Yeung and Ragsdale, 1997; Yeung *et al.*, 1998). *Actinomyces* expresses two distinct filamentous structures, called the Type 1 and Type 2 fimbriae (Yeung, 1999), which are respectively made of FimP and FimA as the major fimbrial subunits (Donkersloot *et al.*, 1985; Yeung *et al.*, 1987) (Fig. 5.2A). The sequences of the genes encoding these proteins revealed that each of them contain the CWSS (Yeung and Ragsdale, 1997; Yeung *et al.*, 1998). Furthermore, each fimbrial subunit gene was found to be associated with a putative sortase

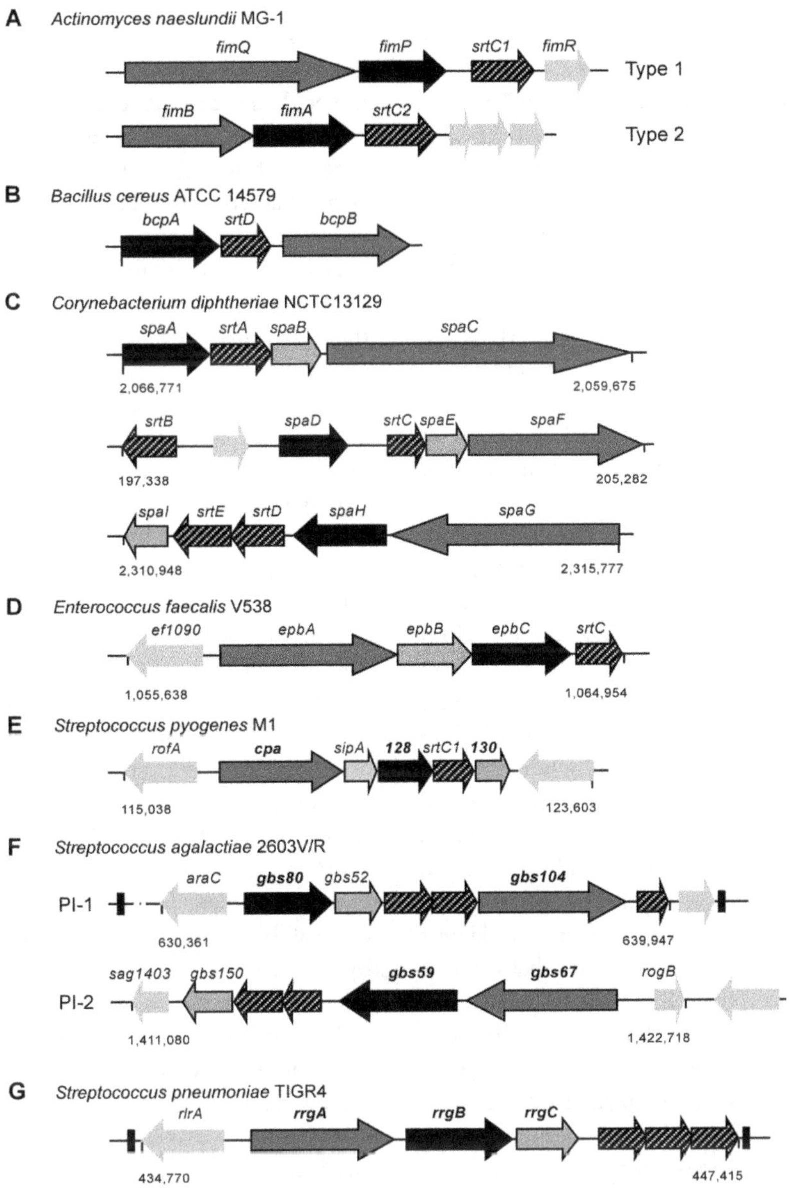

Figure 5.2 Pilus gene clusters of Gram-positive bacteria. Graphic presentation of pilus gene clusters identified in the chromosome of **A** *Actinomyces naeslundii* MG-1, **B** *Bacillus cereus* ATCC 14579, **C** *Corynebacterium diphtheriae* NCTC13129, **D** *Enterococcus fecalis* V538, **E** *Streptococcus pyogenes* M1, **F** *Streptococcus agalactiae* 2603V/R and **G** *Streptococcus pneumoniae* TIGR4. Each cluster contains pilus specific sortase gene(s) (hatched), genes encoding a major subunit (black) and minor pilins (dark grey and light grey arrows with borders). Uncharacterized and accessory genes are coloured in grey. The genes that encode pilins being used in vaccine studies are shown in bold. Numbers below clusters indicate genomic location of the clusters.

gene forming a fimbrial specific gene cluster (Yeung and Ragsdale, 1997; Yeung *et al.*, 1998).

This clustering of pilus/fimbrial genes and putative sortases turned out to be a conserved feature of genome organization as the sequences of various bacterial genomes became available. A systematic bioinformatic analysis was carried out to identify sortases and sortase-associated surface proteins that contain the LPXTG motif in specific Gram-positive organisms, such as *C. diphtheriae,* previously shown to produce pili (Ton-That and Schneewind, 2003). Three separate gene clusters were uncovered from the genome of *C. diphtheriae,* encoding a total of five sortase enzymes (SrtA-E) and nine putative surface proteins with the LPXTG motif, that were named SpaA-I (Spa for sortase-mediated pilus assembly) (Ton-That and Schneewind, 2003) (Fig. 5.2C). Many additional LPXTG containing proteins are found dispersed throughout the genome, and there is a sixth sortase gene (*srtF*) located elsewhere in the corynebacterial chromosome (Ton-That and Schneewind, 2003).

Multiple sortase genes are found within almost every Gram-positive genome (Comfort and Clubb, 2004; Dramsi *et al.,* 2005). These sortases can be grouped into four classes (A, B, C and D) according to their conserved features. Close homologues of *S. aureus* SrtA form class A sortase or the housekeeping sortase, which include SrtF of corynebacteria, each of which anchors the various surface proteins to the cell wall in respective organisms. The same is true for class B sortases, with SrtB of *S. aureus* as the founding member, that are involved in iron acquisition (Mazmanian *et al.,* 2003). Class C sortases form the largest group and the only sortase enzymes that catalyse the formation of pili. Named pilus-specific sortases, these include the SrtA-E of corynebacteria. Whereas the core domain required for catalysis is conserved in all sortase enzymes (Zong *et al.,* 2004a; Zong *et al.,* 2004b), the class C sortase enzymes possess a long C-terminal domain that is presumably required for membrane anchoring of these proteins (Comfort and Clubb, 2004; Dramsi *et al.,* 2005). Although the class D sortases are found in many Gram-positive bacteria with high and low GC% content, their function is not clear.

The prototype SpaA-type pili of *Corynebacterium diphtheriae*

While Gram-negative bacteria typically express long pili that are easily visualized by electron microscopy, the short pilus fibres produced by many Gram-positive bacteria are difficult to analyse by conventional microscopy. Thus, although pili were reported for some of these microbes as early as 1970 (Yanagawa and Otsuki, 1970; Girard and Jacius, 1974), little was known about the molecular mechanism of pilus assembly in Gram-positive bacteria until recently. It was possible to detect distinct pilus structures on the surface of corynebacteria through immuno-gold labelling electron microscopy (IEM) using antibodies raised against recombinant Spa proteins (Ton-That and Schneewind, 2003). With this method, the first pilus gene cluster *spaA-srtA-spaB-spaC* of *C. diphtheriae* was shown to encode the SpaA-type pili made of a major shaft protein SpaA, a tip protein SpaC and another minor pilin SpaB (Ton-That and Schneewind, 2003). This SpaA pilin is closely related to FimA and FimP of *Actinomyces,* the respective shaft forming proteins of the two different fimbriae in this organism (Mishra *et al.,* 2007).

In wild-type corynebacterial cells, short pili are detected by immuno-gold labelling EM in which SpaB is colocalized with SpaA. When SpaA is overproduced, however, corynebacteria assemble long pili, and SpaB pilins are found interspersed along the long pilus shafts (Ton-That and Schneewind, 2003) (Fig. 5.3). Unlike the Gram-negative pili, corynebacterial pili are covalently linked together and anchored to the cell wall by a covalent linkage. This is established by the fact that SpaABC polymers are only released from the cell by treatment with a cell wall hydrolase, not by boiling cell samples with SDS and a reducing agent, and the resistance of the polymers to formic acid, which is known to dissociate noncovalent pilus polymers (Ton-That and Schneewind, 2003).

Recently, it has been shown that SpaB and SpaC are also anchored to the bacterial cell wall apart from pilus structures (Mandlik *et al.,* 2007). This surface anchoring of the minor pilins is mediated by the housekeeping sortase SrtF (Swaminathan *et al.,* 2007). Similar to the *spaA*

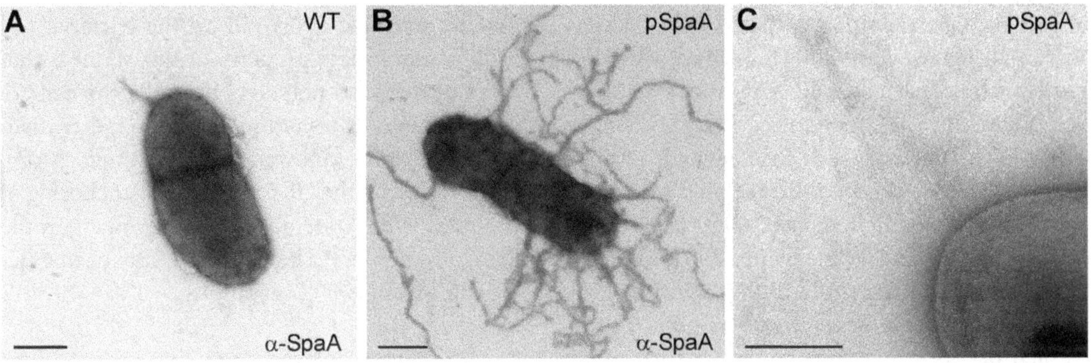

Figure 5.3 Pilus structure of *Corynebacterium diphtheriae*. Wild type corynebacteria (**A**) or its isogenic derivative expressing the major pilin SpaA (**B** and **C**) were used in this experiment. Corynebacteria were stained with a specific antiserum against SpaA (α-SpaA) followed by IgG-conjugated 12-nm gold particles (**A** and **B**). The pili were viewed by an electron microscope after staining with 1% uranyl acetate (**A–C**). Bars indicate a distance of 0.5 μm.

gene cluster, the other two gene clusters encode two distinct pilus structures, named SpaD- and SpaH-type pili (Gaspar and Ton-That, 2006; Swierczynski and Ton-That, 2006), which comprise a major subunit forming the shaft, a tip pilin and another pilin that is found along the structure as well as on the bacterial surface.

Molecular genetics and biochemical studies showed that deletion of *srtA*, the only sortase found in the *spaA* gene cluster, abolished the formation of the SpaA-type pili (Ton-That and Schneewind, 2003). Conversely, in a multiple deletion mutant strain that expresses SrtA as the only sortase, SpaA polymers are still detected on the bacterial surface (Mandlik *et al.*, 2007). Together these findings establish that SrtA is a pilus-specific sortase that is both necessary and sufficient for the assembly of the SpaA pili. Similarly, specific sortases encoded within the respective gene clusters are required for the formation of the SpaD- and SpaH-type pili; while sortases SrtB/SrtC assemble the SpaD-type pilus (Gaspar and Ton-That, 2006), SrtD/SrtE are essential for the SpaH-type pilus assembly (Swierczynski and Ton-That, 2006). The deletion of *srtF* does not diminish the polymerization of any of these pili, nor are any of the pili polymerized in a strain that expresses SrtF as the only sortase. Clearly, the housekeeping sortase SrtF is not involved in pilus polymerization (Swaminathan *et al.*, 2007).

Mechanism of sortase-mediated pilus polymerization

The critical question now arises as to how a pilus-specific sortase catalyses the cross-linking of the three pilins SpaA/B/C into a covalent polymer in an order that assembles a tip protein followed by the major pilin shaft. Since each pilin contains the C-terminal CWSS and their assembly requires sortase SrtA, it has been proposed that pilus assembly occurs by a mechanism similar to the transpeptidation reaction catalysed by the housekeeping sortase that links a surface protein to the cell wall peptidoglycan (Ton-That and Schneewind, 2004) (see Fig. 5.1). Specifically, SrtA cleaves the LPXTG motif between threonine and glycine and forms an acyl-enzyme intermediate with the pilin substrates. For transpeptidation to occur, a nucleophilic attack by the amino group of an element within an adjoining pilin is required to resolve the acyl-enzyme intermediate. Cycles of transpeptidation reactions continue for pilus extension as long as the pilins are abundantly available (Fig. 5.4B).

Several lines of evidence support this model described in Fig. 5.4B. First, in addition to the LPXTG motif, corynebacterial SpaA and its homologues contain the pilin motif with a conserved lysine residue. Consistent with the notion that this lysine serves as a nucleophile for the proposed transpeptidation reaction, a SpaA mutation that changed the conserved lysine 190 to alanine abolishes polymerization of SpaA pilins (Ton-That and Schneewind, 2003). Second, with the same analysis, when the LPXTG motif of SpaA was mutated, SpaA polymers were no longer observed (Ton-That and Schneewind, 2003). Third, SpaA homologues with the pilin motif and the CWSS are found in many Gram-positive bacteria (Ton-That and Schneewind, 2003). Finally, recent work by Kang *et al.* (2007) demonstrated a covalent linkage of *Streptococcus pyogenes* T1 pili formed between threonine of the LPXTG motif and a lysine residue of a neighbouring T1 pilin.

As previously mentioned, the SpaA-type pili contain two minor pilins SpaB and SpaC. The SpaC pilin has been observed primarily at the tip region of the pilus (Ton-That *et al.*, 2004b). How is the tip pilin SpaC linked to SpaA, while SpaC and SpaB are dispensable for the assembly of a SpaA polymer? SpaA and SpaC have the same LPLTG motif, although SpaC lacks the pilin motif (Ton-That and Schneewind, 2003). A plausible scenario for the SpaA-SpaC linkage is that it may be formed via the threonine of the SpaC LPLTG motif and the lysine of the SpaA pilin motif. Consistent with this model, mutation of the SpaC LPLTG motif abrogates surface display of SpaC pilins, but not SpaA polymers (Mandlik *et al.*, 2007). Secondly, in the SpaA K190A mutant, SpaC fails to form pilus polymers (Ton-That and Schneewind, 2003). Nonetheless, this actual linkage should be experimentally addressed in future work.

Another basic question is what determines SpaC as the first pilin that is normally assembled at the pilus tip, and what prevents SpaA from being assembled as the first pilin. It is thus important to examine whether or not SpaA pili always contain SpaC at the tip region. An equally puzzling problem is how the minor pilin SpaB is incorporated into the pilus structure. Although a glutamate residue in the SpaA E box has been identified to be important for the incorporation of SpaB in the pilus (Ton-That *et al.*, 2004b) (Fig. 5.4), this glutamate might contribute to stabilizing the SpaA-SpaB association or the proper SpaA structure (Mandlik *et al.*, 2008b).

Cell wall anchoring of pilus polymers

As mentioned above, a corynebacterial mutant strain that expresses SrtA as the only sortase is able to assemble SpaA pili on the bacterial cell wall. A remarkable property of this strain is that most of the pilus polymers made by the cell are secreted into the culture medium rather than anchored to the cell wall (Mandlik *et al.*, 2007), suggesting that the efficient cell wall anchoring of the pilus is dependent on a factor missing in the SrtA-only strain. A clue to the identity of this factor came from two observations.

1. The SpaA polymers are also secreted into the medium when the housekeeping gene *srtF* was inactivated (Swaminathan *et al.*, 2007).
2. A strain that expresses both SrtA and SrtF assembles wild type levels of the SpaA pili on the cell surface (Swaminathan *et al.*, 2007).

Thus, the housekeeping sortase SrtF is required for the efficient anchoring of SpaA polymers to the cell wall. Importantly, these results also established another fundamental aspect of the mechanism of pilus assembly in Gram-positive bacteria: pilus assembly is a biphasic process in which pilus polymerization precedes the anchoring step of the polymers to the cell wall.

The next logical question is what governs the termination of pilus polymerization and which pilin serves as the terminal subunit for SrtF catalysed cell wall anchoring (Fig. 5.4C). Although SpaA pilin can be anchored to the cell wall, this reaction does not seem to be efficient (Mandlik *et al.*, 2007; Swaminathan *et al.*, 2007). A possible candidate for this termination is the minor pilin SpaB, which could facilitate this process by acting as the preferred substrate for SrtF as previously suggested (Comfort and Clubb, 2004; Dramsi *et al.*, 2005). Recent work from this laboratory supports this contention (Mandlik *et al.*, 2008a). Consistently, an equivalent molecule acting similarly to SpaB has been found in *Streptococcus agalactiae*, namely GBS150 (Nobbs *et al.*, 2008).

General mechanism of pilus assembly

Besides *Actinomyces* and corynebacteria, pili have been identified in many other Gram-positive bacteria including *S. agalactiae*, *Streptococcus*

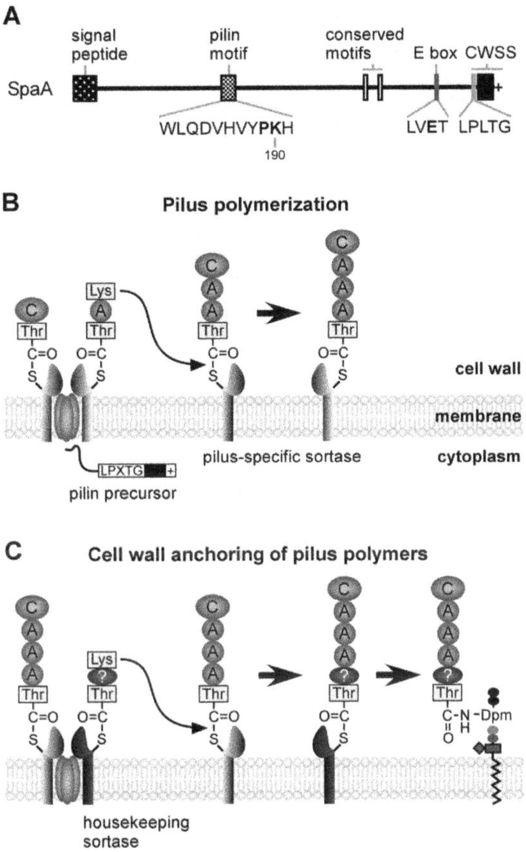

Figure 5.4 Working model of pilus biogenesis. **A** Graphic presentation of SpaA pilin. Its signal peptide, pilin motif, conserved motifs, E box and cell wall sorting signal (CWSS) are indicated. Shown in bold is the conserved lysine 190 in the pilin motif. **B,C** Presented is the biphasic mode of pilus assembly: pilus polymerization (**B**) followed by cell wall anchoring of pilus polymer with SpaB as a terminal subunit (**C**). See text for details.

pyogenes, Streptococcus pneumoniae, Enterococcus faecalis, and *Bacillus cereus* (Lauer *et al.*, 2005; Mora *et al.*, 2005; Barocchi *et al.*, 2006; Dramsi *et al.*, 2006; LeMieux *et al.*, 2006; Nallapareddy *et al.*, 2006; Rosini *et al.*, 2006; Budzik *et al.*, 2007). As with *Actinomyces* and corynebacteria, the pilus and sortase genes in these organisms are clustered together and often flanked by transposon elements, suggesting horizontal gene transfer as the mechanism for their acquisition by various bacteria (Mandlik *et al.*, 2008b). As expected, sortase is the central enzyme in these pathogens for assembling the pili on their surface (Marraffini *et al.*, 2006; Scott and Zahner, 2006; Telford *et al.*, 2006; Mandlik *et al.*, 2008b). Furthermore, the major pilus subunits in all but one of these organisms contain the pilin motif, E box and the CWSS (Mandlik *et al.*, 2008b) (Fig. 5.3A). The notable exception is *S. pyogenes*, whose pilins do not harbour the pilin motif or E box, although their pilin linkage is still mediated by the threonine of the LPXTG motif and a lysine residue (Kang *et al.*, 2007).

Based on these conserved features of the SpaA homologues, it is apparent that sortase-mediated pilus assembly is wide spread in Gram-positive bacteria, including non-pathogenic ones and others that colonize animals (Table 5.1). What about the actual assembly mechanism? Is the biphasic mode of pilus assembly, involving a polymerization step followed by the cell wall anchoring step as described recently in *C. diphtheriae* (Mandlik *et al.*, 2008b), the universal mechanism of pilus assembly? Limited data available so far suggests that it may be so. In *S. agalactiae*, inactivation of the housekeeping sortase gene was reported to diminish pilus formation on the bacterial surface (Dramsi *et al.*, 2006). Furthermore, *B. cereus* pili require a similar enzyme for their anchoring to the cell wall (Budzik *et al.*, 2007); the pili are released into the culture medium when SrtA is knocked out (Budzik *et al.*, 2007). Curiously, as we discuss below, the streptococcal pili are heterotrimeric like the corynebacterial pili (Dramsi *et al.*, 2006; Rosini *et al.*, 2006), whereas the bacillus pili seem to be made of only two components (Budzik *et al.*, 2007).

Heterotrimeric pili of Gram-positive bacteria

Like corynebacterial pili, all streptococcal and enterococcal pili contain three distinct components (Mora *et al.*, 2005; Barocchi *et al.*, 2006; Dramsi *et al.*, 2006; LeMieux *et al.*, 2006; Nallapareddy *et al.*, 2006; Rosini *et al.*, 2006). In *S. agalactiae* (GBS), two pilus islands, PI-1 and PI-2 (see Fig. 5.2), have been identified in various strains of *S. agalactiae* (Telford *et al.*, 2006). Each island encodes a different pilus that consists of a major pilin forming the pilus shaft and two minor pilins associated with the pilus, though the specific location of the two minor (or ancillary) pilins is uncertain, and there is no designated tip protein (Mora *et al.*, 2005; Dramsi *et al.*, 2006). One ancillary pilin is detected along

Table 5.1 Defined and putative pilus gene clusters of Gram-positive bacteria

Species[a]	Pilus-specific sortase[b]	Number of pilus clusters	Reference
Actinomyces naeslundii	1/1	2	–
Actinomyces odontolyticus ATCC 17982	1/1	2	–
Arthrobacter aurescens TC1	1	1	Mongodin *et al.*, 2006
Bacillus cereus ATCC 10987	1/2	2	Rasko *et al.*, 2004
Bacillus thuringiensis str. Al Hakam	1	1	Challacombe *et al.*, 2007
Bifidobacterium dentium ATCC 27678	1	1	–
Bifidobacterium longum NCC2705	1	1	Schell *et al.*, 2002
Brevibacterium linens BL2	1	1	–
Clostridium perfringens str. 13	1	1	Shimizu *et al.*, 2002
Clostridium spiroforme DSM 1552	2	1	–
Corynebacterium diphtheriae NCTC 13129	1/2/2	3	Cerdeno-Tarraga *et al.*, 2003
Corynebacterium efficiens YS-314	2/2	2	Nishio *et al.*, 2003
Corynebacterium glutamicum R	2	1	Yukawa *et al.*, 2007
Corynebacterium jeikeium K411	1	1	Tauch *et al.*, 2005
Corynebacterium urealyticum DSM 7109	2	1	Tauch *et al.*, 2008
Enterococcus faecalis V583	1	1	Paulsen *et al.*, 2003
Enterococcus faecium DO	2	1	–
Eubacterium dolichum DSM 3991	1	1	–
Faecalibacterium prausnitzii M21/2	1	1	–
Lactococcus lactis subsp. *lactis* Il1403	1	1	Bolotin *et al.*, 2001
Leuconostoc mesenteroides subsp. *mesenteroides* ATCC 8293	2	1	–
Ruminococcus obeum ATCC 29174	2	1	–
Ruminococcus torques ATCC 27756	2/1	2	–
Streptococcus agalactiae 2603V/R	3/2	2	Tettelin *et al.*, 2002
Streptococcus pneumoniae TIGR4	3	1	Tettelin *et al.*, 2001
Streptococcus pyogenes M1 GAS	1	1	Ferretti *et al.*, 2001
Streptococcus suis 05ZYH33	3	1	Challacombe *et al.*, 2007

a They all possess the housekeeping sortase.

b Number of pilus-specific sortases in each cluster.

–, references are not available.

the pilus structure with a lesser frequency than the major pilin (Dramsi *et al.*, 2006; Rosini *et al.*, 2006). The other is hardly found on the bacterial surface, although its association with the major pilus polymers is well documented (Dramsi *et al.*, 2006; Rosini *et al.*, 2006). Similar to GBS pili, *S. pyogenes* pili include a backbone pilin and two ancillary proteins (Mora *et al.*, 2005). Immuno-gold labelling against the latter pilins does not detect full-length pili, but gold particles are found on and away from the bacterial surface (Mora *et al.*, 2005). Failure in detecting minor pilins in these species could be explained if antibody recognition is masked within the pilus structure.

In contrast to GAS and GBS pili, the location of *S. pneumoniae* minor pilins RrgA and RrgB on the pilus backbone RrgC is rather unique (Barocchi *et al.*, 2006; Hilleringmann *et al.*, 2008). Immuno- and cryo-electron microscopic studies showed that RrgA and RrgB are interspersed along the pilus backbone, and RrgC is sometimes found colocalized with RrgA (Hilleringmann *et al.*, 2008). In the EbpA/B/C pilus system of *E. faecalis* (Fig. 5.2), immuno-gold labelling with EbpB or EbpC stains long pilus structures, whereas gold particles labelled with EbpA appear predominantly on the bacterial surface (Nallapareddy *et al.*, 2006).

Thus, it appears that although the mechanism of sortase-mediated polymerization of the major pilus subunits is well conserved, the mode of assembly of the minor pilins into the pili is variable. Since some of the minor pilins are recognized by a sortase different than the one that polymerizes the major subunits (Dramsi *et al.*, 2006; Rosini *et al.*, 2006; Swierczynski and Ton-That, 2006), it is conceivable that sortase-mediated incorporation of the minor pilins is independent of polymerization of the major pilin subunits.

Heterodimeric pili of Gram-positive bacteria

Actinomyces naeslundii and *Bacillus cereus* are the two known species that apparently express heretodimeric pili (Budzik *et al.*, 2007; Mishra *et al.*, 2007) (see Fig. 5.2). *Actinomyces* strain MG-1 assembles two types of long fimbriae (Mishra *et al.*, 2007). Type 1 fimbriae consist of the major subunit FimP and a tip fimbria FimQ, whereas Type 2 fimbriae are made of FimA forming the fimbrial shaft and FimB located at the tip (Mishra *et al.*, 2007). Similar to *Actinomyces*, vegetative forms of *B. cereus* produce surface pili that are composed of the pilus shaft BcpA and the tip pilin BcpB (Budzik *et al.*, 2007). Since the minor pilins are dispensable for pilus polymerization (Ton-That *et al.*, 2004b; Dramsi *et al.*, 2006; Rosini *et al.*, 2006), it is plausible that one of the minor pilin genes might have been lost during the course of evolution. On the other hand, there is no proof that the *Actinomyces* and *B. cereus* pili are made of only two components. These pili might contain minor pilins encoded by genes located outside the respective pilus gene clusters.

Regulation of pili

One of the most poorly understood aspects of Gram-positive pili is how pilus expression and assembly are regulated. *C. diphtheriae* expresses all of the three types of pili under laboratory conditions, although their relative level of expression varies and the length of individual pili is short (Ton-That and Schneewind, 2003; Gaspar and Ton-That, 2006; Swierczynski and Ton-That, 2006). By comparison, *Actinomyces* fimbriae are more abundant and considerably longer (Mishra *et al.*, 2007). The expression of certain pilins in various group B streptococcal strains is sporadic. The minor pilin GBS104 of the PI-1 pili was not detected by immunofluorescence in strain 2603V/R (Maione *et al.*, 2005) (Tettelin *et al.*, 2002), yet in some other reported strains, this pilin is abundantly expressed (Maione *et al.*, 2005). This indicates that specific genetic factors are responsible for controlling the expression of pilins. Indeed, a gene encoding a transcriptional regulator is often found within most streptococcal pilus islands (Telford *et al.*, 2006) (see Fig. 5.2), including the PI-1 and PI-2 of GBS, the GAS pilus gene clusters and the *S. pneumoniae rlrA* islet. For the pneumococcal pilus, a regulatory system that involves a positive regulator *rlrA* (*rofA*-like regulator) and a transcriptional repressor *mgrA* has been described (Barocchi *et al.*, 2006). Importantly, the inactivation of *mgrA* results in pilus expression in a pneumococcal strain in which pili are normally undetected (Barocchi *et al.*, 2006). Moreover, RlrA activator is itself controlled by several regulatory networks, including the two component systems CbpR/S and HK/RR03 for adhesion and the transcriptional regulators of divalent cation transporters MerR and PsaR (Rosch *et al.*, 2008).

There is also evidence for environmental control of Gram-positive pili. When grown in tryptic soy broth with glucose supplemented with horse serum, *E. faecalis* expresses high levels of high-molecular-weight Ebp polymers that are typically reduced in the same medium without serum (Nallapareddy *et al.*, 2006). Interestingly, serum increases the amount of the mRNA encoding the pilus-specific sortase *srtC*, while *ebp* mRNA levels remain unchanged (Nallapareddy *et al.*, 2006), suggesting that pilus assembly might be enhanced by upregulating the amount of sortase in the cell. Curiously, in *C. diphtheriae*, a binding site for the transcription regulator DtxR has been identified upstream of the housekeeping sortase gene *srtF* (Swaminathan *et al.*, 2007), but its potential regulatory role has not been tested yet.

Molecular structure of pili

Gram-positive pili represent unique biological protein polymers whose three-dimensional structures are only beginning to be elucidated. GBS52 of *S. agalactiae* is the first Gram-positive pilin whose atomic structure has been determined crystallographically (Krishnan *et al.*, 2007). Although the minor pilin GBS52 was not detected with the pilus structures by immuno-electron microscopy, Western blotting has documented its incorporation into the pilus polymer (Rosini *et al.*, 2006). GBS52 binds specifically to lung epithelial cells and this host–bacterial adherence is significantly reduced with streptococcal mutants lacking *gbs52* (Krishnan *et al.*, 2007). Consistent with the role of GBS52 as a specific bacterial adhesin, GBS52 structure contains two IgG-like fold domains N1 and N2, with N2 acting as the major determinant for specific adherence by GBS52 (Krishnan *et al.*, 2007).

Similarly, the crystal structure of *S. pyogenes* major pilin T1 (antigen 1) comprises two all-β domains, each having the immunoglobulin-like fold (Kang *et al.*, 2007). The T antigens have been known for their antigenicity and resistance to trypsin, and the proteins have been implicated in vaccine development (Mora *et al.*, 2005). Consistent with the model of lysine transpeptidation-mediated pilus assembly proposed above (see Fig. 5.3A), the T1 pilin subunits were shown to be linked by the threonine of one pilin LPXTG motif and the lysine of an adjoining T1 pilin (Kang *et al.*, 2007). Unexpectedly, however, the T1 structure discovered the presence of two intramolecular lysine-asparagine isopeptide bonds, which presumably provide strength and stability to the pili during bacterial colonization (Kang *et al.*, 2007; Yeates and Clubb, 2007). In line with this finding is the recent work in *B. cereus* that shows the covalent linkage between bacillus pilin subunits is formed by a conventional lysine-transpeptidation reaction, i.e. between the LPXTG threonine of one pilin and the pilin motif lysine of an adjoining pilin (Budzik *et al.*, 2008).

A more elaborate view of the pilus structure was provided by the cryo-EM structure of *S. pneumoniae* T4 pili that are composed of the major pilin RrgB and two ancillary proteins RrgA and RrgC (Barocchi *et al.*, 2006). The cryo-EM structure suggests that the T4 pilus contains at least two RrgB 'protofilaments' that form a coiled-coil structure (Hilleringmann *et al.*, 2008). Importantly, RrgA and RrgC are found colocalized and/or interspersed along the protofilament structure (Hilleringmann *et al.*, 2008). A major challenge for future work is to ascertain the rules that determine the ordered assembly of these structures.

Function of pili in bacterial pathogenesis

Pili have long been considered to be key virulent determinants for bacterial pathogenesis (Sauer *et al.*, 2000; Burrows, 2005; Torres *et al.*, 2005). As summarized below, considerable progress has been made recently in dissecting some of the specific functions of Gram-positive pili in adherence, colonization, biofilm formation and the evasion of host immunity.

Bacterial adherence and colonization and host cell receptors for pili

As mentioned above, *C. diphtheriae* produces three types of pili (Ton-That and Schneewind, 2003; Gaspar and Ton-That, 2006; Swierczynski and Ton-That, 2006). Recent work by Mandlik and colleagues, examining the binding properties of a large battery of corynebacterial mutants to a variety of host cells in culture, revealed that the SpaA-type pili mediate the specific adherence of corynebacteria to pharyngeal epithelial cells (Mandlik *et al.*, 2007), the major site of *C. diphtheriae* infections (Holmes, 2000). Most importantly, the two minor pilins SpaB and SpaC are solely responsible for this specific adhesion and they can function as such when present within the pilus or anchored to the cell wall as monomeric forms (Mandlik *et al.*, 2007). The study revealed further that the SpaD- and SpaH-type pili display preferential affinity for laryngeal and lung epithelial cells (Mandlik *et al.*, 2007).

Similarly, certain types of *S. agalactiae* pili mediate adherence to a specific tissue but not many others (Dramsi *et al.*, 2006; Krishnan *et al.*, 2007; Maisey *et al.*, 2007). Like the SpaA pili, the ancillary pilins of *S. agalactiae* are the main contributors for this host tissue tropism (Dramsi *et al.*, 2006; Krishnan *et al.*, 2007). This is not a universal feature, however, since *S. pyogenes* M1 utilizes the pilus backbone as the major adhesin for human tonsil epithelium and primary human keratinocytes (Abbot *et al.*, 2007; Manetti *et al.*, 2007), although one of the two ancillary proteins also contributes to adhesion (Abbot *et al.*, 2007).

So far, fewer reports on the use of animal models of infection to examine the role of pili in bacterial adherence/colonization have emerged. Consistent with *in vitro* cell culture studies, piliated *S. pneumoniae* was shown to be more virulent and better colonizers, since they outcompete a mutant strain that lacks pili (Barocchi *et al.*, 2006). Similarly studies with *E. faecalis* infection in an endocarditis model of rats demonstrated that, compared to the wild type, more than 50% of non-piliated mutant bacteria were cleared from vegetation and kidneys (Nallapareddy *et al.*, 2006). Consistently, studies with a murine intraperitoneal infection model has shown that the survival rate of a *S. agalactiae* mutant with a deletion of the major subunit (PilB) gene is 20% less than that of the wild type bacteria (Maisey *et al.*, 2007). Interestingly, in the systemic challenge model, injection of nonpathogenic Gram-positive *Lactococcus lactis* expressing PilB to mice is lethal, as compared to that of the wild type lactococci considered as avirulent (Maisey *et al.*, 2007). Clearly, expression of PilB presents a unique feature beneficial to GBS colonization and survival.

Pili in the formation of biofilm

A classic example of bacterial biofilm is our dental plaque, which contains more than 500 bacterial species including *Actinomyces* and oral streptococci (Kolenbrander, 2000). *Actinomyces* and oral streptococci are among the earliest bacteria known to colonize the tooth surface, and they are thought to promote the subsequent colonization of other bacteria during early stages of plaque development (Nyvad and Kilian, 1987). The binding of *Actinomyces* to the tooth surface is mediated by the Type 1 fimbriae by their adherence with salivary acidic proline-rich proteins (PRPs) and statherines coating the tooth enamel (Gibbons and Hay, 1988; Gibbons *et al.*, 1988). *Actinomyces* and oral streptococci are also known to coaggregate (Gibbons and Nygaard, 1970; Cisar *et al.*, 1979), and their coaggregation requires the Type 2 fimbriae, which recognize β-linked galactose and N-acetylgalactosamine-containing motifs present in the polysaccharides of oral streptococci (Sandberg *et al.*, 1995). A flow-cell system for biofilm development has demonstrated that coaggregation between *Actinomyces* and streptococci is essential for biofilm formation (Palmer *et al.*, 2001; Palmer *et al.*, 2003). Interestingly, *A. naeslundii* coaggregation with streptococci is blocked by the Fab fragments of polyclonal antibodies raised against the Type 2 fimbriae, but not with the Fab fragments of monoclonal antibodies that were raised against the major fimbrial subunits FimA (Revis *et al.*, 1982). It is thus likely that the tip fimbrial protein FimB mediates bacterial coaggregation (Mishra *et al.*, 2007).

Recent work in *E. faecalis* and *S. pyogenes* also demonstrated the role of pili in biofilm formation (Nallapareddy *et al.*, 2006; Manetti *et al.*, 2007). As discussed above, the *ebpA–ebpB–ebpC–srtC* pilus locus encodes a heterotrimeric pilus that is required for enterococcal infection. Gene disruption in this locus results in a significant defect in biofilm development (Nallapareddy *et al.*, 2006). In *S. pyogenes*, it was shown that the major pilin largely contributes to the formation of biofilm *in vitro* (Manetti *et al.*, 2007). How pili contribute to biofilm development in these systems, however, remains to be elucidated.

Pili and innate immunity: immune modulation and phagocyte survival

Invading microbial pathogens activate the host innate immune system, inducing an inflammatory response by macrophages and neutrophils. As the first line of defense against bacteria, macrophages kill the invading bacteria by phagocytosis and secrete cytokines and chemokines initiating inflammation that is beneficial to fight infection by host cells.

Recent findings support the role of Gram-positive pili in modulating the host immune system and conferring bacterial survival in macrophages and neutrophils. When mice were challenged with *S. pneumoniae* strains that express pili or no pili, high levels of proinflammatory cytokines TNF and IL-6 were detected in the serum of animals infected with piliated pneumoniae only (Barocchi *et al.*, 2006). It is not clear how pneumococcal pili trigger inflammatory responses, however, the high level of cytokines thus generated may elicit tissue damage that allows for a more robust bacterial invasion. In fact, the role of pili in mediating phagocytic survival is evident for GBS. A GBS strain that lacks the major pilin PilB is more susceptible to killing by macrophages and neutrophils *in vitro* as compared to the wild type strain (Maisey *et al.*, 2007). Moreover, the expression of PilB in *L. lactis* confers similar resistance (Maisey *et al.*, 2007). This resistance to phagocytic killing may be due to the ability of pili to sequester the host cathelicidin antimicrobial peptides (Maisey *et al.*, 2007).

Prospect of pili as vaccine candidates

Pili are attractive molecules for vaccine development since they are not only expressed ubiquitously on the bacterial cell surface but they play important roles in pathogenesis. The most attractive vaccines are those that offer broad protection against the vast majority of pathogens. For *S. agalactiae, S. pyogenes* and enterococci, pili may well be an ideal candidate for vaccine development because they are present on the surface of most clinical isolates and are required for bacterial adherence (Sillanpaa *et al.*, 2004; Maione *et al.*, 2005; Telford *et al.*, 2006). Importantly, antibody response to specific Gram-positive pili has been documented in the case of *E. faecalis* infection of humans (Sillanpaa *et al.*, 2004). Moreover, efforts to develop a broad-spectrum protein-based vaccine against GBS and GAS have yielded promising results. Maione and colleagues screened several hundred recombinant proteins for their protective function in an infant mouse model of GBS infection (Maione *et al.*, 2005). Pregnant mice were immunized with individual proteins and their offsprings were challenged with wild-type GBS at a lethal dose that killed 80–90% of the animals. Remarkably, three out of four antigens that conferred protection to the pups in this large screen are pilus proteins expressed by many GBS isolates (Maione *et al.*, 2005).

Similar results have also been reported for the pilus antigens from GAS and *S. pneumoniae*. Immunization of mice with a combination of the three GAS pilus proteins Cpa/128/130 conferred protection against mucosal challenge with wild type bacteria (Mora *et al.*, 2005), with an efficiency comparable to that of the leading vaccine candidate M protein (Dale, 2008). As for *S. pneumoniae,* both active and passive immunization with recombinant pilus proteins protected mice from intraperitoneal challenge with the virulent bacteria (Gianfaldoni *et al.*, 2007). Unfortunately, pili may not be most ideal for developing a multivalent pneumococcal vaccine because only a small number of clinical isolates contain the *rlrA* pilus gene cluster (< 30%) (Basset *et al.*, 2007; Aguiar *et al.*, 2008).

Conclusions and future trends

Upon entering the hosts, bacteria are faced with the perpetual challenge to keep up with changing environments and threats of elimination. To encounter this, the bacteria must develop mechanisms and strategies to survive and persist within the hosts. Clearly, Gram-positive pili represent another form of the bacterial arsenal in this regard. The bacteria orchestrate the assembly of pilus components into filamentous structures as well as surface molecules that mediate bacterial adherence, colonization, and invasion. These surface entities also confer bacterial resistance to elimination by the hosts in the form of bacterial biofilms, or resistance to killing by phagocytes.

Although a great deal of progress has been made recently, there are still many major gaps in our knowledge of the mechanism of pilus biogenesis and the function of pili in pathogenesis by Gram-positive bacteria. In terms of the mechanism, new technology is required to, first of all, elucidate the true architecture of pili in different pathogens as well as the molecular bonds that link the different pilins. Second, it is important to determine whether or not the housekeeping sortase plays a universal role in anchoring pili to the cell wall. Third, we have no knowledge of how pilin secretion is coupled to the ordered assembly of pilus and factors that dictate the decision between pilus polymerization and cell wall anchoring.

Fourth, there is as yet little or no molecular understanding of how sortases select and act on a set of pilin substrates with such exquisite specificity as clearly demonstrated in corynebacteria.

Even greater gaps of knowledge exist for the function of pili in pathogenesis. First, the specific host cell receptors for most pili remain unknown to date. While it is almost certain that pili are involved in signalling host cells, virtually nothing is known about this potentially important step of pathogenesis. Second, we have little understanding of how pilins modulate host inflammatory response. Are pilins secreted during infection to modulate host immunity? Is pilus assembly regulated during bacterial infection? Filling these gaps should aid significantly in our goals to develop new antibacterials for serious Gram-positive pathogens.

Web resources

1. Database for sortase and sortase substrates: http://www.doe-mbi.ucla.edu/Services/Sortase/
2. Sortase substrate database: http://bamics3.cmbi.kun.nl/cgi-bin/jos/sortase_substrates/index.py
3. Prediction of cell wall sorting signals in Gram-positive bacteria: http://bioinformatics.biol.uoa.gr/CW-PRED/

Acknowledgments

We thank the members of our laboratory for their contribution to the studies of pilus assembly and function. The authors' work has been supported by grants from the Charles H. Hood Foundation and the National Institutes of Health AI061381 and DE017382 to H.T.T.

References

Abbot, E.L., Smith, W.D., Siou, G.P., Chiriboga, C., Smith, R.J., Wilson, J.A., Hirst, B.H., and Kehoe, M.A. (2007). Pili mediate specific adhesion of *Streptococcus pyogenes* to human tonsil and skin. Cell Microbiol. *9*, 1822–1833.

Aguiar, S.I., Serrano, I., Pinto, F.R., Melo-Cristino, J., and Ramirez, M. (2008). The presence of the pilus locus is a clonal property among pneumococcal invasive isolates. BMC Microbiol. *8*, 41.

Barocchi, M.A., Ries, J., Zogaj, X., Hemsley, C., Albiger, B., Kanth, A., Dahlberg, S., Fernebro, J., Moschioni, M., Masignani, V., Hultenby, K., Taddei, A.R., Beiter, K., Wartha, F., von Euler, A., Covacci, A., Holden, D.W., Normark, S., Rappuoli, R., and Henriques-Normark, B. (2006). A pneumococcal pilus influences virulence and host inflammatory responses. Proc. Natl. Acad. Sci. U.S.A. *103*, 2857–2862.

Basset, A., Trzcinski, K., Hermos, C., O'Brien, K.L., Reid, R., Santosham, M., McAdam, A.J., Lipsitch, M., and Malley, R. (2007). Association of the pneumococcal pilus with certain capsular serotypes but not with increased virulence. J. Clin. Microbiol. *45*, 1684–1689.

Bolotin, A., Wincker, P., Mauger, S., Jaillon, O., Malarme, K., Weissenbach, J., Ehrlich, S.D., and Sorokin, A. (2001). The complete genome sequence of the lactic acid bacterium *Lactococcus lactis* ssp. lactis IL1403. Genome Res. *11*, 731–753.

Budzik, J.M., Marraffini, L.A., and Schneewind, O. (2007). Assembly of pili on the surface of *Bacillus cereus* vegetative cells. Mol. Microbiol. *66*, 495–510.

Budzik, J.M., Marraffini, L.A., Souda, P., Whitelegge, J.P., Faull, K.F., and Schneewind, O. (2008). Amide bonds assemble pili on the surface of bacilli. Proc. Natl. Acad. Sci. U.S.A. *105*, 10215–10220.

Burrows, L.L. (2005). Weapons of mass retraction. Mol. Microbiol. *57*, 878–888.

Cerdeno-Tarraga, A.M., Efstratiou, A., Dover, L.G., Holden, M.T., Pallen, M., Bentley, S.D., Besra, G.S., Churcher, C., James, K.D., De Zoysa, A., Chillingworth, T., Cronin, A., Dowd, L., Feltwell, T., Hamlin, N., Holroyd, S., Jagels, K., Moule, S., Quail, M.A., Rabbinowitsch, E., Rutherford, K.M, Thomson, N.R., Unwin, L., Whitehead, S., Barrell, B.G, and Parkhill, J. (2003). The complete genome sequence and analysis of *Corynebacterium diphtheriae* NCTC13129. Nucleic Acids Res. *31*, 6516–6523.

Challacombe, J.F., Altherr, M.R., Xie, G., Bhotika, S.S., Brown, N., Bruce, D., Campbell, C.S., Campbell, M.L., Chen, J., Chertkov, O., Cleland, C., Dimitrijevic, M., Doggett, N.A., Fawcett, J.J., Glavina, T., Goodwin, L.A., Green, L.D., Han, C.S., Hill, K.K., Hitchcock, P., Jackson, P.J., Keim, P., Kewalramani, A.R., Longmire, J., Lucas, S., Malfatti, S., Martinez, D., McMurry, K., Meincke, L.J., Misra, M., Moseman, B.L., Mundt, M., Munk, A.C., Okinaka, R.T., Parson-Quintana, B., Reilly, L.P., Richardson, P., Robinson, D.L., Saunders, E., Tapia, R., Tesmer, J.G., Thayer, N., Thompson, L.S., Tice, H., Ticknor, L.O., Wills, P.L., Gilna, P., Brettin, T.S (2007). The complete genome sequence of *Bacillus thuringiensis* Al Hakam. J. Bacteriol. *189*, 3680–3681.

Cisar, J.O., Kolenbrander, P.E., and McIntire, F.C. (1979). Specificity of coaggregation reactions between human oral streptococci and strains of *Actinomyces viscosus* or *Actinomyces naeslundii*. Infect. Immun. *24*, 742–752.

Comfort, D., and Clubb, R.T. (2004). A comparative genome analysis identifies distinct sorting pathways in Gram-positive bacteria. Infect. Immun. *72*, 2710–2722.

Dale, J.B. (2008). Current status of group A streptococcal vaccine development. Adv. Exp. Med. Biol. *609*, 53–63.

Donkersloot, J.A., Cisar, J.O., Wax, M.E., Harr, R.J., and Chassy, B.M. (1985). Expression of *Actinomyces viscosus* antigens in *Escherichia coli*: cloning of a structural gene (fimA) for type 2 fimbriae. J. Bacteriol. *162*, 1075–1078.

Dramsi, S., Caliot, E., Bonne, I., Guadagnini, S., Prevost, M.C., Kojadinovic, M., Lalioui, L., Poyart, C., and Trieu-Cuot, P. (2006). Assembly and role of pili in group B streptococci. Mol. Microbiol. *60*, 1401–1413.

Dramsi, S., Trieu-Cuot, P., and Bierne, H. (2005). Sorting sortases: a nomenclature proposal for the various sortases of Gram-positive bacteria. Res. Microbiol. *156*, 289–297.

Ferretti, J.J., McShan, W.M., Ajdic, D., Savic, D.J., Savic, G., Lyon, K., Primeaux, C., Sezate, S., Suvorov, A.N., Kenton, S., Lai, H.S., Lin, S.P., Qian, Y., Jia, H.G., Najar, F.Z., Ren, Q., Zhu, H., Song, L., White, J., Yuan, X., Clifton, S.W., Roe, B.A., and McLaughlin R. (2001). Complete genome sequence of an M1 strain of *Streptococcus pyogenes*. Proc. Natl. Acad. Sci. U.S.A. *98*, 4658–4663.

Gaspar, A.H., and Ton-That, H. (2006). Assembly of distinct pilus structures on the surface of *Corynebacterium diphtheriae*. J. Bacteriol. *188*, 1526–1533.

Gianfaldoni, C., Censini, S., Hilleringmann, M., Moschioni, M., Facciotti, C., Pansegrau, W., Masignani, V., Covacci, A., Rappuoli, R., Barocchi, M.A., and Ruggiero, P. (2007). *Streptococcus pneumoniae* pilus subunits protect mice against lethal challenge. Infect. Immun. *75*, 1059–1062.

Gibbons, R.J., and Hay, D.I. (1988). Human salivary acidic proline-rich proteins and statherin promote the attachment of *Actinomyces viscosus* LY7 to apatitic surfaces. Infect. Immun. *56*, 439–445.

Gibbons, R.J., Hay, D.I., Cisar, J.O., and Clark, W.B. (1988). Adsorbed salivary proline-rich protein 1 and statherin: receptors for type 1 fimbriae of *Actinomyces viscosus* T14V-J1 on apatitic surfaces. Infect. Immun. *56*, 2990–2993.

Gibbons, R.J., and Nygaard, M. (1970). Interbacterial aggregation of plaque bacteria. Arch. Oral Biol. *15*, 1397–1400.

Girard, A.E., and Jacius, B.H. (1974). Ultrastructure of *Actinomyces viscosus* and *Actinomyces naeslundii*. Arch. Oral Biol. *19*, 71–79.

Hilleringmann, M., Giusti, F., Baudner, B.C., Masignani, V., Covacci, A., Rappuoli, R., Barocchi, M.A., and Ferlenghi, I. (2008). Pneumococcal pili are composed of protofilaments exposing adhesive clusters of Rrg A. PLoS Pathog. *4*, e1000026.

Holmes, R.K. (2000). Biology and molecular epidemiology of diphtheria toxin and the tox gene. J. Infect. Dis. *181 Suppl. 1*, S156–167.

Hu, P., Bian, Z., Fan, M., Huang, M., and Zhang, P. (2008). Sec translocase and sortase A are colocalised in a locus in the cytoplasmic membrane of *Streptococcus mutans*. Arch. Oral Biol. 53, 150–154.

Kang, H.J., Coulibaly, F., Clow, F., Proft, T., and Baker, E.N. (2007). Stabilizing isopeptide bonds revealed in Gram-positive bacterial pilus structure. Science *318*, 1625–1628.

Kolenbrander, P.E. (2000). Oral microbial communities: biofilms, interactions, and genetic systems. Annu. Rev. Microbiol. *54*, 413–437.

Krishnan, V., Gaspar, A.H., Ye, N., Mandlik, A., Ton-That, H., and Narayana, S.V. (2007). An IgG-like domain in the minor pilin GBS52 of *Streptococcus agalactiae* mediates lung epithelial cell adhesion. Structure *15*, 893–903.

Lauer, P., Rinaudo, C.D., Soriani, M., Margarit, I., Maione, D., Rosini, R., Taddei, A.R., Mora, M., Rappuoli, R., Grandi, G., and Telford, J.L. (2005). Genome analysis reveals pili in Group B Streptococcus. Science *309*, 105.

LeMieux, J., Hava, D.L., Basset, A., and Camilli, A. (2006). RrgA and RrgB are components of a multisubunit pilus encoded by the *Streptococcus pneumoniae* rlrA pathogenicity islet. Infect. Immun. *74*, 2453–2456.

Maione, D., Margarit, I., Rinaudo, C.D., Masignani, V., Mora, M., Scarselli, M., Tettelin, H., Brettoni, C., Iacobini, E.T., Rosini, R., D'Agostino, N., Miorin, L., Buccato, S., Mariani, M., Galli, G., Nogarotto, R., Nardi, D.ei V., Vegni, F., Fraser, C., Mancuso, G., Teti, G., Madoff, L.C., Paoletti, L.C., Rappuoli, R., Kasper, D.L., Telford, J.L., and Grandi, G. (2005). Identification of a universal Group B streptococcus vaccine by multiple genome screen. Science *309*, 148–150.

Maisey, H.C., Hensler, M., Nizet, V., and Doran, K.S. (2007). Group B streptococcal pilus proteins contribute to adherence to and invasion of brain microvascular endothelial cells. J. Bacteriol. *189*, 1464–1467.

Mandlik, A., Das, A., and Ton-That, H. (2008a). The molecular switch that activates the cell wall anchoring step of pilus assembly in Gram-positive bacteria. Proc. Natl. Acad. Sci. U.S.A. *105*, 14147–14152.

Mandlik, A., Swierczynski, A., Das, A., and Ton-That, H. (2007). *Corynebacterium diphtheriae* employs specific minor pilins to target human pharyngeal epithelial cells. Mol. Microbiol. *64*, 111–124.

Mandlik, A., Swierczynski, A., Das, A., and Ton-That, H. (2008b). Pili in Gram-positive bacteria: assembly, involvement in colonization and biofilm development. Trends Microbiol. *16*, 33–40.

Manetti, A.G., Zingaretti, C., Falugi, F., Capo, S., Bombaci, M., Bagnoli, F., Gambellini, G., Bensi, G., Mora, M., Edwards, A.M., Musser, J.M., Graviss, E.A., Telford, J.L., Grandi, G., and Margarit, I. (2007). *Streptococcus pyogenes* pili promote pharyngeal cell adhesion and biofilm formation. Mol. Microbiol. *64*, 968–983.

Marraffini, L.A., Dedent, A.C., and Schneewind, O. (2006). Sortases and the art of anchoring proteins to the envelopes of Gram-positive bacteria. Microbiol. Mol. Biol. Rev. *70*, 192–221.

Mazmanian, S.K., Liu, G., Ton-That, H., and Schneewind, O. (1999). *Staphylococcus aureus* sortase, an enzyme that anchors surface proteins to the cell wall. Science *285*, 760–763.

Mazmanian, S.K., Skaar, E.P., Gaspar, A.H., Humayun, M., Gornicki, P., Jelenska, J., Joachmiak, A., Missiakas, D.M., and Schneewind, O. (2003). Passage of heme-iron across the envelope of *Staphylococcus aureus*. Science *299*, 906–909.

Mazmanian, S.K., Ton-That, H., and Schneewind, O. (2001). Sortase-catalysed anchoring of surface proteins to the cell wall of *Staphylococcus aureus*. Mol. Microbiol. *40*, 1049–1057.

Mishra, A., Das, A., Cisar, J.O., and Ton-That, H. (2007). Sortase-catalyzed assembly of distinct heteromeric fimbriae in *Actinomyces naeslundii*. J. Bacteriol. *189*, 3156–3165.

Mongodin, E.F., Shapir, N., Daugherty, S.C., DeBoy, R.T., Emerson, J.B., Shvartzbeyn, A., Radune, D., Vamathevan, J., Riggs, F., Grinberg, V., Khouri, H., Wackett, L.P., Nelson, K.E., and Sadowsky, M.J. (2006). Secrets of soil survival revealed by the genome sequence of *Arthrobacter aurescens* TC1. PLoS Genet. *2*, e214.

Mora, M., Bensi, G., Capo, S., Falugi, F., Zingaretti, C., Manetti, A.G., Maggi, T., Taddei, A.R., Grandi, G., and Telford, J.L. (2005). Group A Streptococcus produce pilus-like structures containing protective antigens and Lancefield T antigens. Proc. Natl. Acad. Sci. U.S.A. *102*, 15641–15646.

Nallapareddy, S.R., Singh, K.V., Sillanpaa, J., Garsin, D.A., Hook, M., Erlandsen, S.L., and Murray, B.E. (2006). Endocarditis and biofilm-associated pili of *Enterococcus faecalis*. J. Clin. Invest. *116*, 2799–2807.

Navarre, W.W., and Schneewind, O. (1999). Surface proteins of Gram-positive bacteria and mechanisms of their targeting to the cell wall envelope. Microbiol. Mol. Biol. Rev. *63*, 174–229.

Nishio, Y., Nakamura, Y., Kawarabayasi, Y., Usuda, Y., Kimura, E., Sugimoto, S., Matsui, K., Yamagishi, A., Kikuchi, H., Ikeo, K., and Gojobori, T. (2003). Comparative complete genome sequence analysis of the amino acid replacements responsible for the thermostability of *Corynebacterium efficiens*. Genome Res. *13*, 1572–1579.

Nobbs, A.H., Rosini, R., Rinaudo, C.D., Maione, D., Grandi, G., and Telford, J.L. (2008). Sortase A utilizes an ancillary protein anchor for efficient cell wall anchoring of pili in *Streptococcus agalactiae*. Infect. Immun. *76*, 3550–3560.

Nyvad, B., and Kilian, M. (1987). Microbiology of the early colonization of human enamel and root surfaces *in vivo*. Scand. J. Dent. Res. *95*, 369–380.

Palmer, R.J., Jr., Gordon, S.M., Cisar, J.O., and Kolenbrander, P.E. (2003). Coaggregation-mediated interactions of streptococci and actinomyces detected in initial human dental plaque. J. Bacteriol. *185*, 3400–3409.

Palmer, R.J., Jr., Kazmerzak, K., Hansen, M.C., and Kolenbrander, P.E. (2001). Mutualism versus independence: strategies of mixed-species oral biofilms *in vitro* using saliva as the sole nutrient source. Infect. Immun. *69*, 5794–5804.

Paterson, G.K., and Mitchell, T.J. (2004). The biology of Gram-positive sortase enzymes. Trends Microbiol. *12*, 89–95.

Paulsen, I.T., Banerjei, L., Myers, G.S., Nelson, K.E., Seshadri, R., Read, T.D., Fouts, D.E., Eisen, J.A., Gill, S.R., Heidelberg, J.F., Tettelin, H., Dodson, R.J., Umayam, L., Brinkac, L., Beanan, M., Daugherty, S., DeBoy, R.T., Durkin, S., Kolonay, J., Madupu, R., Nelson, W., Vamathevan, J., Tran, B., Upton, J., Hansen, T., Shetty, J., Khouri, H., Utterback, T., Radune, D., Ketchum, K.A., Dougherty, B.A., and Fraser, C.M. (2003). Role of mobile DNA in the evolution of vancomycin-resistant *Enterococcus faecalis*. Science *299*, 2071–2074.

Rasko, D.A., Ravel, J., Okstad, O.A., Helgason, E., Cer, R.Z., Jiang, L., Shores, K.A., Fouts, D.E., Tourasse, N.J., Angiuoli, S.V., Kolonay, J., Nelson, W.C., Kolstø, A.B., Fraser, C.M., Read, T.D. (2004). The genome sequence of *Bacillus cereus* ATCC 10987 reveals metabolic adaptations and a large plasmid related to *Bacillus anthracis* pXO1. Nucleic Acids Res. *32*, 977–988.

Revis, G.J., Vatter, A.E., Crowle, A.J., and Cisar, J.O. (1982). Antibodies against the Ag2 fimbriae of *Actinomyces viscosus* T14V inhibit lactose-sensitive bacterial adherence. Infect. Immun. *36*, 1217–1222.

Rosch, J.W., and Caparon, M.G. (2005). The ExPortal: an organelle dedicated to the biogenesis of secreted proteins in *Streptococcus pyogenes*. Mol. Microbiol. *58*, 959–968.

Rosch, J.W., Mann, B., Thornton, J., Sublett, J., and Tuomanen, E. (2008). Convergence of regulatory networks on the pilus locus of *Streptococcus pneumoniae*. Infect. Immun. *76*, 3187–3196.

Rosini, R., Rinaudo, C.D., Soriani, M., Lauer, P., Mora, M., Maione, D., Taddei, A., Santi, I., Ghezzo, C., Brettoni, C., Buccato, S., Margarit, I., Grandi, G., and Telford, J.L. (2006). Identification of novel genomic islands coding for antigenic pilus-like structures in *Streptococcus agalactiae*. Mol. Microbiol. *61*, 126–141.

Sandberg, A.L., Ruhl, S., Joralmon, R.A., Brennan, M.J., Sutphin, M.J., and Cisar, J.O. (1995). Putative glycoprotein and glycolipid polymorphonuclear leukocyte receptors for the *Actinomyces naeslundii* WVU45 fimbrial lectin. Infect. Immun. *63*, 2625–2631.

Sauer, F.G., Mulvey, M.A., Schilling, J.D., Martinez, J.J., and Hultgren, S.J. (2000). Bacterial pili: molecular mechanisms of pathogenesis. Curr. Opin. Microbiol. *3*, 65–72.

Schell, M.A., Karmirantzou, M., Snel, B., Vilanova, D., Berger, B., Pessi, G., Zwahlen, M.C., Desiere, F., Bork, P., Delley, M., Pridmore, R.D., and Arigoni, F. (2002). The genome sequence of *Bifidobacterium longum* reflects its adaptation to the human gastrointestinal tract. Proc. Natl. Acad. Sci. U.S.A. *99*, 14422–14427.

Schneewind, O., Fowler, A., and Faull, K.F. (1995). Structure of the cell wall anchor of surface proteins in *Staphylococcus aureus*. Science *268*, 103–106.

Schneewind, O., Mihaylova-Petkov, D., and Model, P. (1993). Cell wall sorting signals in surface proteins of Gram-positive bacteria. EMBO J. *12*, 4803–4811.

Schneewind, O., Model, P., and Fischetti, V.A. (1992). Sorting of protein A to the staphylococcal cell wall. Cell *70*, 267–281.

Scott, J.R., and Zahner, D. (2006). Pili with strong attachments: Gram-positive bacteria do it differently. Mol. Microbiol. *62*, 320–330.

Shimizu, T., Ohtani, K., Hirakawa, H., Ohshima, K., Yamashita, A., Shiba, T., Ogasawara, N., Hattori, M., Kuhara, S., and Hayashi, H. (2002). Complete genome sequence of *Clostridium perfringens*, an anaerobic flesh-eater. Proc. Natl. Acad. Sci. U.S.A. *99*, 996–1001.

Sillanpaa, J., Xu, Y., Nallapareddy, S.R., Murray, B.E., and Hook, M. (2004). A family of putative MSCRAMMs from *Enterococcus faecalis*. Microbiology *150*, 2069–2078.

Sjoquist, J., Meloun, B., and Hjelm, H. (1972). Protein A isolated from *Staphylococcus aureus* after digestion with lysostaphin. Eur. J. Biochem. *29*, 572–578.

Swaminathan, A., Mandlik, A., Swierczynski, A., Gaspar, A., Das, A., and Ton-That, H. (2007). Housekeeping sortase facilitates the cell wall anchoring of pilus polymers in *Corynebacterium diphtheriae*. Mol. Microbiol. *66*, 961–974.

Swierczynski, A., and Ton-That, H. (2006). Type III pilus of corynebacteria: Pilus length is determined by the level of its major pilin subunit. J. Bacteriol. *188*, 6318–6325.

Tauch, A., Kaiser, O., Hain, T., Goesmann, A., Weisshaar, B., Albersmeier, A., Bekel, T., Bischoff, N., Brune, I., Chakraborty, T., Kalinowski, J., Meyer, F., Rupp, O., Schneiker, S., Viehoever, P., and Pühler, A. (2005). Complete genome sequence and analysis of the multiresistant nosocomial pathogen *Corynebacterium jeikeium* K411, a lipid-requiring bacterium of the human skin flora. J. Bacteriol. *187*, 4671–4682.

Tauch, A., Trost, E., Tilker, A., Ludewig, U., Schneiker, S., Goesmann, A., Arnold, W., Bekel, T., Brinkrolf, K., Brune, I., Götker, S., Kalinowski, J., Kamp, P.B., Lobo, F.P., Viehoever, P., Weisshaar, B., Soriano, F., Dröge, M., and Pühler, A. (2008). The lifestyle of *Corynebacterium urealyticum* derived from its complete genome sequence established by pyrosequencing. J. Biotechnol. *136*, 11–12.

Telford, J.L., Barocchi, M.A., Margarit, I., Rappuoli, R., and Grandi, G. (2006). Pili in Gram-positive pathogens. Nat. Rev. Microbiol. *4*, 509–519.

Tettelin, H., Masignani, V., Cieslewicz, M.J., Eisen, J.A., Peterson, S., Wessels, M.R., Paulsen, I.T., Nelson, K.E., Margarit, I., Read, T.D., Madoff, L.C., Wolf, A.M., Beanan, M.J., Brinkac, L.M., Daugherty, S.C., DeBoy, R.T., Durkin, A.S., Kolonay, J.F., Madupu, R., Lewis, M.R., Radune, D., Fedorova, N.B., Scanlan, D., Khouri, H., Mulligan, S., Carty, H.A., Cline, R.T., Van, A.ken SE., Gill, J., Scarselli, M., Mora, M., Iacobini, E.T., Brettoni, C., Galli, G., Mariani, M., Vegni, F., Maione, D., Rinaudo, D., Rappuoli, R., Telford, J.L., Kasper, D.L., Grandi, G., Fraser, C.M. (2002). Complete genome sequence and comparative genomic analysis of an emerging human pathogen, serotype V *Streptococcus agalactiae*. Proc. Natl. Acad. Sci. U.S.A. *99*, 12391–12396.

Tettelin, H., Nelson, K.E., Paulsen, I.T., Eisen, J.A., Read, T.D., Peterson, S., Heidelberg, J., DeBoy, R.T., Haft, D.H., Dodson, R.J., Durkin, A.S., Gwinn, M., Kolonay, J.F., Nelson, W.C., Peterson, J.D., Umayam, L.A., White, O., Salzberg, S.L., Lewis, M.R., Radune, D., Holtzapple, E., Khouri, H., Wolf, A.M., Utterback, T.R., Hansen, C.L., McDonald, L.A., Feldblyum, T.V., Angiuoli, S., Dickinson, T., Hickey, E.K., Holt, I.E., Loftus, B.J., Yang, F., Smith, H.O., Venter, J.C., Dougherty, B.A., Morrison, D.A., Hollingshead, S.K., Fraser, C.M. (2001). Complete genome sequence of a virulent isolate of *Streptococcus pneumoniae*. Science *293*, 498–506.

Ton-That, H., Marraffini, L.A., and Schneewind, O. (2004a). Protein sorting to the cell wall envelope of Gram-positive bacteria. Biochim. Biophys. Acta *1694*, 269–278.

Ton-That, H., Marraffini, L.A., and Schneewind, O. (2004b). Sortases and pilin elements involved in pilus assembly of *Corynebacterium diphtheriae*. Mol. Microbiol. *53*, 251–261.

Ton-That, H., and Schneewind, O. (2003). Assembly of pili on the surface of *Corynebacterium diphtheriae*. Mol. Microbiol. *50*, 1429–1438.

Ton-That, H., and Schneewind, O. (2004). Assembly of pili in Gram-positive bacteria. Trends Microbiol. *12*, 228–234.

Torres, A.G., Zhou, X., and Kaper, J.B. (2005). Adherence of diarrheagenic *Escherichia coli* strains to epithelial cells. Infect. Immun. *73*, 18–29.

Yanagawa, R., and Otsuki, K. (1970). Some properties of the pili of *Corynebacterium renale*. J. Bacteriol. *101*, 1063–1069.

Yeates, T.O., and Clubb, R.T. (2007). Biochemistry. How some pili pull. Science *318*, 1558–1559.

Yeung, M.K. (1999). Molecular and genetic analyses of *Actinomyces* spp. Crit. Rev. Oral Biol. Med. *10*, 120–138.

Yeung, M.K., Chassy, B.M., and Cisar, J.O. (1987). Cloning and expression of a type 1 fimbrial subunit of *Actinomyces viscosus* T14V. J. Bacteriol. *169*, 1678–1683.

Yeung, M.K., Donkersloot, J.A., Cisar, J.O., and Ragsdale, P.A. (1998). Identification of a gene involved in assembly of *Actinomyces naeslundii* T14V type 2 fimbriae. Infect. Immun. *66*, 1482–1491.

Yeung, M.K., and Ragsdale, P.A. (1997). Synthesis and function of *Actinomyces naeslundii* T14V type 1 fimbriae require the expression of additional fimbria-associated genes. Infect. Immun. *65*, 2629–2639.

Yukawa, H., Omumasaba, C.A., Nonaka, H., Kos, P., Okai, N., Suzuki, N., Suda, M., Tsuge, Y., Watanabe, J., Ikeda, Y., Vertès, A.A., and Inui, M. (2007). Comparative analysis of the *Corynebacterium glutamicum* group and complete genome sequence of strain R. Microbiology *153*, 1042–1058.

Zong, Y., Bice, T.W., Ton-That, H., Schneewind, O., and Narayana, S.V. (2004a). Crystal structures of *Staphylococcus aureus* sortase A and its substrate complex. J. Biol. Chem. *279*, 31383–31389.

Zong, Y., Mazmanian, S.K., Schneewind, O., and Narayana, S.V. (2004b). The structure of sortase B, a cysteine transpeptidase that tethers surface protein to the *Staphylococcus aureus* cell wall. Structure (Camb) *12*, 105–112.

What is Essential for Flagellar Assembly?

6

Shin-Ichi Aizawa

Abstract

The flagellum is a motile organelle for most swimming bacteria. In *Salmonella enterica*, flagellar assembly and function requires about 50 genes, among which 21 gene products are incorporated in the complete flagellar structure as component proteins. In other eubacterial species including Gram-positive bacteria and spirochaetes, the number of genes and component proteins required for flagellar assembly is more or less the same, suggesting that the assembly mechanism for the flagellar structure is universal and that the flagellar machine is essentially the same in all species. Since the flagellum is so well designed and beautifully constructed by an ordered assembly pathway, even I, who am not a creationist, get an awe-inspiring feeling from its 'divine' beauty (Pallen and Matzke, 2006). However, if the flagellum has evolved from a primitive form, where are the remnants of its ancestor? Why don't we see any intermediate or simpler forms of flagella than what they are today? How was it possible that the flagella have evolved without leaving traces in history?

First of all, it should be remembered that we know the flagellar systems of only about 20 bacterial species (Fujii *et al.*, 2008). There must be a vast number of flagellated bacteria in the world still beyond our reach. We will be able to know only a very small fraction of the flagellar world. However, simpler flagellar structures, albeit under abnormal conditions, have been observed in the last few years. In this chapter, I would like to ask a primitive question, 'What is essential for flagellar assembly?', and I will show recent results on flagellar assembly mostly done in my lab in collaboration with Kelly Hughes's lab at the University of Utah.

A conventional image of flagellar assembly

First, let's look at how the flagellum of *S. enterica* serovar Typhimurium is constructed; it is regarded as the paradigm system. The order of the steps toward the assembly of a flagellum (the morphological pathway) was determined by analysing intermediate structures in various flagellar mutants and aligning them by size from small to large. Flagellar assembly starts in the cytoplasmic membrane, progresses through the periplasmic space, and finally extends outside the cell (Aizawa, 2004).

From the structural point of view, we see the flagellum consisting of two parts: the secretion apparatus and the axial structure. The major components of the axial structure are FlgG for the rod, FlgE for the hook, and FliC for the filament, each of which assembles with the aid of its corresponding cap protein. The cap protein for each substructure is FlgJ for the rod, FlgD for the hook, and FliD for the filament; the latter stays at the tip of the filament, whereas the former two serve as transient scaffolding proteins and thus are absent from the completed flagellum. Minor components are FlgB, FlgC, FlgF that connect the MS ring complex and the rod, and FlgK and FlgL that connect the hook and the filament.

The smallest flagellar structure recognizable by electron microscopy is the MS ring complex; therefore, the MS ring complex is regarded as the structural foundation. When two other flagellar substructures, the C ring and the C rod, attach to the cytoplasmic face of the M ring, this complex starts secreting other flagellar proteins to continue flagellar construction.

The first extracellular structure assembled onto the MS ring complex is a rod. The rod grows through the peptidoglycan layer with the aid of the cap protein FlgJ that also has the muramidase activity (Nambu *et al.*, 1999) but is soon stopped by the physical barrier of the outer membrane. The hold-up will be released when the outer ring complex makes a hole in the membrane. Once the physical barrier by the outer membrane has been removed, the hook starts growing beneath the FlgD scaffold until the hook length reaches about 55 nm. The hook length control coincides with the switching of substrates to be secreted from the rod-hook (early gene products) mode to the hook associated proteins (HAPs)-flagellin (late gene products) mode. Upon switching of the modes, FlgD is replaced by HAPs, followed by filament growth. The filament growth proceeds only in the presence of FliD (HAP2 or filament cap protein); without this cap, exported flagellin molecules are lost to the medium.

Now, let's consider flagellar mutants that do not produce complete flagella under normal conditions but do so under abnormal conditions and examine which flagellar proteins are not absolutely necessary for flagellar assembly.

Flagellation by exogenous flagellin: a cap protein FliD

Mutants that are deficient in flagellar assembly do not swarm but remain at the inoculated spot on semi-solid plates. However, some flagellar mutants that do not swarm in plates may swim in liquid. A typical example is the flagellar shape mutants that have filaments of altered shapes such as straight or curly. They cannot make a swarm ring, although the motor is perfectly functional.

FliD-deleted mutant cells cannot form flagellar filaments as mentioned above but produce a small swarm size on semi-solid plates during a prolonged incubation. These cells have complete flagella as observed by electron microscopy and can swim in liquid (Iino *et al.*, 1972). This resurrection of flagellar assembly without FliD occurred due to a high concentration of flagellins that accumulated around the cell bodies. This situation can be reproduced *in vitro* by adding flagellin monomers in the culture media, allowing flagellins to polymerize on the hooks of *fliD* mutant cells (Kagawa *et al.*, 1981). In spirochaetes, flagella are formed in the periplasmic space. Although the *fliD* gene is found in the genome of these organisms, we do not know how efficiently FliD works as the cap protein for periplasmic flagella. It will be interesting to test whether the periplasmic flagella may be able to grow or not in the absence of FliD. In short, FliD may be dispensable for constructing functional flagellum.

Ignoring helpers: chaperones FliS and FliT

Among 21 flagellar structural components, only late gene products require secretion chaperones for normal assembly: FliS for FliC, FliT for FliD, and FlgN for FlgK and FlgL (Evans *et al.*, 2006). These component proteins are secreted into the media in large amounts, suggesting that these proteins may be produced in excess (Komoriya *et al.*, 1999). In the absence of either FliS or FliT, cells produce functional but short flagella (Fig. 6.1). In *flgN* mutants, 10% of the cell population can produce intact flagella and swim (Kutsukake *et al.*, 1994). Therefore, although these chaperones are necessary for efficient flagellar assembly,

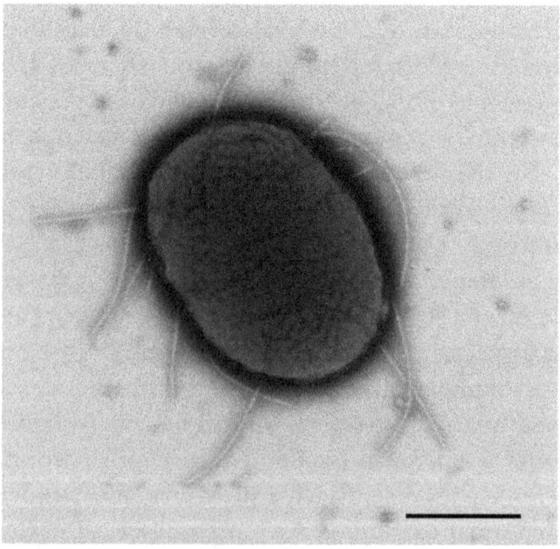

Figure 6.1 Electron micrographs of a *Salmonella enterica fliS* mutant cell (TH7343). Stained with 1% PTA; pH 7.0. Bar indicates 1 μm.

they are dispensable for constructing a functional flagellum.

Leakiness: the hook-length controller FliK

There are a group of mutants whose original phenotypes are Fla$^-$ or deficient in flagellar structure, but produce complete flagella at a low probability under normal conditions; a typical one is a *fliK* mutant. FliK is required for controlling the hook length and substrate-specificity switching in the flagellum-specific type III secretion system (T3SS: see below) from the rod-hook mode to the HAPs-flagellin mode (Shibata *et al.*, 2007). The FliK molecule consists of two regions: a non-structural N-terminal region and a compact C-terminal domain (Williams *et al.*, 1998). The FliK C-terminal domain interacts with FlhB, a gate component of T3SS, to switch the substrate-specificity (Ferris *et al.*, 2005). In a mutant that lacks the C-terminal domain, cells produce polyhooks. In contrast, lack of the N-terminal region gives rise to polyhook-filaments if the C-terminal domain is left intact (Hirano *et al.*, 2005). In *fliK* deletion mutants, 10% of the cell population retain polyhook-filaments and thus are motile (Aizawa, unpublished data). This phenotype is caused by the imperfect (leaky) regulation of FlhB gating. In fact, according to Dr. Shigeru Yamaguchi, who has a large collection of flagellar mutants, many of the flagellar genes give rise to leaky mutants: *flgA, flgD, flgH, flgI, flgJ, fliD, fliS, fliT, fliH, fliI, fliJ, fliK,* and *fliR* (Yamaguchi, unpublished data). These genes, except *fliR*, do not encode a structural component protein but some sort of regulatory protein.

Bypasses for flagellar construction: ATPase FliI

Flagella have been regarded as a self-assembly system, similar to that of bacteriophages. However, flagellar assembly is quite different from phage assembly in several ways. First, the flagellum, being an extracellular structure, assembles not in the cytoplasm but outside the cell. Second, secretion machinery is required for secretion of the component proteins from the cytoplasm to outside the cell. Third, assembly proceeds by addition of one subunit protein at a time at the distal end of the nascent structure.

For this kind of assembly, a protein secretion system must play an important role. As a matter of fact, among the 14 genes required in the very first step of flagellar assembly, more than half of the gene products are necessary for forming a protein complex called the type III export apparatus (Macnab, 2004). One of them, FliI, is an ATPase, indicating that one step in the export process requires ATP hydrolysis as an energy source (Vogler *et al.*, 1991). However, functional flagella can be formed in the absence of FliI under abnormal conditions, indicating, together with other evidence, that protein secretion is energized by proton motive force (Paul *et al.*, 2008; Minamino and Namba, 2008).

The physical body of the export apparatus has not been identified yet; the C rod is a strong candidate, judging from its location within the C ring and from genetic evidence. Genetic analysis indicates that at least five components are required for the C rod: FliP, FliQ, FliR, FlhA, and FlhB, all of which have membrane spanning regions (Aizawa, 2001). How these component proteins are inserted in the small space of the central area of the MS ring complex is still mysterious.

When FliI was overproduced in either *fliH* or *fliN* mutants, the transformed cells produced complete flagella. FliI, the ATPase, is anchored on the periphery of the C ring, which is composed of FliN, by association with the dimer of FliH (McMurry *et al.*, 2006).

The rod is designed to be short: rod protein FlgG

The rod consists of two parts: the proximal half (rod 1) and the distal half of the rod (rod 2). Rod 1 is composed of FlgB, FlgC, and FlgF, while rod 2 is composed of FlgG only. Point mutations in the *flgG* gene give rise to formation of long rods (Chevance *et al.*, 2007). The *flgG* mutants that gave rise to long rods are not specific mutants; there are not the hot spots usually seen in ordinary mutants but rather the mutation sites were scattered on the FlgG molecule. Interestingly, the rod length is under the control of FliK in a similar manner as in the hook-length control. The *fliK*-deletion mutant cells produce long rods with undefined length, thus polyrods. The lengths of polyrods are widely distributed (144.3 ± 79.3 nm), whereas those of long rods are

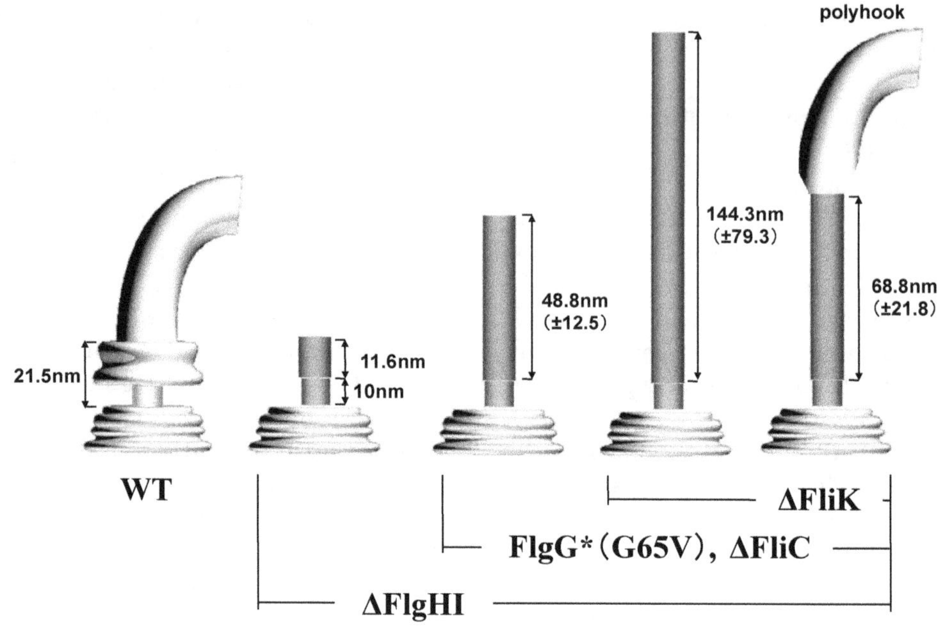

Figure 6.2 Schematic presentation of mutant rods with various lengths. Wild-type rod that is covered by the PL ring complex is exposed in the FlgHI-deleted mutant to allow direct measurements of the length. Additional mutations are shown at the bottom with arrows. The distal rod is distinguished from the proximal rod by a black bar.

a little better controlled (48.8 ± 12.5 nm) but not as good as the hook length (55 ± 6 nm) (Fig. 6.2).

The long rods are often associated with attached stacks of multiple P rings (up to seven rings). A *flgG* (G183R) mutant produces polyrods without rings, while another *flgG* (G183W) mutant produces polyrods with a stack of P rings attached. The L ring is not involved in formation of the P ring stack. When the L ring was attached to the rod, it stops the P ring stacking and concomitantly stops the rod elongation.

Since there are so many mutation points (so far 14 different sites) on FlgG that cause the phenomenon, we wonder if the wild-type rod is designed to be short. How is this short length controlled? The mechanism to maintain a short rod length is different from that for the hook length. FliK plays a major role in the hook-length control, while information intrinsic to the component protein FlgG seems important in the rod-length control. The length of the wild-type rod 2 is 11.6 nm, which corresponds to the thickness of four turns of the FlgG basic helix. This reminds us of tobacco mosaic virus (TMV); the coat proteins form three classes of aggregate, depending on environmental conditions: helical, disk, and A-protein. Four-layer disk is a closed structure consisting of a dimer of a bilayered disk (Caspar, 1963). The wild-type rod may be a closed structure so that polymerization of FlgG monomers on the pre-existing rod is prevented. Then, a mystery remains: How does the hook (helical structure) grow directly from the rod (closed structure) without any adjusting proteins in between?

The hookless filaments

In nature, the hook is observed in flagella in all species so far examined. Even archaeal flagella, which have flagellins completely different from eubacterial flagellin, have a hook-like curved portion at the base of the filament (Chaban *et al.*, 2007). Thus, the hook as a universal joint seems indispensable for flagellar function. However, the hook as a substructure is dispensable for flagellar assembly under abnormal conditions. The player is the rod again.

Among *Salmonella enterica flgG* mutants described above, some cells produced flagella that do not have a characteristic hook-curved portion between the filament and the basal body (Chevance *et al.*, 2007). To test whether the flagellum has a very short hook or no hook, we introduced *flgD* or *flgE* mutations into the *flgG* mutant and found that the filament can, albeit at a very low probability, elongate directly from the rod in the absence of the hook. The hookless filaments are formed and confined in the narrow periplasmic space, because there are no outer rings to penetrate the outer membranes. However, a few filaments are exposed to the outside during the cell division, probably by their penetrating through thin layers of the peptidoglycans in the outer membrane. To our surprise, the hookless filaments can rotate and uncoordinatedly push the cell body. At present, we do not know whether the hookless filaments are functional when they are confined in the periplasmic space. Interestingly, when the cell was elongated by adding a cell division inhibitor, cephalexin, the cell shape became helical due to the filament bundles in a similar manner to that seen in spirochaetes (Fig. 6.3).

Furthermore, to determine if the long rod is required for formation of hookless filaments, we examined the *flgD* or *flgE* mutations in the wild-type background. Most cells produce incomplete basal structures, called rivets, both with and without the outer rings (Kubori *et al.*, 1992), although with very low probability again, we found hookless filaments in the preparation. In summary, the rod can substitute for the hook in flagellar assembly, and the subsequent hookless filaments are functional.

Evolutionary throwback or just a mistake?

A new morphological pathway of flagellar assembly is summarized in Fig. 6.4. Roles that FlgG plays in this figure are intriguing; the hookless flagellum which had disappeared generations ago may reappear by such a simple genetic change as a point mutation. Is this atavism or evolutionary throwback? In general, atavisms occur because genes for previously existing phenotypical features are often preserved in DNA, even though the genes are not expressed in some or most of the organisms possessing them. In the flagellum, the key traits were hidden in a gene, *flgG*.

The amino acid sequences of FlgG and FlgE are homologous, especially at both terminal regions. FlgE (402 a.a.) is twice as large as FlgG (260 a.a.), suggesting that the central region might be responsible for curvature of the hook (Samatey *et al.*, 2004). The ancient flagellum could be constructed without the hook, and later on the hook was added to give flexibility to the flagellum.

Although it takes two generations to complete flagellar assembly in peritrichously flagellated bacteria (Aizawa and Kubori, 1998; Karlinsey *et al.*, 2000), it should be complete in one generation in bacteria with a single, polar flagellum. All flagellar genes seem to have evolved under strong selective pressure for efficiency of assembly. After

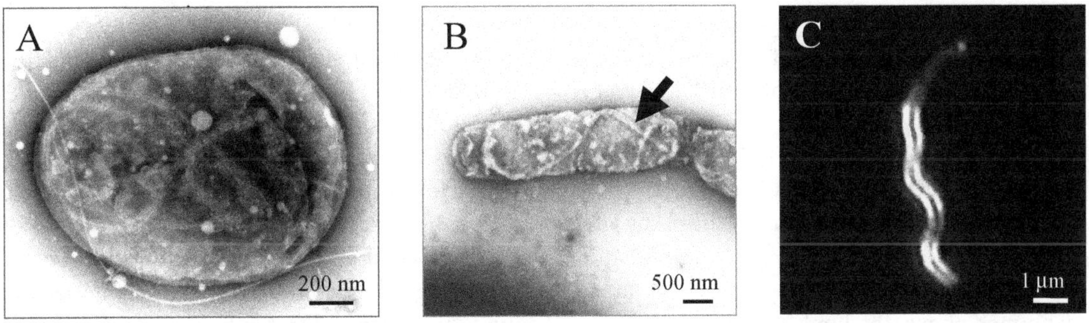

Figure 6.3 Microscopic images of *Salmonella enterica* FlgG* mutants. In *flgG** mutants, flagella were formed in the periplasmic space like periplasmic flagella in spirochaetes (**A**). When the mutant cells were treated with cephalexin (final concentration 10 μg/ml), cells were elongated and became helical as observed by dark-field microscopy (**C**). This helicity of the cell is probably formed by helical flagella in the periplasmic space as shown in a micrograph (**B**).

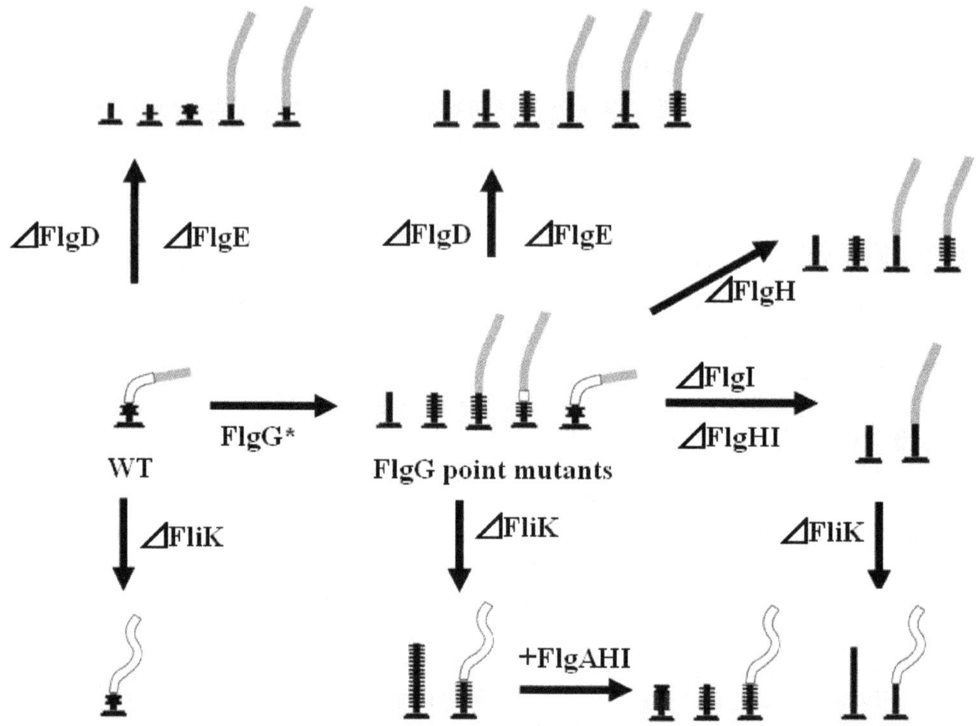

Figure 6.4 New morphological pathway of flagellar assembly. Starting from wild type (WT) at the left centre, we see intermediate flagellar structures in the backgrounds of mutations indicated by arrows. The number of rings on each structure is arbitrary. Rod, black; hook, white; filament, grey.

all, the first one who can build her flagellum and swim to the food passes on her DNA. Flagella are beautiful, ion-powered nanomachines that run with so-called 100% efficiency and their assembly pathway has been streamlined by evolution to minimize the time of the assembly process. The flagellum has acquired its beauty by evolving such a sophisticated, efficient machine.

Discussion and future trends

The morphological pathway of flagellar assembly in *Salmonella enterica* serovar Typhimurium was first shown by Suzuki *et al.* (1978), modified by Kubori *et al.* (1992), and has been revised from time to time. It is still incomplete. One of the large pits to be filled in the pathway is the transition phase from the rod to the hook. In *flgG* mutants mentioned above, the hook can grow directly on the rod distal end without the outer (P- and L-) ring complex. The P-ring formation is subsidiary to the rod formation, since the number of the P-ring attached on the rod increases as the rod elongates. When FlgI, the component protein of the P-ring, was overproduced under the same conditions, the number of the P-ring on the rod simply increased in a denser manner, indicating that the P-ring would be formed at any place of the rod without a special mechanism for positioning. In contrast, the L-ring formation seems to play more important roles in the transition; the L-ring formation coincides with release of the P-ring from the rod and formation of a complex of PL rings (Suzuki *et al.*, 1978). The resulting PL ring complex terminates the rod elongation and starts the hook formation both in the wild type and *flgG* mutants. Therefore, the PL ring complex is essential for the transition from the rod to the hook. How is the PL ring complex formed? Why does the complex that locates but not attaches on the rod change the secretion mode or polymerization mode from the rod to the hook? In this context, what are the roles of HAP1 (FlgK) and HAP3 (FlgL)? These two HAPs are not the cap proteins but essential for filament formation. A filament grows either from the pre-existing rod or hook but is always accompanied by HAP1 and HAP3.

It should be remembered that a mutant HAP3 has a strange effect on filament morphology (Fahner *et al.*, 1994). Finally, we should always keep in mind that archaeal flagellar system contradicts the paradigm established with eubacterial flagella: the component proteins (FlaA, B1, B2, and B3) of the filament are homologous to the pilin (PilA) more than the flagellin (FliC), these Fla proteins are N-terminally cleaved, and thus the filament may grow from the base rather than from the flagellar tip (Bardy *et al.*, 2003). Until we solve the contradiction, we will keep asking, 'What is the flagellum?'

Acknowledgements

I am grateful to Noriko Shibata-Takahashi for her diligent work on flagellar structures of *flgG* mutants and preparing figures, and to Kelly Hughes and David DeRosier for their invaluable comments on the manuscript.

References

Aizawa, S.-I.(2004). Flagella. In The Desk Encyclopedia of Microbiology, Schaechter, M. and Lederberg, J., ed. (Elsevier -Academic Press: San Diego, USA), pp. 470–479.

Aizawa, S.-I. (2001). Bacterial flagella and type III secretion systems. FEMS Microbiol. Lett. *202*, 157–164.

Aizawa S.-I., and Kubori, T. (1998). Bacterial flagellation and cell division. Genes Cells *3*, 1–10.

Bardy, S.L., Ng, S.Y., and Jarrell, K.F. (2003) Prokaryotic motility structures. Microbiology *149*, 295–304.

Caspar, D.L. (1963). Assembly and stability of the tobacco mosaic virus particle. Adv. Protein Chem. *8*, 37–121.

Chaban, B., Ng, S.Y.M., Kanbe, M., Saltzman, I., Nimmo, G., Aizawa, S.-I., and Jarrell, K.F. (2007). Systematic deletion analysis of the *fla* genes in the flagella operon identify several genes essential for proper assembly and function of flagella in the archaeon, *Methanococcus maripaludis*. Mol. Microbiol. *66*, 596–609.

Chevance, F.F., Takahashi, N., Karlinsey, J.E., Gnerer, J., Hirano, T., Samudrala, R., Aizawa, S., and Hughes, K.T. (2007). The mechanism of outer membrane penetration by the eubacterial flagellum and implications for spirochete evolution. Genes Dev. *21*, 2326–2335.

Evans, L.D., Stafford, G.P., Ahmed, S., Fraser, G.M., and Hughes, C. (2006). An escort mechanism for cycling of export chaperones during flagellum assembly. Proc. Natl. Acad. Sci. U.S.A. *103*, 17474–17479.

Fahner, K.A., Block, S.M., Krishnaswamy, S., Parkinson, J.P., and Berg, H.C. (1994) A mutant hook-associated protein (HAP3) facilitates torsionally induced transformations of the flagellar filament of *Escherichia coli*. J. Mol. Biol. *238*, 173–186.

Ferris, H.U., Furukawa, Y., Minamino, T., Kroetz, M.B., Kihara, M., Namba, K., and Macnab, R.M. (2005). FlhB regulates ordered export of flagellar components via autocleavage mechanism. J. Biol. Chem. *280*, 41236–41242.

Fujii, M., Shibata, S., and Aizawa, S.-I. (2008). Polar, peritrichous, and lateral flagella belong to three distinguishable flagellar families. J. Mol. Biol. *379*, 273–283.

Hirano, T., Shibata, S., Ohnishi, K., Tani, T., and Aizawa, S.-I (2005). N-terminal signal region of FliK is dispensable for length control of the flagellar hook. Mol. Microbiol. *56*, 346–360.

Iino, T., Suzuki, H., and Yamaguchi, S. (1972). Reconstitution of *Salmonella* flagella attached to cell bodies. Nat. New Biol. *237*, 238–240.

Kagawa, H., Morishita, H., and Enomoto, M. (1981). Reconstitution *in vitro* of flagellar filaments onto hook structures attached to bacterial cells. J. Mol. Biol. *153*, 467–470.

Karlinsey, J.E., Tanaka, S., Bettenworth, V., Yamaguchi, S., Boos, W., Aizawa, S.-I., and Hughes, K.T., (2000). Completion of the hook–basal body complex of the *Salmonella typhimurium* flagellum is coupled to FlgM secretion and *fliC* transcription Mol. Microbiol. *37*, 1220–1231.

Komoriya, K., Shibano, N., Higano, T., Azuma, N., Yamaguchi, S., and Aizawa, S.-I. (1999). Flagellar proteins and type III-exported virulence factors are the predominant proteins secreted into the culture media of *Salmonella typhimurium*. Mol. Microbiol. *34*, 767–779.

Kubori, T., Shimamoto, N., Yamaguchi, S., Namba, K., and Aizawa, S.-I. (1992). Morphological pathway of flagellar assembly in *Salmonella typhimurium*. J. Mol. Biol. *226*, 433–446.

Kutsukake, K., Okada, T., Yokoseki, T., and Iino, T. (1994). Sequence analysis of the *flgA* gene and its adjacent region in *Salmonella typhimurium*, and identification of another flagellar gene, *flgN*. Gene *143*, 49–54.

Macnab, R.M. (2004). Type III flagellar protein export and flagellar assembly. Biochim Biophys Acta *1694*, 207–217.

McMurry, J.L., Murphy, J.W., and Gonzalez-Pedrajo, B. (2006). The FliN-FliH interaction mediates localization of flagellar export ATPase FliI to the C ring complex. Biochemistry. *45*, 11790–11798.

Minamino, T., and Namba, K. (2008). Distinct roles of the FliI ATPase and proton motive force in bacterial flagellar protein export. Nature *451*, 485–488.

Nambu, T., Minamino, T., Macnab, R.M., and Kutsukake, K. (1999). Peptidoglycan-hydrolyzing activity of the FlgJ protein essential for flagellar rod formation in *Salmonella typhimurium*. J. Bacteriol. *181*, 1555–1561.

Pallen, M.J., and Matzke, N.J. (2006). From the origin of species to the origin of bacterial flagella. Nat. Rev. Microbiol. *4*, 784–790.

Paul, K., Erhardt, M., Hirano, T., Blair, D.F., Hughes, K.T. (2008). Energy source of flagellar type III secretion. Nature *451*, 489–492.

Samatey, F.A., Matsunami, H., Imada, K., Nagashima, S., Shaikh, T.R., Thomas, D.R., Chen, J.Z., DeRosier, D.J., Kitao, A., and Namba, K. (2004). Structure of the bacterial flagellar hook and implication for the molecular universal joint mechanism. Nature *431*, 1062–1068.

Shibata, S., Takahashi, N., Chevance, F.F.V., Karlinsey, J.E., Hughes, K.T., and Aizawa, S.-I. (2007). FliK regulates flagellar hook length as an internal ruler. Mol. Microbiol. *64*, 1404–1415.

Suzuki, T., Iino, T., Horiguchi, T., and Yamaguchi, S. (1978) Incomplete flagellar structures in nonflagellate mutants of *Salmonella typhimurium*. J. Bacteriol. *133*, 904–915.

Vogler, A., Homma, M., Irikura, V.M., and Macnab, R.M. (1991). *Salmonella typhimurium* mutants defective in flagellar filament regrowth and sequence similarity of FliI to F_0F_1, vacuolar, and archaebacterial ATPase subunits. J. Bacteriol. *173*, 3564–3572.

Williams, A.W., Yamaguchi, S., Togashi, F., Aizawa, S.-I., Kawagishi, I., and Macnab, R.M. (1996). Mutations in *fliK* and *flhB* affecting flagellar hook and filament assembly in *Salmonella typhimurium*. J. Bacteriol. *178*, 2960–2970.

The Coordination of Flagellar Gene Expression and the Flagellar Assembly Pathway

7

Jonathon Brown, Alexandra Faulds-Pain and Phillip Aldridge

Abstract

The assembly of a bacterial flagellum requires the coordinated synthesis of over 60 gene products. These genes encode structural and regulatory proteins. A key feature of all flagellar systems is the ability to sense the progression of assembly and the assimilation of this information to allow for the coordination of flagellar gene expression. Our understanding of the regulation of flagellar gene expression encompasses flagellar systems from all corners of the bacterial kingdom. All flagellar systems coordinate flagellar gene expression through a transcriptional hierarchy central to an integrated regulatory network of multiple regulatory components. These networks exhibit a number of conserved circuit architectures reflective of the strong conservation found within the structural components of the flagellum. However, a fascinating level of diversity is evident amongst flagellar systems in how assembly intermediates are sensed and how the perception of these intermediates modulates the activity of the associated transcriptional hierarchy. This chapter explores the conservation and divergence exhibited amongst flagellar system regulatory networks.

Introduction

To assemble one or more flagella per cell is a significant energy burden upon the bacterial cell. For flagellar-mediated motility there is also a need to either rapidly increase flagellar assembly or completely shut down assembly in response to environmental changes. Flagellar systems have thus evolved to regulate the process of assembly to maintain a balance between the need for movement and the energy cost to the cell.

The overall design of the regulatory network that controls flagellar gene expression underpinning flagellar assembly is highly conserved throughout the bacterial kingdom. The analysis of species-specific flagellar systems uncovers a fascinating level of diversity within the conserved design. Most notably there is significant variation in how flagellar systems integrate assembly signals into the network.

The majority of the literature on flagellar gene expression focuses on the response during the initiation of flagellar assembly: when cells pass through a non-flagellated to flagellated transition (Chilcott and Hughes, 2000; Aldridge and Hughes, 2002; McCarter, 2006). For flagellar systems such as the monoflagellated systems of *Vibrio* spp., *Pseudomonas* spp. and *Caulobacter crescentus* where only one flagellum per cell is built, this view is often sufficient (England and Gober, 2001; McCarter, 2001). However, for systems that produce more than one flagellum per cell, whether they are peritrichous or polar, the initiation of flagellar assembly is only part of the system. The network also requires a way to modulate flagellar gene expression during steady state assembly. To complicate matters, some bacteria produce two distinct types of flagella: a polar flagellum and lateral flagella (Stewart and McCarter, 2003; Pallen *et al.*, 2006; Kanbe *et al.*, 2007). These systems can be environmentally regulated where one system is shut down when the other is switched on. In contrast, some bacterial species produce two distinct flagella at the same time

(Kanbe *et al.*, 2007). In these cases the systems must be able to delineate signals from each other to prevent unwanted cross-talk.

Our understanding of coordinated flagellar gene expression and assembly is continually expanding to more and more flagellar systems from all corners of the bacterial kingdom. Through experimental analysis and genome studies, a number of common regulatory networks are becoming evident. The flagellar systems of *Escherichia coli* and *Salmonella enterica* serovar Typhimurium (hereafter described as *Salmonella*), *Vibrio* spp., *Pseudomonas* spp., *C. crescentus* and *Sinorhizobium meliloti* exemplify these prominent systems.

The assembly pathway and its checkpoints

A key feature of coordinated flagellar gene expression is the utilization of a series of intermediate assembly stages as checkpoints (Chilcott and Hughes, 2000) (Fig. 7.1). Flagella are subdivided into three substructures: the basal body, the hook and the long whip-like filament. The basal body consists of the MS-ring, C-ring and rod. The MS-ring assembles first in the inner membrane and acts as a foundation for the rest of the structure. The C-ring assembles onto the cytoplasmic interface of the MS-ring along with the associated Type 3 secretion (T3S) apparatus. After T3S apparatus assembly, the majority of structural subunits are then secreted through the central channel of the MS-ring for incorporation at the tip of the growing structure (Aldridge and Hughes, 2002).

The first non-cytoplasmic structure to be assembled is the rod that will span either the periplasmic space of Gram-negative bacteria or the Gram-positive cell wall (Pallen *et al.*, 2006). Initiation of rod assembly is used by *C. crescentus*, *Vibrio* spp. and *Pseudomonas* spp. as an intermediate assembly checkpoint that coordinates flagellar gene expression with assembly (McCarter, 2001; Muir and Gober, 2001; Dasgupta *et al.*, 2003) (Fig. 7.1). This checkpoint coincides with the activation of a complete T3S apparatus required

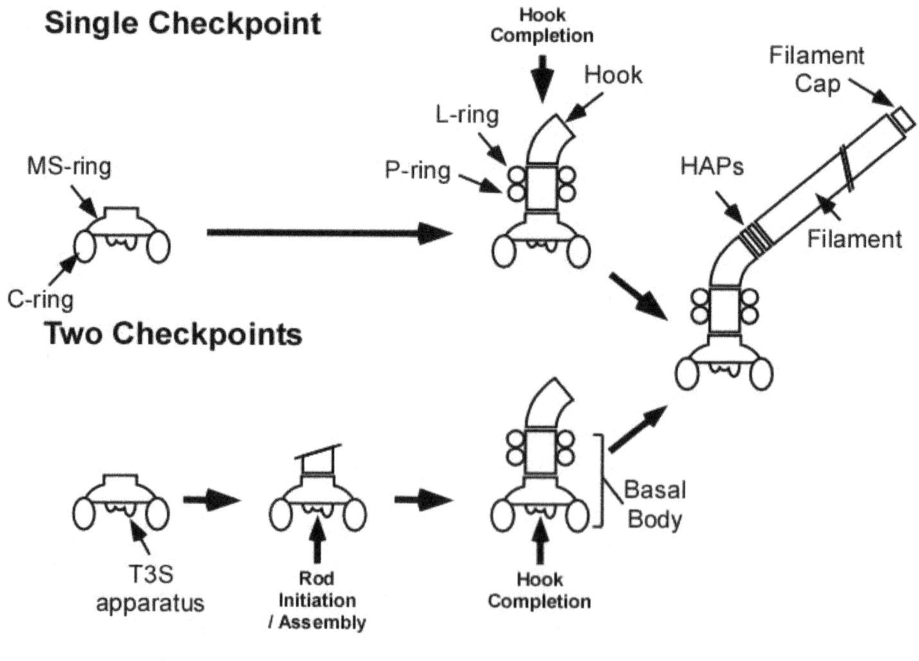

Figure 7.1 The assembly checkpoints utilized by flagellar systems to coordinate the assembly pathway and flagellar gene expression. Assembly starts with the assembly of the MS-ring, C-ring and T3S apparatus. Flagellar systems utilize either a single checkpoint, associated with hook completion, or utilize a second intermediate basal body checkpoint. This second checkpoint coincides with the initiation of rod assembly. After HBB completion, filament assembly will complete the flagellar structure. The individual structural components and substructure nomenclature are highlighted for reference purposes.

for the assembly of the rest of the flagellum. Completion of the rod breaches the cell surface concluding basal body assembly.

Following basal body completion the hook is assembled onto the protruding rod. Hook–basal body (HBB) completion is the second assembly checkpoint utilized to coordinate flagellar gene expression and assembly (Kutsukake *et al.*, 1990; Hughes *et al.*, 1993). This checkpoint plays a pivotal role in coordinating flagellar gene expression in all flagellar systems studied so far (Fig. 7.1). HBB completion is associated with a T3S apparatus substrate specificity switch from HBB-associated subunits to filament-associated subunits. The specificity switch is the result of a self-cleavage event of the integral T3S component FlhB (Williams *et al.*, 1996; Minamino and Macnab, 2000). The average hook length is 55 nm and the conformational changes in FlhB have been shown to be key in terminating hook assembly when its optimal length has been reached (Minamino *et al.*, 1999). A recent structural analysis of the FlhB homologue EscU has elucidated exactly how the cleavage event occurs (Zarivach *et al.*, 2008). Hook length is controlled by the regulation of FlhB cleavage by the protein FliK. Mutants of *fliK* form what have been termed polyhooks (extra long hooks) (Suzuki *et al.*, 1978). Three models have been proposed to explain how FliK regulates hook length. These models are the measuring cup model (Makishima *et al.*, 2001), the molecular ruler (Minamino and Pugsley, 2005; Shibata *et al.*, 2007) and the most recent model: the molecular clock model (Moriya *et al.*, 2006). These models and the regulation of FlhB cleavage have been recently reviewed in some detail (Minamino and Pugsley, 2005; Chevance and Hughes, 2008). Current evidence would suggest that the molecular ruler/clock models are a more accurate description of FliK regulation of FlhB cleavage. Although evidence still exists for the cup model, the spatial considerations of the space found within the C-ring 'cup' cannot account accurately for hook length (Chevance and Hughes, 2008).

After hook completion the hook-associated proteins (HAPs: FlgK and FlgL) assemble to form the hook-filament junction. Once the filament cap protein, FliD, caps the hook-filament junction, filament assembly begins. The filament subunit flagellin is incorporated into the growing filament under FliD (Yonekura *et al.*, 2000). An average filament of *Salmonella* is thought to contain up to 20,000 molecules of flagellin (Samatey *et al.*, 2001). Flagellar filament lengths have been measured in numerous organisms. For example, the average length of the *E. coli* filament is defined as 7 microns although lengths for *E. coli* and *Salmonella* filaments of between 10 to 15 microns have been reported (Yonekura *et al.*, 2000; Darnton *et al.*, 2007). In contrast, the filament of *C. crescentus* is 4.5 to 5 microns (Li and Tang, 2006). It is difficult to define when a bacterial cell determines filament assembly to be complete. No strict filament completion checkpoint has been characterized for any flagellar system to date. This may, however, reflect the nature of the system as it must be left in a state that can respond immediately to any filament damage due to external forces.

The HBB assembly checkpoints have led to flagellar genes being split into classes dependent upon where they come in the assembly pathway. For the single checkpoint systems these classes are known as early, middle and late genes (Chilcott and Hughes, 2000). For the two checkpoint systems, the genes are often grouped into Class I to IV although sometimes classes II and III are defined as IIA and IIB due to both groups being involved in HBB assembly (McCarter, 2006). However, these gene classes can sometimes be confusing, so to avoid confusion we will discuss the genes and checkpoints with respect to the flagellar structure.

The *Salmonella* paradigm

Members of the enteric γ-proteobacteria, especially *E. coli* and *Salmonella*, represent the best-studied flagellar systems. The mechanism by which these bacteria couple flagellar gene expression to assembly is accepted as the paradigm for all systems and is thus discussed in detail in many textbooks.

The flagellar system of *Salmonella* consists of three flagellar promoter classes organized into a transcriptional hierarchy. There is a single Class 1 promoter (P_{class1}) that drives the expression of the enteric master regulator $FlhD_4C_2$ (Arnosti, 1990; Kutsukake *et al.*, 1990). In conjunction with σ^{70}, $FlhD_4C_2$, activates Class 2 promoters

(P_{class2}) that drive HBB subunit gene expression and the regulators of the system (Fig. 7.2). Two of these regulators are the flagellar-specific sigma factor σ^{28} and its anti-sigma factor FlgM (Ohnishi *et al.*, 1990; Gillen and Hughes, 1991). The last promoter class, P_{class3}, strictly require σ^{28} for activation and drive the expression of the flagellins, chemotaxis machinery and the motor force generators. The P_{class3} promoters are activated by σ^{28} only once HBB completion has occurred. Prior to HBB completion σ^{28} activity is inhibited via a tight interaction with FlgM. The FlgM interaction is strong enough to remove σ^{28} from an RNA polymerase holoenzyme complex, thus keeping P_{class3} activity as low as possible prior to when it is really needed (Chadsey and Hughes, 2001). On HBB completion, FlgM is physically secreted from the cell through the growing flagellar structure as a result of the T3S apparatus substrate specificity switch (Hughes *et al.*, 1993) (Figs. 7.2 and 7.3). FlgM secretion leads to P_{class3} activation allowing for completion of the growing flagellum. Interestingly, σ^{28} itself plays an active role in facilitating FlgM secretion, almost as if σ^{28} is itself sensing HBB completion by delivering FlgM to the T3S apparatus (Aldridge *et al.*, 2006b) (Fig. 7.3).

The utilization of alternate sigma factors

Although the *Salmonella* system introduces flagellar gene regulation, it does not really reflect how the majority of bacterial species control flagellar gene expression. Furthermore, a number of regulatory components known to modulate the activity of the *Salmonella* hierarchy are restricted to systems that show extensive similarity to *Salmonella*. The most notable difference between the *Salmonella* hierarchy and other flagellar systems is the utilization of a second alternate sigma factor, σ^{54}, as well as σ^{28}. It is the σ^{54}-dependent flagellar systems that are responsive to two assembly checkpoints: rod assembly and HBB completion (Fig. 7.2) (Dingwall *et al.*, 1992; McCarter, 2001; Dasgupta *et al.*, 2003).

The α-proteobacterium, *C. crescentus,* senses these two HBB intermediates; however, the transcriptional hierarchy is driven by its housekeeping sigma factor, σ^{73}, and σ^{54} only (Dingwall *et al.*, 1992). Bioinformatic analysis of flagellar

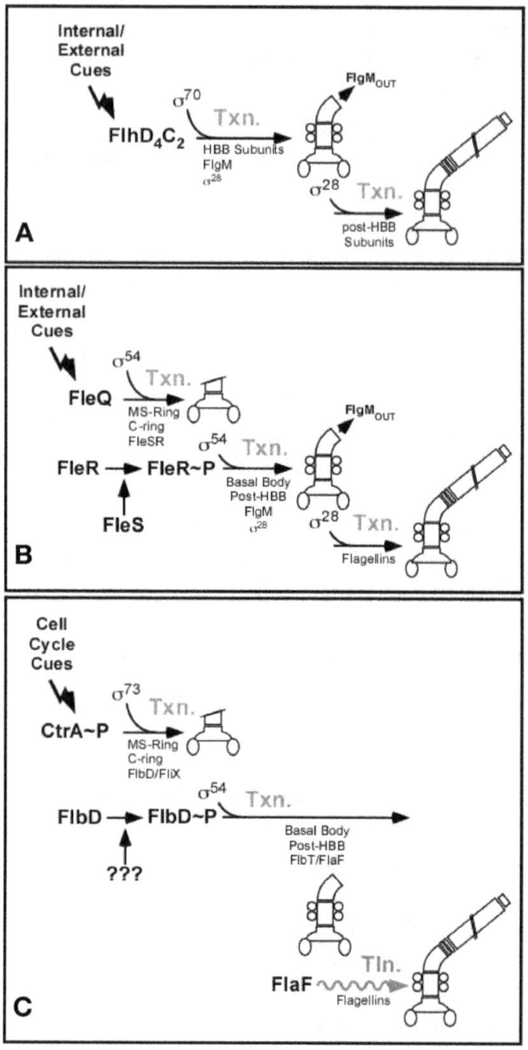

Figure 7.2 Comparison of the regulatory cascades that coordinate flagellar gene expression and assembly in **A** *E. coli/Salmonella,* **B** FleQ-dependent systems and **C** *C. crescentus*. For all three systems the regulatory components necessary to activate the cascade are shown. All systems possess a transcription hierarchy containing either two or three tiers below the master regulators. All systems utilize either two or three sigma factors but this is always a combination of (i) the housekeeping sigma factors, (ii) σ^{54}, (iii) and/or σ^{28}. In the *E. coli/Salmonella* and the FleQ-dependent systems, flagellar gene expression is coupled to HBB completion via FlgM secretion. The signals that activate FleR and FlbD at the rod-assembly checkpoint are not fully understood, although FleR is activated by its cognate sensor kinase FleS. The cascade of *C. crescentus* incorporates control of translation as well as transcription to achieve the third tier of coordinated assembly and flagellar gene expression: this system lacks the FlgM–σ^{28} regulatory circuit. Txn: Transcription; Tln: Translation

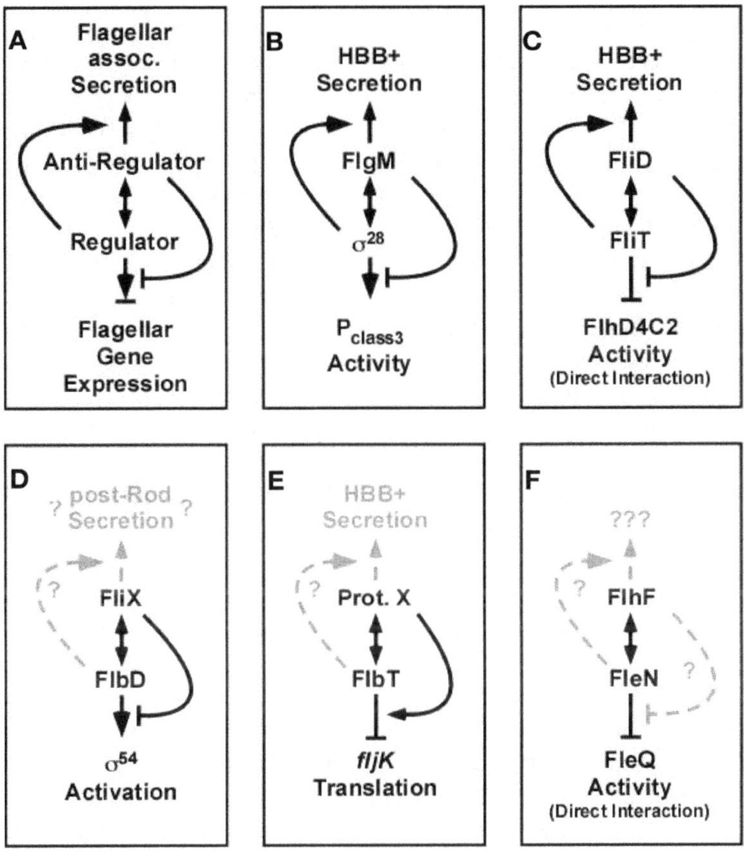

Figure 7.3 Common features of regulatory feedback loops associated with the coordination of flagellar gene expression and assembly. Circuit components shown in Black are known while grey components in D to F are proposed mechanisms based on our understanding of how flagellar systems achieve coordination. **A** The general principle as determined from the *Salmonella* flagellar system. **B** The regulatory feedback loop of FlgM and σ^{28}. **C** The FliD–FliT regulatory feedback loop. This regulatory circuit is identical to that of FlgM–σ^{28}. The same is true for FlgN and the HAPs. **D** The FliX–FlbD circuit. FliX is similar to FliD and FlgM in that it inhibits the activity of FlbD. Unlike the *Salmonella* system, the actual mechanism by which assembly is sensed remains elusive. A plausible scenario is proposed in which FliX is physically secreted during HBB assembly. **E** The FlbT–ProteinX regulatory feedback loop. In contrast to all other circuits, the regulatory effect of protein X is positive. When compared to the other regulatory circuits it is plausible that protein X is a flagellar-associated secretion substrate, secreted on HBB completion. **F** A comparison of the interaction of FlhF and FlhG (FleN) to the other regulatory circuits. In this circuit what FlhF is sensing and how these two proteins coordinate flagellar placement and numbers still remains to be defined.

systems suggests that the absence of σ^{28} is common amongst the α-proteobacteria. When σ^{28} and FlgM are present in an α-proteobacteria genome, there is strong evidence to suggest that horizontal gene transfer has resulted in the acquisition of a second flagellar system. *Rhodobacter sphaeroides* and *Bradyrhizobium japonicum* are two such examples (Kanbe *et al.*, 2007; Poggio *et al.*, 2007). Outside of the α-proteobacteria, all flagellar systems characterized possess σ^{28} and FlgM. Furthermore, all σ^{28}-dependent systems that have been experimentally investigated couple σ^{28} activation to HBB completion via FlgM secretion.

Both σ^{54} and σ^{28} possess contrasting modes of transcriptional activation and regulation. σ^{28} belongs to the σ^{70} family of sigma factors while σ^{54} is a member of its own sigma factor family (Lonetto *et al.*, 1992; Buck *et al.*, 2000). The σ^{70} family all recognize classic but unique −10 and −35 promoter sequences allowing them to recruit RNA polymerase (Lonetto *et al.*, 1992). σ^{28} is one of the smallest members of the σ^{70} family. The activity of a wide range of σ^{70} family members is regulated by anti-sigma factors such as FlgM. For example, the control of sporulation in *Bacillus subtilis* and stress in *E. coli* are both regulated by sigma factor–anti-sigma factor interactions (Hughes and Mathee, 1998).

In contrast to the σ^{70} family, σ^{54} is unable to complete open complex formation without an accessory protein (Buck *et al.*, 2000). The accessory protein or enhancer binding protein (EBP) interacts with a specific DNA sequence upstream of the σ^{54} promoter and, in turn, interacts with the $E\sigma^{54}$ holoenzyme to initiate transcription (Morett and Segovia, 1993; Wigneshweraraj *et al.*, 2008). The DNA binding sites for EBPs are often 70–150 bp upstream of σ^{54} promoters (Wigneshweraraj *et al.*, 2008).

Instead of possessing classic −10 and −35 boxes, σ^{54} utilizes sequences much closer together: −12 and −24 sites (Hunt and Magasanik, 1985). The σ^{54}-EBPs promote open complex formation and transcription initiation through ATP hydrolysis, driven by a conserved ATPase domain possessed by all EBPs (Buck *et al.*, 2000). σ^{54}-dependent promoters are therefore under much tighter control than σ^{28}-dependent promoters. For example if the EBP is not available, or in an inactive state, the σ^{54} promoter will be silent while for σ^{28} its basal activity is dependent on the intracellular FlgM concentration.

One of the first EBPs shown to activate σ^{54} was NtrC, a two-component system response regulator (Morett and Buck, 1989). Two-component systems consist of a sensor kinase and a cognate response regulator. The sensor kinase will sense either an external or internal signal through an input domain and autophosphorylate itself at a conserved histidine residue (Hoch, 2000). The response regulators are activated via transfer of the phosphate groups from their partner sensor kinase to a conserved aspartate in their N-terminal receiver domain. Many response regulators comprise two domains: the receiver domain and an output domain most often containing a helix–turn–helix DNA binding motif (Hoch, 2000). However, some response regulators, like the NtrC family of σ^{54}-EBPs, possess multiple domains: a receiver domain, a central ATPase domain and a C-terminal DNA binding domain (Galperin *et al.*, 2001). Apart from their integral role in the activation of σ^{54}-dependent promoters, two-component regulatory systems play a key role in controlling the activity of non-enteric flagellar associated transcriptional hierarchies.

The ability of σ^{54} to control flagellar gene expression via signal transduction cascades suggests a mechanism that allows a much faster response to environmental changes or flagellar assembly. Thus, σ^{54}-dependent flagellar hierarchies have plausibly evolved as a result of the species-specific requirement for a tighter level of regulation required during activation but also during shut down of flagellar systems.

The master regulators

All flagellar systems possess a master regulator at the top of their transcriptional hierarchy. These transcriptional activators dictate when bacteria will commit to flagellar assembly. Flagellar systems are extremely robust, in that once committed to flagellar assembly, there is very little that can perturb completion of at least the first flagellar structure. When a decision is made to alter the level of flagellar assembly it is invariably fed into the system by the master regulators. As a result, the list of environmental factors and global regulators that influence the expression and activity of the master regulators is immense (Soutourina and Bertin, 2003). For example, in *Salmonella* six transcriptional start sites have been mapped for the master *flhDC* operon, exemplifying the complexity associated with *flhDC* expression (Kutsukake, 1997; McCarter, 2006). McCarter (2006) has proposed that the evident complexity of regulating the expression of flagellar system master regulators is, in part, a result of the energy burden flagellar assembly exerts on the cell. It may also reflect the significant impact flagella have upon the bacterial life cycle.

The $FlhD_4C_2$ hetero-complex, when first described, was an unusual transcriptional activator as most bacterial transcription factors are single proteins that act as monomers, dimers or oligomers (Wang *et al.*, 2006). $FlhD_4C_2$ belongs to the class I transcription factors that activate transcription via a direct interaction with the alpha-subunit of RNA polymerase (Liu *et al.*, 1995). Both FlhC and FlhD can bind DNA. However, FlhD can increase the affinity of FlhC for DNA ten-fold (Claret and Hughes, 2000).

The $FlhD_4C_2$ DNA binding site is always adjacent to the −35 box of P_{class2} promoters. For several divergent P_{class2} promoters, the entire intergenic region is the $FlhD_4C_2$ binding site (Stafford *et al.*, 2005). This is in contrast to σ^{54}-EBPs that activate σ^{54} from a binding site separated from the promoter. The $FlhD_4C_2$–DNA complex is extremely stable and thus once bound to a P_{class2} promoter can potentially promote transcription for a prolonged period (Stafford *et al.*, 2005). The affinity of $FlhD_4C_2$ to all its binding sites is not constant. Consistent with this variation is the observation that HBB subunit gene expression is organized into a subhierarchy (Kalir *et al.*, 2001). Apart from responding to external cues, $FlhD_4C_2$ negatively autoregulates its own expression thus directly modulating its own levels and the basal activity of the dependent transcriptional hierarchy (Kutsukake, 1997).

In contrast to $FlhD_4C_2$, other flagellar systems utilize a diverse range of transcriptional activators as master regulators. These include σ^{54}-EBPs, response regulators and LuxR family members (Aldridge and Hughes, 2002). Alternative master regulators include FleQ from *Pseudomonas* spp. (also known as FlrA and FlaK in *Vibrio* spp. flagellar systems) (Arora *et al.*, 1997), CtrA from *C. crescentus* (Quon *et al.*, 1996) and VisNR from *S. meliloti* (Sourjik *et al.*, 2000) (Fig. 7.2).

The FleQ family of master regulators belong to the NtrC subfamily of σ^{54}-EBP transcriptional activators. The FleQ-family is different from most EBPs, as the phosphorylated aspartate of the receiver domain is not conserved (Jyot *et al.*, 2002). The expression of *fleQ* in the opportunistic pathogen *Pseudomonas aeruginosa* has been shown to respond to Vfr and the AlgT regulon in the lungs of cystic fibrosis (CF) patients (Dasgupta *et al.*, 2002; Tart *et al.*, 2005). Isolates of *P. aeruginosa* from CF patients are commonly mucoid and non-motile while environmental isolates are non-mucoid and motile (Tart *et al.*, 2005). The alternate sigma factor AlgT is responsible for the expression of AmrZ, a transcriptional regulator involved in this phenotypic conversion/switch. AmrZ negatively regulates *fleQ* expression via an interaction with the *fleQ* promoter (Tart *et al.*, 2006). A similar switch is seen between fimbriae and flagellar biogenesis in *Salmonella* (Clegg and Hughes, 2002). This cross-talk requires both $FlhD_4C_2$ and the fimbriae regulatory protein FimZ. When fimbriae are expressed, *flhDC* expression is repressed by FimZ. In contrast, when cells are committed to flagellar assembly, $FlhD_4C_2$ negatively regulates *fim* gene expression (Clegg and Hughes, 2002). This type of regulation is a common theme associated with the regulation of flagellar systems by their master regulators. Certain growth conditions will reciprocally regulate flagellar assembly and a second cellular process, for example, exopolysaccharide (EPS) production or fimbriae biogenesis. This switch is commonly associated with the transition from a motile planktonic life style to a sessile surface-associated life style.

The majority of bacterial species use developmental switches to exploit particular habitats during their life cycle. *C. crescentus* on the other hand goes through such a transition every time it divides. *C. crescentus* divides asymmetrically producing a sessile stalked (ST) cell and a motile swarmer (SW) cell (Jenal, 2000; Laub *et al.*, 2000) (Fig. 7.4). The ST cell is similar to a eukaryotic stem cell in that it can immediately initiate chromosome replication and cell division. In contrast, the SW cell possesses a single polar flagellum but is unable to replicate its chromosome until it passes through a SW to ST cell transition (Jenal, 2000). Assembly of the polar flagellum in the pre-divisional ST cell is tightly integrated into the *C. crescentus* cell cycle.

Transcription profiling of *C. crescentus* genes showed that 19% underwent varying levels of transcription throughout the cell cycle. The master regulator, CtrA, controls the expression of approximately 25% of these cell cycle regulated genes (Laub *et al.*, 2000). CtrA is a member of the response regulator super-family and is essential in *C. crescentus*. Phosphorylated CtrA is key to the coordination of DNA replication, flagellar assembly and cell division (Domian *et al.*, 1997). The activity of CtrA is regulated via its transcription rate, activation by phosphorylation and degradation by the protease ClpXP (Jenal and Fuchs, 1998). The timed synthesis and destruction of CtrA means that the protein is present in SW cells and the early pre-divisional cell but is absent from ST cells and during the initiation of DNA replication (Domian *et al.*, 1997). CtrA activates σ^{73} transcription of the *C. crescentus* genes that

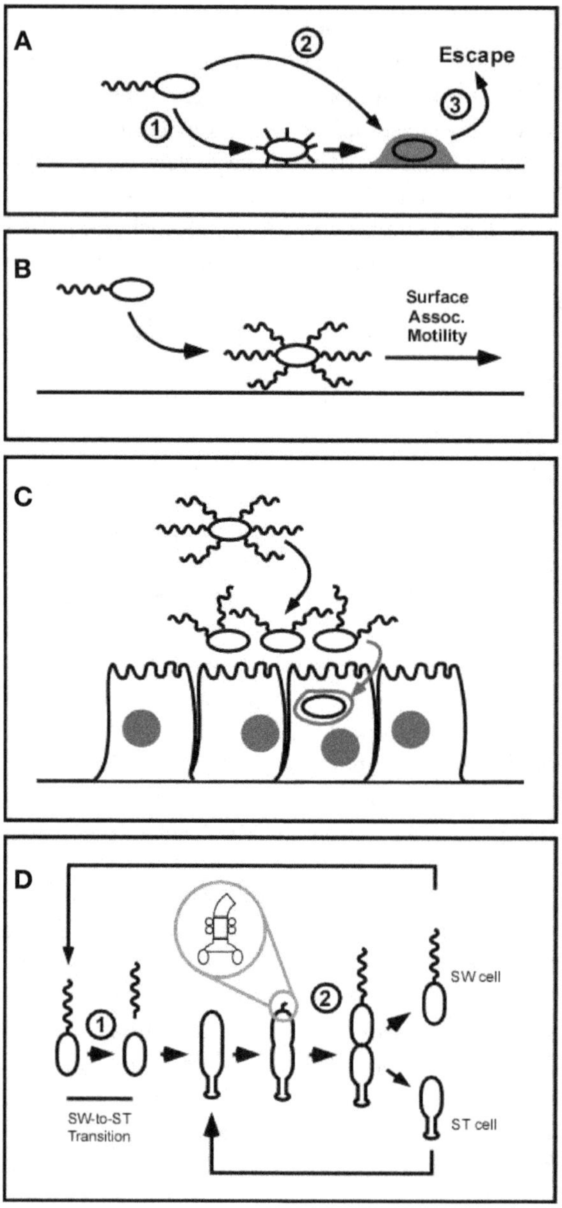

Figure 7.4 Examples of developmental switches associated with a transition from a flagellar-mediated motile lifestyle to a sessile non-motile lifestyle. Notably for many of these developmental events, the secondary signal molecular c-di-GMP is primarily involved in the down regulation of the planktonic flagellar-mediated life style. **A** Seen often associated with biofilm development, planktonic bacteria will associate with a surface either via a flagellar to fimbriae/pili switch (1) or directly (2). Concomitant to this switch is the upregulation of EPS production. Critical for all life cycles is the inverse to occur (3) allowing bacteria to switch back to a planktonic life style to 'escape' in search of other habitats. **B** The switch between flagellar-mediated motility in liquid environments to surface-associated movement. In *V. parahaemolyticus,* this switch is once more influenced by c-di-GMP controlling the ability of the cells to swarm on surfaces. **C** A schematic of host–bacterial interactions. Flagella play a key role in the early stages of host interactions, be it during symbiosis, or pathogen association with both plants and animals. For the symbiont *S. meliloti*, flagellar assembly is downregulated through its master regulators during the establishment of a new symbiotic interaction via quorum sensing. **D** The cell cycle of *C. crescentus*. During the cell cycle c-di-GMP plays a role in regulating flagellar ejection during the SW-to-ST transition (1) while also influencing flagellar assembly and rotation later on during growth of the pre-divisional cell (2).

encode the MS-ring, the C-ring and the flagellar specific regulatory proteins, such as the σ^{54}-EBP FlbD (England and Gober, 2001).

Cell cycle control of flagellar assembly is not unique to *C. crescentus*. There is sufficient evidence to suggest that *flhDC* expression and activity responds to cell cycle cues (Pruss and Matsumura, 1996). A protein that shows significant homology to MinD, involved in regulating cell division, negatively regulates FleQ family members (Dasgupta *et al.*, 2000) (see later). Furthermore, for all systems that produce a single flagellum per cell, a given degree of input from cell cycle progression is essential.

In contrast to *C. crescentus,* the α-proteobacterium *S. meliloti* produces approximately four peritrichous flagella per cell. The *S. meliloti* flagellar system is under the control of two LuxR family members, VisNR, and the OmpR transcriptional activator family member Rem (Sourjik *et al.*, 2000; Rotter *et al.*, 2006). LuxR transcriptional activators are activated by a ligand binding their N-terminal domains (Fuqua and Greenberg, 2002). In contrast, the OmpR family are response regulators with a distinctive winged helix–turn–helix C-terminal DNA binding domain: CtrA also belongs to this family (Kenney, 2002). LuxR family members possess as a ligand a homoserine lactone derivative produced by their associated quorum sensing system (Fuqua and Greenberg, 2002). The activating ligand(s) for VisN or VisR remain elusive. Sourjik *et al* (2000) showed that the ligand binding sites of VisN and VisR are not conserved suggesting that they bind alternative ligands or cooperate to bind one molecule as a dimer. Both VisN and VisR are required for activation of the flagellar system suggesting they act as a hetero complex, leading to a system comparable to $FlhD_4C_2$ (Sourjik *et al.*, 2000).

The expression of *visNR* is negatively regulated by the quorum sensing system ExpR/Sin and the ExoS/ChvI two component system (Clegg and Hughes, 2002; Hoang *et al.*, 2008). Interestingly, ExpR/Sin regulation of VisNR is somewhat similar to that of AlgT and FimZ in *Pseudomonas* spp. and *Salmonella* (Clegg and Hughes, 2002; Tart *et al.*, 2006). ExpR/Sin regulates EPS synthesis in response to high population density essential for establishment of a successful interaction with a host plant. This is a period of the *S. meliloti* life cycle when flagellar-mediated motility is not advantageous (Fig. 7.4C). Therefore, once more, we observe a developmental switch that controls flagellar assembly in response to the transition to an alternative life phase of a bacterium.

Rem is downstream of VisNR in the cascade and has been shown to be the principle activator of flagellar gene transcription (Rotter *et al.*, 2006). Like the FleQ family members, Rem although being a response regulator seems to function in the flagellar system independently of phosphorylation. *S. meliloti* only swims during the exponential phase of growth and the expression of *rem* reflects this. Little is known about what sigma factors are utilized in *S. meliloti* for the hierarchy, although the OmpR-family are known to activate transcription from σ^{70} dependent promoters. *S. meliloti* belongs to the α-proteobacteria that do not encode obvious σ^{28} or FlgM homologues. Importantly, the proteins that couple flagellar gene expression to HBB completion in *C. crescentus* are conserved in *S. meliloti* (Llewellyn *et al.*, 2005). This suggests that, potentially, this flagellar system is exclusively transcribed by σ^{70} and the cascade is achieved independent of the use of alternative sigma factors. However, this still needs to be experimentally proven.

The regulators of flagellar assembly

A significant number of flagellar-associated accessory regulators modulate the activities of their associated transcriptional hierarchies and subunit production rates. These accessory regulators directly or indirectly regulate transcription and translation of flagellar genes (Table 7.1). The main role of these regulators is to sense and feed the status of flagellar assembly into the activity of the transcriptional hierarchy (Aldridge and Hughes, 2002). On first glance the majority of these regulators, apart from homologues in related flagellar systems, show very little similarity to each other. When stripped down to simple regulatory circuits what becomes apparent, however, is that the circuit design involving these regulators is conserved across all flagellar systems (Aldridge and Hughes, 2002) (Fig. 7.3).

The study of the design and dynamics of regulatory networks is becoming an accepted field

Table 7.1 Comparison of the properties of the regulatory components associated with flagellar systems (for references see text)

Bacterial species	Regulator	Other names	Mode of regulation[a]	Target[b]	T3S chaperone activity
Master regulators					
E. coli/Salmonella	$FlhD_4C_2$		±	DNA	
P. aeruginosa	FleQ	FlaK(Vp)	+	DNA	
		FlrA(Vc)			
C. crescentus	CtrA		+	DNA	
S. meliloti	VisNR		+	DNA	
	Rem			DNA	
Accessory regulators					
Many	σ^{28} (FliA)		+	DNA	YES
	FlgM		–	σ^{28}	
E. coli/Salmonella	FliT		–	$FlhD_4C_2$	YES
	FlgN		+	?	YES
	FliZ		+	?	
C. crescentus	FlbD		±	DNA	?
	FliX		–	FlbD	
	FlbT		–	mRNA	?
	FlaF		+	mRNA	
P. aeruginosa	FleS	FlaL (Vp)	+	FleR	?
		FlrB (Vc)			
	FleR	FlaM (Vp)	+	DNA	
		FlrC (Vc)			
	FlhF		+	FlhG	
	FlhG	FleN (Pa)	–	FleQ	
				FlhF	
R. sphaeroides	FleT	FlgR (Hp)	+	FleQ	

Hp, *H. pylori*; Vp, *V. parahaemolyticus*; Vc, *V. cholerae*; Pa, *P. aeruginosa*

[a]The majority of flagellar-associated transcriptional factors have to date only been shown to exert distinct positive or negative regulatory inputs into their associated regulatory networks. However a few, such as $FlhD_4C_2$ and FlbD, have been shown to act as both activators and repressors of transcription.

[b]Proteins that have been shown to interact with DNA or RNA have their targets listed as either DNA or mRNA. Proteins that regulate flagellar gene expression via protein–protein interactions have their interaction partner listed.

of research integrating experimental biology and mathematical modelling of biological systems. This research is changing our perception of how regulatory networks function. Alon and co-workers have demonstrated that regulatory networks are built from a small number of network motifs (Milo *et al.*, 2002; Shen-Orr *et al.*, 2002; Alon, 2007). These can be simple regulatory pathways where X regulates Y, negative auto-regulation, positive auto-regulation, feedback loops and feed-forward loops. For example, the enteric flagellar system possesses a feed-forward loop that controls expression of a subset of flagellar genes. Feed-forward loops (FFL) comprise two transcription factors (in this case $FlhD_4C_2$ and σ^{28}) where one transcription factor, σ^{28}, is dependent on the first, $FlhD_4C_2$, for expression (Kalir *et al.*, 2005). Both factors then drive expression of an overlapping set of promoters. Thus, the activity of the hierarchy is dictated by the feed-forward loop controlling the level of input of $FlhD_4C_2$ and σ^{28}. In contrast, integrated feedback loops assimilate signals obtained from the status of assembly, altering the activity of the FFL accordingly to achieve optimal flagellar assembly rates (Kalir *et al.*, 2005).

For *Salmonella*, the flagellar assembly status is processed by the interaction of a subset of secretion substrates with members of the T3S-chaperone family. The primary role of T3S-chaperones is to increase the efficacy of secretion for their respective secretion substrates (Parsot *et al.*, 2003). T3S-chaperones are also thought to protect their secretion substrate partners from degradation and premature interactions (Bennett and Hughes, 2000). In *Salmonella*, three of the four known T3S-chaperones have a secondary role as regulators of flagellar gene expression. As the T3S-chaperones all interact with filament associated secretion substrates, the role of this group of proteins, as regulators, is to modulate flagellar gene expression after HBB completion (Aldridge and Hughes, 2002).

T3S chaperones are also found amongst non-enteric flagellar systems. The most prominent T3S-chaperone found in non-enteric systems is similar to FliS, which interacts with flagellin, the filament subunit (Ozin *et al.*, 2003). In *V. parahaemolyticus*, it is known as FlaJ (Stewart and McCarter, 1996) while a recent systems biology approach identified a putative T3S-chaperone that interacts with flagellins from *Bacillus subtilis* (Titz *et al.*, 2006). Interestingly, a regulatory role for FliS has yet to be described.

For non-enteric flagellar systems, the regulatory circuits that coordinate gene expression with assembly and the mode of action of many of the accessory regulators are known. Evidence is available to suggest how these regulatory circuits sense the assembly pathway but their mode of detection is still unclear. A property of all T3S-chaperones is that they exhibit very little similarity to each other. It is plausible that a number of the accessory regulators to be described interact directly with flagellar subunits prior to secretion. Alternatively, these regulators may be under the regulation of T3S-chaperones via secondary protein–protein interactions.

The accessory proteins of *Salmonella*

Other than the obvious role of FlgM and σ^{28}, there are three key regulators with respect to flagellar gene expression in the *Salmonella* system: FlgN, FliT and FliZ. FlgN and FliT are both members of the T3S-chaperone family, along with σ^{28} and FliS (Fraser *et al.*, 1999; Bennett and Hughes, 2000; Auvray *et al.*, 2001). FliZ is not known to interact with any secretion substrate but it plays an important role in the regulation of $FlhD_4C_2$ activity (Kutsukake *et al.*, 1999).

FlgN

FlgN is a 16.5 kDa protein originally described as the T3S-chaperone for the HAPs (FlgK and FlgL) (Fraser *et al.*, 1999). Mutants of *flgN* still assemble flagella but their numbers are significantly reduced from six to eight per cell to one or two (Kutsukake *et al.*, 1994). If the expression of the HAPs is increased the Δ*flgN* phenotype is suppressed showing that FlgN is not essential for flagellar assembly (Aldridge *et al.*, 2003). FlgN is responsible for positively regulating *flgM* translation from the *flgMN* P_{class3} transcript (Karlinsey *et al.*, 2000). How FlgN regulates translation is still not understood. Similar to the σ^{28}–FlgM interaction, the HAPs act as anti-FlgN regulators; preventing positive regulation of *flgM* translation until the hook-filament junction has been initiated (Aldridge *et al.*, 2003). This model suggests that the interaction of FlgN and the HAPs is a method of sensing the assembly status after HBB completion. The ability of FlgN to assess assembly through a drop in HAP concentration results in the modulation of σ^{28} activity by counteracting the depletion of FlgM levels. FlgN has recently been identified in *Xanthomonas axonopodis* and shown to interact with at least FlgK (Khater *et al.*, 2007). Bioinformatic studies have suggested that

FlgN is also present in *Campylobacter jejuni* (Pallen *et al.*, 2005). Whether FlgN also regulates *flgM* translation in these systems has still to be shown, although the need to modulate σ^{28} activity during filament assembly is still necessary.

FliT

FliT is the second T3S-chaperone shown to regulate flagellar gene expression in the *Salmonella* system. A 14 kDa protein, FliT interacts with the filament cap FliD (Fraser *et al.*, 1999). FliT is not essential for flagellar assembly as FliT mutants are still fully motile (Yokoseki *et al.*, 1995). Only five subunits of FliD are required per flagellum, a level plausibly below a threshold that requires FliT-assisted secretion (Yonekura *et al.*, 2000). However, the motile phenotype of Δ*fliT* mutants may also be due to its role in negatively regulating $FlhD_4C_2$ activity; Δ*fliT* mutants suppress the Δ*flgN* phenotype (Aldridge *et al.*, 2003). Kutsukake *et al* (1999) showed that FliT inhibits P_{class2} promoter activity independent of σ^{28}. The absence of FliT also leads to a reciprocal increase in P_{class3} activity, presumed to be through increasing σ^{28} levels through its P_{class2} promoter (Kutsukake *et al.*, 1999). Recently, Yamamoto and Kutsukake (2006) have shown that FliT interacts with $FlhD_4C_2$. In doing so, $FlhD_4C_2$ is inhibited from binding to P_{class2} promoters thus reducing their transcription rates, defining FliT as an anti-$FlhD_4C_2$ factor. Furthermore, the regulatory activity of FliT is dependent on the FliD secretion rate. Thus, FliD can be seen as an anti-FliT factor or, more correctly, an anti-anti- $FlhD_4C_2$ factor. Thus, FliT is central to a regulatory feedback loop that couples P_{class2} activity to the assembly status of the growing flagella, post-HBB completion. Interestingly, FlgN, FliT and σ^{28} are all central components of assembly dependent feedback loops of identical architecture (Fig. 7.3).

FliZ

FliZ is a small protein (21.7 kDa) encoded within the *fliAZY* operon; *fliA* encodes σ^{28} (Mytelka and Chamberlin, 1996). FliZ shares very weak homology to phage integrases and has no demonstrated interaction partners (Saini *et al.*, 2008). Its expression is sensitive to secretion, as the *fliAZY* operon is positively autoregulated by σ^{28} itself through the *fliAZY* σ^{28}-dependent promoter and, subsequently, FlgM secretion rates (Saini *et al.*, 2008). Kutsukake *et al* (1999) first demonstrated that FliZ positively regulates P_{class2} activity. Expression of *fliZ* from a multicopy plasmid enhanced transcription of P_{class2} promoters independent of σ^{28}. FliZ also effects expression of the virulence associated SPI-1 T3S system of *Salmonella* (Lucas et al., 2000; Iyoda *et al.*, 2001). This again demonstrates the ability of bacteria to evolve interacting regulatory networks that influence their ability to adapt to alternative life styles. Saini *et al* (2008) recently discovered that FliZ positively regulates the concentration of $FlhD_4C_2$ post-translationally. Their work also showed that the regulation of $FlhD_4C_2$ by FliT and FliZ is independent of each other. This suggests FliZ enhances P_{class2} activity during the early stages of flagellar assembly. In contrast, at a later stage of assembly, when multiple flagella are present, the activity of FliT will downregulate the $FlhD_4C_2$ activity to avoid an unwanted energy burden on the cell (Saini *et al.*, 2008).

In contrast to the results of Saini *et al* (2008), Lanois *et al* (2008) have shown that FliZ, from the insect pathogen *Xenorhabdus nematophila*, acts by increasing the efficacy of $FlhD_4C_2$ binding to DNA. Similar to *Salmonella*, FliZ has been shown to be a key regulator coupling motility and virulence in this pathogen (Lanois *et al.*, 2008). Even though *Xenorhabdus* spp. belongs to the enterics, the regulation of the flagellar hierarchy is somewhat different. While most enterics possess a *fliAZY* operon, *X. nematophila* is more like *Vibrio* and *Pseudomonas* spp. that are annotated to have a *fliAZ* operon. The *X. nematophila* FliZ shares 56% similarity with *Salmonella* FliZ suggesting that in *Salmonella* FliZ is also acting through a DNA interaction of an as yet unknown target (Saini *et al.*, 2008). This target could also explain the cross network communication with the SPI-1 system that requires FliZ. Potential candidates would include global regulators or proteases that target system specific regulators such as $FlhD_4C_2$.

The regulators of the *C. crescentus* flagellar system

The flagellar system of *C. crescentus* is very different than that of *Salmonella* and the FleQ dependent systems. *C. crescentus* does not possess either *fliA* or *flgM* in its genome. In its place is an elegant regulatory network that couples flagellin gene expression to HBB completion. Two key proteins, FlbD and FliX, regulate the activity of σ^{54}-dependent flagellar-specific promoters in *C. crescentus* (Muir and Gober, 2001) (Fig. 7.2). FlbD is the σ^{54}-EBP and drives expression of the majority of subunits and regulatory proteins needed after rod assembly, while FliX regulates FlbD activity (Muir and Gober, 2004). Two other proteins, FlbT and FlaF, in response to HBB completion regulate flagellin expression, negatively and positively, respectively (Mangan *et al.*, 1999; Llewellyn *et al.*, 2005). FlbT and FlaF are conserved in the flagellar system of *S. meliloti* and other α-proteobacterial species suggesting that their mode of regulation overcomes the lack of *fliA* and *flgM* from genomes of α-proteobacterial species.

FlbD

FlbD is the central σ^{54}-EBP of the *C. crescentus* system. Unlike FleQ, FlbD belongs to a group of flagellar-associated σ^{54}-EBPs that do respond to phosphorylation. Mutants of *flbD* are filamentous and produce no flagella (Muir *et al.*, 2005). The filamentous phenotype is not exclusive to Δ*flbD* strains; the majority of pre-rod flagellar mutants exhibit a filamentous phenotype (Dingwall *et al.*, 1992). The initiation of flagellar assembly is a cell cycle checkpoint itself; exemplified by the role CtrA plays in controlling flagellar gene expression at this early stage. The cell cycle associated flagellar checkpoint coincides with FlbD activation of flagellar specific σ^{54} promoters during rod assembly (Benson *et al.*, 1994). After passing the critical stages of assembly necessary for the cell cycle checkpoint, i.e. rod assembly, the assembly pathway is less dependent upon cell cycle progression. As a result the majority of post-rod flagellar subunit mutants exhibit normal cell morphology.

Gober and coworkers originally proposed that the protein FlbE was the FlbD sensor kinase (Wingrove and Gober, 1996). However, mutagenesis of the conserved residues in FlbE did not alter its ability to regulate FlbD activity (Muir and Gober, 2001). The *flbE* gene is located between *fliG* and *fliN*, and through comparisons with other flagellar systems synteny would suggest that *flbE* is in fact *fliH*. The *flbE* gene was first described prior to the genome revolution when the number of flagellar systems that had been sequenced was extremely restricted (Benson *et al.*, 1994). Up-to-date bioinformatic analysis of FlbE confirms similarity to the flagella export protein FliH (Pallen *et al.*, 2006). FliH is a component of the T3S apparatus. Importantly the similarity of FlbE to FliH does suggest a link to how FlbD is activated in response to flagellar assembly. Pre-rod flagellar assembly does not require T3 secretion as this stage is involved in the assembly of the MS-ring, C-ring and the T3S apparatus itself. However, rod assembly does require T3S apparatus activity, as the rod proteins are the first to be secreted through the assembled secretion apparatus. Interestingly, FlbD negatively autoregulates its own expression by binding the *fliF* operon promoter: *flbD* is the last gene in the *fliF* operon (Wingrove and Gober, 1996). This suggests that in order to produce a single flagellum *C. crescentus* has evolved a mechanism to reduce pre-rod gene expression once the rod assembly checkpoint has been reached.

FliX

FliX is a small protein identified through the characterization of mutants that would bypass the rod assembly checkpoint with respect to flagellar gene expression. These mutants were first termed *bfa* (bypass flagellar assembly) (Mangan *et al.*, 1995; Muir *et al.*, 2001). FliX regulates the activity of FlbD via a direct interaction with the FlbD receiver domain (Muir and Gober, 2004). Thus, like FlgM and the T3S-chaperones of the *Salmonella* system, FliX can be defined as an anti-FlbD factor (Fig. 7.3). In contrast to FlgM and the partner proteins for FlgN and FliT, FliX is not secreted from the cell. However, the time point for FliX secretion would coincide with rod assembly where the structure is open to the periplasm and not the external environment. In certain *Salmonella* flagellar mutants FlgM is secreted prematurely into the periplasm and degraded (Aldridge *et al.*, 2006a). Therefore, it is plausible that FliX is secreted into the periplasm where it would be degraded and thus undetectable (Fig. 7.3). This model is consistent with initiation of rod assembly coinciding with the activation of the secretion apparatus: the time at which FlbD activation is necessary. A similar mechanism is used during hook-length control, as the hook-length determinant FliK is also secreted during HBB assembly (Minamino *et al.*, 1999). Even though the rod-assembly checkpoint is prior to HBB completion it is possible that the circuit architecture is similar to that of the T3S-chaperone circuits in *Salmonella* (Fig. 7.3).

FlbT

In *C. crescentus*, the genes encoding the distal rod components, hook, HAPs and flagellins are all transcribed from σ^{54}-dependent promoters after the rod checkpoint has been passed (Wu *et al.*, 1995). However, the flagellins are not translated until the HBB is complete (Mangan *et al.*, 1999). The mRNA binding protein FlbT achieves coupling of the translation of flagellin genes to HBB completion (Mangan *et al.*, 1999). FlbT has been shown to negatively regulate the translation of at least the flagellin gene *fljK* prior to HBB completion: the *C. crescentus* filament is assembled from six flagellins FljJ–FljO (Ely *et al.*, 2000). FlbT regulates *fljK* translation by binding to the 5′ untranslated region (UTR) of the *fljK* mRNA and promoting RNA degradation (Mangan *et al.*, 1999). Whether FlbT also interacts with the UTRs of the other flagellins has not been shown. FlbT is not secreted from the cell once the HBB is complete. This suggests the assembly signal that alleviates FlbT negative regulation is transduced through an unidentified partner. Interestingly, a direct interaction of FlbT and *fljK* mRNA requires an accessory protein as Anderson and Gober (2000) could only demonstrate an interaction using cell lysates, not purified protein. This suggests that the assembly signal recognized by FlbT is transduced via this unidentified protein that is plausibly a flagellar-specific secretion substrate such as a structural subunit (Fig. 7.3).

The mechanism by which FlbT regulates flagellin translation is not unique to *C. crescentus*. *Salmonella* flagellins exhibit phase variation between the flagellins FliC and FljB (Pearce and Stocker, 1967; Aldridge *et al.*, 2006c). FljA and the recombinase Hin regulate the genetic switch dictating which of these two flagellins is utilized (only one is ever found in filaments) (Szekely and Simon, 1983). The *fljBA* operon is located downstream of a 1 kb region that is subject to inversion by Hin (Hughes *et al.*, 1988). This DNA region contains the σ^{28}-dependent promoter needed to drive transcription of the *fljBA* operon. When this region is flipped so that σ^{28} can drive *fljBA* expression, FljA represses *fliC* translation by binding the *fliC* 5′ untranslated region of its mRNA inducing mRNA degradation. In the *fljBA* 'OFF' orientation, *fliC* translation can occur as no FljA is produced (Hughes *et al.*, 1988). Thus, a FlbT-like mode of regulation can be found in other unrelated bacteria but in this case it regulates the switch between two flagellins but not flagellin expression *per se*.

FlaF

A second regulatory protein, FlaF, acts as an antagonist to FlbT after HBB completion in *C. crescentus*. The *fljK* mRNA is predicted to fold into two conformations, which are equally energetically favourable. One formation permits translation while the other blocks the ribosomal binding site thus preventing translation (Anderson and Gober, 2000). FlbT is proposed to maintain the mRNA in the non-permissive conformation promoting its degradation and on HBB completion FlaF interacts with the 5′ UTR altering its conformation to the translational permissive state (Anderson and Gober, 2000). However, the translation of all six flagellins does not respond in a similar manner to FlaF.

A Δ*flaF* mutant results in a complete loss of *fljK* flagellin gene expression. When combined with a Δ*flbT* mutant this expression is restored (Llewellyn *et al.*, 2005). However, Δ*flaF* Δ*flbT* mutants do not produce a flagellar filament suggesting that the function of FlaF is not merely regulatory but is also involved in the export of the flagellins (Llewellyn *et al.*, 2005). The stability of the flagellins does not change in Δ*flaF* mutants suggesting that FlaF is not acting as a T3S-chaperone but is playing a distinct role in the regulation of flagellin secretion in this system.

The regulation of flagellin translation by FlbT/FlaF overcomes the lack of FlgM–σ^{28} regulation. Our current understanding of this system has identified how FlbT/FlaF proteins function but how HBB assembly is sensed and transduced to FlbT/FlaF is still unclear. Interestingly, it would seem that the FlbT regulatory circuit is inverse to those seen for FliX and the T3S-chaperones of the *Salmonella* system; a third component is required for FlbT activity rather than inhibiting its regulatory activity. Thus, *C. crescentus* exploits both transcriptional and translational regulation to achieve coordinated flagellar assembly and gene expression. Importantly, the FlbT/FlaF regulatory circuits in effect create what is better described as an expression cascade rather than a transcriptional hierarchy (Fig. 7.2).

The *Vibrio* spp. and *Pseudomonas* spp. flagellar systems

In the *Vibrio/Pseudomonas* FleQ-dependent systems, a more classical transcription hierarchy couples flagellar gene expression to assembly via the activation of a second σ^{54}-EBP and σ^{28}. Two further regulatory proteins are crucial for efficient assembly of the flagellum in these systems: FlhF and FlhG (Dasgupta *et al.*, 2000).

FleS (FlaL, FlrB).

FleS is the sensor kinase of the flagellar specific two-component system FleSR from *P. aeruginosa.* This two-component system requires FleQ for expression and activates flagellar gene expression in response to the rod assembly checkpoint (Ritchings *et al.*, 1995). Thus, like *C. crescentus,* these systems transduce basal body assembly status through a signal transduction cascade. However, these systems possess an associated sensor kinase while *C. crescentus* does not. FleS homologues include FlaL from *V. parahaemolyticus* (McCarter, 2001) and FlrB from *V. cholerae* (Correa *et al.*, 2000). The structures of the three homologues differ in their N termini input domains. FlrB and FleS contain an N-terminal PAS domain, a domain known to be sensitive to a number of environmental signals (McCarter, 2001). No PAS domain has been identified in FlaL. This suggests that the method of sensing flagellar assembly for these three sensor kinases may be different due to species-specific constraints. Similar to FlbD activation, what these sensor kinases are sensing still remains elusive.

FleR (FlaM, FlrC).

FleR is the cognate response regulator of the FleSR system of *P. aeruginosa* (Ritchings *et al.*, 1995). Its homologues are FlaM in *V. parahaemolyticus* (Correa *et al.*, 2000) and FlrC in *V. cholerae* (McCarter, 2001; Muir and Gober, 2001). FleR is an NtrC-like σ^{54}-EBP that is activated by phosphorylation by FleS. Thus, as with FlbD, post-rod flagellar gene expression is σ^{54}-dependent through activation of σ^{54} by the FleR-like proteins. However, in these systems, σ^{28} drives expression of some of the flagellin genes and late flagellar associated proteins, creating a true transcriptional hierarchy architecture (Fig. 7.2).

FlhF

FlhF is a member of the FtsY GTPase signal recognition protein family (Bange *et al.*, 2007). FlhF has been shown to be associated with flagellar assembly in *Pseudomonas* spp. (Dasgupta *et al.*, 2000), *Vibrio* spp. (Kusumoto *et al.*, 2006), *Helicobacter pylori* (Niehus *et al.*, 2004), *B. subtilis* (Carpenter *et al.*, 1992) and *B. cereus* (Salvetti *et al.*, 2007). Except in the *Bacillus* systems that produce peritrichous flagella, the role of FlhF in flagellar biogenesis is to regulate polar localization of the single polar flagellum produced. FlhF localizes to the cell pole in *P. aeruginosa* and *V. alginolyticus* and directs flagellar assembly to that location. A discrepancy is seen between *flhF*-containing flagellar systems as to whether mutants can or cannot make flagella. For *P. aeruginosa* and *P. putida,* flagella are still made in *flhF* mutants but their location and functionality can be impaired, while in the *Bacillus* systems no obvious change in flagellation occurs. In contrast, for *V. aliginolyticus, V. cholerae* and *H. pylori* Δ*flhF* mutants are aflagellate (Dasgupta *et al.*, 2000; Niehus *et al.*, 2004; Kusumoto *et al.*, 2006). This suggest that in some systems FlhF is essential for flagellar assembly while in others it is only required for accurate placement of the flagella. Interestingly, these systems cluster into those that produce sheathed flagella (*V. alginolyticus, V. cholerae* and *H. pylori*) versus those that produce an unprotected flagellum (*P. aeruginosa* and *P. putida*). However, it is also plausible that the additional FtsY homologues (FtsY itself and Ffh) found in some of these bacteria are able to compensate for the loss of FlhF.

FlhG

A second protein that regulates flagellar number in the FleQ-dependent systems is FlhG or FleN. FlhG/FleN are homologues of MinD, a protein involved in regulating the placement of the cell division plane (Kusumoto *et al.*, 2006). MinD and FlhG possess a conserved ATP-Binding domain; the activity of MinD, in conjunction with its partner MinC, is dependent upon ATP hydrolysis (Raskin and de Boer, 1999). FleN in *P. aeruginosa* negatively regulates FleQ activity via a direct interaction (Dasgupta *et al.*, 2000). The FlhG negative feedback loop is common to all FleQ-family master regulators. A FlhG homologue is also present in *B. subtilis*, however, very little is known about the transcriptional hierarchy in this Gram-positive flagellar system. The presence of a FlhG homologue suggests that the flagellar system is σ^{54}-dependent.

Deletions of *flhG* exhibit increased flagellar gene expression and a hyperflagellated phenotype (Dasgupta *et al.*, 2000). These flagella are still localized to the pole. Consistently, *flhG* overexpression can lead to an aflagellate phenotype (Dasgupta *et al.*, 2000). In *V. alginolyticus*, FlhG has been shown to also interact with FlhF (Kusumoto *et al.*, 2008). Analysis of FlhG-GFP fusions suggests FlhG is a cytoplasmic protein distributed throughout the cell. In contrast, FlhF-GFP fusions localize to the pole but exhibited some cytoplasmic distribution in $flhG^{+}$ cells. In $\Delta flhG$ mutants, however, all FlhF-GFP localized to the pole. This suggests that flagellar placement is regulated via the FlhF–FlhG interaction while the FleQ–FlhG interaction regulates flagellar abundance through flagellar gene expression (Kusumoto *et al.*, 2008). This view is supported by $\Delta flhFG$ double mutants of *V. alginolyticus* which possess multiple, peritrichous flagella.

Regulation of flagellar gene expression by σ^{54} in other systems

Two additional proteins that regulate σ^{54} in flagellar systems are FlgR and FleT from *H. pylori* and *R. sphaeroides*, respectively (Brahmachary *et al.*, 2004; Poggio *et al.*, 2005). FlgR/FleT are novel σ^{54} enhancer proteins that do not possess a DNA binding domain. Furthermore, FleT does not possess an N-terminal receiver domain (Poggio *et al.*, 2005). FlgR is activated by its cognate sensor kinase FlgS and activates σ^{54}-dependent transcription of genes encoding distal basal body components, the hook, the HAPs and one of the two flagellins in *H. pylori* (Brahmachary *et al.*, 2004). The lack of a DNA binding domain questions how such σ^{54}-EBPs discriminate between σ^{54} promoters. This is unnecessary in *H. pylori* as FlgR seems to be the only σ^{54}-EBP and thus σ^{54} is exclusively utilized to drive flagellar gene expression (Brahmachary *et al.*, 2004). In contrast to *H. pylori, R sphaeroides* does possess multiple σ^{54}-EBPs however the issue of specificity is overcome by the presence of multiple *rpoN* genes. The σ^{54}-EBPs are not interchangeable, thus σ^{54}-dependent systems utilize their own specific σ^{54}.

R. sphaeroides activates its transcriptional hierarchy through a FleQ-like protein that lacks an N-terminal receiver domain but does possess a DNA-binding domain (Poggio *et al.*, 2005). FleT is dependent on the *R. spheroides* FleQ protein for expression. The *fleT* operon encodes for FleT, the MS-ring and several T3S apparatus components (Poggio *et al.*, 2005). Transcription of at least the rod genes and other basal body components in this system requires both FleQ and FleT. FleQ alone shows the highest affinity to the *fleT* promoter and FleT was found to enhance FleQ binding to flagellar promoters (Poggio *et al.*, 2005). This has lead Poggio *et al* to propose that FleQ/FleT function as a hetero-oligomer when activating flagellar promoters other than the *fleT* promoter. Thus, the *R. sphaeroides* flagellar system is comparable to $FlhD_4C_2$ and VisNR dependent systems in that a hetero-complex transcriptional activator controls the activity of the hierarchy.

A case for c-di-GMP

A new layer of regulation within flagellar systems has recently become apparent, controlled by the novel secondary messenger molecule: bis-(3′,5′)-cyclic diguanylic acid (also known as cyclic di-GMP or c-di-GMP) (Ross *et al.*, 1991). For the majority of flagellar systems studied, c-di-GMP has been shown to exert an inhibitory effect upon flagellar gene expression, flagellar assembly and/or function. The regulation of flagellar systems by c-di-GMP represents a novel regulatory mechanism that can laterally regulate flagellar systems at multiple stages of the assembly pathway independent of the master regulators.

Cyclic-di-GMP was first described as a positive effector of cellulose synthase activity in *Gluconacetobacter xylinus* (Ross *et al.*, 1991). c-di-GMP regulation of cellular processes is predominantly involved in the transition from a planktonic (motile) free state to a sessile community/biofilm-based life style (Jenal and Malone, 2006). The principle mode of action for c-di-GMP is to modulate the activity of a diverse range of proteins via a direct molecule–protein interaction (Wolfe and Visick, 2008). Two protein domains: diguanylate cyclase (DGC's) and phosphodiesterase (PDE's) synthesize and degrade c-di-GMP, respectively. Both elements work together to maintain a steady state of c-di-GMP within a cell (Ross *et al.*, 1991). Often these domains are found within a single protein but they can also be in separate proteins.

Hickman and Harwood (2008) have recently demonstrated the transcriptional activator FleQ to be responsive to c-di-GMP in *P. aeruginosa*. FleQ acts as a transcriptional repressor of the *pel* operon transcription. The *pel* operon is involved in EPS biosynthesis. The repression of *pel* expression is relieved by c-di-GMP proposed to be via a direct interaction with FleQ. Thus, c-di-GMP/FleQ regulation of the *pel* operon is a further example of cross talk between two systems that are advantageous for two different life-styles.

c-di-GMP also controls the transition from one life style to another via the *scrABC/scrG* system of *V. parahaemolyticus*. *V. parahaemolyticus* utilizes two flagellar systems; one, a polar sheathed flagellum, is used for movement through liquids while the other, a lateral (peritrichous) flagellar system, is essential for surface-associated movement (McCarter, 2001) (Fig. 7.4B). The lateral system resembles very much the flagellar system of *Salmonella,* however its master regulator is a σ^{54}-EBP, LafK. The *scrABC* operon was isolated through screening a cosmid library for clones that exhibited lateral flagellar gene expression in liquid cultures. The three genes of the *scrABC* operon each have an observable effect upon the lateral flagellar gene expression, and all are essential for lateral flagellar mediated swarming (Boles and McCarter, 2002). The *scrC* gene encodes a membrane anchored DGC/PDE protein while *scrG* encodes a PDE showing weak homology to the consensus motifs of this family. Consistent with other observed developmental switches associated with flagellar systems, *scrABC* also negatively regulates EPS synthesis (Boles and McCarter, 2002). Thus, this system will inhibit EPS synthesis, indicative of sessile surface-associated growth, until the bacteria have identified a preferred niche through surface-associated motility.

Flagellar systems are extremely robust and there are very few examples of mechanisms that inhibit flagellar assembly that do not regulate the system through the master regulators. The exception to this is a number of recent reports into c-di-GMP regulation of flagellar assembly. Two examples are TipF from *C. crescentus* (Huitema *et al.*, 2006) and the Mif system from *V. fischeri* (O'Shea *et al.*, 2006). TipF acts as a PDE to control c-di-GMP levels that influence hook translation rates (Huitema *et al.*, 2006). A Δ*tipF* mutant produces a flagellum lacking at least the hook and filament. This suggests that TipF regulates flagellar assembly and plausibly flagellar gene expression during late basal body/hook assembly. Whether TipF is actually involved in FlbD activation of flagellar gene expression has as yet not been determined. Interestingly, Δ*tipF* mutants also exhibit cell division defects. A second protein, TipN, has been proposed to act as a pole birth scar marker that allows TipF to be correctly localized to the new pole to guide flagellar assembly to the correct site.

A second example of c-di-GMP regulation of flagellar assembly is the Mif system of *V. fischeri*. In contrast to TipF, how the Mif system regulates flagellar assembly remains somewhat elusive. *V. fischeri* requires high magnesium concentrations for flagellar assembly but it does not influence flagellar gene transcription. The *mif* genes (*mifA* and *mifB*) were identified using a screen for motility in the absence of Mg^{2+} (O'Shea *et al.*, 2006). Both MifA and MifB are DGC c-di-GMP producing enzymes. The PDE MifD balances the system. Analysis of *mif* mutants suggests that their input into the system is not through transcription but either via translation or inhibition of the assembly process directly. How and why this level of control is required in the *V. fischeri* system requires further research.

Discussion and future trends

In this chapter we have focused on how flagellar gene expression is coupled to assembly. Similarly to our in-depth understanding of the flagellar structure, our knowledge of the regulation of flagellar gene expression is also very detailed. We have shown that a number of common network architectures are evident. The main difference between these networks is how flagellar assembly is sensed and that perception is transduced into the rate of transcription from the underlying transcriptional hierarchy. However, a major gap in our understanding is how the assembly status in non-enteric flagellar systems is being sensed and how that information is being utilized to modulate the activity of the regulators of these systems.

We know from *Salmonella* that the assembly status is fed into the regulatory network by the flagellar-associated T3S-chaperones. We also know that for a number of non-enteric systems putative T3S-chaperones can be identified using innovative bioinformatic tools. However, one feature of the T3S-chaperone family is the low similarity between its members, unless they are obvious homologues from related systems. This suggests that all flagellar systems do utilize T3S-chaperones but the important questions are when and how. A common feature of all *Salmonella* T3S-chaperones is that they interact with filament-associated secretion substrates and regulate flagellar gene expression after HBB completion; when the transcriptional hierarchy has become redundant due to FlgM secretion. Prior to HBB completion, the need for T3S-chaperone facilitated secretion has still to be confirmed. Furthermore, the majority of non-enteric systems exploit two, not one, HBB assembly checkpoints. Thus, the non-enteric rod assembly checkpoint may require a novel mechanism for sensing this time point. Interestingly, these systems control flagellar gene expression in response to this assembly intermediate via signal transduction cascades. How these two-component systems perceive this checkpoint also remains elusive.

A key breakthrough in our understanding of how cells regulate flagellar systems has been our growing appreciation of the perception of a bacterial life cycle as a series of developmental events. Key to a successful transition from one life style to the next is the coordination of gene expression required for successful survival/colonization in a new environment. With respect to movement, many of these developmental cycles are associated with a motile to non-motile transition or vice versa (Fig. 7.4). Studies of flagellar systems have shown them to be extremely robust during assembly of at least the first complete structure. Bacteria will often be at a selective advantage if, when a motile to non-motile developmental switch is needed, the change can be immediate. However, the issue is what to do with the completed flagella? The recent explosion in research on the secondary signalling molecule c-di-GMP leads to the intriguing observation that, with respect to flagellar systems, c-di-GMP is always associated with the inhibition of flagellar gene expression, assembly, and most importantly function. Important questions are now: how prevalent is such control? Does this type of regulation always require input from c-di-GMP? And are there alternative modes of regulating flagellar systems during developmental switches associated with bacterial life cycles?

References

Aldridge, P., and Hughes, K.T. (2002). Regulation of flagellar assembly. Curr. Opin. Microbiol. *5*, 160–165.

Aldridge, P., Karlinsey, J., and Hughes, K.T. (2003). The type III secretion chaperone FlgN regulates flagellar assembly via a negative feedback loop containing its chaperone substrates FlgK and FlgL. Mol. Microbiol. *49*, 1333–1345.

Aldridge, P., Karlinsey, J.E., Becker, E., Chevance, F.F., and Hughes, K.T. (2006a). Flk prevents premature secretion of the anti-sigma factor FlgM into the periplasm. Mol. Microbiol. *60*, 630–643.

Aldridge, P.D., Karlinsey, J.E., Aldridge, C., Birchall, C., Thompson, D., Yagasaki, J., and Hughes, K.T. (2006b). The flagellar-specific transcription factor, σ^{28}, is the Type III secretion chaperone for the flagellar-specific anti-σ^{28} factor FlgM. Genes Dev. *20*, 2315–2326.

Aldridge, P.D., Wu, C., Gnerer, J., Karlinsey, J.E., Hughes, K.T., and Sachs, M.S. (2006c). Regulatory protein that inhibits both synthesis and use of the target protein controls flagellar phase variation in *Salmonella enterica*. Proc. Natl. Acad. Sci. U.S.A. *103*, 11340–11345.

Alon, U. (2007). Network motifs: theory and experimental approaches. Nat. Rev. Genet. *8*, 450–461.

Anderson, P.E., and Gober, J.W. (2000). FlbT, the post-transcriptional regulator of flagellin synthesis in *Caulobacter crescentus*, interacts with the 5′ untranslated region of flagellin mRNA. Mol. Microbiol. *38*, 41–52.

Arnosti, D.N. (1990). Regulation of *Escherichia coli* sigma F RNA polymerase by *flhD* and *flhC* flagellar regulatory genes. J. Bacteriol. *172*, 4106–4108.

Arora, S.K., Ritchings, B.W., Almira, E.C., Lory, S., and Ramphal, R. (1997). A transcriptional activator, FleQ, regulates mucin adhesion and flagellar gene expression in *Pseudomonas aeruginosa* in a cascade manner. J. Bacteriol. *179*, 5574–5581.

Auvray, F., Thomas, J., Fraser, G.M., and Hughes, C. (2001). Flagellin polymerisation control by a cytosolic export chaperone. J. Mol. Biol. *308*, 221–229.

Bange, G., Petzold, G., Wild, K., Parlitz, R.O., and Sinning, I. (2007). The crystal structure of the third signal-recognition particle GTPase FlhF reveals a homodimer with bound GTP. Proc. Natl. Acad. Sci. U.S.A. *104*, 13621–13625.

Bennett, J.C., and Hughes, C. (2000). From flagellum assembly to virulence: the extended family of type III export chaperones. Trends Microbiol. *8*, 202–204.

Benson, A.K., Wu, J., and Newton, A. (1994). The role of FlbD in regulation of flagellar gene transcription in *Caulobacter crescentus*. Res. Microbiol. *145*, 420–430.

Boles, B.R., and McCarter, L.L. (2002). *Vibrio parahaemolyticus scrABC*, a novel operon affecting swarming and capsular polysaccharide regulation. J. Bacteriol. *184*, 5946–5954.

Brahmachary, P., Dashti, M.G., Olson, J.W., and Hoover, T.R. (2004). *Helicobacter pylori* FlgR is an enhancer-independent activator of σ^{54}-RNA polymerase holoenzyme. J. Bacteriol. *186*, 4535–4542.

Buck, M., Gallegos, M.T., Studholme, D.J., Guo, Y., and Gralla, J.D. (2000). The bacterial enhancer-dependent sigma(54) (sigma(N)) transcription factor. J. Bacteriol. *182*, 4129–4136.

Carpenter, P.B., Hanlon, D.W., and Ordal, G.W. (1992). *flhF*, a *Bacillus subtilis* flagellar gene that encodes a putative GTP-binding protein. Mol. Microbiol. *6*, 2705–2713.

Chadsey, M.S., and Hughes, K.T. (2001). A multipartite interaction between Salmonella transcription factor σ^{28} and its anti-sigma factor FlgM: implications for σ^{28} holoenzyme destabilization through stepwise binding. J. Mol. Biol. *306*, 915–929.

Chevance, F.F., and Hughes, K.T. (2008). Coordinating assembly of a bacterial macromolecular machine. Nat. Rev. Microbiol. *6*, 455–465.

Chilcott, G.S., and Hughes, K.T. (2000). Coupling of flagellar gene expression to flagellar assembly in *Salmonella enterica* serovar Typhimurium and *Escherichia coli*. Microbiol. Mol. Biol. Rev. *64*, 694–708.

Claret, L., and Hughes, C. (2000). Functions of the subunits in the FlhD(2)C(2) transcriptional master regulator of bacterial flagellum biogenesis and swarming. J. Mol. Biol. *303*, 467–478.

Clegg, S., and Hughes, K.T. (2002). FimZ is a molecular link between sticking and swimming in *Salmonella enterica* serovar Typhimurium. J. Bacteriol. *184*, 1209–1213.

Correa, N.E., Lauriano, C.M., McGee, R., and Klose, K.E. (2000). Phosphorylation of the flagellar regulatory protein FlrC is necessary for *Vibrio cholerae* motility and enhanced colonization. Mol. Microbiol. *35*, 743–755.

Darnton, N.C., Turner, L., Rojevsky, S., and Berg, H.C. (2007). On torque and tumbling in swimming *Escherichia coli*. J. Bacteriol. *189*, 1756–1764.

Dasgupta, N., Arora, S.K., and Ramphal, R. (2000). *fleN*, a gene that regulates flagellar number in *Pseudomonas aeruginosa*. J. Bacteriol. *182*, 357–364.

Dasgupta, N., Ferrell, E.P., Kanack, K.J., West, S.E., and Ramphal, R. (2002). *fleQ*, the gene encoding the major flagellar regulator of *Pseudomonas aeruginosa*, is σ^{70} dependent and is downregulated by *Vfr*, a homologue of *Escherichia coli* cyclic AMP receptor protein. J. Bacteriol. *184*, 5240–5250.

Dasgupta, N., Wolfgang, M.C., Goodman, A.L., Arora, S.K., Jyot, J., Lory, S., and Ramphal, R. (2003). A four-tiered transcriptional regulatory circuit controls flagellar biogenesis in *Pseudomonas aeruginosa*. Mol. Microbiol. *50*, 809–824.

Dingwall, A., Zhuang, W.Y., Quon, K., and Shapiro, L. (1992). Expression of an early gene in the flagellar regulatory hierarchy is sensitive to an interruption in DNA replication. J. Bacteriol. *174*, 1760–1768.

Domian, I.J., Quon, K.C., and Shapiro, L. (1997). Cell type-specific phosphorylation and proteolysis of a transcriptional regulator controls the G1-to-S transition in a bacterial cell cycle. Cell *90*, 415–424.

Ely, B., Ely, T.W., Crymes, W.B., Jr., and Minnich, S.A. (2000). A family of six flagellin genes contributes to the *Caulobacter crescentus* flagellar filament. J. Bacteriol. *182*, 5001–5004.

England, J.C., and Gober, J.W. (2001). Cell cycle control of cell morphogenesis in *Caulobacter*. Curr. Opin. Microbiol. *4*, 674–680.

Fraser, G.M., Bennett, J.C., and Hughes, C. (1999). Substrate-specific binding of hook-associated proteins by FlgN and FliT, putative chaperones for flagellum assembly. Mol. Microbiol. *32*, 569–580.

Fuqua, C., and Greenberg, E.P. (2002). Listening in on bacteria: acyl-homoserine lactone signalling. Nat. Rev. Mol. Cell. Biol. *3*, 685–695.

Galperin, M.Y., Nikolskaya, A.N., and Koonin, E.V. (2001). Novel domains of the prokaryotic two-component signal transduction systems. FEMS. Microbiol. Lett. *203*, 11–21.

Gillen, K.L., and Hughes, K.T. (1991). Molecular characterization of *flgM*, a gene encoding a negative regulator of flagellin synthesis in *Salmonella typhimurium*. J. Bacteriol. *173*, 6453–6459.

Hoang, H.H., Gurich, N., and Gonzalez, J.E. (2008). Regulation of motility by the ExpR/Sin quorum-sensing system in *Sinorhizobium meliloti*. J. Bacteriol. *190*, 861–871.

Hoch, J.A. (2000). Two-component and phosphorelay signal transduction. Curr. Opin. Microbiol. *3*, 165–170.

Hughes, K.T., Gillen, K.L., Semon, M.J., and Karlinsey, J.E. (1993). Sensing structural intermediates in bacterial flagellar assembly by export of a negative regulator. Science *262*, 1277–1280.

Hughes, K.T., and Mathee, K. (1998). The anti-sigma factors. Annu. Rev. Microbiol. *52*, 231–286.

Hughes, K.T., Youderian, P., and Simon, M.I. (1988). Phase variation in *Salmonella*: analysis of Hin recombinase and *hix* recombination site interaction *in vivo*. Genes Dev. *2*, 937–948.

Huitema, E., Pritchard, S., Matteson, D., Radhakrishnan, S.K., and Viollier, P.H. (2006). Bacterial birth scar proteins mark future flagellum assembly site. Cell *124*, 1025–1037.

Hunt, T.P., and Magasanik, B. (1985). Transcription of *glnA* by purified *Escherichia coli* components: core RNA polymerase and the products of *glnF*, *glnG*, and *glnL*. Proc. Natl. Acad. Sci. U.S.A. *82*, 8453–8457.

Iyoda, S., Kamidoi, T., Hirose, K., Kutsukake, K., and Watanabe, H. (2001). A flagellar gene *fliZ* regulates the expression of invasion genes and virulence phenotype in *Salmonella enterica* serovar Typhimurium. Microb. Pathog. *30*, 81–90.

Jenal, U. (2000). Signal transduction mechanisms in *Caulobacter crescentus* development and cell cycle control. FEMS Microbiol. Rev. *24*, 177–191.

Jenal, U., and Fuchs, T. (1998). An essential protease involved in bacterial cell-cycle control. EMBO J. *17*, 5658–5669.

Jenal, U., and Malone, J. (2006). Mechanisms of cyclic-di-GMP signaling in bacteria. Annu. Rev. Genet. *40*, 385–407.

Jyot, J., Dasgupta, N., and Ramphal, R. (2002). *FleQ*, the major flagellar gene regulator in *Pseudomonas aeruginosa*, binds to enhancer sites located either upstream or atypically downstream of the RpoN binding site. J. Bacteriol. *184*, 5251–5260.

Kalir, S., Mangan, S., and Alon, U. (2005). A coherent feed-forward loop with a SUM input function prolongs flagella expression in *Escherichia coli*. Mol. Syst. Biol. *1*, 2005 0006.

Kalir, S., McClure, J., Pabbaraju, K., Southward, C., Ronen, M., Leibler, S., Surette, M.G., and Alon, U. (2001). Ordering genes in a flagella pathway by analysis of expression kinetics from living bacteria. Science *292*, 2080–2083.

Kanbe, M., Yagasaki, J., Zehner, S., Gottfert, M., and Aizawa, S. (2007). Characterization of two sets of subpolar flagella in *Bradyrhizobium japonicum*. J. Bacteriol. *189*, 1083–1089.

Karlinsey, J.E., Lonner, J., Brown, K.L., and Hughes, K.T. (2000). Translation/secretion coupling by type III secretion systems. Cell *102*, 487–497.

Kenney, L.J. (2002). Structure/function relationships in OmpR and other winged-helix transcription factors. Curr. Opin. Microbiol. *5*, 135–141.

Khater, L., Alegria, M.C., Borin, P.F., Santos, T.M., Docena, C., Tasic, L., Farah, C.S., and Ramos, C.H. (2007). Identification of the flagellar chaperone FlgN in the phytopathogen *Xanthomonas axonopodis* pathovar citri by its interaction with hook-associated FlgK. Arch. Microbiol. *188*, 243–250.

Kusumoto, A., Kamisaka, K., Yakushi, T., Terashima, H., Shinohara, A., and Homma, M. (2006). Regulation of polar flagellar number by the *flhF* and *flhG* genes in *Vibrio alginolyticus*. J. Biochem. *139*, 113–121.

Kusumoto, A., Shinohara, A., Terashima, H., Kojima, S., Yakushi, T., and Homma, M. (2008). Collaboration of FlhF and FlhG to regulate polar-flagella number and localization in *Vibrio alginolyticus*. Microbiology *154*, 1390–1399.

Kutsukake, K. (1997). Autogenous and global control of the flagellar master operon, *flhD*, in *Salmonella typhimurium*. Mol. Gen. Genet. *254*, 440–448.

Kutsukake, K., Ikebe, T., and Yamamoto, S. (1999). Two novel regulatory genes, *fliT* and *fliZ*, in the flagellar regulon of *Salmonella*. Genes Genet. Syst. *74*, 287–292.

Kutsukake, K., Ohya, Y., and Iino, T. (1990). Transcriptional analysis of the flagellar regulon of *Salmonella typhimurium*. J. Bacteriol. *172*, 741–747.

Kutsukake, K., Okada, T., Yokoseki, T., and Iino, T. (1994). Sequence analysis of the *flgA* gene and its adjacent region in *Salmonella typhimurium*, and identification of another flagellar gene, flgN. Gene *143*, 49–54.

Lanois, A., Jubelin, G., and Givaudan, A. (2008). FliZ, a flagellar regulator, is at the crossroads between motility, haemolysin expression and virulence in the insect pathogenic bacterium *Xenorhabdus*. Mol. Microbiol. *68*, 516–533.

Laub, M.T., McAdams, H.H., Feldblyum, T., Fraser, C.M., and Shapiro, L. (2000). Global analysis of the genetic network controlling a bacterial cell cycle. Science *290*, 2144–2148.

Li, G., and Tang, J.X. (2006). Low flagellar motor torque and high swimming efficiency of *Caulobacter crescentus* swarmer cells. Biophys. J. *91*, 2726–2734.

Liu, X., Fujita, N., Ishihama, A., and Matsumura, P. (1995). The C-terminal region of the alpha subunit of *Escherichia coli* RNA polymerase is required for transcriptional activation of the flagellar level II operons by the FlhD/FlhC complex. J. Bacteriol. *177*, 5186–5188.

Llewellyn, M., Dutton, R.J., Easter, J., O'Donnol, D., and Gober, J.W. (2005). The conserved *flaF* gene has a critical role in coupling flagellin translation and assembly in *Caulobacter crescentus*. Mol. Microbiol. *57*, 1127–1142.

Lonetto, M., Gribskov, M., and Gross, C.A. (1992). The sigma 70 family: sequence conservation and evolutionary relationships. J. Bacteriol. *174*, 3843–3849.

Lucas, R.L., Lostroh, C.P., DiRusso, C.C., Spector, M.P., Wanner, B.L., and Lee, C.A. (2000). Multiple factors independently regulate *hilA* and invasion gene expression in *Salmonella enterica* serovar Typhimurium. J. Bacteriol. *182*, 1872–1882.

Makishima, S., Komoriya, K., Yamaguchi, S., and Aizawa, S.I. (2001). Length of the flagellar hook and the capacity of the type III export apparatus. Science *291*, 2411–2413.

Mangan, E.K., Bartamian, M., and Gober, J.W. (1995). A mutation that uncouples flagellum assembly from transcription alters the temporal pattern of flagellar gene expression in *Caulobacter crescentus*. J. Bacteriol. *177*, 3176–3184.

Mangan, E.K., Malakooti, J., Caballero, A., Anderson, P., Ely, B., and Gober, J.W. (1999). FlbT couples flagellum assembly to gene expression in *Caulobacter crescentus*. J. Bacteriol. *181*, 6160–6170.

McCarter, L.L. (2001). Polar flagellar motility of the *Vibrionaceae*. Microbiol. Mol. Biol. Rev. *65*, 445–462.

McCarter, L.L. (2006). Regulation of flagella. Curr. Opin. Microbiol. *9*, 180–186.

Milo, R., Shen-Orr, S., Itzkovitz, S., Kashtan, N., Chklovskii, D., and Alon, U. (2002). Network motifs: simple building blocks of complex networks. Science *298*, 824–827.

Minamino, T., Gonzalez-Pedrajo, B., Yamaguchi, K., Aizawa, S.I., and Macnab, R.M. (1999). FliK, the protein responsible for flagellar hook length control in Salmonella, is exported during hook assembly. Mol. Microbiol. *34*, 295–304.

Minamino, T., and Macnab, R.M. (2000). Domain structure of *Salmonella* FlhB, a flagellar export component responsible for substrate specificity switching. J. Bacteriol. *182*, 4906–4914.

Minamino, T., and Pugsley, A.P. (2005). Measure for measure in the control of type III secretion hook and needle length. Mol. Microbiol. *56*, 303–308.

Morett, E., and Buck, M. (1989). In vivo studies on the interaction of RNA polymerase- σ^{54} with the *Klebsiella pneumoniae* and *Rhizobium meliloti nifH* promoters. The role of NifA in the formation of an open promoter complex. J. Mol. Biol. *210*, 65–77.

Morett, E., and Segovia, L. (1993). The σ^{54} bacterial enhancer-binding protein family: mechanism of action and phylogenetic relationship of their functional domains. J. Bacteriol. *175*, 6067–6074.

Moriya, N., Minamino, T., Hughes, K.T., Macnab, R.M., and Namba, K. (2006). The type III flagellar export specificity switch is dependent on FliK ruler and a molecular clock. J. Mol. Biol. *359*, 466–477.

Muir, R.E., Easter, J., and Gober, J.W. (2005). The trans-acting flagellar regulatory proteins, FliX and FlbD, play a central role in linking flagellar biogenesis and cytokinesis in *Caulobacter crescentus*. Microbiology *151*, 3699–3711.

Muir, R.E., and Gober, J.W. (2001). Regulation of late flagellar gene transcription and cell division by flagellum assembly in *Caulobacter crescentus*. Mol. Microbiol. *41*, 117–130.

Muir, R.E., and Gober, J.W. (2004). Regulation of FlbD activity by flagellum assembly is accomplished through direct interaction with the trans-acting factor, FliX. Mol. Microbiol. *54*, 715–730.

Muir, R.E., O'Brien, T.M., and Gober, J.W. (2001). The *Caulobacter crescentus* flagellar gene, *fliX*, encodes a novel trans-acting factor that couples flagellar assembly to transcription. Mol. Microbiol. *39*, 1623–1637.

Mytelka, D.S., and Chamberlin, M.J. (1996). *Escherichia coli fliAZY* operon. J. Bacteriol. *178*, 24–34.

Niehus, E., Gressmann, H., Ye, F., Schlapbach, R., Dehio, M., Dehio, C., Stack, A., Meyer, T.F., Suerbaum, S., and Josenhans, C. (2004). Genome-wide analysis of transcriptional hierarchy and feedback regulation in the flagellar system of *Helicobacter pylori*. Mol. Microbiol. *52*, 947–961.

O'Shea, T.M., Klein, A.H., Geszvain, K., Wolfe, A.J., and Visick, K.L. (2006). Diguanylate cyclases control magnesium-dependent motility of *Vibrio fischeri*. J. Bacteriol. *188*, 8196–8205.

Ohnishi, K., Kutsukake, K., Suzuki, H., and Iino, T. (1990). Gene *fliA* encodes an alternative sigma factor specific for flagellar operons in *Salmonella typhimurium*. Mol. Gen. Genet. *221*, 139–147.

Ozin, A.J., Claret, L., Auvray, F., and Hughes, C. (2003). The FliS chaperone selectively binds the disordered flagellin C-terminal D0 domain central to polymerisation. FEMS. Microbiol. Lett. *219*, 219–224.

Pallen, M.J., Bailey, C.M., and Beatson, S.A. (2006). Evolutionary links between FliH/YscL-like proteins from bacterial type III secretion systems and second-stalk components of the F_0F_1 and vacuolar ATPases. Protein Sci. *15*, 935–941.

Pallen, M.J., Penn, C.W., and Chaudhuri, R.R. (2005). Bacterial flagellar diversity in the post-genomic era. Trends Microbiol. *13*, 143–149.

Parsot, C., Hamiaux, C., and Page, A.L. (2003). The various and varying roles of specific chaperones in type III secretion systems. Curr. Opin. Microbiol. *6*, 7–14.

Pearce, U.B., and Stocker, B.A. (1967). Phase variation of flagellar antigens in *Salmonella*: abortive transduction studies. J. Gen. Microbiol. *49*, 335–349.

Poggio, S., Abreu-Goodger, C., Fabela, S., Osorio, A., Dreyfus, G., Vinuesa, P., and Camarena, L. (2007). A complete set of flagellar genes acquired by horizontal transfer coexists with the endogenous flagellar system in *Rhodobacter sphaeroides*. J. Bacteriol. *189*, 3208–3216.

Poggio, S., Osorio, A., Dreyfus, G., and Camarena, L. (2005). The flagellar hierarchy of *Rhodobacter sphaeroides* is controlled by the concerted action of two enhancer-binding proteins. Mol. Microbiol. *58*, 969–983.

Pruss, B.M., and Matsumura, P. (1996). A regulator of the flagellar regulon of *Escherichia coli*, *flhD*, also affects cell division. J. Bacteriol. *178*, 668–674.

Quon, K.C., Marczynski, G.T., and Shapiro, L. (1996). Cell cycle control by an essential bacterial two-component signal transduction protein. Cell *84*, 83–93.

Raskin, D.M., and de Boer, P.A. (1999). MinDE-dependent pole-to-pole oscillation of division inhibitor MinC in *Escherichia coli*. J. Bacteriol. *181*, 6419–6424.

Ritchings, B.W., Almira, E.C., Lory, S., and Ramphal, R. (1995). Cloning and phenotypic characterization of *fleS* and *fleR*, new response regulators of *Pseudomonas aeruginosa* which regulate motility and adhesion to mucin. Infect. Immun. *63*, 4868–4876.

Ross, P., Mayer, R., and Benziman, M. (1991). Cellulose biosynthesis and function in bacteria. Microbiol. Rev. *55*, 35–58.

Rotter, C., Muhlbacher, S., Salamon, D., Schmitt, R., and Scharf, B. (2006). Rem, a new transcriptional activator of motility and chemotaxis in *Sinorhizobium meliloti*. J. Bacteriol. *188*, 6932–6942.

Saini, S., Brown, J.D., Aldridge, P.D., and Rao, C.V. (2008). FliZ is a post-translational activator of $FlhD_4C_2$-dependent flagellar gene expression. J. Bacteriol. *190*, 4979–4988.

Salvetti, S., Ghelardi, E., Celandroni, F., Ceragioli, M., Giannessi, F., and Senesi, S. (2007). FlhF, a signal recognition particle-like GTPase, is involved in the regulation of flagellar arrangement, motility behaviour and protein secretion in *Bacillus cereus*. Microbiology *153*, 2541–2552.

Samatey, F.A., Imada, K., Nagashima, S., Vonderviszt, F., Kumasaka, T., Yamamoto, M., and Namba, K. (2001). Structure of the bacterial flagellar protofilament and implications for a switch for supercoiling. Nature *410*, 331–337.

Shen-Orr, S.S., Milo, R., Mangan, S., and Alon, U. (2002). Network motifs in the transcriptional regulation network of *Escherichia coli*. Nat. Genet. *31*, 64–68.

Shibata, S., Takahashi, N., Chevance, F.F., Karlinsey, J.E., Hughes, K.T., and Aizawa, S. (2007). FliK regulates flagellar hook length as an internal ruler. Mol. Microbiol. *64*, 1404–1415.

Sourjik, V., Muschler, P., Scharf, B., and Schmitt, R. (2000). VisN and VisR are global regulators of chemotaxis, flagellar, and motility genes in *Sinorhizobium (Rhizobium) meliloti*. J. Bacteriol. *182*, 782–788.

Soutourina, O.A., and Bertin, P.N. (2003). Regulation cascade of flagellar expression in Gram-negative bacteria. FEMS Microbiol. Rev. *27*, 505–523.

Stafford, G.P., Ogi, T., and Hughes, C. (2005). Binding and transcriptional activation of non-flagellar genes by the *Escherichia coli* flagellar master regulator $FlhD_2C_2$. Microbiology *151*, 1779–1788.

Stewart, B.J., and McCarter, L.L. (1996). *Vibrio parahaemolyticus* FlaJ, a homologue of FliS, is required for production of a flagellin. Mol. Microbiol. *20*, 137–149.

Stewart, B.J., and McCarter, L.L. (2003). Lateral flagellar gene system of *Vibrio parahaemolyticus*. J. Bacteriol. *185*, 4508–4518.

Suzuki, T., Iino, T., Horiguchi, T., and Yamaguchi, S. (1978). Incomplete flagellar structures in nonflagellate mutants of *Salmonella typhimurium*. J. Bacteriol. *133*, 904–915.

Szekely, E., and Simon, M. (1983). DNA sequence adjacent to flagellar genes and evolution of flagellar-phase variation. J. Bacteriol. *155*, 74–81.

Tart, A.H., Blanks, M.J., and Wozniak, D.J. (2006). The AlgT-dependent transcriptional regulator AmrZ (AlgZ) inhibits flagellum biosynthesis in mucoid, nonmotile *Pseudomonas aeruginosa* cystic fibrosis isolates. J. Bacteriol. *188*, 6483–6489.

Tart, A.H., Wolfgang, M.C., and Wozniak, D.J. (2005). The alternative sigma factor AlgT represses *Pseudomonas aeruginosa* flagellum biosynthesis by inhibiting expression of *fleQ*. J. Bacteriol. *187*, 7955–7962.

Titz, B., Rajagopala, S.V., Ester, C., Hauser, R., and Uetz, P. (2006). Novel conserved assembly factor of the bacterial flagellum. J. Bacteriol. *188*, 7700–7706.

Wang, S., Fleming, R.T., Westbrook, E.M., Matsumura, P., and McKay, D.B. (2006). Structure of the *Escherichia coli* FlhDC complex, a prokaryotic heteromeric regulator of transcription. J. Mol. Biol. *355*, 798–808.

Wigneshweraraj, S., Bose, D., Burrows, P.C., Joly, N., Schumacher, J., Rappas, M., Pape, T., Zhang, X., Stockley, P., Severinov, K., and Buck, M. (2008). Modus operandi of the bacterial RNA polymerase containing the σ^{54} promoter-specificity factor. Mol. Microbiol. *68*, 538–546.

Williams, A.W., Yamaguchi, S., Togashi, F., Aizawa, S.I., Kawagishi, I., and Macnab, R.M. (1996). Mutations in *fliK* and *flhB* affecting flagellar hook and filament assembly in *Salmonella typhimurium*. J. Bacteriol. *178*, 2960–2970.

Wingrove, J.A., and Gober, J.W. (1996). Identification of an asymmetrically localized sensor histidine kinase responsible for temporally and spatially regulated transcription. Science *274*, 597–601.

Wolfe, A.J., and Visick, K.L. (2008). Get the message out: cyclic-Di-GMP regulates multiple levels of flagellum-based motility. J. Bacteriol. *190*, 463–475.

Wu, J., Benson, A.K., and Newton, A. (1995). Global regulation of a σ^{54}-dependent flagellar gene family in *Caulobacter crescentus* by the transcriptional activator FlbD. J. Bacteriol. *177*, 3241–3250.

Yokoseki, T., Kutsukake, K., Ohnishi, K., and Iino, T. (1995). Functional analysis of the flagellar genes in the *fliD* operon of *Salmonella typhimurium*. Microbiology *141*, 1715–1722.

Yonekura, K., Maki, S., Morgan, D.G., DeRosier, D.J., Vonderviszt, F., Imada, K., and Namba, K. (2000). The bacterial flagellar cap as the rotary promoter of flagellin self-assembly. Science *290*, 2148–2152.

Zarivach, R., Deng, W., Vuckovic, M., Felise, H.B., Nguyen, H.V., Miller, S.I., Finlay, B.B., and Strynadka, N.C. (2008). Structural analysis of the essential self-cleaving type III secretion proteins EscU and SpaS. Nature *453*, 124–127.

Structure and Mechanism of the Flagellar Rotary Motor

8

David F. Blair

Abstract

The flagellar motor is a biochemical machine that transforms energy stored in the membrane ion gradient into the mechanical work of rotation, at a high rate of speed and with great efficiency. Discrete steps in rotation have recently been detected, attaining a long-sought goal of biophysical measurement and opening a path to detailed physiological dissection of the mechano-chemical cycle of the motor. Structural and biochemical studies of the elements that function in torque generation have provided a fairly detailed picture of the rotor and have begun to reveal crucial features of the stator. A molecular-level understanding of the mechanism appears almost within reach.

Introduction

Rotary flagella provide what may be the most widespread form of cellular motility in nature. Swimming bacteria face the dual challenges of large viscous drag and chaotic thermal motion (Berg, 1993), yet can move at speeds greater than 10 body lengths per second and can direct their movement with exquisite sensitivity toward a wide variety of chemical attractants and other stimuli (Berg and Brown, 1972). The apparatus enabling this movement is a large, multi-component protein machine in the cell membrane that converts energy stored in an ion gradient into the mechanical work of rotation. The flagellum has been studied most extensively in the enteric species *Escherichia coli* and *Salmonella enterica* (serovar Typhimurium). Although little is known of the precise roles and benefits of flagellar motility in natural settings, it is clear that flagella provide a robust and versatile solution to the problems of movement in the micro-world: Flagellar genes occur in about half of the sequenced bacterial genomes, in diverse, deeply branched lineages.

The flagellar motor has been the subject of more than 30 years of investigation, and one of the earliest discoveries was that it is energized not by ATP but by the membrane gradient or ions, usually protons (H^+)(Larsen *et al.*, 1974a) but in some cases sodium ions (Na^+)(Hirota *et al.*, 1981). Thus, the central question regarding mechanism is how ion movement is harnessed to drive rotation. Flagellar motility benefits the cell by allowing migration that is not only fast, but is purposefully directed toward nutrients, temperatures, or other circumstances favouring survival. Directional control relies on the ability of the flagellar motor to reverse, or in some cases to slow, in response to various sensory cues (Larsen *et al.*, 1974b; Armitage and Schmitt, 1997). This involves a set of sensing and information-processing proteins, which together regulate the level of a signalling protein that interacts with the motor to reverse or slow rotation (Parkinson *et al.*, 1983; Welch *et al.*, 1993). Thus, another fundamental question concerns the molecular mechanism of switching.

Rotation of the bacterial flagellum, as contrasted with the whipping action that its name would suggest, was demonstrated 35 years ago (Berg and Anderson, 1973; Silverman and Simon, 1974), and for some years was believed to be the unique example of true rotary motion in biology. Subsequent work showed that the F_1F_0

ATP synthase carries out an essentially similar process, harnessing the H^+ (or Na^+) gradient to power rotation of one protein complex relative to another, in that case to drive conformational changes needed to make ATP (Noji *et al.*, 1997; Zhou *et al.*, 1997). Evidently, ion-driven rotation is an ancient device that has found at least two very different applications. Some components of the flagellum appear, moreover, to be evolutionarily related to subunits of the ATP synthase (Vogler *et al.*, 1991; Pallen *et al.*, 2006).

While the mechanism of the flagellar motor is not yet understood in detail, experimental constraints on the mechanisms have begun to tighten in recent years and a chemically detailed understanding of the motor may soon be in reach. In this chapter, I will summarize recent progress in understanding the motor mechanism, with major emphasis on structural and physiological experiments. Much elegant and important structural work has also been done on the filament and hook (see, for example, Samatey *et al.*, 2001, 2004); for a more detailed treatment of those topics, see Yonekura *et al.* (2003). Numerous hypotheses for the motor mechanism have been proposed; these will not be surveyed extensively, but some recent models will be briefly described and an attempt will be made to formulate a chemically explicit mechanism that accords with present knowledge of motor structure and physiology.

Overview of flagellar structure

The flagellum comprises a long helical filament that functions as propeller, a short hook-shaped structure, and the basal body, a multi-ring structure traversing the cellular membrane(s) (Fig. 8.1). In *Escherichia coli* and *Salmonella*, the filament is formed from a single protein called flagellin (or FliC). The filament is formed from multiple flagellins in some species (Ely *et al.*, 2000; Li *et al.*, 2000), and in some cases is surrounded by an extension of the outer membrane (Sjoblad *et al.*, 1983). Whatever the details of its composition, the flagellar filament is helical and rigid, as required for its propeller function. The hook is also helical, but is shorter and more flexible (Samatey *et al.*, 2004). It is believed to function as a universal joint to allow the several flagella on a cell to come together into one (or a few) coherently rotating bundle(s). The structures of major parts of the hook and filament proteins have been solved and docked into EM maps, providing a clear, structurally grounded account of the helical shapes and contrasting stiffnesses of these structures (Samatey *et al.*, 2001, 2004). The rings of the basal body are named according to their locations relative to the cell membrane: L, Lipopolysaccharide; P, Peptidoglycan; MS, cytoplasmic Membrane and Supramembranous; C, Cytosol (DePamphilis and Adler, 1971; Aizawa *et al.*, 1985; Khan *et al.*, 1992; Francis *et al.*, 1994). Torque generation takes place in the vicinity of the membrane, and so the MS- and C-rings are of greatest interest for understanding the rotation mechanism.

The flagellar basal body has been studied by electron microscopy for more than 30 years (DePamphilis and Adler, 1971; Dimmitt and Simon, 1971), and a series of refinements in specimen preparation and microscopy have led to steady, and lately dramatic, improvements in resolution (Fig. 8.1). A recent single-particle reconstruction by Thomas *et al.* (2006) achieved resolution sufficient to distinguish features the size of small protein domains, providing an effective framework for constraining models of protein organization within the structure. The part of the flagellum that has been imaged at high resolution is just the rotor, because the stator proteins, which are integral to the cytoplasmic membrane, are lost during basal-body purification. The stator consists of a set of protein complexes in the membrane surrounding the basal body, first seen by freeze-fracture EM (Khan *et al.*, 1988; Fig. 8.1). In another very important advance, Jensen and co-workers recently reported the first images of the intact flagellum (i.e. including both the rotor and stator), obtained by high-resolution electron tomography of *Treponema primitia* cells (Murphy *et al.*, 2006). The stator complexes are found roughly where expected on the basis of the freeze-fracture images (Khan *et al.*, 1988) and other considerations (Zhou *et al.*, 1998a), but the connections between stator and rotor appear more extensive than anticipated (Fig. 8.1, panel E). The stator of *T. primitia* contains 16 discrete elements, a value consistent with the earlier freeze-fracture work.

The flagellum also contains a central, membrane-embedded structure that functions during

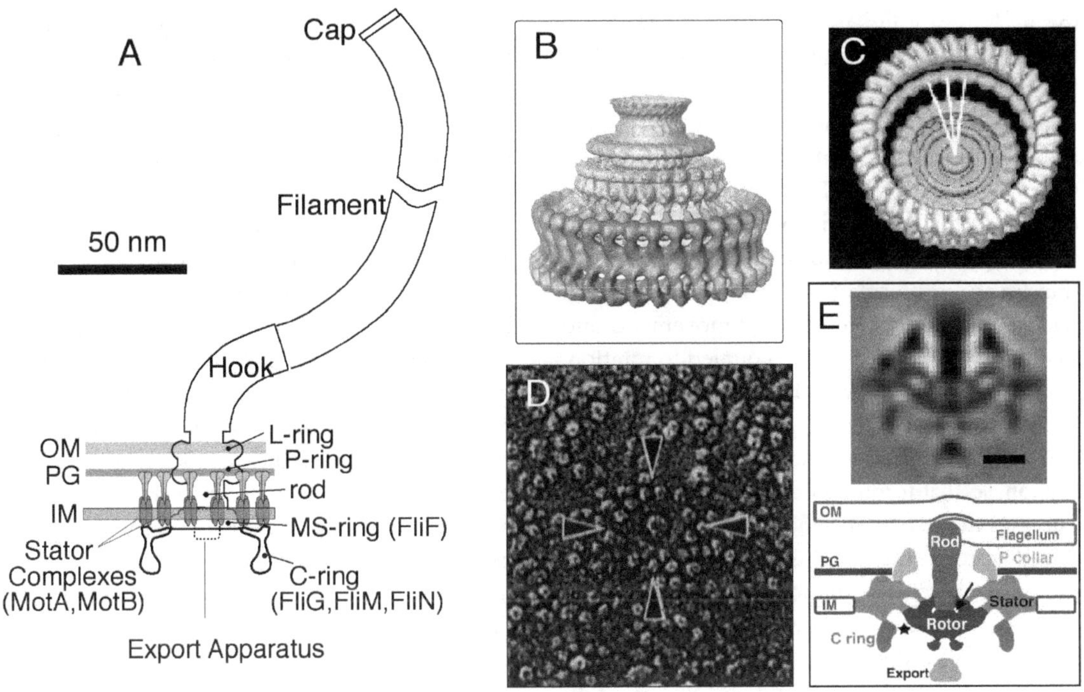

Figure 8.1 Structure of the flagellum. **A** Major structural features of the flagellum and its relationship to the membrane layers in a Gram-negative bacterium. **B** Shape of the lower part of the basal body as determined by electron cryomicroscopy and single-particle reconstruction (Thomas *et al.*, 2006). **C** The basal-body reconstruction viewed from the bottom. The ridges on the bottom of the C-ring are spaced at intervals of 4 nm (Young *et al.*, 2003; Thomas *et al.*, 2006). **D** Freeze-fracture electron micrograph showing rings of particles, which can be assigned to the MotA/MotB stator complexes (Khan *et al.*, 1988). **E** The basal body and stator of the flagellum of *Treponema primitia*, imaged in cells by cryoelectron tomography (Murphy *et al.*, 2006). The electron density map is above (scale bar, 20nm), and an interpretive cartoon assigning the density to various features is shown below. (Images provided by G. Jensen.)

assembly to export the protein subunits needed to form exterior structures. Like the stator, the membrane-associated parts of this apparatus are likely to be disrupted during the detergent treatments used in basal-body purification, and so the export apparatus has not yet been imaged at high resolution. A central protrusion that is probably the membrane-bound part of the apparatus has been seen in images of membranes obtained by a freeze-etch procedure (Katayama *et al.*, 1996).

Motor physiology

Silverman and Simon first visualized the rotation of individual flagellar motors by tethering cells to coverslips by their filaments and observing the resulting rotation of the cell bodies (Silverman and Simon, 1974). At about the same time, Larsen and coworkers reported the discovery of the ion gradient as the energy source for rotation (Larsen *et al.*, 1974a). Since then, the tethered-cell assay and its variations have been used in numerous studies of motor physiology, often in combination with methods for manipulating the energizing-ion gradient. Because the cell body has a larger drag coefficient than the slender flagellar filament, motors turning tethered cells work against an unnaturally large load and consequently turn at low rates. Experiments with tethered cells have led to the following key observations: When motors turn slowly, their torque is proportional to the magnitude of the proton electrochemical potential gradient (the proton-motive force, or pmf), and is the same whether the pmf is present in the form of a concentration gradient or an electrical potential difference (Manson *et al.*, 1980; Berg *et al.*, 1982). Slow-turning motors do not exhibit a significant hydrogen-isotope effect (producing the same torque in D_2O as in H_2O) and their torque is nearly independent of temperature in the range 4–38° C (Khan and Berg, 1983; Chen and Berg, 2000a).

These properties are not expected if the torque depends on the rates of chemical processes occurring in the motor. They can be economically accounted for by postulating that the processes occurring within the motor – for example, ion transfer and conformational changes – are all fast relative to the net rate of rotor movement, and that proton transfer is tightly coupled to rotation. Under these assumptions the slow-turning motor functions to convert energy between chemical and mechanical forms in essentially reversible fashion, so that torque is determined not by rates but by energetics. Specifically, the energy obtained from all the protons flowing through the motor during one revolution (the pmf times the number of protons that flow) must match, or just slightly exceed, the mechanical work performed (2π times the torque). A definitive demonstration of the tight-coupling hypothesis would require more direct measurement of the proton stoichiometry. Just one such measurement has been made (Meister *et al.*, 1987). The results were in approximate agreement with the tight-coupling hypothesis but additional experiment of this kind are needed to provide a stricter test of the tight-coupling proposal. Berry and coworkers recently described a fluorescence-based method for measuring Na^+ concentration in individual cells (Lo *et al.*, 2006), which may allow a new approach to the stoichiometry question based on measurement of ion fluxes through sodium-fuelled motors.

As noted above, the stator consists of a set of discrete protein complexes in the membrane around the basal body. By examining the effects of controlled, low-level expression of stator proteins, Berg and co-workers (Block and Berg, 1984; Blair and Berg, 1988) demonstrated that the motor contains several stator elements that function independently to generate torque. Initial estimates of the number of torque generators were as high as 16 (Block and Berg, 1984), later revised downward to eight (Blair and Berg, 1988), settling most recently on a value around 11 on the basis of improved measurements by Berry and coworkers (Reid *et al.*, 2006). Whatever the number of torque generators present (which may vary with species and circumstances), it seems clear that the stator units function independently to generate torque. Thus, the rotation mechanism must involve processes occurring within each stator complex, and does not appear to depend in any essential way on interactions between the complexes.

After several years of fruitful study with tethered cells, it was found that lightly loaded motors – those turning filaments rather than cell bodies – exhibit quite different properties. By measuring rates of bundle rotation in freely swimming cells, Lowe *et al.* found that lightly loaded motors can reach speeds as great as 300 Hz (Lowe *et al.*, 1987). Under this condition of light load, the motor speed was found to depend strongly on both temperature and hydrogen isotope, as expected if chemical processes in the motor, including proton dissociation, are rate-limiting. Since this discovery, numerous studies have been undertaken to measure motor performance over a range of loads and speeds. These studies have used various means for applying external torque to the cell (Berg and Turner, 1993; Washizu *et al.*, 1993; Berry *et al.*, 1995; Berry and Berg, 1999), or have used beads attached to flagellar stubs to monitor rotation and to adjust the viscous load (for example, Chen and Berg, 2000b). The resulting torque-vs.-speed characteristic (Fig. 8.2) places useful constraints on the mechanism. It exhibits a plateau of nearly constant torque until reaching a 'knee' value above which torque drops fairly steeply. The 'knee' speed, and the limiting, zero-torque speed, vary with the temperature (Chen and Berg, 2000b). The shape of the torque-speed characteristic is indicative of a 'powerstroke' mechanism in which rotor movement is driven actively, in preference to a 'Brownian ratchet' model in which the energy of protons is used to rectify thermal movements (Berry *et al.*, 1995). Further studies extended the bead approach to examine the low-load, high-speed performance of motors containing only one or a few torque generators (Ryu *et al.*, 2000). Most recently, very small gold beads were used to decrease the viscous load even further. Remarkably, motors with only a single torque generator were found to turn at the full ~300-Hz rate, provided the load is very light (Fig. 8.2) (Yuan and Berg, 2008). This implies that the stator elements operate with a high duty cycle, meaning that they remain engaged with the rotor

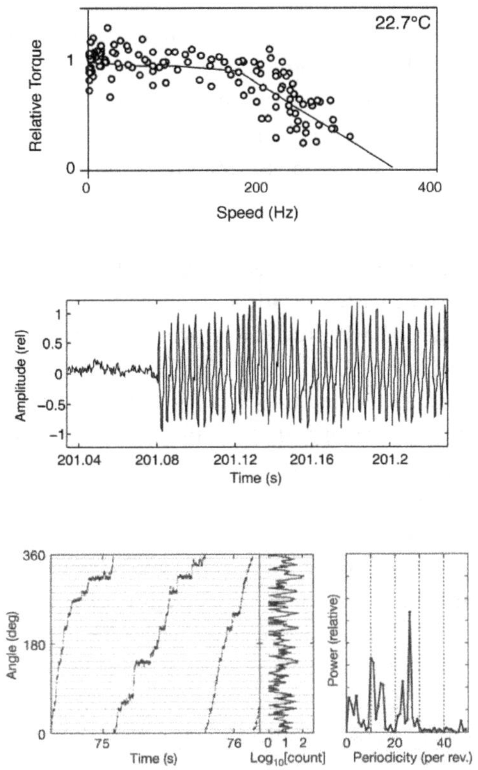

Figure 8.2 Key physiological properties of the motor. **Top:** The torque-vs.-speed characteristic of the *Escherichia coli* motor, measured using small beads attached to flagellar stubs (Chen and Berg, 2000b). The wide plateau of nearly constant torque is indicative of a motor that uses an active power stroke, rather than a 'Brownian ratchet' mechanism based on rectification of thermal movement. **Middle:** A recent torque-restoration experiment that monitored rotation of the motor during installation of the first torque-generating element, in a situation of very low load. A stator protein was expressed at a low level in stator-deficient cells, and rotation was visualized by attachment of a small gold bead. A single torque generator can drive the motor at 250 Hz or faster when the load is light (Yuan and Berg, 2008). **Bottom:** Detection of individual steps in rotation. Rotation was visualized by attachment of small beads, and the motor of *E. coli* was genetically engineered to use Na^+, using chimeric stator proteins with portions borrowed from *V. alginolyticus*. Na^+ concentration, and thus the motor torque, was kept low to facilitate step detection. Traces of the angular position of the bead vs. time are shown on the left, and the observed periodicity in the motion is represented in terms of a power spectrum on the right (Sowa *et al.*, 2005).

the majority of the time (Ryu *et al.*, 2000; Yuan and Berg, 2008).

As mentioned above, the deuterium/hydrogen isotope effect observed in lightly loaded motors suggests that proton transfer is a rate-limiting factor in the cycle of the motor. Sodium-fuelled motors provide a further indication that ion movements restrict the overall motor rate under light load. The sodium-driven motor of *Vibrio alginolyticus* has been clocked at 1700 Hz, in a measurement that used laser dark-field microscopy to monitor rotation of the polar filaments on cells attached to a surface (Magariyama *et al.*, 1994). A straightforward interpretation of this finding is that ion movements are the major rate limitation, and that sodium ions, because they can be present at higher concentration than protons, can support more rapid rotation.

Studies using chimeric proteins have shown that the motor components functioning in rotation are largely interchangeable between H^+- and Na^+-driven motors (see, for example, Asai *et al.*, 1999, 2000; Gosink and Hase, 2000). The two classes of motor therefore appear to work in essentially the same way. The sodium-fuelled motor provides some advantages for physiological measurements, the most important being the ability to adjust sodium-ion concentration over a wide range without harm to cellular constituents. This property was exploited in a recent study of *E. coli* motors containing stators re-engineered to utilize Na^+. By using motors with just a single, Na^+-using torque generator, and low Na^+ concentrations to keep the driving force low, Sowa *et al.* recently succeeded in observing discrete steps in rotation (Sowa *et al.*, 2005). The search for steps has been under way since the earliest days of flagellar motor physiology but had been hampered by the presence of the compliant hook, which acts to dampen any stepwise character in the motion. Rotation of the Na^+-using chimeras was visualized by means of beads attached to flagellar stubs. This remarkable measurement gave an average step size of about 14°, corresponding to 26 steps per revolution, assuming all steps are of the same size. This value is in about the range predicted from a previous analysis of motor-speed fluctuations (Samuel and Berg, 1995), and appears also to agree with the emerging structural picture of the rotor (see below). Backward steps, of somewhat smaller size, were also seen. The step size provides an important constraint on the mechanism of the motor, and even more importantly, the ability to detect steps should enable new kinds of physiological experiment to dissect the mechanism.

Proteins that function in rotation

The assembled flagellum contains about two dozen proteins and about as many more are needed to regulate its assembly and operation (Macnab, 1996; Aizawa, 2001; Aldridge and Hughes, 2002; Macnab, 2003). The rotation process itself is thought to involve just five proteins. These are FliG, FliM, and FliN, which form a large complex on the rotor dubbed the 'switch complex,' (Yamaguchi *et al.*, 1986a) and MotA and MotB, membrane proteins that form the stator complexes (see Fig. 8.1) (Dean *et al.*, 1984; Stader *et al.*, 1986; Khan *et al.*, 1988).

Switch complex

The switch-complex proteins function (as the name implies) to regulate CW/CCW reversals, and are also essential for flagellar assembly and rotation (Yamaguchi *et al.*, 1986b). While each of them is multifunctional, mutational studies of the switch-complex proteins indicate some degree of functional specialization: FliG is involved most extensively in rotation (Irikura *et al.*, 1993; Lloyd *et al.*, 1996), and interacts directly with the stator protein MotA (Zhou *et al.*, 1998a; Yakushi *et al.*, 2006). FliG is the target of a recently discovered 'clutch' protein that functions in *Bacillus* to disengage the rotor from the stator in cells entering the biofilm state (Blair *et al.*, 2008). FliG was also just found to bind the enzyme fumarate reductase, providing a rationale for the observed requirement of fumarate for flagellar direction reversal (Cohen-Ben-Lulu *et al.*, 2008). FliM is most closely involved in CW/CCW direction control (Sockett *et al.*, 1992), and is a binding target for the chemotactic signalling protein CheY-P (CheY phosphate) (Welch *et al.*, 1993). CheY-P promotes CW rotation of the motor (in *E. coli* or *Salmonella*). FliN is involved in flagellar assembly (Vogler *et al.*, 1991; Irikura *et al.*, 1993; Tang *et al.*, 1995), and also appears to have an important role in switching (Paul *et al.*, 2006).

Crystal structures are now known for major parts of all three switch-complex proteins (Lloyd *et al.*, 1999; Brown *et al.*, 2002; Brown *et al.*, 2005; Park *et al.*, 2006) (Fig. 8.3). The solved portion of FliG, which is roughly the C-terminal two-thirds of the protein, consists of two compact domains joined by an alpha helix and a short extended segment that should be flexible by virtue of two well conserved Gly residues. A prominent ridge on the 'top' of the C-terminal domain (see Fig. 8.3) displays a set of conserved charged residues that are essential for rotation and that form a site of interaction with the stator protein MotA (Lloyd and Blair, 1997; Zhou *et al.*, 1998a; Yakushi *et al.*, 2006). At the end opposite the charged ridge, the FliG C-terminal domain displays a well conserved surface hydrophobic patch that has been shown to interact with FliM (Brown *et al.*, 2002; Brown *et al.*, 2007). The N-terminal domain has a highly conserved surface motif termed EHPQR (for its constituent residues), which is also implicated in binding FliM (Brown *et al.*, 2007). The structure is not yet solved for a sizable (ca. 115-residue) piece at the N-terminus of FliG, which includes the segments that attach to the MS-ring protein FliF (see Fig. 8.1).

The structurally characterized part of FliN, comprising about the C-terminal two-thirds of the protein, forms a compact, saddle-shaped dimer (Brown *et al.*, 2005). The FliN dimers are further associated into a doughnut-shaped tetramer in the crystal, and cross-linking experiments indicate a similar, ring-shaped organization of FliN in cells (Paul and Blair, 2006). A conserved hydrophobic patch on the two-fold axis of the FliN dimer is important for flagellar assembly and was shown to interact with FliH, a component of the flagellar export machinery (Gonzalez-Pedrajo *et al.*, 2006; McMurry *et al.*, 2006; Paul *et al.*, 2006). Regions in and around the hydrophobic patch are also implicated in switching, and might form a site of interaction with the CW-signalling molecule CheY-P: Mutations in this region cause a CCW motor bias that can be suppressed in most cases by overexpression of CheY (Paul *et al.*, 2006). An intriguing feature of the FliN doughnut is that the interfaces between the FliN dimers are slightly bent, giving the doughnut a puckered shape. The puckering gives rise to two states, which are structurally and energetically equivalent in the isolated FliN tetramer but which would be different in the context of the flagellum. Accordingly, Paul *et al.* suggested that CW/CCW switching may involve

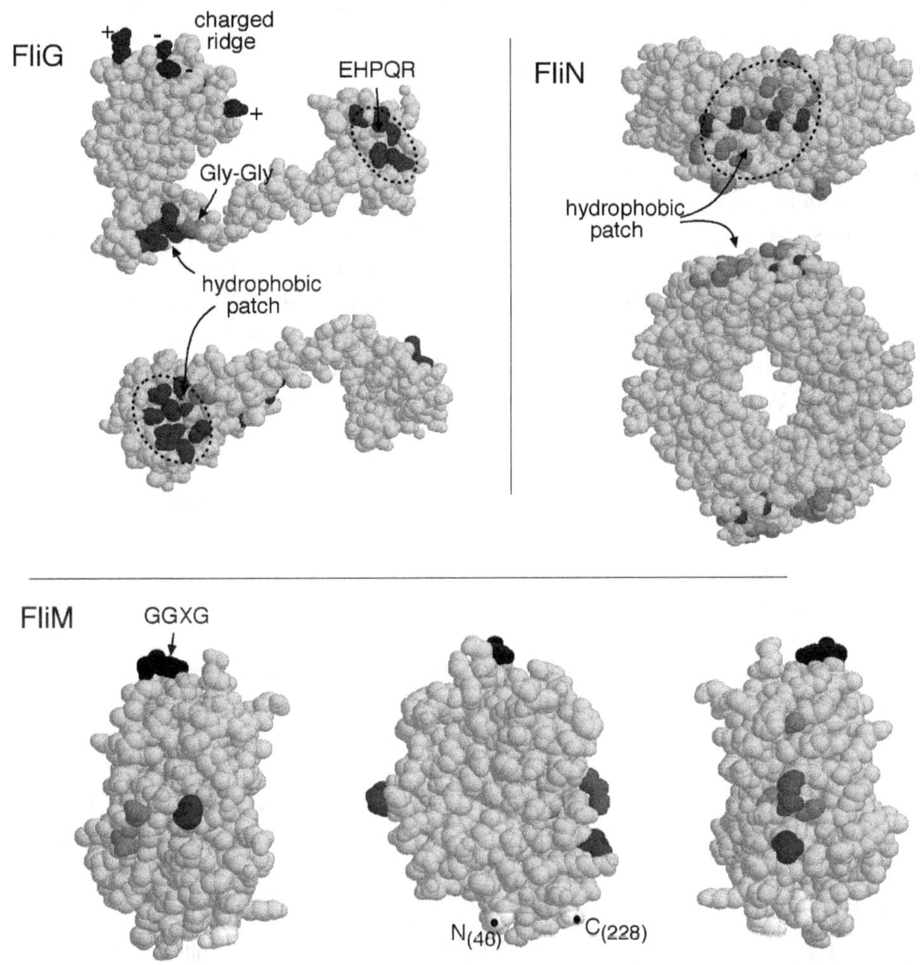

Figure 8.3 Structures of the switch-complex proteins. Conserved features discussed in the text are indicated. In FliG, these include the charged ridge that interacts with the stator protein MotA (Lloyd and Blair, 1997; Zhou *et al.*, 1998a), the conserved EHPQR motif and hydrophobic patch that interact with FliM (Brown *et al.*, 2007), and a Gly-Gly linker that is hypothesized to function as a hinge between the domains (Brown *et al.*, 2002). In FliN, positions coloured light grey identify a candidate binding site for the CW-signalling protein CheY; mutations at these positions give a CCW motor bias that can be suppressed by CheY overexpression (Paul *et al.*, 2006). Positions coloured dark grey on FliN are important for assembly and are involved in its interaction with FliH (Gonzalez-Pedrajo *et al.*, 2006; Paul *et al.*, 2006). In FliM, the conserved GGXG motif and residues nearby have been implicated in binding FliG (Mathews *et al.*, 1998; Passmore *et al.*, 2008). The N- and C-termini of the crystallized domain, which spans residues 46–228, are indicated. A sizable C-terminal domain on FliM, not yet characterized structurally, is involved in binding FliN. Positions coloured dark grey on the edges of FliM gave high-yield cross-linking between subunits (Lowder *et al.*, 2005), and thus identify regions of FliM-FliM contact in the flagellum. Positions coloured light grey on FliM gave a strong CW motor bias (Sockett *et al.*, 1992), and the position coloured medium-grey gave both high-yield cross-linking and strong CW bias.

transitions between the two oppositely puckered states (Paul and Blair, 2006; Paul *et al.*, 2006). The structure has not yet been solved for the N-terminal one-third of FliN. This part of FliN is relatively poorly conserved, and was found not to be critical for flagellar assembly, rotation, or switching (Tang *et al.*, 1995).

The crystallographically characterized part of FliM encompasses residues 46–228, or about 60% of the ~330-residue protein (Park *et al.*,

2006). The structure shows a single globular domain with approximate dimensions 3 nm × 4 nm × 5 nm. A conserved GGXG motif at the top of the domain (see Fig. 8.3) has been implicated in binding FliG (Mathews *et al.*, 1998; Passmore *et al.*, 2008). The missing N- and C-terminal parts would both attach at the end opposite the GGXG motif (at the bottom in Fig. 8.3). The N-terminal segment functions to bind CheY-phosphate, and a structure has been solved for a complex of CheY with a FliM-derived peptide (Lee *et al.*, 2001). The C-terminal domain of FliM contains the determinants for binding FliN (Toker *et al.*, 1996; Mathews *et al.*, 1998). The structure of the C-terminal domain has not yet been solved, but some aspects of its folding can be deduced from its sequence similarity to FliN (Mathews *et al.*, 1998).

Mutational analyses, pulldown assays, and targeted cross-linking have been used to identify interacting surfaces on the switch complex proteins and to deduce features of their arrangement. Further guidance to the subunit organization comes from the EM reconstructions, which as mentioned above have now resolved features the size of small protein domains (Thomas *et al.*, 2006). Together, the various approaches have led to the proposal of a fairly detailed structural model of the switch complex (Fig. 8.4) (Brown *et al.*, 2007). FliG is positioned at the top where it can interact with the stator. FliM is just below, and is proposed to occupy two distinct locations in order to account for its interactions with both the hydrophobic patch and the EHPQR motif on FliG. This two-site organization of FliM can also account for the different copy numbers of FliM (~34 subunits) and FliG (~26 subunits) in the flagellum (Fig. 8.4B). The FliN doughnuts are at the bottom, where ring-shaped density is seen in the micrographs (see Fig. 8.1). The FliM C-terminal domain, which is known to interact with FliN, can be accommodated in this model if it is sandwiched between adjacent FliN tetramers. Modelling of the domain based on its homology to FliN indicates that it would fit between adjacent FliN doughnuts and would give rise to a

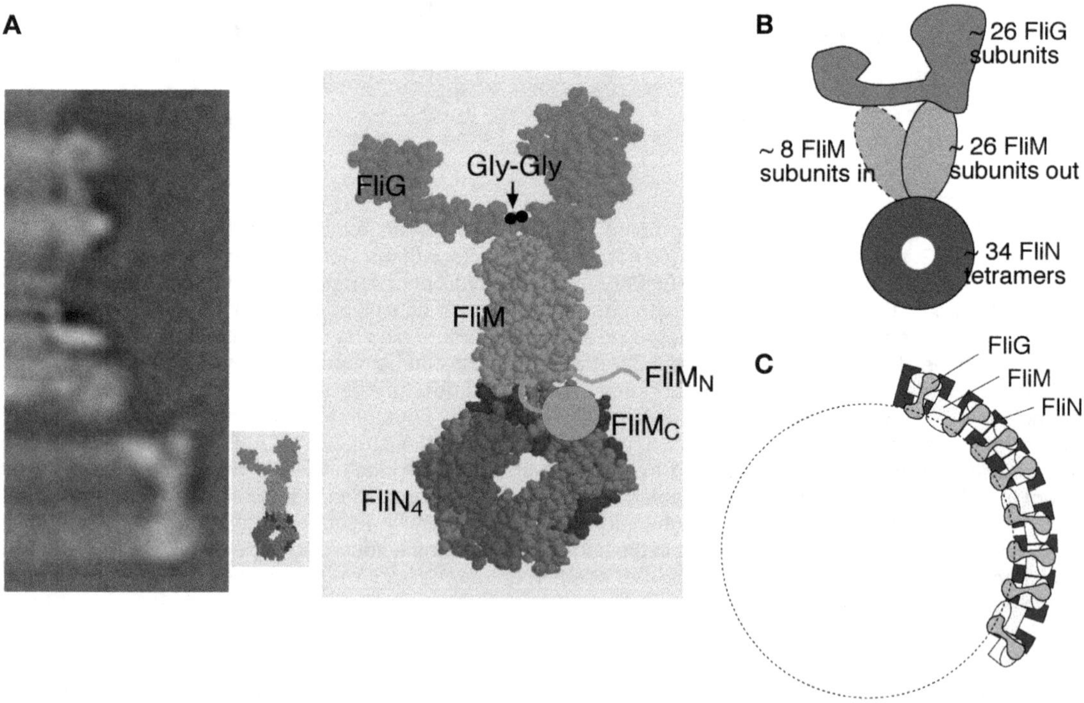

Figure 8.4 Structural model of the switch complex. **A** A proposed arrangement of the three switch-complex proteins and correspondence between the proposed structure and an electron micrograph of the basal body (Thomas *et al.*, 2001). **B** The two orientations proposed for the FliM subunits, and a proposed scheme to account for observed differences in subunit copy number. **C** Top view of the proposed subunit arrangement.

spacing of around 4 nm (M. Sarkar and D.F. Blair, unpublished results), in accordance with the EM (Young *et al.*, 2003; Thomas *et al.*, 2006).

This structural model for the switch complex, together with mutational results on its components, provides the outlines of an hypothesis for switching. The initial step is likely to be the capture of CheY-P by the N-terminal segment of FliM. The CheY-P-binding segment on FliM is joined to the rest of FliM protein by a segment that is poorly conserved and predicted to have non-regular secondary structure (Mathews *et al.*, 1998). This segment might serve as a flexible linker to facilitate the capture of CheY-P and to allow the captured protein to sample other protein surfaces in the neighbourhood, including the hypothesized CheY-binding surface on FliN. Binding of CheY-P to this surface on FliN could reverse the pucker of the FliN tetramer, inducing a twisting or rocking movement that is then transmitted, via FliM, to the C-terminal domain of FliG. The Gly-Gly linker in FliG would function as a hinge to allow movement of the C-terminal domain relative to the rest of the protein. In support of this proposal, Manson and coworkers reported that certain mutations in the Gly-Gly motif exhibit a slow-switching, 'rusty hinge' phenotype (Van Way *et al.*, 2004). Movement of the FliG C-terminal domain would then alter the topography and other features of the rotor stator interface so that the forces there are redirected toward CW rotation.

Stator

The stator complexes function to conduct protons across the membrane and to couple ion flow to rotation (Blair and Berg, 1990). As mentioned, the motor contains about 11 stator complexes that function independently to generate torque (Reid *et al.*, 2006). The complexes are formed from the MotA and MotB proteins and have subunit composition $MotA_4MotB_2$ (Sato and Homma, 2000; Braun *et al.*, 2004). MotB has just a short N-terminal segment in the cytosol, a single membrane segment,, and a large domain in the periplasm that includes segments believed to bind peptidoglycan (Stader *et al.*, 1986; Chun and Parkinson, 1988; DeMot and Vanderleyden, 1994). This attachment to peptidoglycan would ensure that forces generated at the rotor-stator interface are transmitted effectively to the rotor. An invariant aspartate residue is found in the membrane segment of MotB, near the inner end (Fig. 8.5). This aspartate, residue 32 in the MotB protein of *E. coli*, is the only conserved titratable residue in the motor that has been shown to be essential for rotation, and has accordingly been suggested to function directly in H^+ conduction (Sharp *et al.*, 1995; Zhou *et al.*, 1998b). An Asp residue is also found at this position in Na^+-driven motors and may have a role in Na^+ binding and conduction. MotA has four membrane segments, sizable parts in the cytoplasm, and relatively short segments in the periplasm (Fig. 8.5) (Dean *et al.*, 1984; Zhou *et al.*, 1995). Functionally important residues of MotA include two conserved Pro residues near the inner ends of membrane segments 3 and 4 (Braun *et al.*, 1999), and conserved charged resides in the cytoplasmic domain, which form the site of interaction with FliG (Lloyd and Blair, 1997; Zhou *et al.*, 1998a; Yakushi *et al.*, 2006). The stator complexes must presumably remain plugged until being installed in the motor, to prevent wasteful dissipation of the ion gradient. A segment in the periplasmic domain of MotB, just exterior to the membrane segment, has been shown to function in such a gatekeeping capacity (Hosking *et al.*, 2006).

Although there is as yet no high-resolution structural information on the MotA/MotB complex, cross-linking studies and systematic mutagenesis have shed light on the organization of its membrane segments. The membrane segments of the two MotB subunits are together at the centre of the complex, surrounded by membrane segments 3 and 4 of the four MotA subunits, in turn surrounded by an outermost layer formed from membrane segments 1 and 2 of the MotA subunits (Fig. 8.5C) (Braun and Blair, 2001; Braun *et al.*, 2004; Kim *et al.*, 2008). The two Asp32 residues are positioned apart from each other, where they would function within two separate proton channels rather than a single central channel. Conserved residues in the membrane segments are clustered almost exclusively along the linings of the hypothesized channels, and are distinguished by their small size. The channel is predicted to contain few hydrophilic groups. These include just Thr 209, an invariant residue of TM4, and possibly also (depending on the precise helix

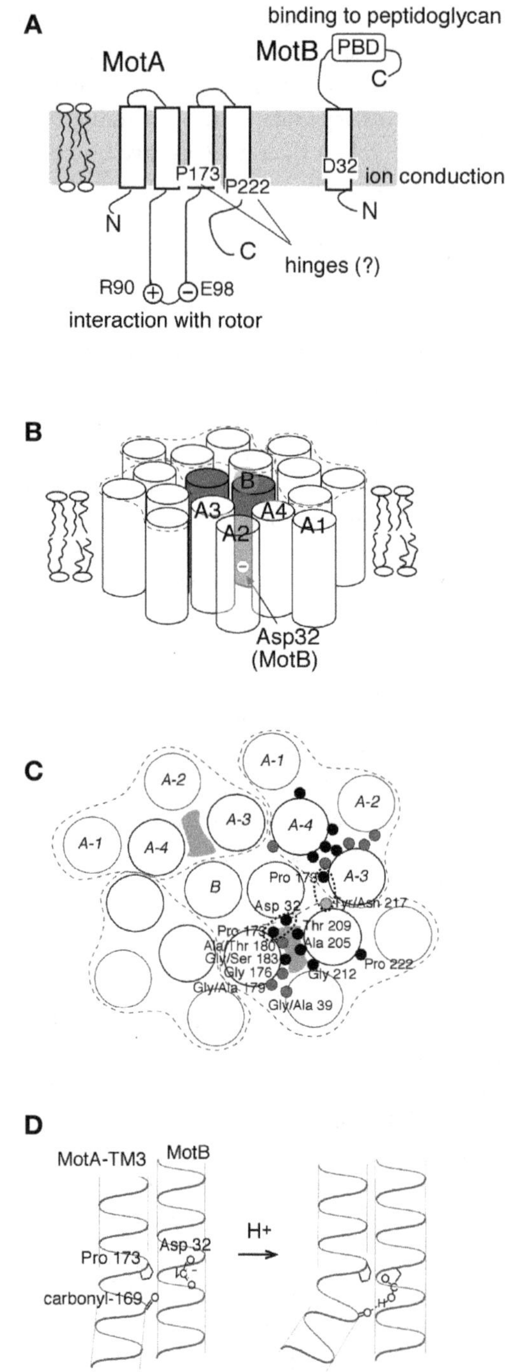

Figure 8.5 Features of the stator. **A** Topology of the MotA and MotB proteins, and functionally important residues identified by mutagenesis. **B** Overall organization of the membrane segments in the $MotA_4MotB_2$ complex. **C** Detailed organization of the transmembrane segments in the complex, as viewed from the 'top' (the outside of the cell) (Braun and Blair, 2001; Braun *et al.*, 2004; Eun A. Kim and D. Blair, unpublished). Invariant positions are shown as black circles, and positions where two fairly similar residues occur are coloured grey. Position 217 is shown as a light-grey circle; this contains either Tyr or Asn, which are of different size but can both provide H-bonding groups (see the text). Close residue relationships that might be important for function are indicated by the dashed ovals. **D** Proposed mechanism for coupling protonation of Asp32 to a conformational change in MotA. Proximity of Asp32 to Pro17 was established in a cross-linking study (Braun *et al.*, 2004). H-bond formation to the carbonyl just below Pro173 would be expected to alter the helix conformation in this region.

orientations) Ser35 of MotB, which is conserved as either Ser or Thr. The scarcity of hydophilic groups and the small size of the channel-lining residues suggests that the conduction pathway is formed from water molecules. In a channel model based on the cross-linking results, a clear pathway is discernible from the outer mouth of the channel down to Asp 32, but below Asp 32 the path appears to be blocked by the side-chains of larger residues. Thus, the structural model suggests that H^+ release into the cell interior would require movements that open a gate in the lower (cytoplasm-proximal) part of the channel.

Some models for the motor mechanism have postulated steps in which the proton is transferred to the rotor at some step, or moves along sites formed jointly by groups on the rotor and stator. Because no single protonatable residue of the rotor is critical for rotation, the actual proton pathway appears more likely to be confined to the stator. In this case, proton movement must be coupled to rotor movement by some means other than direct binding. It was proposed that H^+ association/dissociation at Asp32 might induce conformational changes in the stator, which would cause movements in the cytoplasmic domain of MotA that interacts with the rotor (Zhou *et al.*, 1998b). This conformation-change hypothesis was tested by examining the effects of Asp32 mutations on protease susceptibility of MotA (Kojima and Blair, 2001). Mutations that neutralized Asp32 (thus mimicking the charge-neutralizing effect of protonation) altered the protease susceptibility of MotA, in the cytoplasmic domain that interacts with FliG.

A chemically explicit model for rotation

Many models for the motor mechanism have been proposed; for detailed surveys see, for example, Armitage and Berry (1999), Berry (2000), and Berg (2003). Most models were developed in the absence of much detailed information on the torque-generating components, and have since been ruled out. One recent model appears consistent with the major findings to date, and has also been analysed in sufficient detail to establish its feasibility in quantitative terms (Xing *et al.*, 2006). The model is based on protonation-induced conformational changes in the stator, thus extending an earlier, more basic proposal by Kojima and Blair (Kojima and Blair, 2001). Here, I will describe a chemically detailed model that integrates recent information on the structures of the rotor and stator.

In the working model for the stator complex, the proton-binding residue Asp32 is in close proximity to the conserved residue Pro 173 of MotA (Fig. 8.5). This suggests a specific chemical mechanism for the proton-induced conformational change. Proline residues are known to be important in the function of diverse membrane proteins , where their major effect is to introduce bends, or to increase susceptibility to bending, in helical segments (Brandl and Deber, 1986; Tieleman *et al.*, 2001). A Pro residue in an alpha helix disrupts the normal hydrogen-bonding of the backbone carbonyl four positions before in the sequence. The geometry of a Pro-containing helix can therefore be influenced by the presence of any nearby hydrogen-bond donors that might interact with the unsatisfied carbonyl. In the channel model derived from cross-linking, the side-chain of Asp32 is within hydrogen-bonding distance of carbonyl-169 of MotA, and so the protonated form of Asp32 could form a hydrogen bond to this carbonyl to cause bending of the MotA helix (Fig. 8.5D). The other conserved Pro residue, Pro222, could serve as a second hinge to allow the other cytoplasmic domain to follow.

The stator-complex model (Fig. 8.5) indicates that just two of the MotA subunits contribute directly to the proton channels. In the other two MotA subunits, the residue-169 carbonyl groups are not positioned near Asp32, but the structural model indicates that they are near another potential H-bond donor, the side-chain of Tyr217 (Eun A. Kim and D.F. Blair, unpublished) (Fig. 8.5). While this position is not invariant, it is either Tyr or Asn, thus always providing an H-bond donor. The relevant Tyr217 side-chain extends toward carbonyl-169 of the non-channel MotA subunit, but arises from TM-4 of the channel-forming subunit. Thus, in the context of the conformation-change model, Tyr217 could provide a linkage that coordinates action in the two classes of MotA subunit (channel and non-channel). Specifically, it could ensure that a similar conformational change occurs in both the channel and non-channel subunits, either synchronously or in alternating fashion. Such similarity in the action of the two sets of MotA subunits might help to rationalize the use of the same protein in such seemingly dissimilar structural contexts.

The cytoplasmic domain of MotA has not yet been characterized structurally and remains the least understood element in the motor. The domain does not exhibit strong sequence conservation except in the regions that interact with the rotor. Secondary-structure predictions indicate a largely helical character, which is confirmed by circular dichroism spectroscopy on purified fragment constructs (D. Blair and T. Braun, unpublished results). A mutational study of the domain showed that most missense mutations that eliminate function involve the loss or gain or Pro residues, suggesting that the overall structure of the domain is important but that functionally critical residues are restricted to the region that interacts with the rotor (Zhou and Blair, 1997). Together, the data suggest that the domain functions as a mechanical element to transmit protonation-induced movements to the rotor, possibly serving as a lever arm to amplify the movement.

A hypothesis for torque generation can be constructed by joining this model for the stator to the structural model of the rotor presented above. The model postulates simple in-and-out movements of the MotA cytoplasmic domains, which would take place alternately in the channel- and non-channel subunits (Fig. 8.6A). The movements are driven by cycles of protonation/deprotonation at Asp32, following the H-bond-based mechanism pictured (Fig. 8.5D). Besides driving the rotor, the conformational change would also regulate proton flow, by opening an exit path to the cell interior as the cytoplasmic domain swivels about the Pro173 hinge. At the same time, the upper part of the channel might close, to

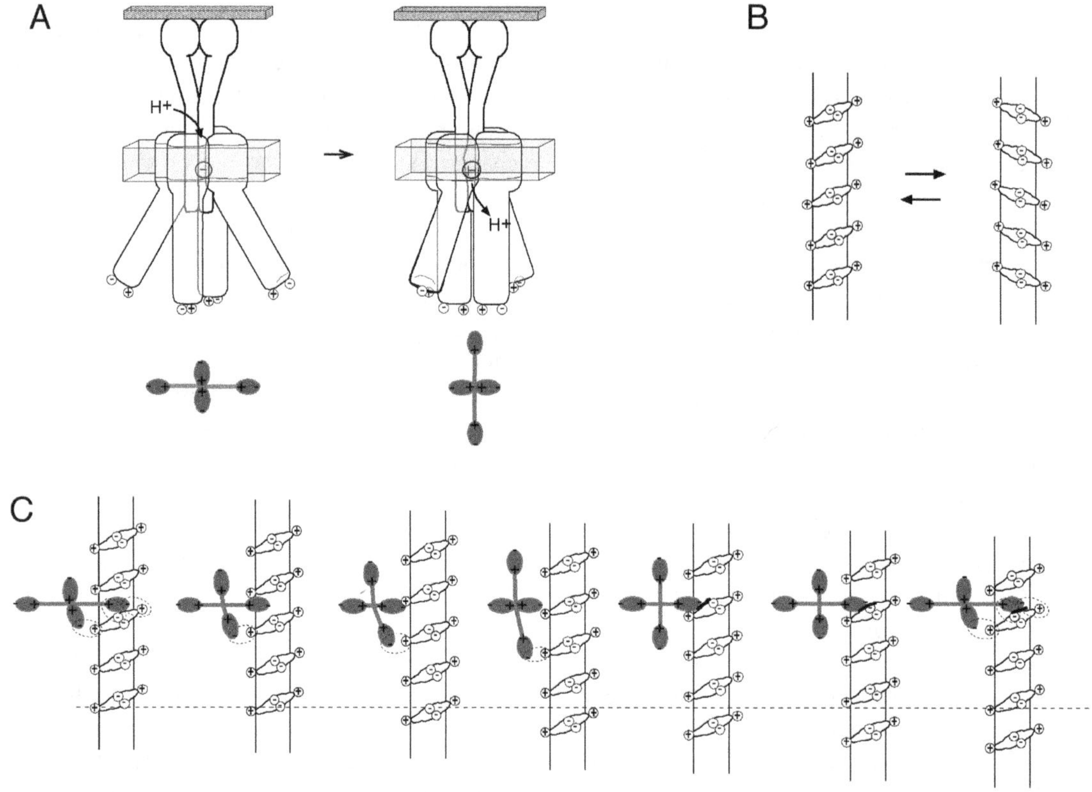

Figure 8.6 Hypothesis for rotation. **A** A specific proposal for the conformational change in MotA. Cytoplasmic domains of two subunits move out, as the other two move in. The charged residues that interact with the rotor are suggested to be at the bottom, possibly at the end of an extended domain that serves as a lever arm to amplify the movement. Cycles of conformational change are driven by proton flow, as indicated. The bottom part of the channel is opened by the conformational change; the upper part could be simultaneously closed by small bending movements in the channel-forming helices. **B** Schematic rendering of the rotor ridges, based on the structural model for the switch complex described above. The important feature is a set of charge-bearing ridges, whose angular orientations would change upon switching, as illustrated. **C** Hypothesized cycle of events in rotation. The model is constructed by positioning the stator of panel A in an appropriate relationship to the rotor of B. Rotation is driven by two adjacent MotA subunits, one pushing inward against the angled face of the charged ridge, and the other moving tangentially and engaging the rotor by an attractive electrostatic interaction. The direction of rotation is determined by the orientation of the ridges, and would reverse when the angle is reversed.

ensure that the channel is not open to both sides at once. Note that proton flow through this part of the channel could be easily blocked by subtle changes in the shape of the channel, if it depends on a precise arrangement of water molecules as proposed. (Threonine is thought to potentiate helix bending (Ballasteros *et al.*, 2000), and the conserved residue Thr209 may have a role here.) Movement of the rotor is driven by the action of two adjacent MotA subunits, each applying force in a different way. One subunit is oriented to push directly inward toward the rotor and would drive rotation by pressing against the angled faces of the ridges on FliG, while the adjacent subunit moves tangentially, engaging the rotor through electrostatic interactions (Fig. 8.6C). The alternating action of the two subunits would produce unidirectional movement of the rotor, and the direction of rotation would be reversed just by reversing the pitch of the ridges on the rotor, as illustrated (Fig. 8.6B and C).

Future directions

Previous physiological measurements on the motor have provided useful global constraints on the mechanism but have not disclosed the details of events occurring within the motor. The ability to detect steps provides a cornerstone for new, potentially very informative, physiological studies. By measuring the rates and sizes of steps, at various ion concentrations and loads, it should be possible to dissect the sequence of events (ion association, conformational change, ion dissociation) in the motor cycle. This approach should be especially powerful when joined to the new methods for measuring Na^+ concentration in cells (Lo *et al.*, 2006).

While the specific model proposed is unlikely to prove correct in all details, it suggests some additional lines of experiment, mainly involving the stator. Previous mutational studies have focused on conserved residues indicated in Fig. 8.4. The structural model of the stator suggests that future mutational studies should focus on the putative linkage residue Tyr217, residues in the vicinity of Pro173 that might influence the bending propensity of TM3, and residues below Asp32 that might function to gate the channel or facilitate proton exit. Structure determination on the stator would obviously be helpful, but the model suggests that some aspects of mechanism could also be addressed through spectroscopic studies. The movements executed by the rotor-interaction sites in MotA are a key feature of the mechanism, and aspects of this movement might be probed by spectroscopic studies of isolated stator complexes, using spin labels or fluorophores attached at key sites.

Another important area for future work will be the flagellar motors of species besides *Escherichia coli* and *Salmonella*. While the literature on flagella of other genera is growing, even basic features of flagellar motor structure and function remain relatively unexplored in species outside the enterics. Given the similarities in flagellar genes of all species one might assume that the core features of mechanism will be similar. Work in other species is nevertheless likely to provide fresh insights. As already noted, studies in *Bacillus* led to the discovery of a clutch-based mechanism for inactivating the motor in the biofilm (Blair *et al.*, 2008), a finding that is likely to prompt searches for analogous clutching mechanisms in other species, including *E. coli*. The switch-complex protein FliN is replaced in *Bacillus* by a much larger protein called FliY (an apparent fusion between FliN and a large piece of FliM), and the default direction of rotation is CW, with CheY being needed for CCW rotation. How does the architecture of the switch complex in Gram-positive species differ from that in *E. coli*, and what is the molecular basis for the different switching properties? Answers to such questions should be uncovered in the next few years, and promise to enlarge our understanding of flagellar motility in the broader microbial world.

References

Aizawa, S.-I. (2001). Flagella. in Bacterial Structure (Academic Press), pp. 155–175.

Aizawa, S.-I., Dean, G.E., Jones, C.J., Macnab, R.M., and Yamaguchi, S. (1985). Purification and characterization of flagellar hook-basal body complex of *Salmonella typhimurium*. J. Bacteriol. *161*, 836–849.

Aldridge, P., and Hughes, K.T. (2002). Regulation of flagellar assembly. Curr. Opin. Microbiol. *5*, 160–165.

Armitage, J.P., and Berry, R.M. (1999). The bacterial flagella motor. Adv. Microb. Physiol. *41*, 291–337.

Armitage, J.P., and Schmitt, R. (1997). Bacterial chemotaxis: *Rhodobacter sphaeroides* and *Sinorhizobium meliloti*--variations on a theme? Microbiology *143*, 3671–3682.

Asai, Y., Kawagishi, I., Sockett, R.E., and Homma, M. (1999). Hybrid motor with H^+ and Na^+-driven components can rotate *Vibrio* polar flagella by using Na^+ ions. J. Bacteriol. *181*, 6332–6338.

Asai, Y., Kawagishi, K., Sockett, R.E., and Homma, M. (2000). Coupling ion specificity of chimeras between H^+-and Na^+-driven motor proteins, MotB and PomB, in *Vibrio* polar flagella. EMBO J. *19*, 3639–3648.

Ballasteros, J.A., Deupi, X., Olivella, M., Kaaksma, E.E.J., and Pardo, L. (2000). Serine and threonine residues bend α-helices in the $\chi_1 = g^-$ conformation. Biophys. J. *79*, 2754–2760.

Berg, H., and Brown, D. (1972). Chemotaxis in *Escherichia coli* analysed by three-dimensional tracking. Nature *239*, 500–504.

Berg, H.C. (1993). Random Walks in Biology (2nd ed.) (Princeton, Princeton Univ. Press.).

Berg, H.C., and Anderson, R.A. (1973). Bacteria swim by rotating their flagellar filaments. Nature *245*, 380–382.

Berg, H.C., Manson, M.D., and Conley, M.P. (1982). Dynamics and energetics of flagellar rotation in bacteria. Symp. Soc. Exp. Biol. *35*, 1–31.

Berg, H.C., and Turner, L. (1993). Torque generated by the flagellar motor of *Escherichia coli*. Biophys. J. *65*, 2201–2216.

Berry, R.M. (2000). Theories of rotary motors. Philos. Trans. R. Soc. Lond. B Biol. Sci. *355*, 503–509.

Berry, R.M., and Berg, H.C. (1999). Torque generated by the flagellar motor of *Escherichia coli* while driven backward. Biophys. J. *76*, 580–587.

Berry, R.M., Turner, L., and Berg, H.C. (1995). Mechanical limits of bacterial flagellar motors probed by electrorotation. Biophys. J. *69*, 280–286.

Blair, D.F., and Berg, H.C. (1988). Restoration of torque in defective flagellar motors. Science *242*, 1678–1681.

Blair, D.F., and Berg, H.C. (1990). The MotA protein of *E. coli* is a proton-conducting component of the flagellar motor. Cell *60*, 439–449.

Blair, K., Berg, H.C., and Kearns, D. (2008). A molecular clutch disables flagella in the *Bacillus subtilis* biofilm. Science *320*, 1636–1638.

Block, S.M., and Berg, H.C. (1984). Successive incorporation of force-generating units in the bacterial rotary motor. Nature *309*, 470–472.

Brandl, C.J., and Deber, C.M. (1986). Hypothesis about the function of membrane-buried proline residues in transport proteins. Proc. Natl. Acad. Sci. U.S.A. *83*, 917–921.

Braun, T.F., al-Mawsawi, L.Q., Kojima, S., and Blair, D.F. (2004). Arrangement of core membrane segments in the MotA/MotB proton-channel complex of *Escherichia coli*. Biochemistry *43*, 35–45.

Braun, T.F., and Blair, D.F. (2001). Targeted disulfide cross-linking of the MotB protein of *Escherichia coli*: Evidence for two H^+ channels in the stator complex. Biochemistry *40*, 13051–13059.

Braun, T.F., Poulson, S., Gully, J.B., Empey, J.C., Van Way, S., Putnam, A., and Blair, D.F. (1999). Function of proline residues of MotA in torque generation by the flagellar motor of *Escherichia coli*. J. Bacteriol. *181*, 3542–3551.

Brown, P.N., Hill, C.P., and Blair, D.F. (2002). Crystal structure of the middle and C-terminal domains of the flagellar rotor protein FliG. EMBO J. *21*, 3225–3234.

Brown, P.N., Mathews, M.A. A., Joss, L.A., Hill, C.P., and Blair, D.F. (2005). Crystal structure of the flagellar rotor protein FliN from *Thermotoga maritima*. J. Bacteriol. *187*, 2890–2902.

Brown, P.N., Terrazas, M., Paul, K., and Blair, D.F. (2007). Mutational analysis of the flagellar protein FliG: Sites of interaction with FliM and implications for organization of the switch complex. J. Bacteriol. *189*, 305–312.

Chen, X., and Berg, H.C. (2000a). Solvent-isotope and pH effects on flagellar rotation in *Escherichia coli*. Biophys. J. *78*, 2280–2284.

Chen, X., and Berg, H.C. (2000b). Torque-speed relationship of the flagellar rotary motor of *Escherichia coli*. Biophys. J. *78*, 1036–1041.

Chun, S.Y., and Parkinson, J.S. (1988). Bacterial motility: membrane topology of the *Escherichia coli* MotB protein. Science *239*, 276–278.

Cohen-Ben-Lulu, G.N., Francis, N.R., Shimoni, E., Noy, D., Davidov, Y., Prasad, K., Sagi, Y., Cecchini, G., Johnstone, R.M., and Eisenbach, M. (2008). The bacterial flagellar switch complex is getting more complex. EMBO J. *27*, 1134–1144.

Dean, G.E., Macnab, R.M., Stader, J., Matsumura, P., and Burke, C. (1984). Gene sequence and predicted amino-acid sequence of the MotA protein, a membrane-associated protein required for flagellar rotation in *Escherichia coli*. J. Bacteriol. *143*, 991–999.

DeMot, R., and Vanderleyden, J. (1994). The C-terminal sequence conservation between OmpA-related outer membrane proteins and MotB suggests a common function in both Gram-positive and Gram-negative bacteria, possibly in the interaction of these domains with peptidoglycan. Mol. Microbiol. *12*, 333–334.

DePamphilis, M.L., and Adler, J. (1971). Fine structure and isolation of the hook-basal body complex of flagella from *Escherichia coli* and *Bacillus subtilis*. J. Bacteriol. *105*, 384–395.

Dimmitt, K., and Simon, M. (1971). Purification and thermal stability of intact *Bacillus subtilis* flagella. J. Bacteriol. *105*, 369–375.

Ely, B., Ely, T.W., Crymes, W.B., and Minnich, S.A. (2000). A family of six flagellin genes contributes to the *Caulobacter crescentus* flagellar filament. J. Bacteriol. *182*, 5001–5004.

Francis, N.R., Sosinsky, G.E., Thomas, D., and DeRosier, D.J. (1994). Isolation, characterization and structure of bacterial flagellar motors containing the switch complex. J. Mol. Biol. *235*, 1261–1270.

Gonzalez-Pedrajo, B., Minamino, T., Kihara, M., and Namba, K. (2006). Interactions between C-ring proteins and export apparatus components: a possible mechanism for facilitating type III protein export. Mol. Microbiol. *60*, 984–998.

Gosink, K.K., and Hase, C.C. (2000). Requirements for conversion of the Na(+)-driven flagellar motor of *Vibrio cholerae* to the H(+)-driven motor of *Escherichia coli*. J. Bacteriol. *182*, 4234–4240.

Hirota, N., Kitada, M., and Imae, Y. (1981). Flagellar motors of alkalophilic *Bacillus* are powered by an electrochemical potential gradient of Na^+. FEBS Lett. *132*, 278–280.

Hosking, E.R., Vogt, C., Bakker, E.P., and Manson, M.D. (2006). The *Escherichia coli* MotAB proton channel unplugged. J. Mol. Biol. *364*, 921–937.

Irikura, V.M., Kihara, M., Yamaguchi, S., Sockett, H., and Macnab, R.M. (1993). *Salmonella typhimurium fliG* and *fliN* mutations causing defects in assembly, rotation, and switching of the flagellar motor. J. Bacteriol. *175*, 802–810.

Katayama, E., Shiraishi, T., Oosawa, K., Baba, N., and Aizawa, S.-I. (1996). Geometry of the flagellar motor in the cytoplasmic membrane of *Salmonella typhimurium* as determined by stereo-photogrammetry of quick-freeze deep-etch replica images. J. Mol. Biol. *255*, 458–475.

Khan, I.H., Reese, T.S., and Khan, S. (1992). The cytoplasmic component of the bacterial flagellar motor. Proc. Natl. Acad. Sci. U.S.A. *89*, 5956–5960.

Khan, S., and Berg, H.C. (1983). Isotope and thermal effects in chemiosmotic coupling to the flagellar motor of *Streptococcus*. Cell *32*, 913–919.

Khan, S., Dapice, M., and Reese, T.S. (1988). Effects of *mot* gene expression on the structure of the flagellar motor. J. Mol. Biol. *202*, 575–584.

Kim, E.A, Price-Carter, M., and Blair, D.F. (2008). Membrane segment organization in the stator complex of the flagellar motor: implications for proton flow and proton-induced conformational change. Biochemistry *47*, 11332–11339.

Kojima, S., and Blair, D.F. (2001). Conformational change in the stator of the bacterial flagellar motor. Biochemistry *40*, 13041–13050.

Larsen, S.H., Adler, J., Gargus, J.J., and Hogg, R.W. (1974a). Chemomechanical coupling without ATP:

the source of energy for motility and chemotaxis in bacteria. Proc. Natl. Acad. Sci. U.S.A. *71*, 1239–1243.

Larsen, S.H., Reader, R.W., Kort, E.N., Tso, W.-W., and Adler, J. (1974b). Change in direction of flagellar rotation is the basis of the chemotactic response in *E. coli*. Nature *249*, 74–77.

Lee, S.-Y., Cho, H.S., Pelton, J.G., Yan, D., Henderson, R.K., King, D.S., Huang, L.-S., Kustu, S., Berry, E.A., and Wemmer, D.E. (2001). Crystal structure of an activated response regulator bound to its target. Nat. Struct. Biol. *8*, 52–56.

Li, C., Corum, L., Morgan, D., Rosey, E.L., Stanton, T.B., and Charon, N.W. (2000). The spirochete FlaA periplasmic flagellar sheath protein impacts flagellar helicity. J. Bacteriol. *182*, 6698–6706.

Lloyd, S.A., and Blair, D.F. (1997). Charged residues of the rotor protein FliG essential for torque generation in the flagellar motor of *Escherichia coli*. J. Mol. Biol. *266*, 733–744.

Lloyd, S.A., Tang, H., Wang, X., Billings, S., and Blair, D.F. (1996). Torque generation in the flagellar motor of *Escherichia coli*: Evidence of a direct role for FliG but not for FliM or FliN. J. Bacteriol. *178*, 223–231.

Lloyd, S.A., Whitby, F.G., Blair, D.F., and Hill, C.P. (1999). Structure of the C-terminal domain of FliG, a component of the rotor in the bacterial flagellar motor. Nature *400*, 472–475.

Lo, C.-J., Leake, M.C., and Berry, R.M. (2006). Fluorescence measurement of intracellular sodium concentration in single *Escherichia coli* cells. Biophys. J. *90*, 357–365.

Lowder, B.J., Duyvesteyn, M.D., and Blair, D.F. (2005). FliG subunit arrangement in the flagellar motor probed by targeted cross-linking. J. Bacteriol. *187*, 5640–5647.

Lowe, G., Meister, M., and Berg, H.C. (1987). Rapid rotation of flagellar bundles in swimming bacteria. Nature *325*, 637–640.

Macnab, R.M. (1996). Flagella and motility. In *Escherichia coli* and *Salmonella typhimurium*: cellular and molecular biology, 2nd ed. ASM Press, Washington, D.C. R.C. F.C. Neidhardt, C.A. Gross, J.L. Ingraham, E.C. C. Lin, K. Brooks Low, Jr., B. Magasanik, W. Reznikoff, M. Riley, M. Schaechter, and H.E. Umbarger, eds.

Macnab, R.M. (2003). How bacteria assemble flagella. Ann. Rev. Microbiol. *57*, 77–100.

Magariyama, Y., Sugiyama, S., Muramoto, K., Maekawa, Y., Kawagishi, I., Imae, Y., and Kudo, S. (1994). Very fast flagellar rotation. Nature *371*, 752.

Manson, M.D., Tedesco, P., and Berg, H.C. (1980). Energetics of flagellar rotation in bacteria. J. Mol. Biol. *138*, 541–561.

Mathews, M.A. A., Tang, H.L., and Blair, D.F. (1998). Domain analysis of the FliM protein of *Escherichia coli*. J. Bacteriol. *180*, 5580–5590.

McMurry, J.L., Murphy, J.W., and Gonzalez-Pedrajo, B. (2006). The FliN-FliH interaction mediates localization of the flagellar export ATPase FliI to the C ring complex. Biochemistry *45*, 11790–11798.

Meister, M., Lowe, G., and Berg, H.C. (1987). The proton flux through the bacterial flagellar motor. Cell *49*, 643–650.

Murphy, G.E., Leadbetter, J.R., and Jensen, G.J. (2006). *In situ* structure of the complete *Treponema primitia* flagellar motor. Nature *442*, 1062–1064.

Noji, H., Yasuda, R., Yoshida, M., and Kinosita, K. (1997). Direct observation of the rotation of F_1-ATPase. Nature *386*, 299–302.

Pallen, M.J., Bailey, C.M., and Beatson, S.A. (2006). Evolutionary links between FliH/YscL-like proteins from bacterial type III secretion systems and second-stalk components of the F_oF_1 and vacuolar ATPases. Protein Sci. *15*, 935–941.

Park, S.-Y., Lowder, B., Bilwes, A.M., Blair, D.F., and Crane, B.R. (2006). Structure of FliM provides insight into assembly of the switch complex in the bacterial flagella motor. Proc. Natl. Acad. Sci. U.S.A. *103*, 11886–11891.

Parkinson, J.S., Parker, S.R., Talbert, P.B., and Houts, S.E. (1983). Interactions between chemotaxis and flagellar genes in *Escherichia coli*. J. Bacteriol. *155*, 265–74.

Passmore, S.E., Meas, R., and Marykwas, D.L. (2008). Analysis of the FliM/FliG motor protein interaction by two-hybrid mutation suppression analysis. Microbiology *154*, 714–724.

Paul, K., and Blair, D.F. (2006). Organization of FliN subunits in the flagellar motor of *Escherichia coli*. J. Bacteriol. *188*, 2502–2511.

Paul, K., Harmon, J.G., and Blair, D.F. (2006). Mutational analysis of the flagellar rotor protein FliN: Identification of surfaces important for flagellar assembly and switching. J. Bacteriol. *188*, 5340–5248.

Reid, S.W., Leake, M.C., Chandler, J.H., Lo, C.-J., Armitage, J.P., and Berry, R.M. (2006). The maximum number of torque-generating units in the flagellar motor of *Escherichia coli* is at least 11. Proc. Natl. Acad. Sci. U.S.A. *103*, 8066–8071.

Ryu, W.S., Berry, R.M., and Berg, H.C. (2000). Torque-generating units of the flagellar motor of *Escherichia coli* have a high duty ratio. Nature *403*, 444–447.

Samatey, F.A., Imada, K., Nagashima, S., Vonderviszt, F., Kumasaka, T., Yamamoto, M., and Namba, K. (2001). Structure of the bacterial flagellar protofilament and implications for a switch for supercoiling. Nature *410*, 331–337.

Samatey, F.A., Matsunami, H., Imada, K., Nagashima, S., Shaikh, T.R., Thomas, D.R., Chen, J.Z., DeRosier, D.J., Kitao, A., and Namba, K. (2004). Structure of the bacterial flagellar hook and implication for the molecular universal joint mechanism. Nature *431*, 1062–1068.

Samuel, A.D., and Berg, H.C. (1995). Fluctuation analysis of rotational speeds of the bacterial flagellar motor. Proc. Natl. Acad. Sci. U.S.A. *92*, 3502–3506.

Sato, K., and Homma, M. (2000). Functional reconstitution of the Na(+)-driven polar flagellar motor component of *Vibrio alginolyticus*. J. Biol. Chem. *275*, 5718–5722.

Sharp, L.L., Zhou, J., and Blair, D.F. (1995). Tryptophan-scanning mutagenesis of MotB, an integral membrane protein essential for flagellar rotation in *Escherichia coli*. Biochemistry *34*, 9166–9171.

Silverman, M., and Simon, M. (1974). Flagellar rotation and the mechanism of bacterial motility. Nature *249*, 73–74.

Sjoblad, R.D., Emala, C.W., and Doetsch, R.N. (1983). Bacterial flagellar sheaths: Structures in search of a function. Cell Motil. Cytoskeleton *3*, 93–103.

Sockett, H., Yamaguchi, S., Kihara, M., Irikura, V.M., and Macnab, R.M. (1992). Molecular analysis of the flagellar switch protein FliM of *Salmonella typhimurium*. J. Bacteriol. *174*, 793–806.

Sowa, Y., Rowe, A.D., Leake, M.C., Yakushi, T., Homma, M., Ishijima, A., and Berry, R.M. (2005). Direct observation of steps in rotation of the bacterial flagellar motor. Nature *437*, 916–919.

Stader, J., Matsumura, P., Vacante, D., Dean, G.E., and Macnab, R.M. (1986). Nucleotide sequence of the *Escherichia coli motB* gene and site-limited incorporation of its product into the cytoplasmic membrane. J. Bacteriol. *166*, 244–252.

Tang, H., Billings, S., Wang, X., Sharp, L., and Blair, D.F. (1995). Regulated underexpression and overexpression of the FliN protein of *Escherichia coli* and evidence for an interaction between FliN and FliM in the flagellar motor. J. Bacteriol. *177*, 3496–3503.

Thomas, D., Morgan, D.G., and DeRosier, D.J. (2001). Structures of bacterial flagellar motors from two FliF-FliG gene fusion mutants. J. Bacteriol. *183*, 6404–6412.

Thomas, D.R., Francis, N.R., Xu, C., and DeRosier, D.J. (2006). The three-dimensional structure of the flagellar rotor from a clockwise-locked mutant of *Salmonella enterica* serovar Typhimurium. J. Bacteriol. *188*, 7039–7048.

Tieleman, D.P., Shrivastava, I.H., Ulmschneider, M.R., and Sansom, M.S. P. (2001). Proline-induced hinges in transmembrane helices: Possible roles in ion channel gating. Proteins: Struct., Func. Bioinform. *44*, 63–72.

Toker, A.S., Kihara, M., and Macnab, R.M. (1996). Deletion analysis of the FliM flagellar switch protein of *Salmonella typhimurium*. J. Bacteriol. *178*, 7069–7079.

Van Way, S., Millas, S.G., Lee, A.H., and Manson, M.D. (2004). Rusty, Jammed, and Well-Oiled Hinges: Mutations affecting the interdomain region of FliG, a rotor element of the *Escherichia coli* flagellar motor. J. Bacteriol. *186*, 3173–3181.

Vogler, A.P., Homma, M., Irikura, V.M., and Macnab, R.M. (1991). *Salmonella typhimurium* mutants defective in flagellar filament regrowth and sequence similarity of FliI to F_0F_1, vacuolar, and archaebacterial ATPase subunits. J. Bacteriol. *173*, 3564–3572.

Washizu, M., Kurahashi, Y., Iochi, H., Kurosawa, O., Aizawa, S.-I., Kudo, S., Magariyama, Y., and Hotani, H. (1993). Dielectrophoretic measurement of bacterial motor characteristics. IEEE Trans. on Indust. Apps. *29*, 286–294.

Welch, M., Oosawa, K., Aizawa, S.-I., and Eisenbach, M. (1993). Phosphorylation-dependent binding of a signal molecule to the flagellar switch of bacteria. Proc. Natl. Acad. Sci. U.S.A. *90*, 8787–8791.

Xing, J., Bai, F., Berry, R.M., and Oster, G. (2006). Torque-speed relationship of the bacterial flagellar motor. Proc. Natl. Acad. Sci. U.S.A. *103*, 1260–1265.

Yakushi, T., Yang, J., Fukuoka, H., Homma, M., and Blair, D.F. (2006). Role of charged residues of rotor and stator in flagellar rotation: Comparative study using H^+-driven and N^+-driven motors in *Escherichia coli*. J. Bacteriol. *188*, 1466–1472.

Yamaguchi, S., Aizawa, S.-I., Kihara, M., Isomura, M., Jones, C.J., and Macnab, R.M. (1986a). Genetic evidence for a switching and energy-transducing complex in the flagellar motor of *Salmonella typhimurium*. J. Bacteriol. *168*, 1172–1179.

Yamaguchi, S., Fujita, H., Ishihara, A., Aizawa, S.-I., and Macnab, R.M. (1986b). Subdivision of flagellar genes of *Salmonella typhimurium* into regions responsible for assembly, rotation, and switching. J. Bacteriol. *166*, 187–193.

Yonekura, K., Maki-Yonekura, S., and Namba, K. (2003). Building the atomic model for the bacterial flagellar filament by electron cryomicroscopy and image analysis. Structure *13*, 407–412.

Young, H.S., Dang, H., Lai, Y., DeRosier, D.J., and Khan, S. (2003). Variable symmetry in *Salmonella typhimurium* flagellar motors. Biophys. J. *84*, 571–577.

Yuan, J., and Berg, H., C. (2008). Resurrection of the flagellar rotary motor near zero load. Proc. Natl. Acad. Sci. U.S.A. *105*, 1182–1185.

Zhou, J., and Blair, D.F. (1997). Residues of the cytoplasmic domain of MotA essential for torque generation in the bacterial flagellar motor. J. Mol. Biol. *273*, 428–439.

Zhou, J., Fazzio, R.T., and Blair, D.F. (1995). Membrane topology of the MotA protein of *Escherichia coli*. J. Mol. Biol. *251*, 237–242.

Zhou, J., Lloyd, S.A., and Blair, D.F. (1998a). Electrostatic interactions between rotor and stator in the bacterial flagellar motor. Proc. Natl. Acad. Sci. U.S.A. *95*, 6436–6441.

Zhou, J., Sharp, L.L., Tang, H.L., Lloyd, S.A., Billings, S., Braun, T.F., and Blair, D.F. (1998b). Function of protonatable residues in the flagellar motor of *Escherichia coli*: a critical role for Asp 32 of MotB. J. Bacteriol. *180*, 2729–2735.

Zhou, Y., Duncan, T.M., and Cross, R.L. (1997). Subunit rotation in *Escherichia coli* F_oF_1-ATP synthase during oxidative phosphorylation. Proc. Natl. Acad. Sci. U.S.A. *94*, 10583–10587.

Flagellum Structure

9

Fadel A. Samatey

Abstract

Since its discovery, the flagellum has fascinated researchers. The scientific community had been wondering about the composition, the construction and the function, at the molecular level, of such an organelle. All available techniques in genetics, molecular biology, electron microscopy, X-ray diffraction, electron cryotomography and molecular dynamics simulation have been used, quite successfully, independently or in a combined manner to answer many of the questions. It is now well established that the flagellum is a complex multi-component organelle that spans from the cell membrane to outside of the cell, with the exception of the spirochaete flagellum that stays in the periplasmic space. The flagellum self-assembles to form a helical propeller that enables prokaryotic cells to swim in their living environments. Diverse structural studies are done to understand the molecular interactions relative to its function. The nature of the flagellum makes it suitable for structural studies using electron microscopy and three-dimensional image reconstruction techniques. Therefore, electron microscopy has given many structural models of flagella from different species of Archaea and Bacteria. Structural investigations of flagellar systems have been more successful by combining X-ray diffraction and electron microscopy. The combination of these techniques produced high-resolution models of the filament from *Salmonella enterica* serovar Typhimurium.

Introduction

The flagellum is an organelle that enables Archaea and Bacteria to swim in their living environment. It is widely spread throughout the archaeal and bacterial world. Flagella come in different sizes; their length and diameter vary among species. They can be polar, peritrichous, and, in some species, sheathed. The first observation of a flagellum obtained by electron microscopy in the early 1940's brought lots of excitement and questions on its structure and from where on the cell it originated (Gray, 1926; Knaysi, 1942; Boltjes, 1948). Many, if not all, of the questions from that time have been answered. As it is often the case, with those answers came new questions that the scientific community is still trying to answer.

Gram-positive and Gram-negative bacteria have essentially identical flagella. In Bacteria, such as *Escherichia* and *Salmonella,* the flagellum can be divided in three main parts: the filament, the hook and the basal body (Fig. 9.1) (Abram *et al.*, 1965; Abram *et al.*, 1966). The filament is made by the polymerization of tens of thousands of copies of the flagellin (FliC). It is the most visible part of the flagellum. The bacterial filament is, in some cases, made of more than one type of flagellin (Minnich *et al.*, 1988; Ely *et al.*, 2000). The hook is a short, highly curved segment made by the polymerization of about one hundred copies of the protein, FlgE. The basal body contains different parts of the flagellum: the motor, the switch complex and the rod (Kojima and Blair, 2004). The

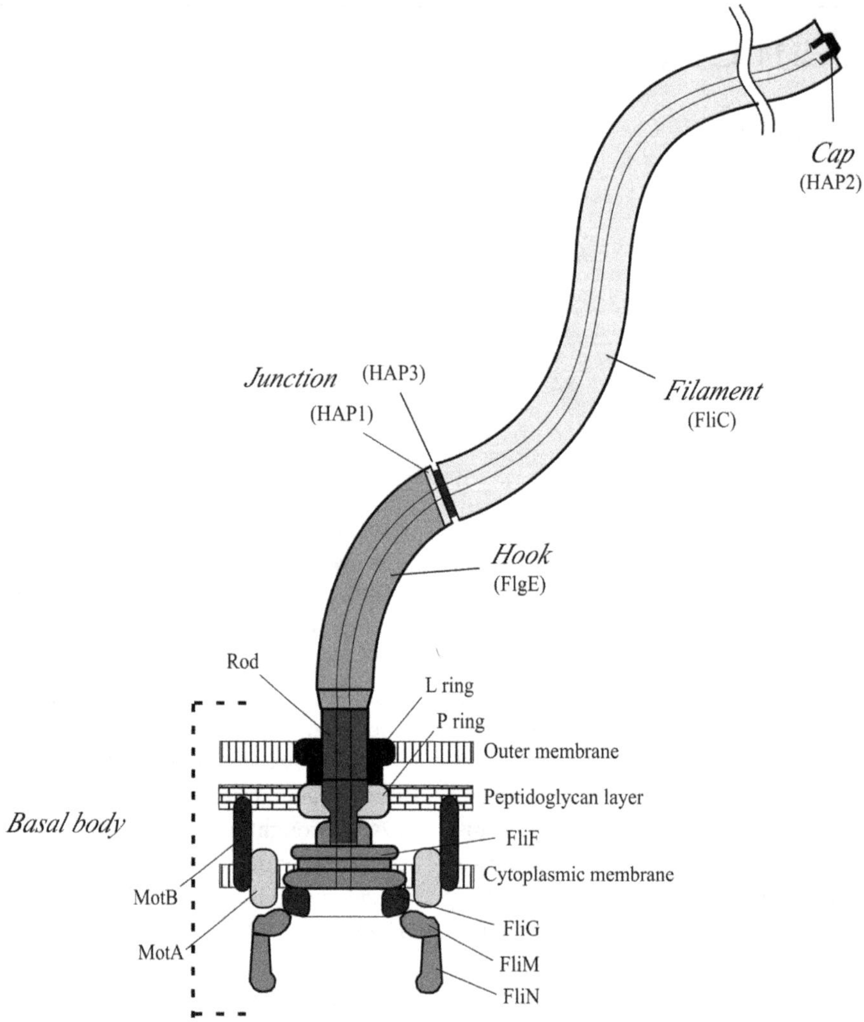

Figure 9.1 Schematic representation of a bacterial flagellum. Reproduced from Namba, K. and Vonderviszt F. (1997). Quart. Rev. Biophys. 30: 1–65, with permission.

rod connects the hook to the motor. It transmits the torque of the motor directly to the hook. The rod is made by the association of five proteins: FliE, FlgB, FlgC, FlgF and FlgG. The stoichiometry and arrangement of the proteins that make the rod is not well established (Jones *et al.*, 1990). The motor is made of a stator, MotA and MotB, and the rotor, FliG, and the MS ring, FliF. All of these proteins are present in multiple copies. The switch complex, that controls the rotation of the motor clockwise or counter-clockwise, consists of two proteins, FliM and FliN (Yamaguchi *et al.*, 1986; Francis *et al.*, 1994). Both FliM and FliN interact with FliG, however their precise stoichiometry is not known. To complete the description of the bacterial flagellum, some additional parts need to be mentioned. Between the hook and the filament there are two junction proteins named HAP1 and HAP3 (hook-associated proteins 1 and 3) and at the tip of the filament, there is a capping structure made by HAP2. The proteins making the rod, the hook, the filament and the junction between the hook and the filament are often referred to as the flagellar axial proteins (Furukawa *et al.*, 2002). The size of each of these proteins varies from species to species. A particular feature in the bacterial flagellum is the existence of a channel, running along the length of the flagellum. Through this channel, the flagellar axial proteins will be exported to assemble at the distal end and form the flagellum. The export process of these proteins is directed by the flagellar type III export apparatus (Macnab, 2003).

Although archaeal and bacterial flagella have a similar function and polymerize to form a filament, their growth, morphology and structure are very different. The archaeal flagellar filaments are, in most cases, made of more than one flagellin protein (Jarrell *et al.*, 1996), while the bacterial flagellar filament is often made of one type of protein. Another important difference between the archaeal and bacterial flagella is their growth mechanism. Because of the lack of a central channel in archaeal flagella, it is difficult to imagine the export of substrates to the distal end during the flagellum growth. Therefore, contrary to bacterial flagella, archaeal flagella have been proposed to grow from the proximal end. Its growth is not dependent on a Type III export apparatus. Archaeal flagella biogenesis machinery is more similar to the bacterial Type IV pili and to the Type II secretion system (Peabody *et al.*, 2003).

High-resolution structure determination of flagellar proteins by X-ray crystallography is difficult to achieve. Many proteins of the flagella are membrane proteins that form complexes of multiple copies. Membrane proteins are challenging for crystallization because they are not water-soluble. Both solubilization and purification of these complexes or of their respective monomers for crystallographic studies are not easy. Flagellar proteins that are not located in the membrane are designed to polymerize by interaction of the N- and C- terminal chains (Vonderviszt *et al.*, 1989; Aizawa *et al.*, 1990; Vonderviszt *et al.*, 1992). These terminal chains are disordered when in solution. Chemicals used for crystallization often induce the aggregation or the polymerization of the flagellar axial proteins. Electron microscopy seems to be the technique of choice for much of the structural work on the flagellum. The helical organization of the different parts making up the flagellum, combined with the large size of the object, makes it appropriate for producing low-resolution structures that show organization of the domains of the flagellum (Mimori *et al.*, 1995) or high-resolution structures that reveal atomic details (Yonekura *et al.*, 2005).

I will mainly describe the structure of the bacterial flagella because there have been many successful structural studies done on this system and some of these led to the atomic model of some parts of the flagellum. I would also like to point out that there have been less structural investigations done on archaeal flagella compared to the bacterial counterpart. Revealing these structures will bring invaluable information in our understanding of the flagellar world. The large diversity of habitats of Archaea, often encompassing 'extreme conditions', brings additional interest to the study of their flagella. The methodology and the techniques used for the studies of bacterial flagella are similar to those being used for archaeal flagella, so it is only a matter of time before we have a high-resolution model of an archaeal flagellum.

The bacterial filament

Early structural studies on the filament

The filament can have different sizes depending on the bacterium. Early structural studies of the filament were done at the start of the 1940s. Experiments in those days first established the method of purification of bacterial flagella and its *in vivo* polymerization ability. These experiments laid the foundation of the structural work done on the filament. X-ray diffraction experiments done on filament fibres from *Proteus vulgaris* and *Bacillus subtilis* showed that the filament has an α-type characteristic and belongs to the group of 'keratin-myosin-epidermis-fibrinogen' (Astbury and Weibull, 1949) with a meridional reflexion at about 5.1 Å. At the beginning of the 1960's, studies of several species of bacteria postulated the polymorphism of the filament (Lowy and McDonough, 1964; Abram and Koffler, 1964; Abram *et al.*, 1966). A decade later, flagella from bacteria, such as *B. subtilis*, *Caulobacter crescentus*, *Escherichia coli*, *P. vulgaris*, *Rhizobium lupini* and *Salmonella enterica* serovar Typhimurium, had been investigated by electron microscopy and by X-ray diffraction. These early studies established the α-helical properties of the filament. New mutant strains of these bacteria, with properties that would later be very useful for the structural studies, were being discovered. Electron microscopy experiments on straight filaments of *S.* Typhimurium and *E. coli* showed that the filament consists of 11 protofilaments that run parallel to the filaments (Asakura, 1970; Kondo and Yanagida, 1975; Calladine, 1975). The study of the filament of wild-type bacteria, if feasible, is nonetheless difficult, as it is a mixture of different conformations of the flagellin monomer (Asakura, 1970). Further studies focused on two types of straight flagellar filament with distinct helical symmetries: the L-type filament in which the longitudinal 11-start helix is left-handed and the R-type filament in which it is right-handed (Kamiya *et al.*, 1980; Fujita *et al.*, 1981). These filaments were isolated from *S.* Typhimurium strains SJW1660 and SJW1655, respectively, both of which are point mutants in flagellin of the wild-type laboratory strain SJW1103. The point mutations G426A and A449V in FliC of strain SJW1103 generated the strains SJW1660 and SJW1655, respectively. With X-ray diffraction data available for well-oriented flagellar filament fibres samples, Namba and colleagues calculated a 20-Å density map and assigned the whole flagellin subunit into it. They provided a better estimation of the central channel through which the export of subunits occurs during the building of the flagellum (Namba *et al.*, 1989).

Towards a 9-Å resolution filament structure

Advancements in electron microscopy combined with new developments in data analysis opened up new possibilities for the improvement of the resolution of density maps of the flagellar filament by electron microscopy (Fujiyoshi, 1989). The use of liquid helium in the cryostage to reduce the radiation damage, combined with a stable field emission source to increase the image quality, contributed to the collection of better data. By averaging about 8000 molecules, Mimori and colleagues produced a high quality density map of the R-type filament from *S.* Typhimurium (Mimori *et al.*, 1995; Mimori-Kiyosue *et al.*, 1996). On the density map they obtained, four distinct domains, D0, D1, D2 and D3, were identified (Fig. 9.2A and B). Domain D0, that makes the inner tube, has a radius from 15 to 30 Å. Domain D1 forms the outer tube that goes from a radius of 35 Å to 55 Å. There are also 11 spoke-like features that seem to connect domains D0 to domain D1 (Mimori *et al.*, 1995). This map also enables a better estimation of the diameter of the central channel of the filament that is between 25 to 30 Å. Previously the central channel had been estimated to be about 60 Å (Namba *et al.*, 1989). Furthermore, by combining new data obtained from X-ray fibre diffraction and from electron microscopy, an improved density map of the filament was obtained (Yamashita *et al.*, 1995; Mimori *et al.*, 1995). On this new map, the assignment of the sequence of the flagellin (FliC) to the four previously found domains was possible. The assignment of the flagellin sequence to domains D0, D1, D2 and D3 was further refined by analysing images taken from filaments lacking some of the terminal chains. The roles of the terminal segments of the flagellin in the polymerization and in the polymorphism of the filament were also revealed (Mimori-Kiyosue *et al.*, 1997; Mimori-Kiyosue *et al.*, 1998). With improvements in the methods of preparation of solutions of well-ordered liquid crystalline fibres of filaments, for X-ray fibre diffraction, Yamashita and colleagues obtained better data on the R-type and the L-type filaments and compared the differences in the packing of these two filaments. They precisely calculated the repeat distance along the protofilament of 51.9 Å and 52.7 Å for the R-type and L-type filaments, respectively (Yamashita *et al.*, 1998).

Structure of a major fragment of FliC

While gradual improvements on the structure of the filament by electron microscopy and image reconstruction were going on, attempts to solve the structure by X-ray crystallography were also tried. Crystallization trials of FliC were not successful. FliC is designed to polymerize into a filament by interaction of the N- and C- terminal chains. Thus, removing the terminal chains will inhibit any propensity of the protein to polymerize. Early studies to map the region of the sequence of FliC involved in the polymerization showed that about 50 amino acid residues from the terminal chains were responsible for the polymerization of FliC. A 41 kDa fragment of FliC (51 kDa) from *S.* Typhimurium was produced by proteolytic digestion of FliC (Vonderviszt *et al.*, 1989; Aizawa *et al.*, 1990; Vonderviszt *et al.*, 1990). This fragment, in which 52 and 44 amino acid residues were missing from the N- and C- termini, respectively, was named F41 (53–450) because of its mo-

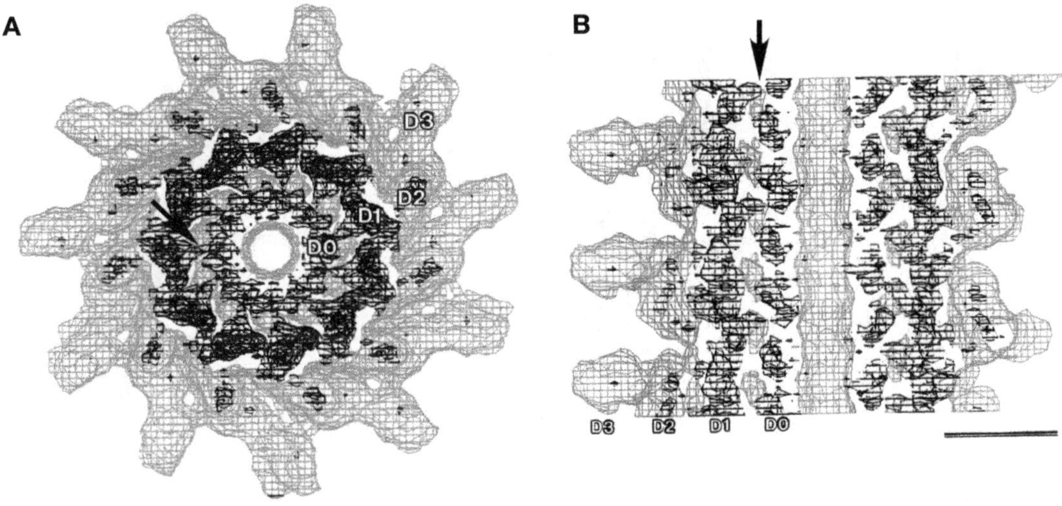

Figure 9.2 The bacterial flagellum: from electron microscopy to protein crystallography. **A** Wire frame representation of 50 Å thick cross-section of the filament in axial view. The density map, obtained by electron cryomicroscopy, was calculated by including data up to 9 Å resolution. The lower contour lines in grey represent the correct volume of the filament and the contour lines in black represent the density level twice as high as those in blue. The identified structural domains are labelled D0, D1, D2 and D3. There are 11 spoke-like features (one of them is marked by an arrow), each of which connects the domains D0 and D1. **B** Side view of the longitudinal section of the filament. The axial length is 147 Å. A pair of concentric tubes of high density is clearly shown with radially tilted spoke-like densities marked by arrows. The scale bar represents 50 Å. Reproduced from Mimori *et al*. (1995). J. Mol. Biol. 249: 69–87, with permission.

lecular weight. The missing amino acid residues correspond to the inner core, domain D0, in the filament. FliC contains four domains: D0, D1, D2 and D3, but F41 has only three domains: D1, D2 and D3. Three different F41 samples of FliC were produced: one from wild type FliC (strain SJW1103), the second from the mutant G426A (strain SJW1660) that gives the L-type filament and the third from the mutant A449V (strain SJW1655) that gives the R-type filament. The mutations G426A and A449V are both present in F41. The different F41 samples were successfully crystallized (Samatey *et al.*, 2000) and a refined atomic model of F41 at 2 Å resolution structure was obtained by X-ray crystallography (Samatey *et al.*, 2001). F41 can be divided in three domains: D1, D2 and D3. Domain D1 extends from Asn 56 to Gln 176 and residues Thr 402 to Arg 450 in N- and C- terminal chains, respectively. Domain D2 consists of a small N-terminal segment, Lys 177 to Gly 189, and segment Ala 284–Ala 401 in the C-terminal chain. Domain D3 is made by segment Tyr 190 – Val 283. The Cα backbone trace of F41 (Fig. 9.3A) shows the secondary structure of the different domains. Domain D1 forms a bundle made by three α-helices and a peculiar motif made of a 'β-turn–β-hairpin–α-helix'. In domain D1, the N-terminal chain forms a long α-helix (residues 57 – 99), followed by a shorter one (104 – 129) and a β-hairpin. The β-hairpin (140–160) is completed by two distorted β-turns (131–134 and 135–138) on the N-terminal side, and slightly more than one turn of α-helix (163–168) on the C-terminal side. The C-terminal chain forms an α-helix (406–447). Domain D2 is mostly made of randomly oriented β-strands that are completed with an α-helix (288–298). Domain D2 could be separated into two sub-domains, D2a and D2b. Domain D3, like D2, is made up of mostly β-strand and one short α-helix. But unlike D2, Domain D3 has a more compact and bulky shape.

Structure of a protofilament

The three different samples of F41, wild type, G426A and A449V, were all crystallized in the space group $P2_1$, with identical cell parameters of a = 51.9 Å, b = 36.5 Å, c = 119.1 Å and β = 91.2° (Samatey *et al.*, 2000). The cell parameter of

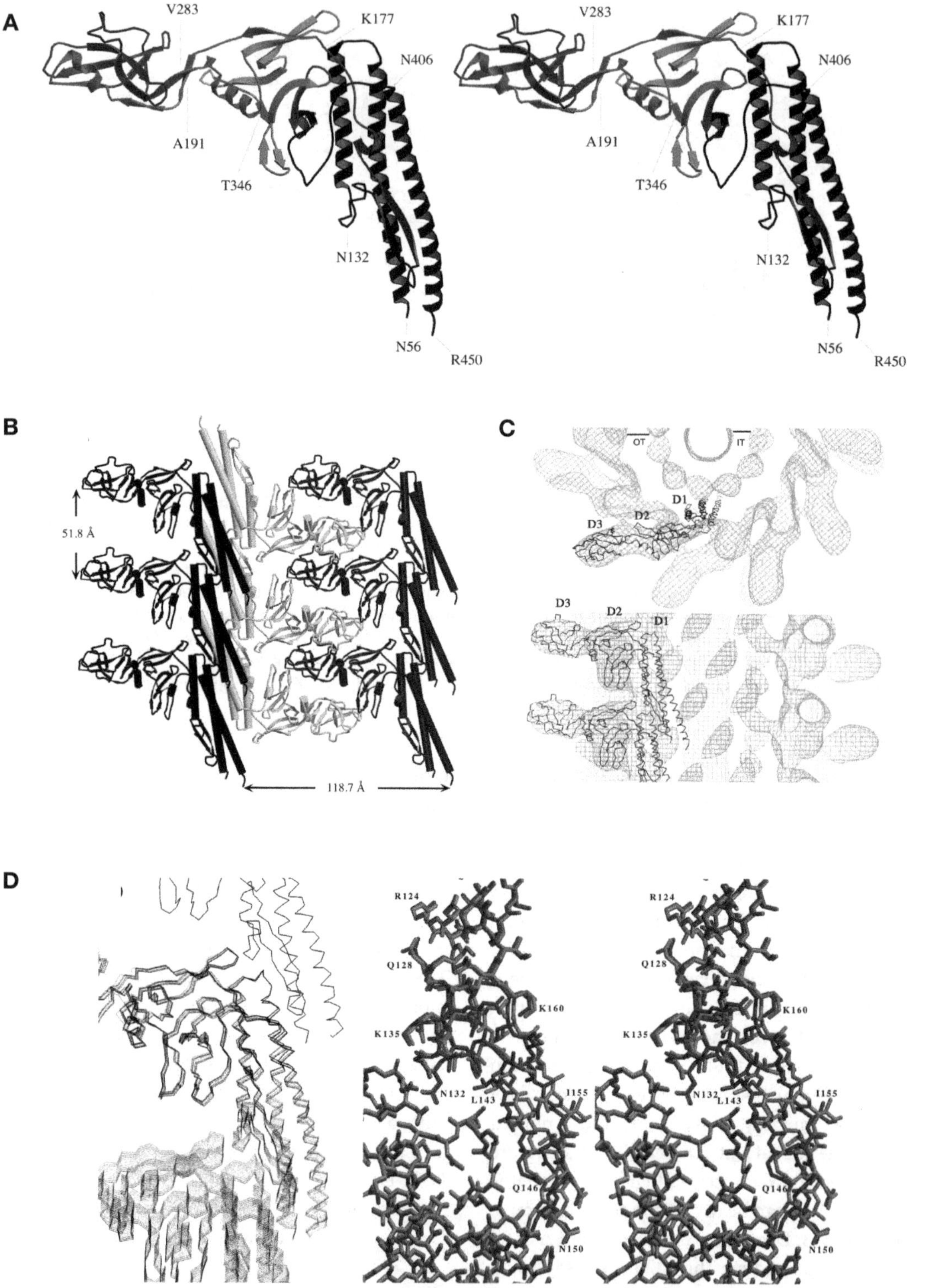

Figure 9.3 Model of the bacterial flagellar filament. **A** Stereo diagram of the Cα backbone trace of F41 with the three domains, D1, D2 and D3. The chain is colour coded from blue to red from the N to the C terminus. **B** Crystal packing of F41 in the a-c plane viewed down the b axis. Each array of the F41 molecules along the 'a' axis, which has been identified as the protofilament of the flagellar filament, has a different colour. The repeat distances are labelled. Antiparallel packing of the protofilaments by the P21 symmetry of the crystal is clearly shown. **C** Docking of a single protofilament model into the density map of the filament that was obtained by

a = 51.9 Å is identical to the repeat distance of two molecules of FliC in the R-type filament (Yamashita *et al.*, 1998). In the crystal, F41 molecules are packed along the a axis just as it would be in a protofilament taken out of the R-type filament (Fig. 9.3B); two protofilaments are packed antiparallel and this pair repeats along the b axis every 36.5 Å, forming a flat sheet. The protofilament from the crystal was docked into the low-resolution density map of the R-type filament obtained by electron microscopy (Mimori *et al.*, 1995). The quality of the docking indicates that the protofilament in the crystal is almost identical to that of the R-type filament (Fig. 9.3C). The three domains, D1, D2 and D3, of F41 correspond to domains D1, D2 and D3, respectively, previously assigned in the density map of the filament (Mimori *et al.*, 1995; Yamashita *et al.*, 1995). Domain D0 in the density map of the filament, that corresponds to the inner-tube, is absent from the crystal structure because of the truncation of the N- and C- terminal chains that are unfolded in the monomeric state of FliC. Domain D1 of the F41 atomic model is located in the outer-tube region of the filament. The outer-tube domain is known to be solely responsible for the formation of two distinct helical lattices, L-type and R-type (Mimori-Kiyosue *et al.*, 1998). This means that the protofilament of the R-type repeat is partly held in the crystal by axial interactions between D1 domains, indicating that each protofilament is an independent, cooperatively switching unit and that the axial inter-subunit interactions between D1 domains along the protofilament are responsible for the two-state switching to produce the two distinct repeat distances of the L-type and the R-type straight filaments. The axial interaction found in the protofilament from the crystal should be similar to that found in the filament. The intersubunit interactions in the axial direction are formed between the first N-terminal α-helix (Asn 56 to Asp 69) and the β-turns/β-hairpin motif (Phe 132 to Asp 151) in domain D1 of the upper subunit and by short segments of two N-terminal α-helices with a short loop connecting these α-helices (Gln 89 to Asp 107), Leu 408 and Gln 409 of the C-terminal α-helix, and Asn 315 of domain D2 of the lower subunit. The nature of the interactions is mainly charge-polar formed between Asp 69 and Asn 100, Asn 132 and Asp 107, the carbonyl oxygen of Ala 149 and Arg 92, and Asp 151 and Gln 409, of the upper and lower subunit, respectively.

Because the docking of a protofilament from the crystal structure into the density map is very good, questions arose. Could a R-type protofilament be converted to a L-type protofilament? Does a protofilament have the ability to switch between the two states of the flagellin represented by the L-type and R-type straight filament structures, and if it does what were the conformational changes induced by the switch? The difference in the repeat distance of the protofilament in the R-type and L-type filament is only 0.8 Å. A simulation of the extension of the crystal protofilament was done on a protofilament made of three molecules of F41 (Samatey *et al.*, 2001). The top subunit of the protofilament was fixed while the bottom subunit was pulled in the axial direction by 0.1 Å steps with its Cα backbone

electron cryomicroscopy. The top panel shows an end-on view from the top, with a 50-Å-thick cross-section of the map. The map is painted where the density is higher than the contour level. Bottom panel shows a side view representation of the docking, with a 50-Å-thick longitudinal section of the map. The resolution of the density map was limited to 20 Å for easy evaluation of the model fitting. D0, D1, D2 and D3 indicate the domains of flagellin; IT and OT represent the inner- and outer-tube regions, respectively. **D** Superimposition of 12 sampled stages of the simulated extension. Domains D1 and D2 of three F41 molecules are shown (bottom half of the top, whole of the middle, and top half of the bottom) and the 12 stages are colour coded from blue to red. Each step of the extension was 0.1 Å, but every five steps is sampled and displayed here; therefore, the bottom subunits in different stages are equally separated by 0.5 Å. The whole β-hairpin at the bottom of domain D1 of the middle subunit shows a discrete jump in its conformation at an extension from 4.5 to 5.0 Å, whereas other parts show only gradual deformation of an elastic nature. On the right panel is a close-up view of the axial contact region in stereo, including the β-hairpin showing the conformational switch. The structures at two sampled stages, at an extension of 4.5 Å in cyan and of 4.7 Å in dark pink, are superimposed to show the conformational switch of the β-hairpin, of which the start and the end are marked by yellow arrowheads. Reproduced from Samatey *et al.* (2001) Nature *410*, 331–337. A colour version of this figure is located in the plate section at the back of the book.

as a rigid body. The energy of the model was minimized after each pulling step. This simulation showed that up to a pulling of 4.5 Å everything is just being translated without a conformational change. However, after that stage, further translation induced an abrupt change in the conformation of the β-hairpin in domain D1. After this conformational change, continued extension caused gradual distortions of the molecules as observed before the transition (Fig. 9.3D). Based on this simulation, the β-hairpin in domain D1 seems to be the key in the filament ability to switch from a R-type to a L-type filament and *vice versa*. Because there was no difference on the atomic structure of three samples of F41, besides the side chain of the mutated amino acid, the effect of the mutations G426A and A449V that produce the L-type and the R-type straight filaments, respectively, could not be explained from the crystal structure. In the crystal structure of the protofilament, amino acid residues Gly 426 and Ala 449 do not have any interacting partner.

These results of the atomic structure of F41 gave valuable information on the interactions within the protofilament and provided a model to explain the switch, from 51.9 Å to 52.7 Å, in the repeat distance of the protofilament of the L-type and the R-type filament. However, it was clear that for a complete understanding of the filament mechanism, it was necessary to have the structure of the different types of filament.

Structure of the filament

New structural studies of the R-type straight filament by electron cryomicroscopy and helical image reconstruction enabled the calculation of a density map from image data up to 4 Å. This big leap forward was made possible by averaging about 41,000 molecular images (Yonekura *et al.*, 2003). In the density map at 4 Å resolution, α-helices and β-sheets could be resolved throughout the FliC molecule but individual strands of β-sheets were difficult to identify in domain D2 and D3. The F41 model, obtained by X-ray crystallography, was docked into the density map as a rigid body. Some reorientations of domain D3 and of the terminal chains were necessary before the docking. The missing terminal chains were traced with large side chains serving as guidelines. The complete atomic model of the R-type filament was built and refined. The model also shows that FliC is made of four domains, D0, D1, D2 and D3, going from the inner core of the filament to the outer part, respectively. The terminal chains in domain D0, which was missing in F41, form an α-helical coiled coil. This structure was explained in detail by Yonekura and his colleagues (Yonekura *et al.*, 2003). The N-terminal α-helix starts from Gln 2 and extends up to Ser 32, whereas the C-terminal α-helix starts from Ala 459 and extends down to Ser 491 (Fig. 9.4A). The spoke region consists of two chains, the first one, NS, from Ser 32 to Ala 44 in the N-terminal chain and the second one, CS, from Glu 454 to Ala 459 in the C-terminal chains. A part from domain D0 that is missing in the crystal structure of F41, is very similar in both the FliC and F41 atomic structure. The superimposition of the structure of FliC from the filament and the crystal structure of F41, relative to both domains D2, shows a slight displacement of domain D3 and the C-terminal chain of F41 is more upright. The β-hairpin motif in domain D1, which was found to play an important role in the switch mechanism of the filament, is perfectly aligned (Fig. 9.4B).

The atomic model of the R-type filament reveals the different inter-subunit interactions. Along the 11-start helix, interactions described earlier with the structure of F41 are completed with contacts between the middle portion of the C-terminal a-helix in domain D0 and the end of the NS spoke. Along the 5-start helix, the two N-terminal α-helices in domain D1 of the subunit 0 interact with the subunit 5 at the level of the β-hairpin of domain D1 and with the lower half of the C-terminal α-helix in domain D1. Contacts are also found in the 16-start helix between the two N-terminal α-helices in domain D1 of subunits -5 and the end of the NS spoke and the first N-terminal α-helix in domain D1 of the subunit 11 (Fig. 9.4C). The interactions in the inner tube are extensive between D0 domains in all directions of the 5, 6- and 11-start helices. These interactions are mostly hydrophobic, giving a high stability to the filament structure. In the outer tube, most of the intersubunit interactions are polar-polar or charge-polar. Based on early structural studies on the filament done by electron microscopy, the flagellar central channel, through which the proteins are transported to

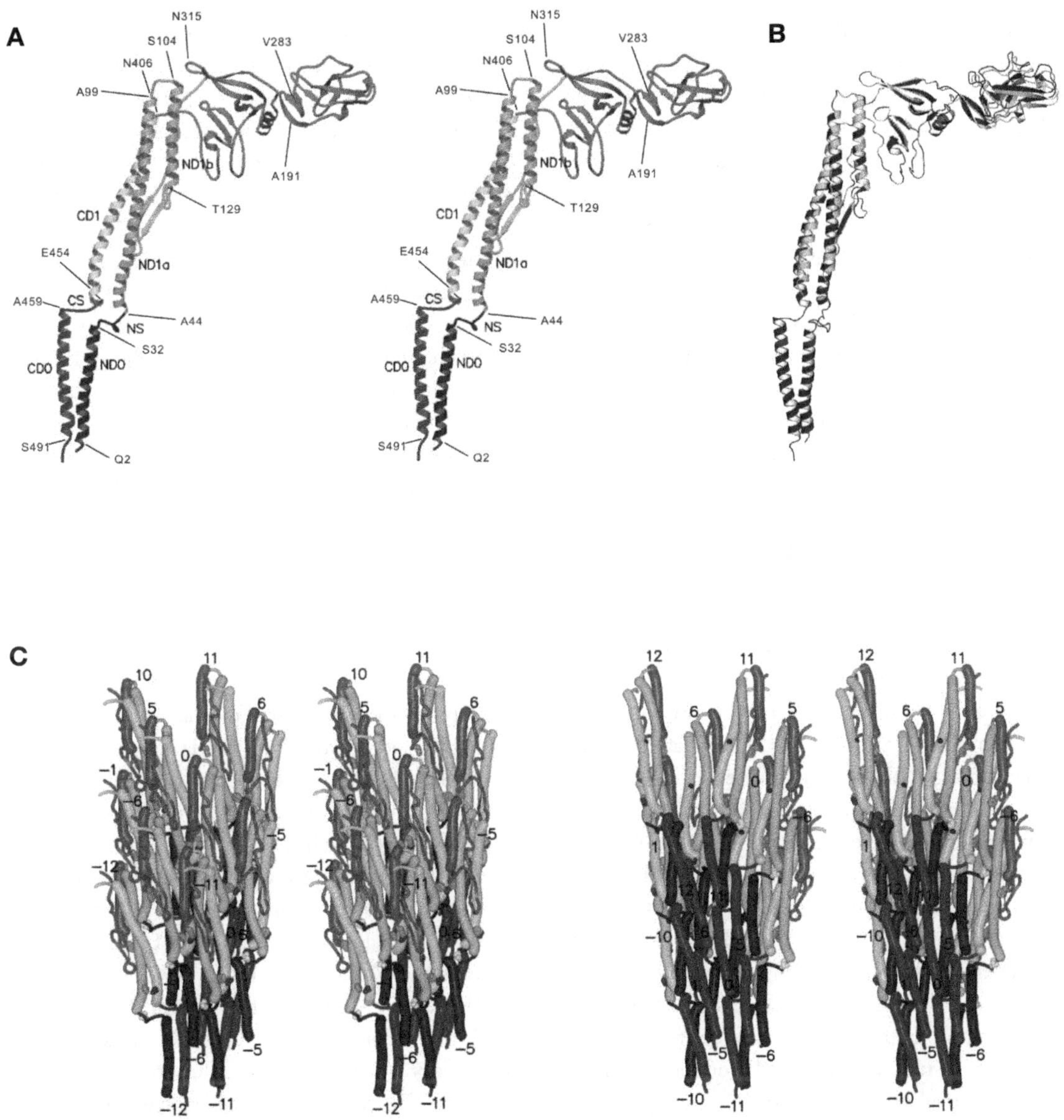

Figure 9.4 Complete model of the bacterial flagellar filament. **A** Stereo diagram of the Cα backbone trace of FliC. **B** Superposition of crystal structure of F41 and FliC. Cα backbone traces of FliC in dark and F41 in grey. The superimposition was done on domain D2. Domain D3 of F41 is slightly shifted. **C** Intersubunit interactions in the inner and outer tubes of the filament in stereo: side view from outside the filament (left panel), side view from inside the filament (right). For more clarity, domains D2 and D3 have been removed. Ten subunits in four protofilaments are displayed. Each protofilament is made of either two or three subunits. Subunits are numbered along the 1-start (right-handed single stranded) helix with a reference subunit numbered 0 and subunits in the distal and proximal directions are given positive and negative numbers, respectively. Various helical lines can be identified easily from these numbers. Thick tubes represent α-helices. Side chains involved in the interactions with amino acid residues that undergo point mutation causing the formation of the R-type straight filament (A449V) and the L-type straight filament (G426A) are presented by space-filling representation. Reproduced from Yonekura *et al.* (2003) Nature *424*, 643–650, with permission.

the distal end, was estimated to be about 30 Å in diameter. With such a diameter, the transport of partly folded proteins would have been possible (Mimori *et al.*, 1995; Mimori-Kiyosue *et al.*, 1996; Mimori-Kiyosue *et al.*, 1997). The complete atomic model of the R-type filament reveals that the diameter of the flagellar central channel is 20 Å. This narrow diameter would not accommodate a partly folded protein for its transport to the distal end. It is more likely that proteins will have to be transported unfolded to the distal end where they will fold and polymerize.

Most of the mutations found on the wild-type strain of *S.* Typhimurium that affect the filament morphology are found in domains D0 and D1. In the structure of the R-type filament, the amino acid residue Ala 449 is located at the bottom of domain D1 before the spoke region. Its neighbours are Phe 53, in the N-terminal α-helix in domain D1 of the same subunit, and Phe 131 in the first of two β-turns in domain D1 of the 5-start subunit. The β-hairpin, previously identified as being responsible for the conformational switch when going from a L-type to an R-type filament, is located just after these two β-turns. The mutation A449V stabilizes its hydrophobic interactions with two surrounding phenylalanine side chains, and therefore stabilizes the lateral disposition of the protofilaments in the R-type structure. Residue Gly 426, the mutation of which into an Ala produces the formation of the L-type straight filament, is located in the middle of the C-terminal α-helix in domain D1. The C_α atom of Gly 426 faces the carbonyl oxygen of Ile 37 in the spoke region of the upper subunit in the protofilament. In a G426A mutation, the C_β atom of alanine would displace the spoke region and change the repeat along the protofilament to 52.7 Å. Complete models of the wild-type filament and of the L-type filament would be necessary to fully appreciate the effects of all the mutations in domains D0 and D1.

The bacterial hook

Low-resolution structure of the hook

The hook is made of the assembly of multiple copies of a single protein called FlgE. FlgE molecular weight varies considerably from species to species. The hook normally has a well-defined length that is regulated by the flagellar regulatory system (Williams *et al.*, 1996; Minamino *et al.*, 1999). Genetic mutations in the regulon produced an abnormally long hook termed polyhook (Kagawa *et al.*, 1979; Johnson *et al.*, 1979). One of the first low-resolution structures of the hook was obtained from structural studies of the hook of *Caulobacter crescentus* by electron microscopy followed by a three-dimensional image reconstruction. *C. crescentus* has a polar flagellum with a short curved hook but these experiments were done on a polyhook mutant (Wagenknecht *et al.*, 1981). The molecular weight of FlgE of *C. crescentus* is about 61 kDa. Structural analysis showed that the hook, like the filament, is made of 11 protofilaments that run parallel to its length and has a central channel for the export of the substrate during the flagellum construction. Similar experiments that validated the results from the polyhook, were done on the wild-type hook of *C. crescentus* and on *S.* Typhimurium (Wagenknecht *et al.*, 1982). *S.* Typhimurium has peritchous flagella with each having a short and highly curved hook made up of about 120 copies of FlgE (41 kDa). These experiments showed that both hooks have very similar structures in spite of the difference in their molecular weight. The flagellar hooks from both species possess continuous interaction in the 6-start helix. This interaction seems to be important for the function of the hook as a universal joint (Wagenknecht *et al.*, 1981; Wagenknecht *et al.*, 1982). A structure of higher-resolution would be necessary to have a complete understanding of the function of the hook.

Experiments of the far-UV circular dichroism spectrum of the hook protein of *S.* Typhimurium showed that it was composed of random coil structures (Vonderviszt *et al.*, 1995). However, calorimetric experiments demonstrated that FlgE has a compact ordered core. Subsequently, secondary structure prediction, combined with far-UV circular dichroism, has shown that monomeric FlgE contains mainly β-sheeted structures. Further NMR experiments, still on the hook protein of *S.* Typhimurium, and proteolytic experiments showed that the terminal parts of FlgE are disordered in solution. About 70 and 30 amino-acid residues at the N and the C-terminal regions, respectively, were making a coiled-coil interaction to form the hook.

Structure of a fragment of FlgE

The disordered terminal chains of FlgE in solution make any attempt to crystallize it very difficult. Full length FlgE has a tendency to polymerize or aggregate when mixed with chemicals used for protein crystallization. To overcome the polymerization of FlgE, 70 and

50 amino acid residues were removed from the N- and C-termini, respectively, producing a 31kDa fragment named FlgE31. FlgE31 was cloned, overexpressed in *E. coli* and crystallized (Samatey *et al.*, 2004). The atomic structure of FlgE31 was determined by X-ray diffraction at a resolution of 1.8 Å. The structure of FlgE31 reveals two of the three domains of FlgE, D1 and D2. Domain D0, that makes the central core of the hook, is absent due to the missing amino acid residues that were removed to prevent polymerization of the full-length protein. In the crystallographic structure, domains D1 and D2 are very distinct. The overall structure of FlgE31 is made of β-strands, but there is a sharp contrast in the packing of the β-strands in domain D1 and in domain D2. In domain D1, the β-strands seem to be randomly oriented and in domain D2 they form an eight-strand β-barrel (Fig. 9.5A). Domain D1 is made by the amino-terminal segment Gly71–Ala144 and the carboxyl-terminal segment Pro285–Ser363. Domain D2 is made by segment Ala145–Lys284. The chain starts with Gly71 in domain D1, makes a succession of four horizontal β-hairpins one above another (Asn79–Leu115 and Gly324–Gln337) followed by a loop that makes an ampersand motif (Thr116–Pro135), a four-stranded (Leu288–Ile314 and Asn357–Ser363) and a two-stranded (Val315–Asn321 and Ser339–Thr346) β sheet (in the upper and lower half, respectively); two consecutive β-turns (Thr346–Phe352, behind the triangular loop); and a vertically extended chain (Pro135–Ala144, in the centre front of the upper half). As predicted by Vonderviszt and co-workers, FlgE31 structure is mainly made of β-sheets. The structure of FlgE31, on its own, brings little to the understanding of the function of the hook.

Model of a straight hook

While the work on FlgE31 was going on, Shaikh and colleagues improved the map of the hook by electron cryo-microscopy. By aligning images of pictures taken from frozen-hydrated polyhooks, they produced a 10 Å resolution map of the hook (Shaikh *et al.*, 2005). The density map of the hook clearly revealed the three domains of FlgE. The inner core domain, D0, is about 1 nm in diameter and 2.5 nm long. Just like in the filament, the inner core is likely to be a α-helical coiled coil made of both terminal chains of FlgE. This domain is absent from the crystal structure of FlgE31. The middle domain lies between radii of 5 and 6 nm and the outermost domain on the surface of the hook at a radius of 7.5 nm.

The docking of FlgE31 into the hook density map was initially done manually and with the use of a real-space refinement program (Chen *et al.*, 2003). The crystallographic structure of FlgE31 could not be docked directly as it appeared in the crystal structure. The docking could only be done by fitting separately each domain directly into the density map, with D1 in the middle part and D2 in the outer feature. Domains D1 and D2 were connected once the docking was done. The docking resulted in a change of the relative orientations of the major axes of domains D1 and D2, an indication of the flexibility in D1-D2 connection (Fig. 9.5B and C). The high-resolution model of the hook shows that all the interactions found in the hook model made from FlgE31 are either polar-polar or polar-charged interactions. Along the 6-start helix there are strong 'head-to-tail' interactions between D2 domains, just as it was predicted in the early structural studies done by electron microscopy. The top and bottom portions of the stacked b-hairpins have interactions along the −5-start and 6-start, respectively, with the C-terminal chains on the lower and upper sides of domain D1. Along the 11-start helix, the main interactions are between the inner side of domain D2 of the lower subunit and the triangular loop (Thr116–Pro135) in the domain D1 on the upper subunit. This D1–D2 interaction is very intimate and seems to hold the protofilament (Fig. 9.5D).

A curved hook and its implications

The hook model was built based on the density map obtained from the straight hook, but the wild-type hook is highly curved, as shown by electron microscopy (Abram *et al.*, 1970). Study of the polymorphism of the hook also showed that, depending on pH and salt concentrations the hook can adopt a different range of curvature and pitches, and in some cases the hook take the form of a right-handed coil with a pitch of 950 Å and a diameter of 350 Å (Kato *et al.*, 1984). These data imply that protofilaments of a curved hook, taken from the inner side of the curve must have

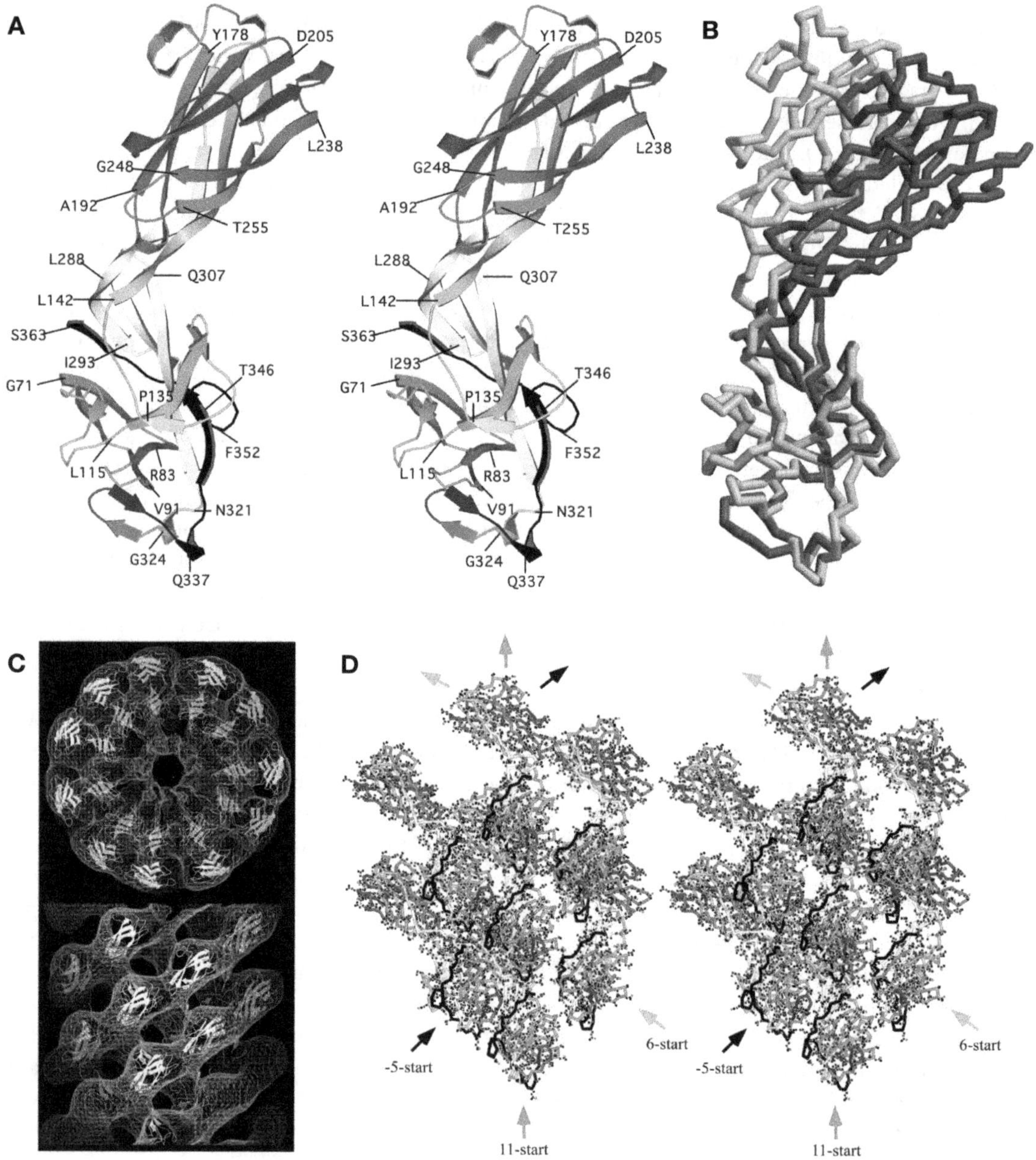

Figure 9.5 Atomic model of the bacterial flagellar straight hook. **A** Stereo view of the Cα backbone trace of FlgE31. The chain is colour-coded from blue to red, going through the rainbow colours, from the N terminus to the C terminus. The model is oriented with domain D1 at the bottom and D2 at the top. **B** Relative domain rearrangement necessary for the docking of FlgE31 into the density map of the straight hook obtained by electron microscopy. The cyan model represents the crystal structure; the dark pink model represent the refined model in the hook. The molecules were aligned respective to the domain D1. **C** Docking of the atomic model of FlgE31 into the outer two domains of the hook. The top panel shows an end-on view and the bottom panel a side view. In the side view, slightly more than half of the hook density on the back is trimmed off for clarity. **D** Stereo view of three protofilaments in the hook with seven subunits viewed from the inside. Atomic models are represented with the Cα backbone in stick form and side chains in ball-and-stick form, and colour-coded as in **A**. Arrows in three different colours indicate the directions of three representative helical lines: the bottom three for D1 domains and the top three for D2 domains. Reproduced from Samatey *et al.* (2004) Nature *431*, 1062–1068. A colour version of this figure is located in the plate section at the back of the book.

a shorter repeat distance then those on the outer side the curve. To reflect these results, a model of a curved hook was created by continuously deforming the initial model of the straight hook into a right-handed helical line with a pitch of 950 Å and a diameter of 350 Å. In the hook, domain D2 makes the outer region, domain D1 the middle region and domain D0 the inner core (Fig. 9.6A and B). Due to the curvature of the wild type hook, the repeat distance between domains of two consecutive molecules of FlgE in the same protofilament will be different. For the same reason, the repeat distance between two similar domains on the most inner protofilament will be different from the repeat distance between two similar domains on the most outer protofilament (Fig. 9.6C). In the case of a right-handed helical hook with a pitch of 950 Å and a diameter of 350 Å, the repeat distance between two D2 of two consecutive molecules of FlgE on the most inner protofilament is 35 Å and is 59 Å in the most outer protofilament (Fig. 9.6B). In the case of domain D1 the distances are 39 Å and 54 Å for the most inner and most outer protofilament, respectively. We remember that, in the case of a straight hook, these distances are all the same and are equal to 46 Å. During the rotation of the flagellum around its central axis, a protofilament position will gradually change, the most inner protofilament taking the place of the most outer and vice versa. It will continuously switch from a compressed to an extended position with the repeat distance between two D2 of two consecutive molecules of FlgE going from 35 Å to 59 Å, respectively. The change in the protofilament position will happen 300 times per second in the case of the proton-driven flagellum of *S.* Typhimurium. To fully understand the behavior of the main axial intermolecular interactions that hold the protofilament structure, Kitao and colleagues did a molecular dynamic simulation of the extension and the compression of a protofilament of the hook. A three-molecule protofilament was surrounded with water molecules to fill the gaps between domains of interacting subunits. The coordinates of domain D1 of the subunit at the bottom and domain D2 of the top subunit were shifted in an opposite direction along the protofilament while positional restraints were applied to the main chain atoms of these domains to compress or extend by 5 Å and 15 Å, respectively. In these simulations, the conformation of the ampersand loop shows little changes after axial compression or extension (Samatey *et al.*, 2004; Furuta *et al.*, 2007). However, the bonding interactions between amino acid residues of the ampersand loop and the inner face of D2 show multiple steps of changes in bonding partners that induces a large slippage at the D1–D2 interface (Fig. 9.6D). The flexibility in the relative domain orientation between D1 and D2, which was indicated in the docking process of the crystal structure of FlgE31 into the hook density map, seems to have a function in the changes in bonding partners. On the hook model, the different subunits are arranged like tiles. This arrangement enables the subunits to slide against each other, giving to the hook its flexibility and helping it to function as a perfect universal joint.

The basal body

The basal body is probably the most complex component of the bacterial flagellum due to the fact that it is made up of completely different proteins which assemble in multi-copies that are intricately connected. To understand the whole system, one would need to be able to dock the atomic structure of each protein into a low-resolution map of the complete basal body. There had been successful X-ray crystallography experiments from which the structures of fragments of proteins, that are part of the motor, were revealed (Lloyd *et al.*, 1999; Brown *et al.*, 2002). These experiments brought valuable information on the structure of the motor but its full mechanism at the molecular level remained to be understood.

Image reconstructions from electron cryomicrographs of the rotor of the flagellar motor enable the calculation of a low-resolution density map of the C ring and M ring from a clockwise locked mutant of *S.* Typhimurium (Thomas *et al.*, 2006). These data showed that the symmetry of individual M rings varies from 24-fold to 26-fold and that of individual C rings varies from 32-fold to 36-fold. Another set of experiments of electron cryotomography done on *Treponema primitia* revealed a 70-Å resolution structure of the flagellar motor, *in situ* (Murphy *et al.*, 2006). On this model, the stator is seen as having 16-fold symmetry and is directly connected to the rotor.

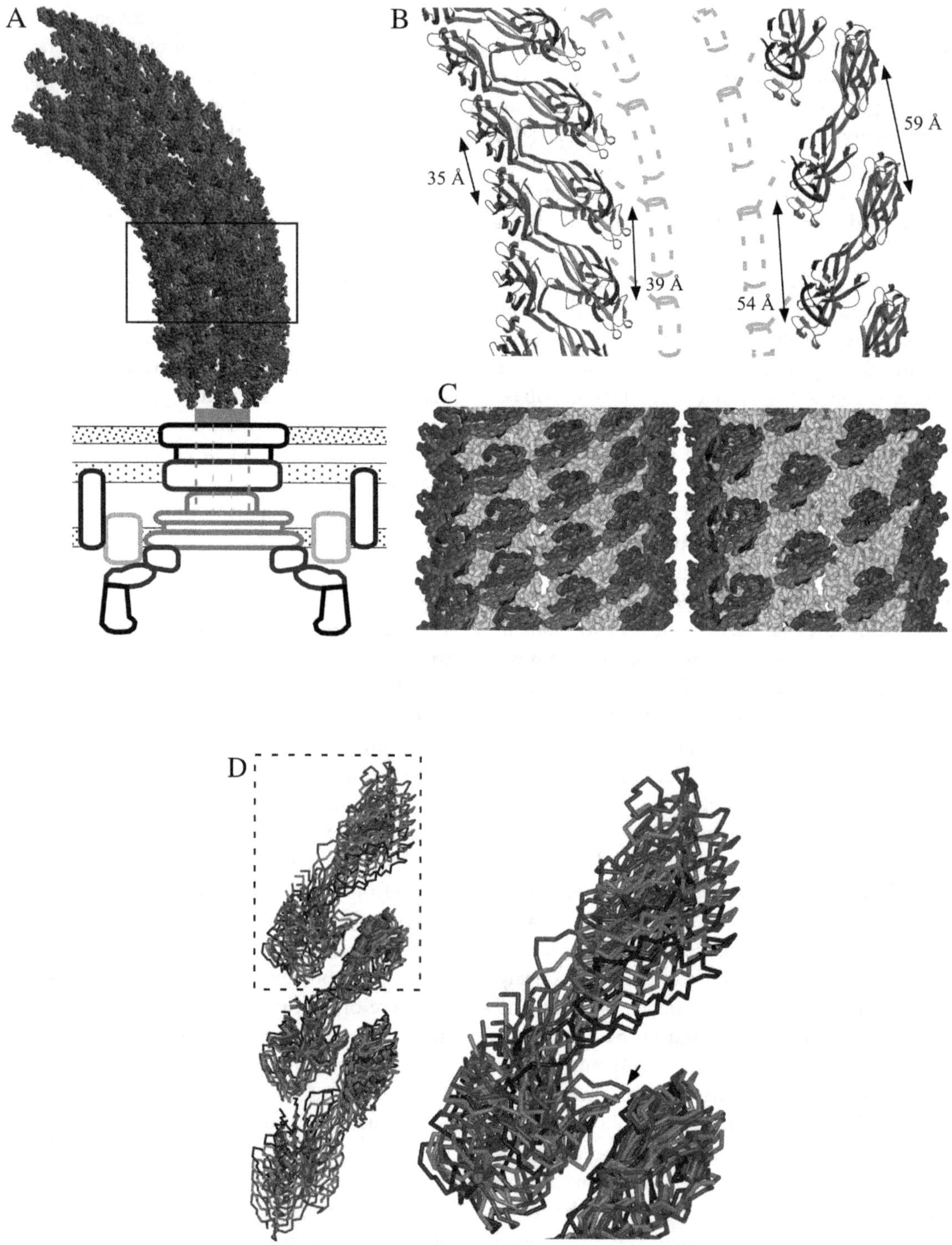

Figure 9.6 Model of the bacterial flagellar coiled hook. **A** Atomic model of the coiled hook with a schematic diagram of the basal body spanning the outer membrane and the cytoplasmic membrane. This coiled hook model is part of a supercoiled polyhook with a helical pitch of 950 Å and a diameter of 350 Å. **B** Magnified image of the coiled hook with the innermost and outermost protofilaments on the left and right, respectively. The inner core domains formed from both terminal chains and the central channel are represented by dotted grey lines. **C** Intermolecular packing arrangements of D2 domains on the inner side (left panel) and on the outer side (right panel) of the coiled hook surface. Domain D1 is coloured in grey. Domain D2 is colour-coded from blue to red, going through the rainbow colours, from the N-terminus to the C-terminus. **D** Simulated extension and compression of hook protofilament. Three subunits were used, and domain D2 at

With both the crystal structure and the electron microscopy models being gradually upgraded, it is only a question of time until we have a high-resolution model of the basal body of the bacterial flagellum. We will then be able to understand the different interactions between the proteins that make up this complex substructure.

Future trends

Many parts in the bacterial flagella such as the hook-associated proteins, HAP1, HAP2 and HAP3, which are essential for the formation and the function of the flagellum (Homma *et al.*, 1984), have not been described. At the distal end of the flagellum, HAP2, forming a pentamer, is found capping the filament (Maki *et al.*, 1998), where it works as a scaffold for the filament. There have been low-resolution structures of the pentameric HAP2 but it is still difficult to see in detail the interaction between HAP2 and the filament (Yonekura *et al.*, 2000; Maki-Yonekura *et al.*, 2003). With the efforts put on the crystallization of HAP2, the atomic model and a detailed map of the interaction between the filament and HAP2 will soon come to light.

Both the flagellar filament and the hook are made of 11 protofilaments that run parallel to the flagellum. However, the atomic structures of the proteins that make up the filament and the hook, FliC and FlgE, respectively, show that these proteins have a completely different folding. They have no homology in sequence nor in structure. The packing and the interaction of subunits in the filament and in the hook are also very different. These differences make us realize the importance of the two junction proteins, HAP1 (FlgK) and HAP3 (FlgL), which connect the hook to the filament. Major fragments of FlgK and FlgL were recently crystallized and their atomic structure solved at a resolution of 2.1 Å and 1.9 Å, respectively. These structures, which will soon be published (Imada *et al*, personal communication), reveal all the interactions in the 'hook–HAP1–HAP3–filament' junction.

The rod, which is made by five different proteins, has evaded, up to now, all attempts to reveal its structure (Saijo-Hamano *et al.*, 2004). However, with the newly described polyrod (Chevance *et al.*, 2007), we now have an adequate sample for structural investigation of the distal end of the rod by electron cryomicroscopy. With all these in hand, in the near future, a high-resolution model of the axial part of the flagellum from *S.* Typhimurium will be available. This model will start from FlgG, located at the distal end of the rod, to HAP2 at the distal end of the flagellum.

Important questions still remain. Are the important folds and motifs in the axial proteins of the flagellum, such as the β-hairpin in domain D1 of FliC, conserved from species to species? Does the functional mechanism of the hook and the filament hold valid for all species? More important, do we have to assume that bacterial flagellar filaments always contain eleven protofilaments? A recent report shows that it is not the case as some bacterial flagellar filaments may have a different quaternary structure (Galkin *et al.*, 2008). More structural studies will be necessary to answer these questions.

Now that the structures of most of the flagellar axial proteins have been solved, groups have started to focus a little more on the membrane components of the flagellum. With gradual improvements in preparation of samples and the use of electron cryotomography, we will most likely see a high quality structure sometime, *in situ*, of the motor embedded in the membrane.

the top subunit was moved upwards, whereas domain D1 of the bottom subunit was moved downwards. Protofilament models at five different stages at 5-Å intervals are superimposed with different colours: dark blue, light blue, green, yellow and red, from the most compressed to the most extended state. The left panel shows the whole three subunits while the right panel shows a magnified view of the upper half. D2 domains at the top and D1 domains at the bottom are at equal intervals (2.5 Å) along the vertical axis. It should be noted that the top corner of the triangular loop of domain D1 of the top subunit (indicated by arrows for the most compressed and the most extended, respectively) shows stepwise movements (a jump from dark blue to light blue, very small movements from light blue to green and to yellow, and another jump to red), whereas domain D2 of the middle subunit stays more or less in the same position. This indicates that the triangular loop of domain D1 and the surface of domain D2 can have distinct side-chain bonding partners depending on the state of extension or compression. Reproduced from Samatey *et al.* (2004) Nature *431*, 1062–1068. A colour version of this figure is located in the plate section at the back of the book.

References

Abram, D., and Koffler, H. (1964). In vitro formation of flagella-like filaments and other structures from flagellin. J. Mol. Biol. *9*, 168–185.

Abram, D., Koffler, H., and Vatter, A.E. (1965). Basal structure and attachment of flagella in cells of *Proteus vulgaris*. J. Bacteriol. *90*, 1337–1354.

Abram, D., Vatter, A.E., and Koffler, H. (1966). Attachment and structural features of flagella of certain bacilli. J. Bacteriol. *91*, 2045–2068.

Abram, D., Mitchen, J.R., Koffler, H., and Vatter, A.E. (1970). Differentiation within the bacterial flagellum and isolation of the proximal hook. J. Bacteriol. *101*, 250–261.

Aizawa, S.I., Vonderviszt, F., Ishima, R., and Akasaka, K. (1990). Termini of *Salmonella* flagellin are disordered and become organized upon polymerization into flagellar filament. J. Mol. Biol. *211*, 673–677.

Asakura, S. (1970). Polymerization of flagellin and polymorphism of flagella. Adv. Biophys. *1*, 99–155.

Astbury, W.T., and Weibull, C. (1949). X-ray diffraction study of the structure of bacterial flagella. Nature *163*, 280–282.

Boltjes, T.Y. (1948). Function and arrangement of bacterial flagella. J. Pathol. Bacteriol. *60*, 275–287.

Brown, P.N., Hill, C.P., and Blair, D.F. (2002). Crystal structure of the middle and C-terminal domains of the flagellar rotor protein FliG. EMBO J. *21*, 3225–234.

Calladine, C.R. (1975). Construction of bacterial flagella. Nature *255*, 121–124.

Chen, J.Z., Fürst, J., Chapman, M.S., and Grigorieff, N. (2003). Low-resolution structure refinement in electron microscopy. J. Struct. Biol. *144*, 144–151.

Chevance, F.F., Takahashi, N., Karlinsey, J.E., Gnerer, J., Hirano, T., Samudrala, R., Aizawa, S., and Hughes, K.T. (2007). The mechanism of outer membrane penetration by the eubacterial flagellum and implications for spirochete evolution. Genes Dev. *21*, 2326–2335.

Ely, B., Ely, T.W., Crymes, W.B., and Minnich, S.A. (2000). A family of six flagellin genes contributes to the *Caulobacter crescentus* flagellar filament. J. Bacteriol. *182*, 5001–5004.

Francis, N.R., Sosinsky, G.E., Thomas, D., and DeRosier, D.J. (1994). Isolation, characterization and structure of bacterial flagellar motors containing the switch complex. J. Mol. Biol. *235*, 1261–1270.

Fujita, H., Yamaguchi, S., Taira, T., and Iino, T. (1981). A simple method for the isolation of flagellar shape mutants in *Salmonella*. J. Gen. Microbiol. *125*, 213–216.

Fujiyoshi, Y. (1989). High resolution cryo-electron microscopy for biological macromolecules. J. Electron Microsc. (Tokyo) *38 Suppl*, S97–101.

Furukawa, Y., Imada, K., Vonderviszt, F., Matsunami, H., Sano, K., Kutsukake, K., and Namba, K. (2002). Interactions between bacterial flagellar axial proteins in their monomeric state in solution. J. Mol. Biol. *318*, 889–900.

Furuta, T., Samatey, F.A., Matsunami, H., Imada, K., Namba, K., and Kitao, A. (2007). Gap compression/extension mechanism of bacterial flagellar hook as the molecular universal joint. J. Struct. Biol. *157*, 481–490.

Galkin, V.E., Yu, X., Bielnicki, J., Heuser, J., Ewing, C.P., Guerry, P., and Egelman, E.H. (2008). Divergence of quaternary structures among bacterial flagellar filaments. Science *320*, 382–385.

Gray, P.H. (1926). A method of staining bacterial flagella. J. Bacteriol. *12*, 273–274.

Homma, M., Kutsukake, K., Iino, T., and Yamaguchi, S. (1984). Hook-associated proteins essential for flagellar filament formation in *Salmonella typhimurium*. J. Bacteriol. *157*, 100–108.

Jarrell, K.F., Bayley, D.P., and Kostyukova, A.S. (1996). The archaeal flagellum: a unique motility structure. J. Bacteriol. *178*, 5057–5064.

Johnson, R.C., Walsh, M.P., Ely, B., and Shapiro, L. (1979). Flagellar hook and basal complex of *Caulobacter crescentus*. J. Bacteriol. *138*, 984–989.

Jones, C.J., Macnab, R.M., Okino, H., and Aizawa, S. (1990). Stoichiometric analysis of the flagellar hook-(basal-body) complex of *Salmonella typhimurium*. J. Mol. Biol. *212*, 377–387.

Kagawa, H., Aizawa, S.I., and Asakura, S. (1979). Transformations in isolated polyhooks. J. Mol. Biol. *129*, 333–336.

Kamiya, R., Asakura, S., and Yamaguchi, S. (1980). Formation of helical filaments by copolymerization of two types of 'straight' flagellins. Nature *286*, 628–630.

Kato, S., Okamoto, M., and Asakura, S. (1984). Polymorphic transition of the flagellar polyhook from *Escherichia coli* and *Salmonella typhimurium*. J. Mol. Biol. *173*, 463–476.

Knaysi, G. (1942). On the width and origin of bacterial flagella. Science *95*, 406–407.

Kojima, S., and Blair, D.F. (2004). The bacterial flagellar motor: structure and function of a complex molecular machine. Int. Rev. Cytol. *233*, 93–134.

Kondo, H., and Yanagida, M. (1975). Structure of straight flagellar filaments from a mutant of *Escherichia coli*. J. Mol. Biol. *96*, 641–652.

Lloyd, S.A., Whitby, F.G., Blair, D.F., and Hill, C.P. (1999). Structure of the C-terminal domain of FliG, a component of the rotor in the bacterial flagellar motor. Nature *400*, 472–475.

Lowy, J., and McDonough, M.W. (1964). Structure of filaments produced by re-aggregation of *Salmonella* flagellin. Nature *204*, 125–127.

Macnab, R.M. (2003). How bacteria assemble flagella. Annu. Rev. Microbiol. *57*, 77–100.

Maki, S., Vonderviszt, F., Furukawa, Y., Imada, K., and Namba, K. (1998). Plugging interactions of HAP2 pentamer into the distal end of flagellar filament revealed by electron microscopy. J. Mol. Biol. *277*, 771–777.

Maki-Yonekura, S., Yonekura, K., and Namba, K. (2003). Domain movements of HAP2 in the cap-filament complex formation and growth process of the bacterial flagellum. Proc. Natl. Acad. Sci. U.S.A. *100*, 15528–15533.

Mimori, Y., Yamashita, I., Murata, K., Fujiyoshi, Y., Yonekura, K., Toyoshima, C., and Namba, K. (1995). The structure of the R-type straight flagellar filament of *Salmonella* at 9 A resolution by electron cryomicroscopy. J. Mol. Biol. *249*, 69–87.

Mimori-Kiyosue, Y., Vonderviszt, F., Yamashita, I., Fujiyoshi, Y., and Namba, K. (1996). Direct interaction of flagellin termini essential for polymorphic ability of flagellar filament. Proc. Natl. Acad. Sci. U.S.A. *93*, 15108–15113.

Mimori-Kiyosue, Y., Vonderviszt, F., and Namba, K. (1997). Locations of terminal segments of flagellin in the filament structure and their roles in polymerization and polymorphism. J. Mol. Biol. *270*, 222–237.

Mimori-Kiyosue, Y., Yamashita, I., Fujiyoshi, Y., Yamaguchi, S., and Namba, K. (1998). Role of the outermost subdomain of *Salmonella* flagellin in the filament structure revealed by electron cryomicroscopy. J. Mol. Biol. *284*, 521–530.

Minamino, T., González-Pedrajo, B., Yamaguchi, K., Aizawa, S.I., and Macnab, R.M. (1999). FliK, the protein responsible for flagellar hook length control in *Salmonella*, is exported during hook assembly. Mol. Microbiol. *34*, 295–304.

Minamino, T., and Namba, K. (2004). Self-assembly and type III protein export of the bacterial flagellum. J. Mol. Microbiol. Biotechnol. *7*, 5–17.

Minnich, S.A., Ohta, N., Taylor, N., and Newton, A. (1988). Role of the 25-, 27-, and 29-kilodalton flagellins in *Caulobacter crescentus* cell motility: method for construction of deletion and Tn5 insertion mutants by gene replacement. J. Bacteriol. *170*, 3953–3960.

Murphy, G.E., Leadbetter, J.R., and Jensen, G.J. (2006). In situ structure of the complete *Treponema primitia* flagellar motor. Nature *442*, 1062–1064.

Namba, K., Yamashita, I., and Vonderviszt, F. (1989). Structure of the core and central channel of bacterial flagella. Nature *342*, 648–654.

Namba, K. and Vonderviszt, F. (1997). Molecular architecture of the bacterial flagellum. Quart. Rev. Biophys. *30*, 1–65.

Peabody, C.R., Chung, Y.J., Yen, M.R., Vidal-Ingigliardi, D., Pugsley, A.P., and Saier, M.H. (2003). Type II protein secretion and its relationship to bacterial Type IV pili and archaeal flagella. Microbiology *149*, 3051–3072.

Saijo-Hamano, Y., Uchida, N., Namba, K., and Oosawa, K. (2004). In vitro characterization of FlgB, FlgC, FlgF, FlgG, and FliE, flagellar basal body proteins of *Salmonella*. J. Mol. Biol. 339, 423–435.

Samatey, F.A., Imada, K., Vonderviszt, F., Shirakihara, Y., and Namba, K. (2000). Crystallization of the F41 fragment of flagellin and data collection from extremely thin crystals. J. Struct. Biol. *132*, 106–111.

Samatey, F.A., Imada, K., Nagashima, S., Vonderviszt, F., Kumasaka, T., Yamamoto, M., and Namba, K. (2001). Structure of the bacterial flagellar protofilament and implications for a switch for supercoiling. Nature *410*, 331–337.

Samatey, F.A., Matsunami, H., Imada, K., Nagashima, S., and Namba, K. (2004). Crystallization of a core fragment of the flagellar hook protein FlgE. Acta Crystallogr. D Biol. Crystallogr. *60*, 2078–2080.

Samatey, F.A., Matsunami, H., Imada, K., Nagashima, S., Shaikh, T.R., Thomas, D.R., Chen, J.Z., Derosier, D.J., Kitao, A., and Namba, K. (2004). Structure of the bacterial flagellar hook and implication for the molecular universal joint mechanism. Nature *431*, 1062–1068.

Shaikh, T.R., Thomas, D.R., Chen, J.Z., Samatey, F.A., Matsunami, H., Imada, K., Namba, K., and Derosier, D.J. (2005). A partial atomic structure for the flagellar hook of *Salmonella typhimurium*. Proc. Natl. Acad. Sci. U.S.A. *102*, 1023–1028.

Thomas, D.R., Francis, N.R., Xu, C., and DeRosier, D.J. (2006). The three-dimensional structure of the flagellar rotor from a clockwise-locked mutant of *Salmonella enterica* serovar *typhimurium*. J. Bacteriol. *188*, 7039–7048.

Vonderviszt, F., Kanto, S., Aizawa, S., and Namba, K. (1989). Terminal regions of flagellin are disordered in solution. J. Mol. Biol. *209*, 127–133.

Vonderviszt, F., Uedaira, H., Kidokoro, S., and Namba, K. (1990). Structural organization of flagellin. J. Mol. Biol. *214*, 97–104.

Vonderviszt, F., Ishima, R., Akasaka, K., and Aizawa, S. (1992). Terminal disorder: a common structural feature of the axial proteins of bacterial flagellum? J. Mol. Biol. *226*, 575–579.

Vonderviszt, F., Závodszky, P., Ishimura, M., Uedaira, H., and Namba, K. (1995). Structural organization and assembly of flagellar hook protein from *Salmonella typhimurium*. J. Mol. Biol. *251*, 520–532.

Wagenknecht, T., DeRosier, D., Shapiro, L., and Weissborn, A. (1981). Three-dimensional reconstruction of the flagellar hook from *Caulobacter crescentus*. J. Mol. Biol. *151*, 439–465.

Wagenknecht, T., DeRosier, D.J., Aizawa, S., and Macnab, R.M. (1982). Flagellar hook structures of *Caulobacter* and *Salmonella* and their relationship to filament structure. J. Mol. Biol. *162*, 69–87.

Williams, A.W., Yamaguchi, S., Togashi, F., Aizawa, S.I., Kawagishi, I., and Macnab, R.M. (1996). Mutations in fliK and flhB affecting flagellar hook and filament assembly in *Salmonella typhimurium*. J. Bacteriol. *178*, 2960–2970.

Yamaguchi, S., Fujita, H., Ishihara, A., Aizawa, S., and Macnab, R.M. (1986). Subdivision of flagellar genes of *Salmonella typhimurium* into regions responsible for assembly, rotation, and switching. J. Bacteriol. *166*, 187–193.

Yamashita, I., Vonderviszt, F., Mimori, Y., Suzuki, H., Oosawa, K., and Namba, K. (1995). Radial mass analysis of the flagellar filament of *Salmonella*: implications for the subunit folding. J. Mol. Biol. *253*, 547–558.

Yamashita, I., Hasegawa, K., Suzuki, H., Vonderviszt, F., Mimori-Kiyosue, Y., and Namba, K. (1998). Structure and switching of bacterial flagellar filaments studied by X-ray fiber diffraction. Nat. Struct. Biol. *5*, 125–132.

Yonekura, K., Maki, S., Morgan, D.G., DeRosier, D.J., Vonderviszt, F., Imada, K., and Namba, K. (2000). The bacterial flagellar cap as the rotary promoter of flagellin self-assembly. Science *290*, 2148–2152.

Yonekura, K., Maki-Yonekura, S., and Namba, K. (2003). Complete atomic model of the bacterial flagellar filament by electron cryomicroscopy. Nature *424*, 643–650.

Yonekura, K., Maki-Yonekura, S., and Namba, K. (2005). Building the atomic model for the bacterial flagellar filament by electron cryomicroscopy and image analysis. Structure *13*, 407–412.

Glycosylation of Flagellins

Susan M. Logan

10

Abstract

In recent years it has become clearly established that a number of bacteria produce flagellins which are posttranslationally modified with novel glycan structures via *O*-linkage. In contrast, the flagella of Archaea, which closely resemble bacterial Type IV pili in structure, can also have glycosylated flagellin monomers, although the glycan moiety in this system is attached in *N*-linkage. The unique mechanisms of assembly of these distinct flagellar structures play an integral role in the respective glycosylation processes. The glycosylation of flagellin has been shown to affect flagellar assembly or play a role in virulence of a number of pathogenic species. This chapter will focus on reviewing the current knowledge of structural diversity of flagellar glycans and mechanisms of glycosylation, and, in addition, describe the unique biological roles of some of these flagellar glycans.

Introduction

The bacterial flagellum is recognized as playing an important role in the ability of many organisms to adapt to diverse biological niches including soil and aquatic environments, as well as the unique microenvironments which bacterial pathogens colonize within their respective protozoal, plant or animal hosts. An extensive body of work directed towards the flagellar nanomachine has provided comprehensive detail on both structure and function and has utilized a range of disciplines including genetics, physiology, bioenergetics and structural biology (Macnab, 2004). More recently, extensive structural analysis has revealed that post-translational modification of the flagellin structural proteins by glycosylation occurs in many bacterial species (Logan, 2006). While the molecular mechanism whereby these flagellin structural proteins are glycosylated is as yet poorly understood, the role these glycans play in either assembly of flagella filaments or in unique biological interactions is becoming clearly established.

Bacterial flagellar structures

The most extensively studied flagellar systems are from *Escherichia coli* and *Salmonella enterica* serovar Typhimurium, which produce peritrichous flagella composed of a single flagellin monomer (FliC). Peritrichous flagella are distributed uniformly over the surface of the cell and are commonly found on many members of the family Enterobacteriaceae. Currently, there is no evidence in the literature indicating that the peritrichous flagellins from Enterobacteriaceae are glycosylated.

In contrast, a number of organisms produce polar flagella (Table 10.1). Polar flagella may be found as either a single filament (monotrichous) or in a bundle (multitrichous) and can extend from either one (polar) or both (bipolar) ends of the cell and, in some cases, these flagella may be covered with a sheath (for example *Helicobacter pylori*, *Vibrio parahaemolyticus*). Glycosylation of polar flagella appears to be the norm rather than the exception (Table 10.1). Types of polar flagella include simple flagella which are composed of only a single flagellin protein monomer and are typical of *Legionella pneumophila* and *Pseudomonas aeruginosa*. Alternatively, some

Table 10.1 Flagellar structures and flagellin glycoproteins

Organism	Flagella type	Glycan linkage	Number of sites	Glycan Structure	Role in flagellar assembly
Gram-negative					
Aeromonas caviae Sch3	P, C LAF, C	ND ND	ND ND	ND ND	Y ND
Aeromonas caviae UU 51	P, C LAF, C	*O*-linkage ND	FlaA 6 sites, FlaB 7 sites ND	Pse5Ac7Ac8Ac ND	Y ND
Aeromonas hydrophila AH3	P, C LAF, C	ND ND	ND ND	ND ND	Y Y
Campylobacter jejuni 81-176	P, C	*O*-linkage	FlaA 19 sites	Pse5Ac7Ac, Pse5Ac7Am, Pse8OAc Pse5Am7AcGlnAc	Y
Campylobacter coli VC167	P, C	*O*-linkage	FlaA 16 sites	Pse5Ac7Ac, Leg5Am7Ac, Leg5AmNMe7Ac	Y
Caulobacter crescentus CB15	P, C	ND	ND	ND	Y
Helicobacter pylori 1061	P, C, SH	*O*-linkage	FlaA 7 sites, FlaB 10 sites	Pse5Ac7Ac	Y
Helicobacter pylori NCTC11637	P, C, SH	*O*-linkage	FlaA 7 sites, FlaB 10 sites	Pse5Ac7Ac	Y
Pseudomonas aeruginosa PAK	P, S	*O*-linkage	FliC 2 sites	Rhamnose linking sugar, heterogeneous glycan with novel, conserved trisaccharide capping	N

Pseudomonas aeruginosa JJ692	P, S	*O*-linkage	FliC 2 sites	Rhamnose	N
Pseudomonas aeruginosa PAO	P, S	*O* linkage	FliC 2 sites	Deoxyhexose linking sugar, 209 Da capping structure	N
Pseudomonas syringae	P, S	*O*-linkage	FliC 6 sites	β-D-Qui*p*4N(3-hydroxy-1-oxobutyl)2Me-(1,3)-α-L-Rha*p*-(1,2)-α-L-Rha*p*	N
Gram-positive					
Clostridium botulinum FE9909	Pt, C	*O* linkage	FlaA 7 sites	αLeg5GluNMe7Ac	ND
Listeria monocytogenes	Pt, S	*O*-linkage	FliC 6 sites	β-GlcNAc	N
Archaea					
Halobacterium halobium	Pt, C	*N*-linkage	ND	Glc(4–1)GlcASO_4(4–1)GlcASO_4(4–1)GlcASO_4	Y
Methanococcus voltae	Pt, C	*N*-linkage	FlaB1 5 sites, FlaB2 5 sites, FlaB3 2 sites, FlaA 1 site	β-ManpNAcA6Thr-(1–4)-β-GlcpNAc3NAcA-(1–3)-β-GlcpNAc	Y

ND, not determined; P, polar; LAF, lateral; Pt, peritrichous; Pp, periplasmic; S, simple; C, complex; SH, sheathed; Y, yes; N, no.

organisms produce complex flagella which are composed of more than one flagellin protein monomer, for example *Caulobacter crescentus, H. pylori* and *Campylobacter jejuni*.

In addition to polar flagella, many bacterial species produce lateral flagella (LAF). These are visible as a limited number of filaments extending at random points on the cell surface but not from either pole and they are not as abundant as peritrichous flagella of Enterobacteriaceae. These filaments also usually comprise more than one structural protein and are often produced when cells are grown on solid surfaces and facilitate a swarming type of motility. Glycosylation of lateral flagellin has been described in a limited number of cases.

Another unique flagellar structure is the periplasmic flagellum of spirochaetes (Charon and Goldstein, 2002). These flagellar organelles are attached subterminally to each end of the cell cylinder and the flagella rotate within the periplasmic space conferring a unique form of motility to spirochaete cells. The flagellar filament is a complex type composed of core FlaB flagellin proteins which may be enveloped by a sheath-like structure composed of a FlaA protein. The FlaA protein monomer shows no sequence similarity to other bacterial flagellins (Ge *et al.*, 1998). To date, only indirect evidence for glycosylation of spirochaete flagellins has been presented.

The flagellar structure produced by Archaea is quite distinct from bacterial flagella and is more closely related structurally to bacterial Type IV pili. The flagellar filaments from the archaeal species studied to date are complex in nature, with the exception of the crenarchaeon, *Sulfolobus solfataricus* (Szabo *et al.*, 2007) and glycosylation of flagellin monomers has been reported.

Flagellar glycan structures

Archaea

The first structural characterization of a flagellar glycan was completed for the flagellin of *Halobacterium halobium* and was shown to comprise *N*-linked oligosaccharides of sulphated 1,4-glucuronic acid (Wieland *et al.*, 1985). More recently, a detailed structural characterization of the flagellins from *Methanococcus voltae* revealed that these proteins were glycosylated with an *N*-linked trisaccharide with a mass of 779 Da. Mass spectrometry examination of a proteolytic digest of flagella identified a total of 14 sites of linkage on the four flagellin structural proteins (FlaA, FlaB1, FlaB2 and FlaB3) and MS/MS fragmentation of a glycopeptide demonstrated that the novel glycan was composed of three sugar residues of masses 318, 258 and 203 Da. A typical *N*-linked eukaryotic sequon (Asn-Xaa-Ser/Thr) was identified as the site of glycan attachment rather than the more recently characterized bacterial *N*-linked sequon (Asp/Glu-Xaa-Asn-Xaa-Ser/Thr) (Kowarik 2006). The novel glycan structure was elucidated by NMR analysis and shown to be composed of βMan*p*NAcA6Thr-(1,4)-β-Glc*p*NAc3NacA-(1,3)-βGlc*p*NAc linked to Asn (Voisin *et al.*, 2005). In both *M. voltae* and *H. halobium*, the same glycan was found to decorate the respective S-layer protein, demonstrating a common *N*-linked pathway for these two major surface structures in Archaea.

Genetic characterization of *M. voltae* has identified genes involved in biosynthesis of the flagellar glycan as well as in transfer of the glycan to the flagellin protein. An STT3 oligosaccharyltransferase homologue, *aglB*, has been shown to encode a protein which is involved in the transfer of the glycan from a dolichol carrier to the flagellin protein while *aglA* encodes a glycosyltransferase which is responsible for transfer of the terminal sugar to the glycan (Chaban *et al.*, 2006). A third gene which has been functionally characterized through complementation in yeast, *aglH*, appears to be responsible for the attachment of GlcNAc to the lipid-linked carrier (Shams-Eldin *et al.*, 2008). The identification of β-GlcNAc as the linking sugar to asparagine is of significance as this is the same sugar as that used in higher organisms for the attachment of complex and polymannose oligosaccharides. The similarity of these systems points to a potential to utilize archaeal enzymes for glycoengineering of eukaryotic glycoproteins (Schaffer and Messner, 2004).

Gram-negative bacteria

Pseudomonas *spp.*

While Pseudomonads are found in many natural environments, there are also numerous examples where these organisms have been shown to be opportunistic pathogens of both plants and animals. Motility has been recognized as critical to the pathogenic process for the plant pathogen *P. syringae,* as well as the human pathogen *P. aeruginosa,* and in both organisms the flagellin structural protein has been shown to be glycosylated.

Strains of *P. syringae* produce a tuft of polar flagella which is composed of a single flagellin protein monomer, FliC. Six sites of glycosylation have been identified on flagellin from *P. syringae* pv tabaci and mutation of these sites demonstrated that glycosylation was essential for adhesion, swarming and virulence on host tobacco leaves but was not required for flagellar assembly (Taguchi *et al.,* 2006b). The pathovar classification of *P. syringae* is due to virulence towards different host plant species and flagellin has been shown to be an elicitor of a hypersensitive reaction (HR) towards non-host plants for *P. syringae* pv tabaci, *P. syringae* pv glycinea and *P. syringae* pv tomato infections. Glycosylation of each flagellin has been shown to be critical for the induction of this response (Taguchi *et al.,* 2003a, Taguchi *et al.,* 2003b; Takeuchi *et al.,* 2003). Recent studies have also revealed that glycosylation plays a role in stabilization of the filament structure (Taguchi *et al.,* 2008).

Structural analysis of the flagellin from *P. syringae* pv tabaci 6605 revealed that the flagellin was glycosylated with a novel trisaccharide consisting of two rhamnosyl (Rha) residues and one modified 4-amino-4,6-dideoxyglucosyl (Qui4N) residue; β-D-Qui*p*4N(3-hydroxy-1-oxobutyl)2Me-(1,3)-α-L-Rha*p*-(1,2)-α-L-Rha*p*. In contrast, although the flagellin proteins from *P. syringae* pv tabaci 6605 and *P. syringae* pv glycinea are identical in amino acid sequence, the trisaccharide isolated from *P. syringae* pv. glycinea race 4 flagellin was distinct in that it contained both L-Rha and D-Rha at a molar ratio of 4:1 (Takeuchi *et al.,* 2007). It remains to be determined if this change in chirality of the Rha residues in the glycan structure contributes to the unique elicitor activity of each flagellin with respect to non-host plants. At the genetic level, a glycosylation island (GI) has been identified between *flgL* and *fliC* which contains three ORFs; *orf1* and *orf2* which show homology to *orfN* a glycosyltransferase from the *P. aeruginosa* flagellar GI, and orf3 which shares homology to a putative 3-oxoacyl-(acyl carrier protein) synthase of *P. putida.* Deletion of these genes reduced both virulence for respective host plants and HR inducing activity in non-host plants (Taguchi *et al.,* 2003a; Taguchi *et al.,* 2003b; Ishiga *et al.,* 2005, Taguchi *et al.,* 2006a).

P. aeruginosa also produces a simple polar flagellum and flagellins are classified as type-a or type-b based on amino acid sequence, antigenicity and molecular weight. Structural analysis of both flagellin types has identified two sites of glycosylation in each monomer where a novel glycan is attached (Schirm *et al.,* 2004a; Verma *et al.,* 2006). Flagellin from *P. aeruginosa* PAK (type–a) is modified with a heterogeneous glycan of up to 11 monosaccharide units and the sugar linking the glycan to protein has been identified as rhamnose. In contrast, the glycan from *P. aeruginosa* PAO (type b) flagellin is simpler and also appears to be linked to protein via a deoxyhexose monosaccharide to which is attached distally a unique modification of 209 Da. The structure of this 209 Da residue has yet to be determined.

A GI has also been identified in both *P. aeruginosa* strains in a similar location to that identified in *P. syringae,* between *flgL* and *fliC.* The GI of strain PAK consists of 14 ORFs, many of which encode proteins which show homology to enzymes involved in carbohydrate biosynthesis. The genetic content of this island has been shown to be variable amongst other type-a producing strains indicative of variability in glycan structure of type-a flagellins (Arora *et al.,* 2001; Arora *et al.,* 2004). In contrast, the GI of strain PAO contains only four ORFs, likely reflective of the simpler glycan structure found on strains producing type-b flagellins. Both GIs contain a gene encoding a glycosyltransferase (PA1091, *orfN*) which is critical for the addition of glycan to the protein. Studies of GI mutants have demonstrated that glycosylation is not required for flagellar assembly and motility in *P. aeruginosa,* although the presence of glycan was critical to virulence. In the burned mouse model of infection, genetically defined mutants producing non-glycosylated flagella were significantly attenuated (Arora *et al.,* 2005). Additionally, glycosylation of the flagellin protein appears to play a role in the proinflammatory action of *P. aeruginosa* flagellin. IL8 release from A549 cells stimulated with nonglycosylated flagellin was significantly reduced when compared to wild-type flagellin (Verma *et al.*, 2005).

Campylobacter jejuni *and* Campylobacter coli

C. jejuni and *C. coli* produce bipolar, complex flagella composed of two structural proteins, FlaA and FlaB. Motility has been shown to be essential for colonization of the viscous intestinal mucosa in the human host (Black *et al.*, 1988) and in contrast to *Pseudomonas* spp, glycosylation of *Campylobacter* flagellins has been shown to be essential for flagellar assembly, subsequent motility and virulence (Guerry *et al.*, 1996; Linton *et al.*, 2000; Thibault *et al.*, 2001; Logan *et al.*, 2002; Guerry *et al.*, 2006). As one of the most heavily glycosylated prokaryotic proteins described to date, the glycans from *Campylobacter* flagellin contribute 10% to the weight of the protein. Flagellin from *C. jejuni* 81–176 was shown to be glycosylated at 19 sites with 5,7-diacetamido-3,5,7,9-tetradeoxy-L-glycero-L-manno-nonulosonate acid (pseudaminic acid, Pse5Ac7Ac) and an acetamidino derivative of this novel sugar, Pse5Ac7Am (Thibault *et al.*, 2001; McNally *et al.*, 2006). In addition, minor amounts of related derivatives have been found on the flagellin protein. Flagellin from a second strain, *C. coli* VC167, was modified with Pse5Ac7Ac as well as with two derivatives of the related nonulosonate sugar, legionaminic acid (5,7-diacetamido-3,5,7,9-tetradeoxy-D-glycero-D-galacto-nonulosonic acid, Leg5Ac7Ac), Leg5Am7Ac and Leg5AmNMe7Ac (Logan *et al.*, 2002; McNally *et al.*, 2007). The structures of these sialic acid-like sugars which are found both on the flagellin protein and in the cell metabolome of *Campylobacter* as CMP-linked sugars are presented in Fig. 10.1.

Genomic comparisons have revealed that the *Campylobacter* GI is one of the most variable loci in the genome, containing up to 50 genes in some strains. A number of genes encoding proteins with homology to carbohydrate biosynthetic enzymes, including some with similarity to the sialic acid biosynthetic enzymes NeuB, NeuC and NeuA are present within this locus. Extensive mutational analyses, in addition to novel metabolomic analyses and functional studies on recombinant enzymes, have defined the precise function of a number of GI genes in

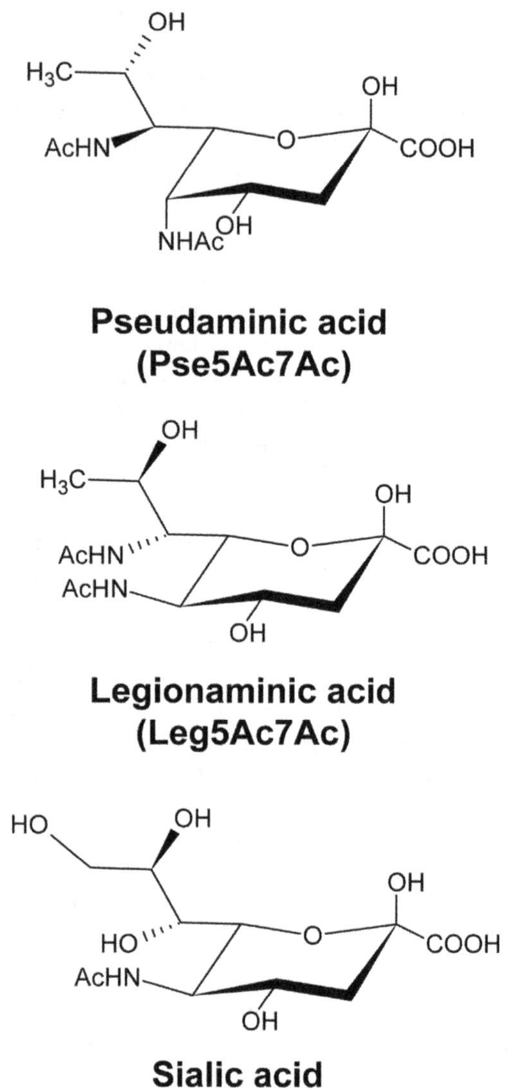

Figure 10.1 Structures of sialic acid-like sugars found in *Campylobacter*. The two nonulosonate sugars, legionaminic acid (Leg5Ac7Ac) and pseudaminic acid (Pse5Ac7Ac), are displayed and compared to sialic acid to demonstrate similarity in overall configuration. In addition to Leg5Ac7Ac and Pse5Ac7Ac, other derivatives of these nonulosonates, which have been structurally characterized by NMR and shown to also be present on *Campylobacter* flagellin monomers, are Pse5Ac7Am, Leg5Am7Ac and Leg5AmNMe7Ac (not shown).

the Pse5Ac7Ac biosynthetic pathway and provided preliminary evidence for a role of the *ptm* genes from this locus in the legionaminic acid biosynthetic pathway (Goon *et al.*, 2003; Soo *et* al., 2004; McNally *et al.*, 2006; Schoenhofen *et al.*, 2006a; Schoenhofen *et al.*, 2006b; Schoenhofen *et al.*, 2006c; McNally *et al.*, 2007). The Pse pathway requires the action of six biosynthetic enzymes; a C5 epimerase/C4,6 dehydratase, aminotransferase, *N*-acetyltransferase, UDP-sugar hydrolase, Pse5Ac7Ac synthase or aldolase, and CMP-Pse5Ac7Ac synthetase, designated PseB, PseC, PseH, PseG, PseI and PseF, respectively and utilizes UDP-GlcNAc as the initial substrate. This pathway exhibits feedback regulation involving the first pathway enzyme, PseB and the final product CMP-Pse5Ac7Ac (McNally *et al.*, 2008), a process which closely resembles the regulation of CMP-Neu5Ac in eukaryotes (Vimr and Troy, 1985). Although early studies had identified *ptmA* and *ptmB* genes from *C. coli* VC167 as involved in flagellar glycosylation, the precise role of each was not known. Targeted metabolomics analysis of *C. coli* VC167 *ptm* isogenic mutants more recently established a role for the *ptmA-F* genes, as well as two new *ptm* genes (*ptmG* and *ptmH*) from the flagellar GI, in the biosynthesis of the novel legionaminic acid derivatives found on the flagellin monomers (McNally *et al.*, 2007).

In addition to carbohydrate biosynthetic genes, the *Campylobacter* GI contains multiple copies of hypothetical genes encoding proteins which belong to the motility accessory family (MAF, Cj1318 family) of proteins (Karlyshev *et al.*, 2002). The homologues of this family in *C. jejuni* 11168 share between 52 and 87% identity at the protein level and many members have homopolymeric tracts indicative of phase variable expression. Insertional inactivation of individual copies of these genes has provided evidence for a role in motility or glycosylation although the precise function of each MAF gene remains to be determined (Schirm *et al.*, 2003, Karlyshev *et al.*, 2005; Guerry *et al.*, 2006). Homologues of these genes are found in only a limited number of other bacterial species including *Wolinella*, *Helicobacter* and *Clostridium* spp.

A second paralogous gene family in the *Campylobacter* genome, the 617 family, includes a number of members from the flagellin GI (Cj1305, Cj1306, Cj1310, Cj1342). As with the MAF family, these genes appear to be capable of slip-strand misrepair which would lead to phase variable gene expression. The function of these genes is also currently unknown although signature-tagged mutagenesis studies demonstrated that Cj0618 played a role in colonization in chickens and motility (Hendrixson and DiRita, 2004). Mutation of Cj1305 and Cj1342 in *C. jejuni* 81–176 had no effect on motility in this strain however (Guerry *et al.*, 2006). No homologues of the 617 family have been found in other bacterial species. In comparative phylogenomics studies of *Campylobacter* isolates, a cluster of genes (Cj1321–Cj1326) within the flagellar GI which includes some of the genes shown to be involved in legionaminic acid biosynthesis have been identified as a specific marker for a 'livestock' clade. The corresponding structure of the flagellar glycan made by these genes and its relevance to colonization of livestock remains to be determined (Champion *et al.*, 2005).

In addition to being essential for flagellar assembly, flagellar glycans of *Campylobacter* have also been shown to contribute significantly to key biological properties of cells which are relevant to the disease process. Autoagglutination (AGG) has been described for a number of bacterial pathogens and is often a critical step in the formation of microcolonies and biofilms (Chiang *et al.*, 1995, Misawa and Blaser, 2000; Golden and Acheson, 2002; Boddey *et al.*, 2006). Flagellar glycans of *Campylobacter* have been shown to play a role in AGG and it is the composition of the glycans found on the flagellar filament which appears to be critical to the process. In *C. jejuni* 81–176, a mutant strain lacking Pse5Ac7Am but producing flagella modified with only Pse5Ac7Ac was AGG$^-$. In a similar fashion, a mutant strain of *C. coli* VC167 which could no longer produce Leg5Am7Ac but only Pse5Ac7Ac was also defective in the AGG phenotype (Guerry *et al.*, 2006). Additionally, *C. jejuni* 81–176 cells lacking PseAm on the flagella filament were also affected in their ability to adhere and invade INT407 cells, pointing to a critical role for the glycans in facilitating host pathogen interactions (Guerry *et al.*, 2006; Guerry, 2007).

Helicobacter pylori

H. pylori produces a polar tuft of sheathed flagella composed of two flagellin monomers, FlaA and FlaB, at one pole of the cell. As with *Campylobacter*, motility is essential for colonization of the mucosal surface of the stomach prior to the development of gastritis and duodenal ulcers (Eaton *et al.*, 1996). Both structural proteins are glycosylated with Pse5Ac7Ac (FlaA seven sites, FlaB 10 sites) and the process of glycosylation is necessary for filament formation and motility (Schirm *et al.*, 2003). In contrast to the intestinal pathogens *C. jejuni* and *C. coli* which produce a number of novel derivatives of Pse5Ac7Ac and Leg5Ac7Ac and incorporate them onto the flagellin glycoproteins, only Pse5Ac7Ac has been found as the *O*-linked glycan on *Helicobacter* flagellin. The presence of a flagellar sheath which covers the filament in *H. pylori* may obviate the need for *H. pylori* cells to produce the diverse array of novel glycosylated structures which is found on *Campylobacter* flagella.

Only a limited number of homologues from the *Campylobacter* flagellar GI were identified in the genome of *H. pylori* (Hp0840, Hp0366, Hp0327, Hp0326B, Hp0178, Hp0326A) and this simplicity in genetic content greatly facilitated the functional elucidation of the Pse5Ac7Ac biosynthetic pathway from both organisms (Schoenhofen *et al.*, 2006b). A related organism *H. felis* which is found naturally in the gastric mucosa of cats and dogs also produces glycosylated flagella (Josenhans *et al.*, 1999).

Aeromonas *spp.*

Like the pseudomonads, aeromonads are ubiquitous aquatic bacteria which can cause disease in fish, amphibians and reptiles. A number of members of this species have been identified as significant human pathogens causing gastrointestinal infections as well as wound infections and septicemia (Thornley *et al.*, 1997). The human pathogens *A. caviae* and *A. hydrophila* produce polar, unsheathed flagella as well as LAF depending on the growth conditions (Rabaan *et al.*, 2001; Kirov *et al.*, 2002; Canals *et al.*, 2006b). FlaA and FlaB of the polar flagella of *A. caviae* were shown to be glycosylated, at six and seven sites respectively, with a novel nonulosonate sugar derivative of mass 373 Da and glycosylation was required for flagellar assembly (Gryllos *et al.*, 2001; Schirm *et al.*, 2005). While bacterial adherence to Hep-2 cells requires polar flagella, it is not yet known if the novel glycan plays a role in this process (Rabaan *et al.*, 2001). The flagellar GI (*flm* locus) contains homologues of a number of genes found in the *H. pylori* and *C. jejuni* GIs.

Studies on LAF from *A. hydrophila* and *A. caviae* has indicated that these flagellins are also glycosylated although the structure of the glycan has yet to be determined (Gavin *et al.*, 2002). Of significance is the identification of a homologue (*maf-5*) of the *Campylobacter* MAF family of genes, in the LAF structural locus of *A. hydrophila*. The product of this gene is required for LAF flagellar production and provides the first preliminary evidence that glycosylation may also be required for LAF assembly in this species (Canals *et al.*, 2006a).

Caulobacter crescentus

C. crescentus cells undergo a unique developmental life cycle and only during the production of swarmer cell progeny is a unipolar, sheathed, complex flagellum produced. While the precise structural characterization of the glycan decorating the flagellin monomers has yet to be made, a GI containing orthologues from a number of bacterial species which do produce glycosylated flagellins, has been characterized. Inactivation of some of these genes affected flagellar assembly indicating, in a similar fashion to those organisms producing complex polar flagella, that glycosylation appears to be critical for assembly (Johnson *et al.*, 1983; Leclerc *et al.*, 1998).

Spirochaete periplasmic flagellins

To date, no detailed structural analysis of a flagellar glycan has been made from this group of organisms, although a number of studies have provided preliminary evidence for glycosylation of the flagellin structural proteins from *Spirochaeta aurantia* (Brahamsha and Greenberg, 1988), *Serpolina hyodysenteriae* (Li *et al.*, 1993), *Treponema pallidum* (Wyss, 1998), and *Borrelia burgdorferi* (Ge *et al.*, 1998). A recent structural analysis of both FlaA and FlaB proteins from *B. burgdorferi* B31 demonstrated that neither proteins carried an *N*-linked glycan, although earlier work had indicated sensitivity of flagellin to *N*-glycosidase F (Sterba *et al.*, 2008).

Gram-positive bacteria

Listeria monocytogenes

L. monocytogenes, when grown at <30°C, produces 4–6 peritrichous flagella which are composed of a single flagellin monomer, FlaA. Structural characterization revealed that the flagellin was glycosylated at up to six sites with *O*-linked β-GlcNAc (Schirm *et al.*, 2004b). This modification is typically found in eukaryotes on many nuclear and cytoplasmic proteins. Recent genetic characterization identified *lmo0688* which lies close to the flagellin structural gene as encoding a protein, an *O*-GlcNAc transferase (OGT) which is responsible for glycosylation. Elegant studies also showed that this protein functions as an anti-repressor for MogR (a flagellar DNA binding transcriptional repressor) and so in addition to its function as an OGT, it also plays a role in transcriptional regulation of its enzymatic substrate (Shen *et al.*, 2006). This is the first flagellin *O*-linked glycosyltransferase to be functionally characterized. The product of this gene has been designated GmaR to indicate its dual role as a glycosyltransferase and motility anti-repressor. As with other organisms producing flagella composed of a single structural monomer, glycosylation is not required for flagellar assembly.

L. monocytogenes is a food-borne facultative intracellular pathogen which causes life-threatening infections in fetuses, newborns and immunocompromised individuals but, as flagella production is down regulated at 37°C, motility is not considered a virulence factor in human infection (Peel *et al.*, 1988). Additionally, although motility is known to be critical for biofilm formation, glycosylation of flagellin does not play a role in this process (Lemon *et al.*, 2007). Not withstanding these observations, the presence of this common eukaryotic posttranslational modification (*O*-linked GlcNAc) on the surface of flagellar filaments of *L. monocytogenes* grown at environmental temperatures may still have significant biological relevance.

Clostridium *spp.*

A number of members of this genus, including the important human pathogens *C. botulinum* and *C. difficile,* produce flagella which have been reported via indirect methods to be glycosylated (Arnold *et al.*, 1998, Bedouet *et al.*, 1998; Lyristis *et al.*, 2000; Tasteyre *et al.*, 2000). While taxonomically clostridia are reported to produce peritrichous flagella, bioinformatics analysis of *C. botulinum* genomes has identified multiple flagellin structural genes suggesting alternate structures may be produced by this organism. In addition, a recent analysis of flagella and flagellin proteins of *C. botulinum* has indicated that the flagellar filament is complex and composed of two highly related flagellin protein monomers, FlaA1 and FlaA2 (Paul *et al.*, 2007). In contrast to *C. botulinum*, the genome sequence of *C. difficile* revealed only a single flagellin structural gene. Significantly, all clostridial genomes examined to date appear to contain a flagellar GI of variable genetic content, lying in close proximity to the flagellar structural genes (Sebaihia *et al.*, 2007). It remains to be established if glycosylation is an integral part of flagellar motility systems of these anaerobic, Gram-positive organisms and whether it is required for assembly of a flagellar filament. Recent structural analysis of the complex flagella from *C. botulinum* revealed that flagellins were glycosylated in *O*-linkage at seven sites. NMR and MS characterization of a glycopeptide from *C. botulinum* FE9909ACS-Alberta revealed that flagellin from this strain was glycosylated with a novel derivative of legionaminic acid, αLeg5GluNMe7Ac (417Da). The biological role of this novel sialic acid-like sugar on flagellin of *C. botulinum* is currently unknown (Twine *et al* 2008).

N- vs *O*-linked glycans

While bacteria have been shown to exclusively glycosylate flagellins with glycans in *O*-linkage, to date only *N*- linked glycans have been found on archaeal flagellins. This is most likely a consequence of the distinct mechanisms of assembly for the respective flagellar structures found in these two groups.

N-linked glycosylation of archaeal flagellins occurs at the outer face of the cytoplasmic membrane (Ng *et al.*, 2006). The similarities of the archaeal *N*-glycosylation process with the eukaryotic and bacterial processes have recently been reviewed (Yurist-Doutsch *et al.*, 2008). Protein *N*-glycosylation is a well-ordered process and extensively characterized in yeast and higher eukarya and more recently in bacteria (for review, (Szymanski *et al.*, 2003). Lipid linkage of glycans and association of assembly components with a membrane (ER in eukarya, cytoplasmic membrane in bacteria and archaea) are critical features of the biosynthetic process. While the identification of glycosylation machinery components has been successfully initiated in archaea, the molecular basis for targeting only the S-layer and flagellin proteins from the larger pool of secreted proteins as well as the identification of an archaeal 'flippase' remains unknown.

In contrast to archaeal flagella, all bacterial glycosylated flagellins examined to date have glycans attached to the protein via *O*-linkage at either serine or threonine residues. In a similar fashion to *O*-glycosylation in eukaryotes, no consensus sequence has been identified for the sites of attachment in this unique bacterial process. Both the synthesis of glycan and assembly appears to occur in the more soluble environment of the cytoplasm and *O*-linked glycosylation substrates are thought to be nucleotide-linked sugars rather than lipid-linked sugars. The mechanistic basis of the glycosylation process for bacterial flagellins is poorly defined yet appears to be highly specific. No reports of other cellular proteins carrying the same *O*-glycan has yet to be made with the exception of the flagellar hook protein from *H. pylori* (S.M. Logan, unpublished) which suggests that the respective flagellar glycosyltransferases have a unique specificity for flagellin and/or flagellin related proteins.

Detailed studies on bacterial flagellar assembly has revealed that the newly synthesized flagellin monomer is prevented from premature polymerization in the cytoplasm through interaction with FliS, the cytosolic chaperone protein (Evdokimov *et al.*, 2003; Ozin *et al.*, 2003). Targeting of the flagellin monomers to the export apparatus is mediated by specific flagellin N-terminus recognition sequences and the protein is secreted through the basal body structure along the hollow core of the filament to the tip where it is then incorporated (Macnab, 2003). The precise step in flagellin synthesis whereby the flagellin monomer becomes glycosylated, either cotranslationally or posttranslationally, remains to be determined. Of significance to this process however, is the finding that the bifunctional flagellar OGT from *L. monocytogenes* which glycosylates the flagellin protein as well as binds to MogR (which itself binds to the *flaA* promoter on the chromosome) may be spatially localized to the site of production of FlaA immediately following translation of mRNA. In this cytosolic location, the flagellar OGT GmaR would be able to glycosylate newly translated FlaA protein, as well as interact with MogR and allow transcription of the *flaA* gene (Shen *et al.*, 2006). A schematic illustrating this putative process of flagellar *O*-linked glycosylation in *L. monocytogenes* is presented in Fig. 10.2.

Flagellar sites of glycosylation

Mass spectrometry has been an invaluable tool for glycoprotein structural characterization. To precisely assign the location of each *O*-linked glycosylation site, β-elimination reactions which involve base catalysed hydrolysis and result in the incorporation of a newly formed amino group of distinct mass at the linkage site on each glycopeptide has been used (Rademaker and Thomas-Oates, 1996). To identify sites of *N*-linked glycosylation, multiple rounds of tandem mass spectrometry (MS^n) was performed on individual glycopeptides (Voisin *et al.*, 2005).

For the four archaeal flagellin structural proteins from *M. voltae* which have been examined, glycosylation of 14/15 predicted *N*-linked sequons were confirmed. The final site was in a peptide of the minor flagellin FlaA and was never identified in mapping experiments. All *N*-linked sequons appear to be accessible to the glycosylation machinery and sites of glycosylation do not appear to be localized to a particular region of each flagellin monomer.

In contrast, the sites of glycosylation of bacterial flagellin are obviously restricted to the central variable region of the flagellin monomer rather than at the more conserved N- and C-terminals (Table 10.2). X-ray crystallography and cryo-EM has provided a three dimensional structure of *S.* Typhimurium flagellin and identified four structural domains D0, D1, D2 and D3 (Samatey *et al.*, 2000). While the D0 and D1 domains found to reside in the N- and C-terminals are highly conserved amongst bacterial flagellins and have been shown to be critical for subunit interactions,

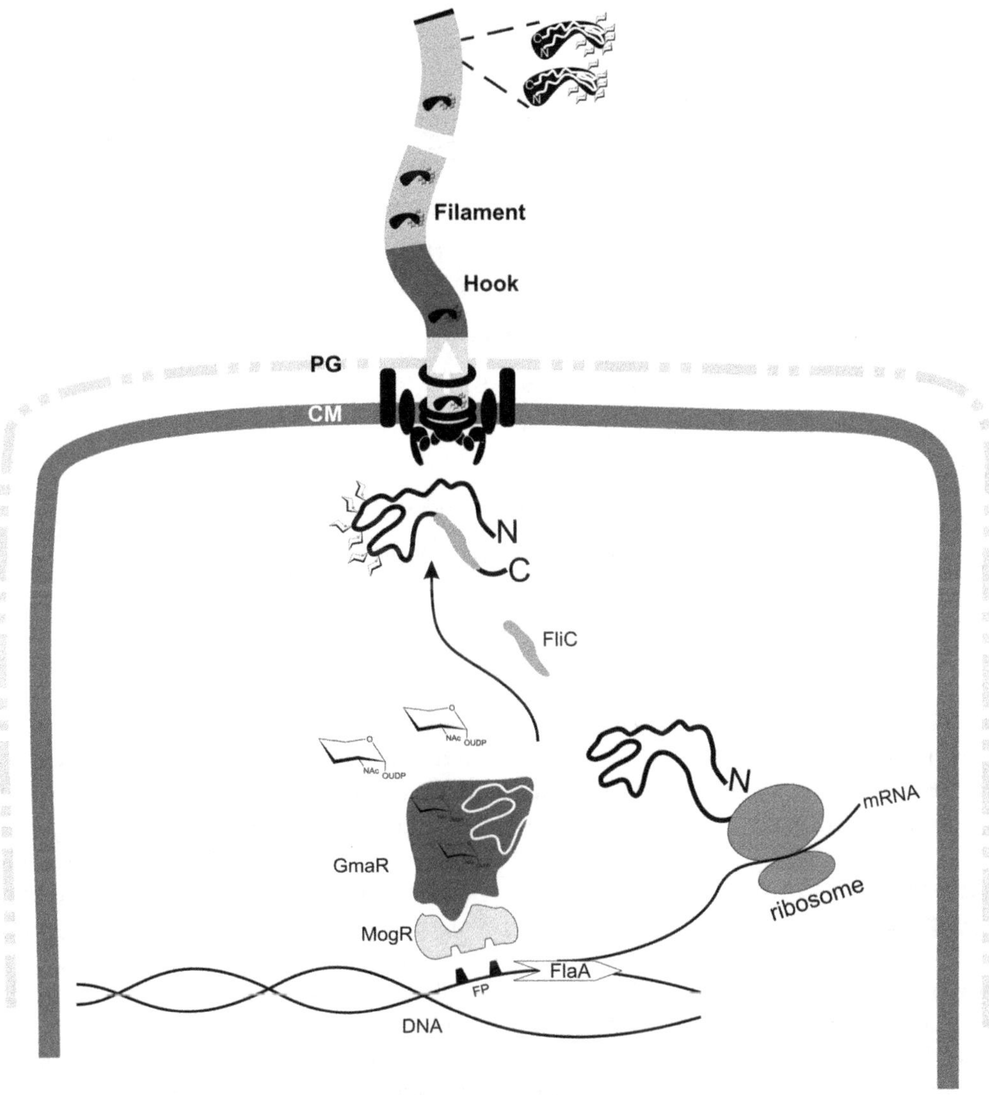

Figure 10.2 Schematic representation of a putative model for cytosolic *O*-linked FlaA glycosylation process in *L. monocytogenes*. When cells are grown at low temperature (<30°C), transcription of *gmaR* occurs. GmaR protein binds to MogR and prevents its binding to target sequences in the promoter of *flaA* (FP) where it represses transcription. *FlaA* is transcribed and the translated FlaA protein is glycosylated at six sites with β-GlcNAc by GmaR using cytosolic pools of UDP-GlcNAc. The chaperone FliC binds to the flagellin monomer preventing premature folding and the glycosylated monomers are secreted through the basal body and hook regions to the filament tip where they are incorporated into the filament. It has not yet been determined if flagellin is cotranslationally or posttranslationally modified by GmaR. However, the bifunctional nature of GmaR as both anti-repressor which binds the transcriptional repressor MogR, as well as a FlaA glycosyltransferase may bring these two activities spatially close together within the cytoplasm. Adapted from Shen *et al.* (2006).

the central variable regions which comprise the D2 and D3 domains do not appear to be critical to subunit interactions in the assembly of *S.* Typhimurium flagella. From the elegant structural models of flagellin and assembled flagella filaments (Yonekura *et al.*, 2003; Yonekura *et al.*, 2005), the D2 and D3 domains are clearly surface exposed and only limited subunit interaction has been described for the D2 domain. These D2 and D3 domains vary in sequence and length amongst other bacterial flagellins and are where the sites of glycosylation have been found on all bacterial

Table 10.2 Sites of glycosylation

Organism	Flagellin (number of aa)	Sites of glycosylation
C. jejuni 81–176	FlaA (576)	S^{206} S^{342} $S^{347}$$T^{393}$ S^{397} S^{400} S^{404} S^{408} S^{417} S^{425} S^{429} S^{436} S^{440} S^{448} S^{451} S^{454} S^{457} S^{460}
H. pylori 1061	FlaA (509)	T^{180} S^{207} S^{245} S^{353} T^{363} S^{394} S^{404}
L. monocytogenes CLIP23485	FlaA (287)	T^{141} T^{145} T^{173} T^{175} S^{180}
P. syringae pv tabaci 6605	FliC (282)	S^{143} S^{164} S^{176} S^{183} S^{193} S^{201}
P. aeruginosa PAO	FliC (488)	T^{189} S^{260}
P. aeruginosa PAK	FliC (394)	S^{191} S^{195}
C. botulinum FE9909ACS	FlaA (373)	S^{126} S^{138} S^{142} S^{175} S^{181} S^{186}
M. voltae	FlaA (207)	N^{163}
	FlaB1 (206)	N^{26} N^{59} N^{65} N^{103} N^{124}
	FlaB2 (204)	N^{26} N^{60} N^{65} N^{101} N^{160}
	FlaB3 (228)	N^{104} N^{117}

flagellins examined to date. The surface exposed location of these regions within the assembled filament would facilitate a role for glycan moieties in biological interactions.

Conclusion and future perspectives

It is now well established that species of both archaea and bacteria are able to produce glycosylated flagellins. Although the flagellar structures produced by each are quite distinct, the necessity for glycosylation in flagellar assembly for archaea and for bacteria which produce complex flagella is clearly documented. In contrast, while glycosylation of simple, bacterial flagella is not required for assembly, the glycans present on these flagellins appear to be critical for other biological processes including contributing to the virulence of a number of bacterial pathogens. The significance of novel derivatives of sialic-acid like nonulosonate sugars on the flagella of *H. pylori, C.jejuni/coli, A. hydrophila, A. caviae,* and *C. botulinum,* all of which are known to colonize the gastrointestinal tract, is likely significant. The presence of these sugars may facilitate binding to eukaryotic receptors or lectins via these sialic acid analogs, or the glycan modifications may contribute to immune evasion by mimicking the host cell surface and masking other highly immunogenic epitopes. The biological role of *O*-GlcNAc on the flagellin of the intracellular pathogen *L. monocytogenesis* is also unclear. Motility is not considered a virulence factor in human infections as flagellar expression is downregulated at 37°C. However, it cannot be overlooked that the same modification is present on many eukaryotic nuclear and cytoplasmic proteins. It has been shown that *L. monocytogenes* cells are ingested and can multiply in protozoa and it remains to be established if flagellin glycosylated with *O*-GlcNAc plays a role in this process (Ly and Muller, 1990).

A recent structural comparison of *C. jejuni* complex filaments, obtained by cryo-EM reconstruction, with the atomic model for *S.* Typhimurium simple filaments indicated these two filament classes exhibit distinctive topographical features (Galkin *et al.*, 2008). By cross-sectional examination, it was shown that *C. jejuni* flagellar filaments had seven protofilaments surrounding the central lumen, as opposed to the eleven protofilaments observed in filaments from *S.* Typhimurium. It appears that the distinctive morphology observed for each filament types results from alternative packing of the D1 domain. The D1 domains of *C. jejuni* and related organisms have diverged from those of *S.* Typhimurium, such that they no longer contain a TLR5 signalling motif (Andersen-Nissen *et al.*,

2005), and these changes may cause alterations in folding and overall ultra-structure. Mapping of the *C. jejuni* glycosylation sites and alignment with *S.* Typhimurium flagellin domains shows that the sites are predominantly localized to the D2 domain (Fig. 10.3). It remains to be determined if the presence of these sialic acid-like sugar modifications on the *C. jejuni* filaments affect the structural organization of *Campylobacter* protofilaments. Whatever role the glycans may play in overall protofilament organization, incorporation has become obligatory for assembly of the *Campylobacter* flagellar filament.

Post-translational modification of flagellins is now clearly established as an integral part of the flagellar machinery in many bacterial species. The biological roles of diverse flagellar glycans may become more clearly defined with the continued addition of new examples of organisms possessing glycosylated flagellins. As a novel therapeutic target, the glycosylation pathways offer significant potential. Motility or the presence of glycan on flagellin has been shown to be critical to the infective process and so small molecule inhibitors of glycan biosynthetic enzymes or of flagellar glycosyltransferases may be of use as therapeutic agents in a manner similar to that recently undertaken for bacterial type III secretion systems (Kauppi *et al.*, 2003; Nordfelth *et al.*, 2005).

From a glycoengineering perspective, elucidation of the mechanistic basis of the glycosylation process could also provide an opportunity to express a recombinant glycopeptide within an immunodominant bacterial antigen present on the surface of bacterial cells. Flagellin fusion proteins have been shown to induce strong and specific immune responses against peptides of interest and the potential of harnessing the glycosylation machinery to express a glycoconjugate fusion protein to which a specific immune response may be raised remains to be explored (Cuadros *et al.*, 2004; Majander *et al.*, 2005a; Majander *et al.*, 2005b).

Acknowledgements

I am grateful to Tom Devecseri for help with figure preparation.

References

Andersen-Nissen, E., Smith, K.D., Strobe, K.L., Barrett, S.L., Cookson, B.T., Logan, S.M., and Aderem, A. (2005). Evasion of Toll-like receptor 5 by flagellated bacteria. Proc. Natl. Acad. Sci. U.S.A. *102*, 9247–9252.

Arnold, F., Bedouet, L., Batina, P., Robreau, G., Talbot, F., Lecher, P., and Malcoste, R. (1998). Biochemical and immunological analyses of the flagellin of *Clostridium tyrobutyricum* ATCC 25755. Microbiol. Immunol. *42*, 23–31.

Arora, S.K., Bangera, M., Lory, S., and Ramphal, R. (2001). A genomic island in *Pseudomonas aeruginosa* carries the determinants of flagellin glycosylation. Proc. Natl. Acad. Sci. U.S.A. *98*, 9342–9347.

Arora, S.K., Neely, A.N., Blair, B, Lory, S., and Ramphal, R. (2005). Role of motility and flagellin glycosylation in the pathogenesis of *Pseudomonas aeruginosa* burn wound infections. Infect. Immun. *73*, 4395–4398.

Arora, S.K., Wolfgang, M.C., Lory, S., and Ramphal, R. (2004). Sequence polymorphism in the glycosylation island and flagellins of *Pseudomonas aeruginosa*. J. Bacteriol. *186*, 2115–2122.

Bedouet, L., Arnold, F., Robreau, G., Batina, P., Talbot, F., and Binet, A. (1998). Evidence for an heterogeneous glycosylation of the *Clostridium tyrobutyricum* ATCC 25755 flagellin. Microbios *94*, 183–192.

Black, R.E., Levine, M.M., Clements, M.L., Hughes, T.P., and Blaser, M.J. (1988). Experimental *Campylobacter jejuni* infection in humans. J. Infect. Dis. *157*, 472–479.

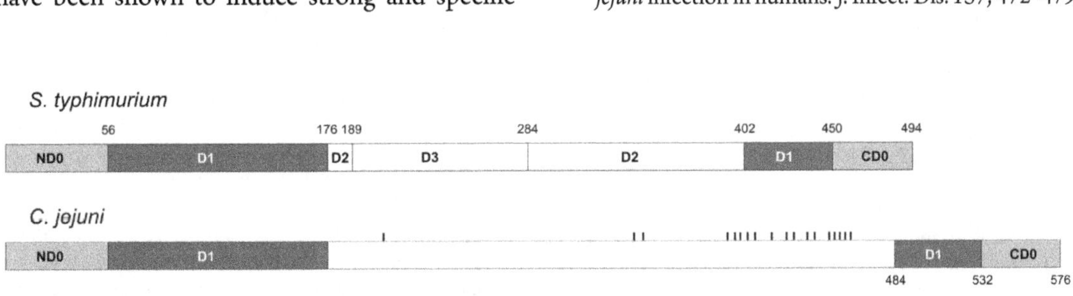

Figure 10.3 Mapping of *O*-linked sites of glycan attachment on the *Campylobacter* flagellin monomer. The flagellins from *C. jejuni* 81–176 (FlaA) and *S.typhimurium* (FliC) are drawn to scale based on the number of amino acid residues. Start and end positions of flagellin structural domains of *S. typhimirium* FliC are indicated (ND0, D1, D2, D3, CD0). Regions of sequence homology are identified with shading (N and C terminals) while non-homologous regions of sequence are identified with open box. Sites of glycosylation on the *Campylobacter* flagellin monomer are indicated by vertical lines.

Boddey, J.A., Flegg, C.P., Day, C.J., Beacham, I.R., and Peak, I.R. (2006). Temperature-regulated microcolony formation by *Burkholderia pseudomallei* requires *pilA* and enhances association with cultured human cells. Infect. Immun. *74*, 5374–5381.

Brahamsha, B. and Greenberg, E.P. (1988). Biochemical and cytological analysis of the complex periplasmic flagella from *Spirochaeta aurantia*. J. Bacteriol. *170*, 4023–4032.

Canals, R., Altarriba, M., Vilches, S., Horsburgh, G., Shaw, J.G., Tomas, J.M., and Merino, S. (2006a). Analysis of the lateral flagellar gene system of *Aeromonas hydrophila* AH-3. J. Bacteriol. *188*, 852–862.

Canals, R., Ramirez, S., Vilches, S., Horsburgh, G., Shaw, J.G., Tomas, J.M., and Merino, S. (2006b). Polar flagellum biogenesis in *Aeromonas hydrophila*. J. Bacteriol. *188*, 542–555.

Chaban, B., Voisin, S., Kelly, J., Logan, S.M., and Jarrell, K.F. (2006). Identification of genes involved in the biosynthesis and attachment of *Methanococcus voltae* N-linked glycans: insight into N-linked glycosylation pathways in Archaea. Mol. Microbiol. *61*, 259–268.

Champion, O.L., Gaunt, M.W., Gundogdu, O., Elmi, A., Witney, A.A., Hinds, J., Dorrell, N., and Wren, B.W. (2005). Comparative phylogenomics of the foodborne pathogen *Campylobacter jejuni* reveals genetic markers predictive of infection source. Proc. Natl. Acad. Sci. U.S.A. *102*, 16043–16048.

Charon, N.W. and Goldstein, S.F. (2002). Genetics of motility and chemotaxis of a fascinating group of bacteria: the spirochetes. Annu. Rev. Genet. *36*, 47–73.

Chiang, S.L., Taylor, R.K., Koomey, M., and Mekalanos, J.J. (1995). Single amino acid substitutions in the N-terminus of *Vibrio cholerae* TcpA affect colonization, autoagglutination, and serum resistance. Mol. Microbiol. *17*, 1133–1142.

Cuadros, C., Lopez-Hernandez, F.J., Dominguez, A.L., McClelland, M., and Lustgarten, J. (2004). Flagellin fusion proteins as adjuvants or vaccines induce specific immune responses. Infect. Immun. *72*, 2810–2816.

Eaton, K.A., Suerbaum, S., Josenhans, C., and Krakowka, S. (1996). Colonization of gnotobiotic piglets by *Helicobacter pylori* deficient in two flagellin genes. Infect. Immun. *64*, 2445–2448.

Evdokimov, A.G., Phan, J., Tropea, J.E., Routzahn, K.M., Peters, H.K., Pokross, M., and Waugh, D.S. (2003). Similar modes of polypeptide recognition by export chaperones in flagellar biosynthesis and type III secretion. Nat. Struct. Biol. *10*, 789–793.

Galkin, V.E., Yu, X., Bielnicki, J., Heuser, J., Ewing, C.P., Guerry, P., and Egelman, E.H. (2008). Divergence of quaternary structures among bacterial flagellar filaments. Science *320*, 382–385.

Gavin, R., Rabaan, A.A., Merino, S., Tomas, J.M., Gryllos, I., and Shaw, J.G. (2002). Lateral flagella of *Aeromonas* species are essential for epithelial cell adherence and biofilm formation. Mol. Microbiol. *43*, 383–397.

Ge, Y., Li, C., Corum, L., Slaughter, C.A., and Charon, N.W. (1998). Structure and expression of the FlaA periplasmic flagellar protein of *Borrelia burgdorferi*. J. Bacteriol. *180*, 2418–2425.

Golden, N.J. and Acheson, D.W. (2002). Identification of motility and autoagglutination *Campylobacter jejuni* mutants by random transposon mutagenesis. Infect. Immun. *70*, 1761–1771.

Goon, S., Kelly, J.F., Logan, S.M., Ewing, C.P., and Guerry, P. (2003). Pseudaminic acid, the major modification on *Campylobacter* flagellin, is synthesized via the Cj1293 gene. Mol. Microbiol. *50*, 659–671.

Gryllos, I., Shaw, J.G., Gavin, R., Merino, S., and Tomas, J.M. (2001). Role of *flm* locus in mesophilic *Aeromonas* species adherence. Infect. Immun. *69*, 65–74.

Guerry, P (2007). The role of flagella in *Campylobacter* virulence. In Virulence mechanisms of bacterial pathogens, Brogden KA, ed. (Washington: American Society for Microbiology), pp. 197–206.

Guerry, P., Doig, P., Alm, R.A., Burr, D.H., Kinsella, N., and Trust, T.J. (1996). Identification and characterization of genes required for post-translational modification of *Campylobacter coli* VC167 flagellin. Mol. Microbiol. *19*, 369–378.

Guerry, P., Ewing, C.P., Schirm, M., Lorenzo, M., Kelly, J., Pattarini, D., Majam, G., Thibault, P., and Logan, S.M. (2006). Changes in flagellin glycosylation affect *Campylobacter* autoagglutination and virulence. Mol. Microbiol. *60*, 299–311.

Hendrixson, D.R. and DiRita, V.J. (2004). Identification of *Campylobacter jejuni* genes involved in commensal colonization of the chick gastrointestinal tract. Mol. Microbiol. *52*, 471–484.

Ishiga, Y., Takeuchi K, Taguchi, F., Inagaki, Y., Toyota, K., Shiraishi, T., and Ichinose, Y. (2005). Defense responses of *Arabidopsis thaliana* inoculated with *Pseudomonas syringae* pv *tabaci* wild type and defective mutants for flagellin (Δ*fliC*) and flagellin glycosylation (Δ*orf1*). J. Gen. Plant Pathol. *71*, 302–307.

Johnson, R.C., Ferber, D.M., and Ely, B. (1983). Synthesis and assembly of flagellar components by *Caulobacter crescentus* motility mutants. J. Bacteriol. *154*, 1137–1144.

Josenhans, C., Ferrero, R.L., Labigne, A., and Suerbaum, S. (1999). Cloning and allelic exchange mutagenesis of two flagellin genes of *Helicobacter felis*. Mol. Microbiol. *33*, 350–362.

Karlyshev, A.V., Ketley, J.M., and Wren, B.W. (2005). The *Campylobacter jejuni* glycome. FEMS Microbiol. Rev. *29*, 377–390.

Karlyshev, A.V., Linton, D., Gregson, N.A., and Wren, B.W. (2002). A novel paralogous gene family involved in phase-variable flagella-mediated motility in *Campylobacter jejuni*. Microbiology *148*, 473–480.

Kauppi, A.M., Nordfelth, R., Uvell, H., Wolf-Watz, H., and Elofsson, M. (2003). Targeting bacterial virulence: inhibitors of type III secretion in *Yersinia*. Chem. Biol. *10*, 241–249.

Kirov, S.M., Tassell, B.C., Semmler, A.B., O'Donovan, L.A., Rabaan, A.A., and Shaw, J.G. (2002). Lateral flagella and swarming motility in *Aeromonas* species. J. Bacteriol. *184*, 547–555.

Leclerc, G., Wang, S.P., and Ely, B. (1998). A new class of *Caulobacter crescentus* flagellar genes. J. Bacteriol. *180*, 5010–5019.

Lemon, K.P., Higgins, D.E., and Kolter, R. (2007). Flagellar motility is critical for *Listeria monocytogenes* biofilm formation. J. Bacteriol. *189*, 4418–4424.

Li, Z., Dumas, F., Dubreuil, D., and Jacques, M. (1993). A species-specific periplasmic flagellar protein of *Serpulina (Treponema) hyodysenteriae*. J. Bacteriol. *175*, 8000–8007.

Linton, D., Karlyshev, A.V., Hitchen, P.G., Morris, H.R., Dell, A., Gregson, N.A., and Wren, B.W. (2000). Multiple N-acetyl neuraminic acid synthetase (neuB) genes in *Campylobacter jejuni*: identification and characterization of the gene involved in sialylation of lipo-oligosaccharide. Mol. Microbiol. *35*, 1120–1134.

Logan, S.M. (2006). Flagellar glycosylation – a new component of the motility repertoire? Microbiology *152*, 1249–1262.

Logan, S.M., Kelly, J.F., Thibault, P., Ewing, C.P., and Guerry, P. (2002). Structural heterogeneity of carbohydrate modifications affects serospecificity of *Campylobacter* flagellins. Mol. Microbiol. *46*, 587–597.

Ly, T.M. and Muller, H.E. (1990). Ingested *Listeria monocytogenes* survive and multiply in protozoa. J. Med. Microbiol. *33*, 51–54.

Lyristis, M., Boynton, Z.L., Petersen, D., Kan, Z., Bennett, G.N., and Rudolph, F.B. (2000). Cloning, sequencing and characterisation of the gene encoding FlaC and the post-translational modification of flagellin from *Clostridium acetobutylicum* ATCC824. Anaerobe *6*, 69–79.

Macnab, R.M. (2003). How bacteria assemble flagella. Annu. Rev. Microbiol. *57*, 77–100.

Macnab, R.M. (2004). Type III flagellar protein export and flagellar assembly. Biochim. Biophys. Acta *1694*, 207–217.

Majander, K., Anton, L., Antikainen, J., Lang, H., Brummer, M., Korhonen, T.K., and Westerlund-Wikstrom, B. (2005a). Extracellular secretion of polypeptides using a modified *Escherichia coli* flagellar secretion apparatus. Nat. Biotechnol. *23*, 475–481.

Majander, K., Korhonen, T.K., and Westerlund-Wikstrom, B. (2005b). Simultaneous display of multiple foreign peptides in the FliD capping and FliC filament proteins of the *Escherichia coli* flagellum. Appl. Environ. Microbiol. *71*, 4263–4268.

McNally, D.J., Aubry, A.J., Hui, J.P., Khieu, N.H., Whitfield, D., Ewing, C.P., Guerry, P., Brisson, J.R., Logan, S.M., and Soo, E.C. (2007). Targeted metabolomics analysis of *Campylobacter coli* VC167 reveals legionaminic acid derivatives as novel flagellar glycans. J. Biol. Chem. *282*, 14463–14475.

McNally, D.J., Hui, J.P., Aubry, A.J., Mui, K.K., Guerry, P., Brisson, J.R., Logan, S.M., and Soo, E.C. (2006). Functional characterization of the flagellar glycosylation locus in *Campylobacter jejuni* 81–176 using a focused metabolomics approach. J. Biol. Chem. *281*, 18489–18498.

McNally, D.J., Schoenhofen, I.C., Houliston, R.S., Khieu, N.H., Whitfield, D.M., Logan, S.M., Jarrell, H.C., and Brisson, J.R. (2008). CMP-Pseudaminic Acid is a natural potent inhibitor of PseB, the first enzyme of the pseudaminic acid pathway in *Campylobacter jejuni* and *Helicobacter pylori*. ChemMedChem. 3, 55–59.

Misawa, N. and Blaser, M.J. (2000). Detection and characterization of autoagglutination activity by *Campylobacter jejuni*. Infect. Immun. *68*, 6168–6175.

Ng, S.Y., Chaban, B., and Jarrell, K.F. (2006). Archaeal flagella, bacterial flagella and Type IV pili: a comparison of genes and posttranslational modifications. J. Mol. Microbiol. Biotechnol. *11*, 167–191.

Nordfelth, R., Kauppi, A.M., Norberg, H.A., Wolf-Watz, H., and Elofsson, M. (2005). Small-molecule inhibitors specifically targeting type III secretion. Infect. Immun. *73*, 3104–3114.

Ozin, A.J., Claret, L., Auvray, F., and Hughes, C. (2003). The FliS chaperone selectively binds the disordered flagellin C-terminal D0 domain central to polymerisation. FEMS Microbiol. Lett. *219*, 219–224.

Paul, C.J., Twine, S.M., Tam, K.J., Mullen, J.A., Kelly, J.F., Austin, J.W., and Logan, S.M. (2007). Flagellin diversity in *Clostridium botulinum* groups I and II: a new strategy for strain identification. Appl. Environ. Microbiol. *73*, 2963–2975.

Peel, M., Donachie, W., and Shaw, A. (1988). Temperature-dependent expression of flagella of *Listeria monocytogenes* studied by electron microscopy, SDS-PAGE and western blotting. J. Gen. Microbiol. *134*, 2171–2178.

Rabaan, A.A., Gryllos, I., Tomas, J.M., and Shaw, J.G. (2001). Motility and the polar flagellum are required for *Aeromonas caviae* adherence to HEp-2 cells. Infect. Immun. *69*, 4257–4267.

Rademaker, G.J. and Thomas-Oates, J. (1996). Analysis of glycoproteins and glycopeptides using fast-atom bombardment. Methods Mol. Biol. *61*, 231–241.

Samatey, F.A., Imada, K., Vonderviszt, F., Shirakihara, Y., and Namba, K. (2000). Crystallization of the F41 fragment of flagellin and data collection from extremely thin crystals. J. Struct. Biol. *132*, 106–111.

Schaffer, C. and Messner, P. (2004). Surface-layer glycoproteins: an example for the diversity of bacterial glycosylation with promising impacts on nanobiotechnology. Glycobiology *14*, 31R-42R.

Schirm, M., Arora, S.K., Verma, A., Vinogradov, E., Thibault, P., Ramphal, R., and Logan, S.M. (2004a). Structural and genetic characterization of glycosylation of type a flagellin in *Pseudomonas aeruginosa*. J. Bacteriol. *186*, 2523–2531.

Schirm, M., Kalmokoff, M., Aubry, A., Thibault, P., Sandoz, M., and Logan, S.M. (2004b). Flagellin from *Listeria monocytogenes* is glycosylated with beta-O-linked N-acetylglucosamine. J. Bacteriol. *186*, 6721–6727.

Schirm, M., Schoenhofen, I.C., Logan, S.M., Waldron, K.C., and Thibault, P. (2005). Identification of unusual bacterial glycosylation by tandem mass spectrometry analyses of intact proteins. Anal. Chem. *77*, 7774–7782.

Schirm, M., Soo, E.C., Aubry, A.J., Austin, J., Thibault, P., and Logan, S.M. (2003). Structural, genetic and functional characterization of the flagellin glycosylation process in *Helicobacter pylori*. Mol. Microbiol. *48*, 1579–1592.

Schoenhofen, I.C., Lunin, V.V., Julien, J.P., Li, Y., Ajamian, E., Matte, A., Cygler, M., Brisson, J.R., Aubry, A., Logan, S.M., Bhatia, S., Wakarchuk, W.W., and Young, N.M. (2006a). Structural and functional characterization of PseC, an aminotransferase involved in the biosynthesis of pseudaminic acid, an essential flagellar modification in *Helicobacter pylori*. J. Biol. Chem. *281*, 8907–8916.

Schoenhofen, I.C., McNally, D.J., Brisson, J.R., and Logan, S.M. (2006b). Elucidation of the CMP-pseudaminic acid pathway in *Helicobacter pylori*: synthesis from UDP-N-acetylglucosamine by a single enzymatic reaction. Glycobiology *16*, 8C-14C.

Schoenhofen, I.C., McNally, D.J., Vinogradov, E., Whitfield, D., Young, M., Dick, S., Wakarchuk, W.W., Brisson, J.-R., and Logan, S.M. (2006c). Functional characterisation of dehydratase/aminotransferase pairs from *Helicobacter* and *Campylobacter*: enzymes distinguishing the pseudaminic acid and bacillosamine biosynthetic pathways. J. Biol. Chem. *281*, 723–732.

Sebaihia, M., Peck, M.W., Minton, N.P., Thomson, N.R., Holden, M.T., Mitchell, W.J., Carter, A.T., Bentley, S.D., Mason, D.R., Crossman, L., Paul, C.J., Ivens, A., Wells-Bennik, M.H., Davis, I.J., Cerdeno-Tarraga, A.M., Churcher, C., Quail, M.A., Chillingworth, T., Feltwell, T., Fraser, A., Goodhead, I., Hance, Z., Jagels, K., Larke, N., Maddison, M., Moule, S., Mungall, K., Norbertczak, H., Rabbinowitsch, E., Sanders, M., Simmonds, M., White, B., Whithead, S., and Parkhill, J. (2007). Genome sequence of a proteolytic (Group I) *Clostridium botulinum* strain Hall A and comparative analysis of the clostridial genomes. Genome Res. *17*, 1082–1092.

Shams-Eldin, H., Chaban, B., Niehus, S., Schwarz, R.T., and Jarrell, K.F. (2008). Identification of the archaeal alg7 gene homologue (encoding N-acetylglucosamine-1-phosphate transferase) of the N-linked glycosylation system by cross-domain complementation in *Saccharomyces cerevisiae*. J. Bacteriol. *190*, 2217–2220.

Shen, A., Kamp, H.D., Grundling, A., and Higgins, D.E. (2006). A bifunctional O-GlcNAc transferase governs flagellar motility through anti-repression. Genes Dev. *20*, 3283–3295.

Soo, E.C., Aubry, A.J., Logan, S.M., Guerry, P., Kelly, J.F., Young, N.M., and Thibault, P. (2004). Selective detection and identification of sugar nucleotides by CE-electrospray-MS and its application to bacterial metabolomics. Anal. Chem. *76*, 619–626.

Sterba, J., Vancova, M., Rudenko, N., Golovchenko, M., Tremblay, T.L., Kelly, J.F., MacKenzie, C.R., Logan, S.M., and Grubhoffer, L. (2008). Flagellin and outer surface proteins from *Borrelia burgdorferi* are not glycosylated. J. Bacteriol. *190*, 2619–2623.

Szabo, Z., Sani, M., Groeneveld, M., Zolghadr, B., Schelert, J., Albers, S.V., Blum, P., Boekema, E.J., and Driessen, A.J. (2007). Flagellar motility and structure in the hyperthermoacidophilic archaeon *Sulfolobus solfataricus*. J. Bacteriol. *189*, 4305–4309.

Szymanski, C.M., Logan, S.M., Linton, D., and Wren, B.W. (2003). *Campylobacter*--a tale of two protein glycosylation systems. Trends Microbiol. *11*, 233–238.

Taguchi, F., Ogawa, Y., Takeuchi, K., Suzuki, T., Toyoda, K., Shiraishi, T., and Ichinose, Y. (2006a). A homologue of the 3-oxoacyl-(acyl carrier protein) synthase III gene located in the glycosylation island of *Pseudomonas syringae* pv. *tabaci* regulates virulence factors via N-acyl homoserine lactone and fatty acid synthesis. J. Bacteriol. *188*, 8376–8384.

Taguchi, F., Shibata, S., Suzuki, T., Ogawa, Y., Aizawa, S., Takeuchi, K., and Ichinose, Y. (2008). Effects of glycosylation on swimming ability and flagellar polymorphic transformation in *Pseudomonas syringae* pv. *tabaci* 6605. J. Bacteriol. *190*, 764–768.

Taguchi, F., Shimizu, R., Ikeda, Y., Inagaki, Y., Toyoda, K., Shiraishi, T., and Ichinose, Y. (2003a). Post-translational modification of flagellin determines the specific induction of HR. Plant Physiol. Biochem. *44*, 342–349.

Taguchi, F., Shimizu, R., Nakajima, R., Toyoda, K., Shiraishi, T., and Ichinose, Y. (2003b). Differential effects of flagellins from *Pseudomonas syringea* pvs. *tabaci, tomato* and *glycinea* on plant defence response. Plant Physiol. Biochem. *41*, 165–174.

Taguchi, F., Takeuchi, K., Katoh, E., Murata, K., Suzuki, T., Marutani, M., Kawasaki, T., Eguchi, M., Katoh, S., Kaku, H., Yasuda, C., Inagaki, Y., Toyoda, K., Shiraishi, T., and Ichinose,Y. (2006b). Identification of glycosylation genes and glycosylated amino acids of flagellin in *Pseudomonas syringae* pv. *tabaci*. Cell Microbiol. *8*, 923–938.

Takeuchi, K., Taguchi, F., Inagaki, Y., Toyoda, K., Shiraishi, T., and Ichinose, Y. (2003). Flagellin glycosylation island in *Pseudomonas syringea* pv glycinea and its role in host specificity. J. Bacteriol. *185*, 6658–6665.

Takeuchi, K., Ono, H., Yoshida, M., Ishii, T., Katoh, E., Taguchi, F., Miki, R., Murata, K., Kaku, H., and Ichinose, Y. (2007). Flagellin glycans from two pathovars of *Pseudomonas syringae* contain rhamnose in D and L configurations in different ratios and modified 4-amino-4,6-dideoxyglucose. J. Bacteriol. *189*, 6945–6956.

Tasteyre, A., Barc, M.C., Karjalainen, T., Dodson, P., Hyde, S., Bourlioux, P., and Borriello, P. (2000). A *Clostridium difficile* gene encoding flagellin. Microbiology *146*, 957–966.

Thibault, P., Logan, S.M., Kelly, J.F., Brisson, J.R., Ewing, C.P., Trust, T.J., and Guerry, P. (2001). Identification of the carbohydrate moieties and glycosylation motifs in *Campylobacter jejuni* flagellin. J. Biol. Chem. *276*, 34862–34870.

Thornley, J.P., Shaw, J.G., Gryllos, I.A., and Eley, E. (1997). Virulence properties of clinically significant *Aeromonas* species: evidence for pathogenicity. Rev. Med. Microbiol. *8*, 61–72.

Twine, S.M., Paul, C.J., Vinogradov, E., McNally, D.J., Brisson, J-R., Mullen, J.A., McMullin, D.R., Jarrell, H.C., Austin, J.W., Kelly, J.F., and Logan, S.M. (2008) Flagellar glycosyaltion in *Clostridium botulinum* FEBS J. *275*, 4428–4444.

Verma, A., Arora, S.K., Kuravi, S.K., and Ramphal, R. (2005). Roles of specific amino acids in the N terminus of *Pseudomonas aeruginosa* flagellin and of flagellin glycosylation in the innate immune response. Infect. Immun. 73, 8237–8246.

Verma, A., Schirm, M., Arora, S.K., Thibault, P., Logan, S.M., and Ramphal, R. (2006). Glycosylation of b-Type flagellin of *Pseudomonas aeruginosa*: structural and genetic basis. J. Bacteriol. *188*, 4395–4403.

Vimr, E.R. and Troy, F.A. (1985). Regulation of sialic acid metabolism in *Escherichia coli*: role of

N-acylneuraminate pyruvate-lyase. J. Bacteriol. *164*, 854–860.

Voisin, S., Houliston, R.S., Kelly, J., Brisson, J.R., Watson, D., Bardy, S.L., Jarrell, K.F., and Logan, S.M. (2005). Identification and characterization of the unique N-linked glycan common to the flagellins and S-layer glycoprotein of *Methanococcus voltae*. J. Biol. Chem. *280*, 16586–16593.

Wieland, F., Paul, G., and Sumper, M. (1985). Halobacterial flagellins are sulfated glycoproteins. J. Biol. Chem. *260*, 15180–15185.

Wyss, C. (1998). Flagellins, but not endoflagellar sheath proteins, of *Treponema pallidum* and of pathogen-related oral spirochetes are glycosylated. Infect. Immun. *66*, 5751–5754.

Yonekura, K., Maki-Yonekura, S., and Namba, K. (2003). Complete atomic model of the bacterial flagellar filament by electron cryomicroscopy. Nature *424*, 643–650.

Yonekura, K., Maki-Yonekura, S., and Namba, K. (2005). Building the atomic model for the bacterial flagellar filament by electron cryomicroscopy and image analysis. Structure *13*, 407–412.

Yurist-Doutsch, S., Chaban, B., VanDyke, D.J., Jarrell, K.F., and Eichler, J. (2008). Sweet to the extreme: protein glycosylation in Archaea. Mol. Microbiol. *68*, 1079–1084.

Lateral Flagella Systems

11

Susana Merino and Juan M. Tomás

Abstract

Flagella-dependent locomotion is usually attributed to only one flagella system, but some bacteria are able to produce dual flagella systems responsible for different types of motility. These bacteria are able to express constitutively polar or subpolar flagella, required for swimming motility, and an inducible, second (lateral) flagella system in viscous environments or on surfaces, essential for swarming motility. Lateral flagella also contribute to the virulence of pathogenic bacteria through adhesion and biofilm formation to host surfaces. Since flagella synthesis and motility have a high metabolic cost for the bacterium, lateral flagella expression is highly regulated by a number of environmental factors and regulators. Although inducible lateral flagella systems were considered uncommon, comparative bioinformatics analysis indicate the presence of homologous non-functional lateral systems in different bacteria that only express one flagella system. These findings show that the presence of dual flagella systems within the same species is more common than was previously thought.

Introduction

Motility is widespread throughout prokaryotes and is advantageous, when nutrients are limited, as a mechanism to migrate to more favourable environments and to compete with other microorganisms. In 1972, J. Henrichsen (Henrichsen, 1972) identified six different motility categories: swimming, swarming, gliding, twitching, sliding and darting. Swimming and swarming were correlated with the presence of flagella systems (Harshey, 1994; Manson *et al.*, 1998); twitching and social gliding were associated with Type IV pili (Matticks, 2002); gliding required either extrusion of copious amounts of slime through a junctional pore complex, or a ratchet structure or contractile cytoskeleton (Bardy *et al.*, 2003); sliding is a form of passive translocation (Martinez *et al.*, 1999); and nothing is known about darting.

Flagella-dependent motility systems are the main mode of locomotion for bacteria, including *Archaea* whose flagellum is distinct in composition and assembly (Jarrell *et al.*, 1996b; Thomas *et al.*, 2001). Furthermore, flagella systems also play an important role in adhesion to substrates, biofilm formation and host invasion in pathogenic bacteria (Otteman *et al.*, 1997; Josenhans *et al.*, 2002).

The bacterial flagellum is a complex rotary structure composed of three main components: the basal body, localized within the cell wall, and the hook and the filament that extend outside the cell surface, except in the case of the periplasmic flagella of spirochaetes. The number and arrangement of flagella on the bacterial surface varies among species. Many bacterial species express only one flagella system which produces single/multiple polar flagella, for example *Pseudomonas aeruginosa*, *Vibrio cholerae* and *Helicobacter pylori*, or peritrichous flagella, such as *Escherichia coli*, *Salmonella enterica* and *Proteus mirabilis*. A limited number of bacteria express two entirely distinct flagella systems which produce polar flagella and lateral induced flagella, such as *Vibrio parahaemolyticus* (Shinoda *et al.*, 1977), *Vibrio alginolyticus* (Kawagishi *et al.*, 1995), *Aeromonas* spp. (Shimada *et al.*, 1985), *Azospirillum* spp. (Tarrand *et al.*, 1978; Hall *et al.*, 1983; Khamman *et al.*,1989), *Rhodospirillum centenum* (McClain *et al.*, 2002), *Helicobacter mustelae* (O'Rouke *et al.*, 1992) and *Plesiomonas shigelloides* (Inoue *et al.*, 1991). Recently, it was reported that *Bradyrhizobium japonicum*, a nitrogen-fixing species, expresses a dual flagella system which produce a subpolar flagellum and lateral induced flagella (Kanbe *et al.*, 2007).

Swimming motility is defined as individual flagella-dependent movement that propels bacteria through liquids while swarming motility is a coordinated flagella-dependent movement of a group of bacteria through viscous environments or over surfaces. Usually swarming motility requires a greater number of flagella than swimming, perhaps because surface friction or higher viscosity reduces flagella rotation. In this sense, many peritrichous flagellated bacteria such as *E. coli*, *S. enterica* serovar Typhimurium, *Serratia marcescens* and *Proteus sp.* use one flagella system for both swimming and swarming motility. In these bacteria, spreading over solid surfaces requires cell multinucleation, elongation, hyperflagellation and secretion of surfactants (Harshey *et al.*, 1994). Bacteria with dual flagella systems, such as *V. parahaemolyticus* (McCarter, 1999) or *Aeromonas hydrophila* (Gavín *et al.*, 2002), express a polar flagellum for swimming and a mixed flagellation consisting of a constitutive polar flagellum and inducible lateral flagella for swarming in high viscous media or on surfaces.

Characteristics of lateral flagella systems

Functional lateral flagella systems, whose flagella are randomly distributed over the cell surface in a manner similar to the peritrichous flagella of *Enterobacteriaceae*, were reported in polar flagellated bacteria with dual flagella systems (Merino *et al.*, 2006). Lateral flagella are synthesized when bacteria are grown in viscous media or on surfaces. Nevertheless, in some bacteria, they also can be induced by other environmental conditions, such as increasing growth temperature in liquid media (*R. centenum*; Berleman *et al.*, 2005) or growth in static liquid media (*A. brasilense* Cd; Madi *et al.*, 1988). Although all bacteria with dual flagella systems express both flagella types in viscous media or on surfaces, only functional lateral flagella are needed for swarming motility, except in *R. centenum*. This bacterium requires the presence of a functional polar flagellum for swarming, because, although lateral flagella rotation occurs in the absence of polar flagella, surface movement is impeded (Jiang *et al.*, 1998). Lateral flagella have distinct properties, morphology and immunology characteristics compared with polar flagella (Shinoda *et al.*, 1977; Gavín *et al.*, 2002). *V. parahaemolyticus*, *V. alginolyticus* and other *Vibrio* species (Shinoda *et al.*, 1992), as well as the purple photosynthetic bacterium *R. centenum*, express a constitutive sheathed polar flagellum, and when grown in viscous media or on solid surfaces, they differentiate into a swarmer cell (in which septation ceases and the cell elongates, producing many non-sheathed lateral flagella that allow the bacteria to swarm) (Ulitzur, 1975; McCarter *et al.*, 1990; Ragatz *et al.*, 1994). In contrast, 50–60% of mesophilic *Aeromonas* isolated from clinical samples have a constitutive polar flagellum and inducible lateral flagella. Both types of flagella are unsheathed but glycosylated (Raaban *et al.*, 2001; *et al.*, 2002; Kirov *et al.*, 2002). *B. japonicum* have one constitutive thick flagellum and a few thin flagella, both unsheathed, and growing at subpolar locations (Kanbe *et al.*, 2007). Furthermore, swarmer cell differentiation in *Aeromonas* spp. and the nitrogen-fixing rhizobacteria *Azospirillum lipoferum* and *B. japonicum* does not result in multinucleated or elongated cells (Alexandre *et al.*, 1999; Kanbe *et al.*, 2007).

The best studied bacteria with dual functional flagella systems are *V. parahaemolyticus* and *A. hydrophila*. The *V. parahaemolyticus* polar flagellum requires approximately 60 genes distributed in five discontinuous regions on chromosome I (Yun-Kyeong *et al.*, 2000) and the lateral flagella are encoded by 38 genes distributed in two discontinuous regions on chromosome II (Stewart *et al.*, 2003). The *A. hydrophila* polar flagellum requires 55 genes distributed in five discontinuous chromosomal regions (Altarriba *et al.*, 2003; Canals *et al.*, 2006b) whereas lateral flagella are encoded by 38 genes distributed in a unique chromosomal region (Gavín *et al.*, 2002; Canals *et al.*, 2006a). In these two bacterial species, the polar and lateral flagella systems do not share either structural or regulatory genes (Kirov,

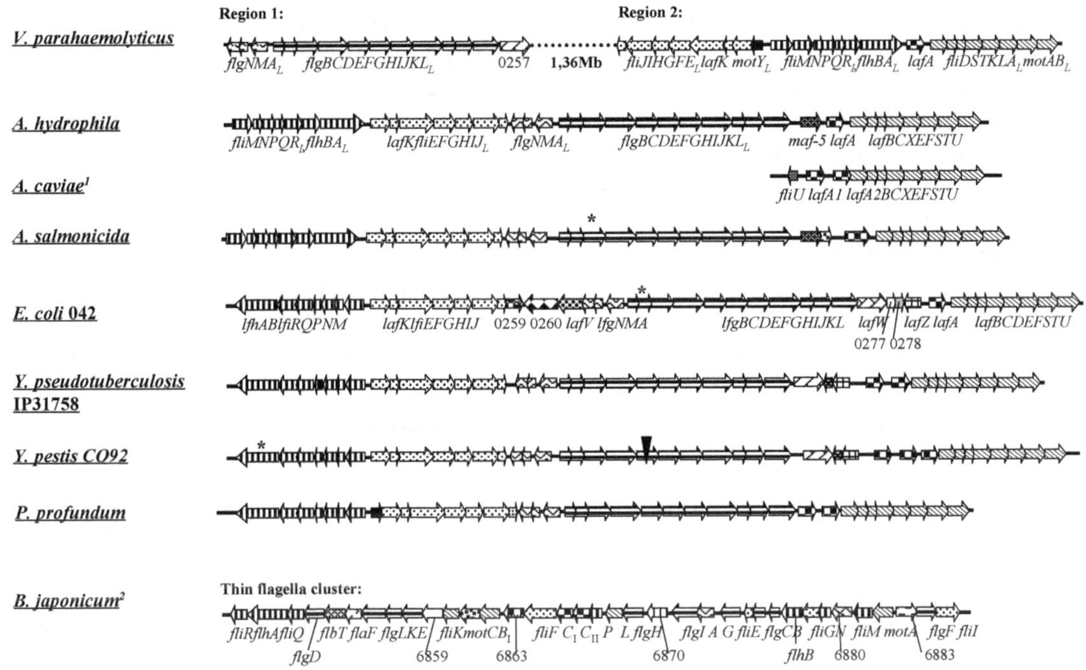

Figure 11.1 Comparative schematic representation of functionally and predicted lateral flagella regions. Arrows showing the same effects correspond to homologous genes among these bacteria, asterisks indicate frameshift mutated genes and black triangles indicate insertion elements. *A. caviae*[1], partially sequenced lateral flagella genes; *B. japonicum*[2], genes of the thin flagella locus.

2003; McCarter, 2004), whereas in *Azospirillum* spp. and *R. centenum*, the two flagellar systems have some structural and/or regulatory overlapping genes (Jiang *et al.*, 1998; Scheludko *et al.*, 1998; McClain *et al.*, 2002). Moreover, in *V. parahaemolyticus* and *V. alginolyticus*, both types of flagella are powered by different ion sources: the polar flagella stator-motor is composed for four proteins ($MotA_P$, $MotB_P$, MotX and $MotY_P$) that generate torque by coupling with the sodium ion flux (Atsumi *et al.*, 1992; Kawagishi *et al.*, 1995), while the lateral flagella stator motor has three proteins ($MotA_L$, $MotB_L$ and $MotY_L$) (Stewart *et al.*, 2003) that generate torque by coupling with the proton ion flux. A similar situation has been observed in *A. hydrophila* (unpublished data) but the protein composition of polar and lateral flagella are different (Canals *et al.*, 2006a; Canals *et al.*, 2006b). However, both *B. japonicus* flagella types are powered by proton-driven motors (Kanbe *et al.*, 2007). As described in peritrichous (lateral non-induced) and polar flagella systems, the number of flagellins varies among species. *V. parahaemolyticus* lateral flagella have one lateral flagellin (Stewart *et al.*, 2003), *B japonicus* has two (Kanbe *et al.*, 2007), while *Aeromonas* spp. has one or two (Gavín *et al.*, 2002; Gavín *et al.*, 2003).

Chromosomal organization of lateral flagella systems

The complete set of genes involved in formation of a functional lateral flagella system was described in two bacterial species: *V. parahaemolyticus* and *A. hydrophila*. However, a partial set of genes was described in other bacteria with functional dual flagella, such as *A. brasilense*, *A. caviae*, *R. centenum* and *B. japonicum* (Merino *et al.*, 2006). *V. parahaemolyticus* lateral flagella genes are arranged in two discontinuous chromosome II regions (region 1 and 2), showing little or no intergenic spacing among large groups of genes (Stewart *et al.*, 2003). Region 1 contains 14 genes which encode the anti-σ^{28} factor ($flgM_L$) and many of the structural proteins involved in the basal body rod, the hook and the P- and L-rings assembly. Region 2 contains 24 genes which encode the master regulator protein (*lafK*), the σ^{28} factor (*fliA*$_L$) and structural proteins involved in switch, motor, export-assembly apparatus and filament assembly (Table 11.1). The 38 *A. hydrophila* lateral flagella genes are arranged in a unique chromosomal region (Canals *et al.*, 2006a).

V. parahaemolyticus region 1 genes are divided among two divergent transcribed set of genes: *flgAMN*$_L$ and *flgBCDEFGHIJKL*$_L$. Homologous *A. hydrophila* genes exhibit the same distribution and direction of transcription. *V. parahaemolyticus* region 2 genes are arranged in four clusters: *fliMNPQR*$_L$*flhBA*$_L$, *lafA* and *fliDSTKLA*$_L$*motAB*$_L$ transcribed in the same direction, and *motY*$_L$-*lafKfliEFGHIJ*$_L$ transcribed in the opposite direction. Homologous *A. hydrophila* genes are transcribed in the same direction and do not contain any homologous gene to *V. parahaemolyticus motY*$_L$, which encodes a protein similar to the *V. alginolyticus* outer membrane sodium-drive stator motor protein MotY$_P$ (Okabe *et al.*, 2002). Furthermore, the *A. hydrophila* lateral flagella region contains, between *flgL*$_L$ and *lafA*, a modification accessory factor gene, *maf-5*, (Karlinshev *et al.*, 2002) which is independently transcribed (Fig. 11.1). The partial lateral flagella sequence of *A. caviae* Sch3N shows, upstream of the capping flagella gene *lafB*, two flagellin subunit genes (*lafA1* and *lafA2*) transcribed in the same direction, and a *fliU* gene transcribed divergently, which encodes a N-lysine methylase involved in flagella biosynthesis (LafV) (Gavín *et al.*, 2002).

Analysis of the *B. japonicum* genome sequence (Kaneko *et al.*, 2002) showed that flagella genes are arranged in two large chromosomal clusters that contain redundant genes as well as several small groups containing one or a few genes. Most of the genes involved in thin flagella (lateral flagella) formation appear to be arranged in one of the large chromosomal regions, but at less two more small flagella clusters are involved in their formation. In contrast to other lateral flagella systems, no homologous genes to the master regulators, *flgJMN* and *fliADHJST,* were found in the *B. japonicum* genome (Kanbe *et al.*, 2007). Curiously, thin flagella genes organization is very different from those of other lateral flagella systems in that many typical pairs of neighbouring genes, such as *motAB, flhAB, fliQR, flgDE, fliFG,* and *flgFG,* are not adjacent (Table 11.1, Fig. 11.1).

The *R. centenum* lateral flagella gene regions have not been completely sequenced, but a second chemotactic operon (*che*$_2$), involved in polar and lateral flagella formation, has been reported (Berleman *et al.*, 2005), in addition to the *che*$_1$ operon that controls the chemotactic and phototactic behaviour of both swimming and swarming cells (Jiang *et al.*, 1997). This second chemotactic operon, organized like those of *R. meliloti* and *R. sphaeroides,* contains eight genes whose encoded proteins are homologues of MCP, CheA, CheY, CheB, CheR, CheW and two ORFs of unknown function.

The comparative bioinformatic analysis of sequenced genomes showed the presence of secondary flagella systems, whose gene product homologues are similar to those of *V. parahaemolyticus* and *A. hydrophila* lateral flagella, in different bacteria such as *Escherichia coli* 042, *Yersinia pestis, Y. pseudotuberculosis, Photobacterium profundum* and *Chromobacterium violaceum* (Fig. 11.1). Vestiges of this system appear in bacteria such as *E. coli* K12 and *Shigella* spp. (Ren *et al.*, 2005). These findings suggest that the presence of dual flagella systems within the same species is more common than was previously thought. Thus, in the enteroaggregative *E. coli* strain 042, the second flagella locus, designed Flag-2, has forty-four genes arranged in a unique chromosomal region, as in the case of *A. hydrophila* lateral flagella genes. Potentially, all gene products required for a functional lateral flagella system are encoded within this locus, but a frameshift mutation in *lfgC,* which encodes a proximal rod protein, appears to inactivate the system in this strain (Ren *et al.*, 2005). Most of the second flagella systems determined by genomic analysis are also arranged in a unique chromosomal region. An exception is found in *C. violaceum* where the second flagella system genes appear to be located in different chromosomal regions (Pereira *et al.*, 2004). *Y. pestis* strains containing a Flag-2 like cluster show frameshift mutations, deletions or insertion elements in key flagellar genes, while in *C. violaceum, Y. pseudotuberculosis* and *P. profundum* lateral flagella clusters do not display any inactivated gene. None of these clusters contain genes homologous to *A. hydrophila maf-5* gene, and only the *P. profundum*

Table 11.1 Lateral flagella genes names and predicted protein functions

Gene name				
A. hydrophila[a]	*V. parahaemolyticus*[b]	*E. coli* 042[c]	*B. japonicum*[d]	Protein function
	motY$_L$			Proton motor
lafT	*motA*$_L$	*lafT*	*motA*	
lafU	*motB*$_L$	*lafU*	*motB*$_l$	
fliF$_L$	*fliF*$_L$	*lfiF*	*fliF*	M-ring
fliP$_L$	*fliP*$_L$	*lfiP*	*fliP*	Type III protein export (integral inner-membrane proteins)
fliQ$_L$	*fliQ*$_L$	*lfiQ*	*fliQ*	
fliR$_L$	*fliR*$_L$	*lfiR*	*fliR*	
flhB$_L$	*flhB*$_L$	*lfhB*	*flhB*	
flhA$_L$	*flhA*$_L$	*lfhA*	*flhA*	
fliH$_L$	*fliH*$_L$	*lfiH*		Type III protein export (membrane-associated proteins)
fliJ$_L$	*fliJ*$_L$	*lfiJ*		
fliI$_L$	*fliI*$_L$	*lfiI*	*fliI*	Export ATP synthase
fliM$_L$	*fliM*$_L$	*lfiM*	*fliM*	Rotor/Switch (C ring)
fliN$_L$	*fliN*$_L$	*lfiN*	*fliN*	
fliG$_L$	*fliG*$_L$	*lfiG*	*fliG*	
fliE$_L$	*fliE*$_L$	*lfiE*	*fliE*	Basal body component
flgI$_L$	***flgIL***	*lfgI*	*flgI*	P-ring
flgA$_L$	***flgAL***	*lfgA*	*flgA*	P-ring assembly
flgJ$_L$	***flgJL***	*lfgJ*		Peptidoglycan hydrolase
flgH$_L$	***flgHL***	*lfgH*	*flgH*	L-ring
flgB$_L$	***flgBL***	*lfgB*	*flgB*	Rod
flgC$_L$	***flgCL***	*lfgC*	*flgC*	
flgD$_L$	***flgDL***	*lfgD*	*flgD*	
flgF$_L$	***flgFL***	*lfgF*	*flgF*	
flgG$_L$	***flgGL***	*lfgG*	*flgG*	
flgE$_L$	***flgEL***	*lfgE*	*flgE*	Hook
flgK$_L$	***flgKL***	*lfgK*	*flgK*	Hook-associated protein 1
flgL$_L$	***flgLL***	*lfgL*	*flgL*	Hook-associated protein 3
lafB	*fliD*$_L$	*lafB*		Hook-associated protein 2
		lafW		Possible hook-associated protein
lafE	*fliK*$_L$	*lafE*	*fliK*	Hook length control
lafA	*lafA*	*lafA*	*fliC*$_I$ and *fliC*$_{II}$	Flagellin
flgN$_L$	***flgNL***	*lfgN*		Chaperones
lafC	*fliS*$_L$	*lafC*		
lafX	*fliT*$_L$	*lafD*		
lafS	*fliA*$_L$	*lafS*		σ^{28}

Table 11.1 Continued

Gene name				
A. hydrophila[a]	*V. parahaemolyticus*[b]	*E. coli* 042[c]	*B. japonicum*[d]	Protein function
flgM$_L$	*flgM*$_L$	*lfgM*		Anti-σ^{28}
lafK	*lafK*	*lafK*		Regulator σ^{54}-dependent
		Ec042–0278	*flaF*	Regulator
		lafZ		Transmembrane regulator
			flbT	Flagella repressor
		Ec042–0259		Cytidylyl transferase
		Ec042–0260		Glycosyl transferase
		lafV		Lysine-N-methylase
maf-5				Motility accesory factor
			bll6859	Lytic transglycosylase
			motC	Possible chemotaxis protein precursor
lafF	*fliL*$_L$	*lafF* Ec042–0277	*fliL* bll6863 bll6870 bll6880 bll6883	Unknown

a *A. hydrophila* gene designation according to accession number DQ124694–5 (Canals *et al*., 2006).

b *V. parahaemolyticus* gene designation according to the nomenclature suggested by Stewart and McCarter, 2003. Boldface print genes are located in lateral flagella region 1 and non-boldface print genes are located in region 2.

c *E. coli* 042 gene designation according to the nomenclature suggested by Ren *et al*., 2005. Underlined genes contain a frameshift mutation.

d *B. japonicum* genes localized in the thin flagella locus and designated according to the nomenclature suggested by Kanbe *et al*. (2007).

cluster contains, upstream of a *lafK*-like gene, a gene homologous to *V. parahaemolyticus motY*$_L$ gene (Fig. 11.1).

E. coli 042 Flag-2 genes, as well as *Y. pestis*, *Y. pseudotuberculosis* and *P. profundum* Flag-2-like genes, exhibit the same distribution and direction of transcription as the homologous genes of *A. hydrophila* lateral flagella, except for those genes homologous to *fliMNPQR*$_L$*flhBA*$_L$, which are transcribed divergently. These second flagella clusters show two variable regions in relation to *A. hydrophila*: between *fliJ*$_L$ (*lfiJ*) and *flgN*$_L$ (*lfgN*) and between *flgL*$_L$ (*lfgL*) and *lafA* (Fig. 11.1). In *A. hydrophila fliJ*$_L$ and *flgN*$_L$ are adjacent genes but in *E. coli* 042 they are separated by three additional genes. Downstream of *lfgN*, and transcribed in the same direction, is *lafV* which is predicted to encode a lysine-N-methylase required for post-translational methylation of lysine residues in some flagellins (Burnens *et al.*, 1997). A *lafV* homologous gene (*fliU*) was reported upstream of *lafA1* in *A. caviae*, although the lack of this gene did not affect lateral flagella (Gavín *et al.*, 2002). Downstream of *E. coli* 042 *lafV*, two genes that are transcribed divergently (Ec042–0259 and Ec042–0260) are located. Homologues of these genes are found as part of capsule polysaccharide biosynthesis clusters. It is possible that these two genes may be involved in post-translational modification of flagella proteins. *Y. pestis*, *Y. pseudotuberculosis* and *P. profundum* strains do not show additional genes between homologous *fliJ*$_L$ and *flgN*$_L$. In *A. hydrophila*, the modification accessory factor gene *maf-5* is located between *flgL*$_L$ and *lafA*. Nevertheless, in *E. coli* 042, this region contains four additional genes: the three first genes, *lafW* and two contiguous *orfs* (Ec042–0277 and Ec042–0278), are transcribed in the same direction, whereas the last one *lafZ*, is transcribed divergently. The *lafW* encoded protein may represent a novel hook-associated protein, and a homologous gene was found in *V. parahaemolyticus* lateral flagella region 1 in an identical position. The Ec042–0277 gene encodes a protein with unknown function and Ec042–0278 encodes a protein that contains a helix–turn–helix domain and exhibits high amino acid identity with several other putative transcriptional regulators. The *lafZ* gene encodes a putative transmembrane transcriptional regulator (Table 11.1). *Y. pestis* and *Y. pseudotuberculosis* strains only contain three genes between homologous *flgL*$_L$ and *lafA*: *lafW*, a gene which encodes a protein with unknown function, and *lafZ*; the last two transcribed divergently to *lafW*. No additional genes were found in *P. profundum*. Furthermore, *Y. pestis*, *Y. pseudotuberculosis* and *P. profundum* have different numbers of flagellin genes located upstream of a homologue of *lafB*, as reported in *A. caviae*.

The finding of two variable regions, *flgL*$_L$ (*lfgL*)-*lafA* and *fliJ*$_L$ (*lfiJ*)-*flgN*$_L$ (*lfgN*), as well as the *V. parahaemolyticus* lateral flagella genes distributed in two chromosomal regions, suggest that there may be recombination points in these two variable areas.

Surprisingly, the genomic analysis of *A. salmonicida*, a non-flagellated bacterium, show the presence of two flagella systems identical to the polar and lateral flagella systems of *A. hydrophila*. The *A. salmonicida* lateral flagella locus show a

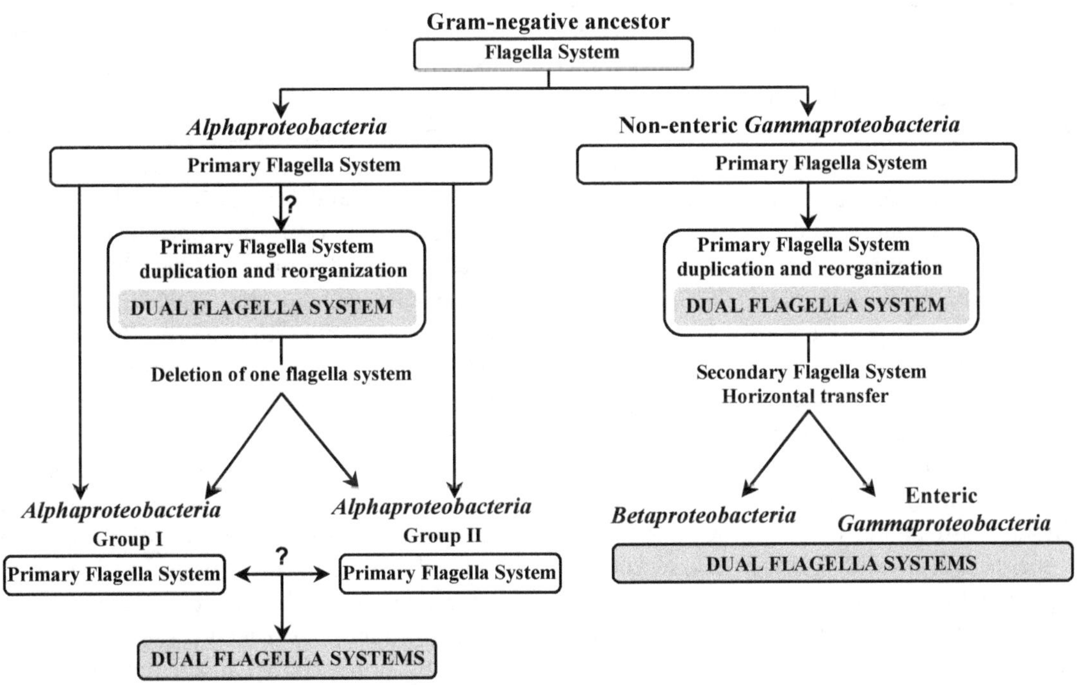

Figure 11.2 Schematic representation of lateral flagella system's origin.

frameshift mutation in *lfgD*, which encodes a proximal rod protein, and a transposase 8 insertion element located in the flagellin gene *lafA* (Merino *et al.*, 2003) that inactivate the system in this strain (Fig. 11.1).

In contrast with the polar or peritrichous flagella systems, none of the lateral flagella systems described contains the export-assembly gene *fliO*. The role of FliO is poorly understood, even in *S. enterica* serovar Typhimurium and *E. coli* (Schoenhals *et al.*, 1998), although it has been postulated their role is in flagellar export.

Origin of lateral flagella systems

Comparison of the complete genome sequences of flagellated bacteria revealed that flagellar structural proteins are based on an ancient core set of 24 flagella genes that were present in the common ancestor to all Bacteria. These core genes originated from a single or few ancestral sequences by a series of duplications that allow individual gene diversification before the shared ancestor of Bacteria (Liu *et al.*, 2007b). Despite the early origin of the core structural genes, flagella systems were subjected to multiple gains, losses, and gene rearrangements, which distributed flagella genes in several clusters. The organization, number, and gene content differ greatly among bacterial lineage. A phylogenetic tree based on gene order distance shows flagella systems clustered into three broad groups (Liu *et al.*, 2007b). Group 1 contains the primary flagella system of several *Alphaproteobacteria* and the *B. japonicum* secondary flagella system (lateral flagella system). This group does not possess many typical gene pairs, like *motAB*, *flhAB*, *fliQR*, *flgDE*, *fliFG*, and *flgFG*. However, they do possess some unique gene pairs like *fliNG-flhB*, *fliM-motA*, *flgAI*, *flgF-fliI*, and *flgD-fliQ*, which are not present in the other groups. Group 2 contains the primary and secondary flagella systems of *Beta-* and *Gammaproteobacteria* and the *Zymomonas mobilis* flagella system. Group 3 also differed in some gene pairs, like the association of *fliK* with *flgD* and the association of *flgA* with *flgG* and *flgH*.

Complete lateral flagella systems, without disrupting genes, were found in six proteobacterial species: four *Gammaproteobacteria* (*V. parahaemolyticus*, *A. hydrophila*, *Y. pseudotuberculosis* and *P. profundum*), one *Betaproteobacteria* (*C. violaceum*) and one *Alphaproteobacteria* (*B. japonicus*). These bacteria have two flagellar systems encoded by physically non-overlapping gene clusters, except that a probable translocation interspersed the two systems in *C. violaceum*. Phylogenetic analysis and cluster organizations revealed that lateral flagella systems were originated twice from the duplication or horizontal transfer of primary flagella systems (Fig. 11.2): once in the *Alphaproteobacteria* and another in the common ancestor of the *Beta-* and *Gammaproteobacteria* (Liu *et al.*, 2007a).

The origin of the *B. japonicum* lateral flagella system is most closely related to the *Alphaproteobacteria* group II primary system rather than to *Beta-* or *Gammaproteobacteria* lateral flagella, whereas its primary flagella systems is closely related to the *Alphaproteobacteria* group I single flagellar system. Therefore, lateral flagella may have been acquired by horizontal transfer of the *Alphaproteobacteria* group II primary system, or maybe the *Alphaproteobacteria* ancestor possessed a second flagellar system and that one system was subsequently lost in both *Alphaproteobacteria* group I and II (Liu *et al.*, 2007a).

The lateral flagella of *Beta-* and *Gammaproteobacterias* appears to have originated from a duplication of the entire flagellar gene complex in the non-enteric gammaproteobacterial lineage, which was then independently horizontally transferred to the *Betaproteobacteria* and to an ancestral lineage of enteric bacteria (Liu *et al.*, 2007a). In contrast to polar or peritrichous flagella systems (primary systems), lateral flagella systems lack *fliO* and the *fliEFGHIJKLMNPQR* gene cluster is split into two gene clusters (*fliEFGHIJ*$_L$ and *fliMNPQR*$_L$) transcribed in the same direction in *A. hydrophila* and in the opposite direction in the other species, except in *C. violaceum* whose *fliJ*$_L$ location is not determined. The *fliKL*$_L$ (*lafEF*) genes are arranged in the *fliD*$_L$-*motB*$_L$ (*lafB-lafU*) cluster. Although some bacterial species maintain both flagella systems, in most species, such as *E. coli*, *Shigella* spp. and *Y. pestis* strains, the lateral flagella genes were either lost or inactivated. It is tempting to speculate that immune selection pressures rendered these flagella systems non-functional in order to be less visible to the innate immune system, since these systems do not provide a selective advantage for survival in a specific habitat. The different genome arrangement of *E. coli/Shigella* and *Yersinia* secondary flagella systems suggests that the lateral gene transfer occurred independently in these two species. Moreover, the absence of remnants of genes in *S. enterica* suggests that the lateral flagella cluster was acquired after the *Escherichia* divergence to *Salmonella* (Ren *et al.*, 2005).

Regulation of lateral flagella systems

Flagella morphogenesis is a complex cascade of events that entails an important cost for the bacterium in terms of resources and energy. Regulation of flagellar assembly involves a combination of transcriptional, translational and post-translational regulatory mechanisms that ensure a careful regulation, proper assembly and quantity of flagellar components. In relation to transcriptional hierarchy, the flagellar clusters are divided into classes of genes temporally expressed during the assembly process by specific transcriptional regulators and sigma factors (Aldridge *et al.*, 2002; Soutourina *et al.*, 2003; McCarter, 2006). All genes of a particular class must be expressed before expressions of the next class to sequentially assemble the flagella proteins (Macnab, 1996).

Bacteria with dual flagella systems have a peritrichous or polar flagella as a primary system and lateral induced flagella as a secondary system. Peritrichous flagella genes of *Gammaproteobacteria*, such as *E. coli* and *S. enterica* serovar Typhimurium, are divided into three transcriptional levels. At the top of the hierarchy (class I) is the *flhDC* master operon that encode the $FlhD_2C_2$ transcriptional activator of class II σ^{70}-dependent promoters. Class II gene products include structural components of the flagellar export system, the basal body and the hook and the regulatory proteins FliA and FlgM. Class III genes encode the flagella filament,

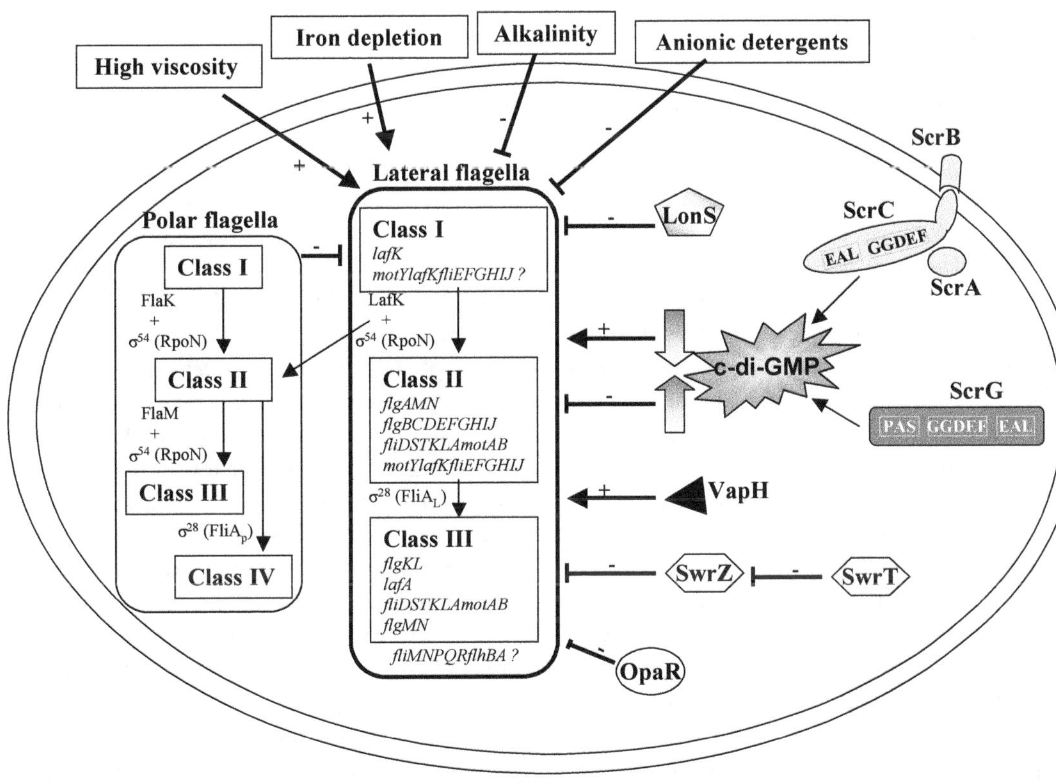

Figure 11.3 *Vibrio parahaemolyticus* lateral flagella hierarchy and factors controlling its expression. Transcription of some lateral flagella clusters (fliMNPQRflhBA?) remain to be elucidated.

hook-associated, motor and chemotaxis proteins, are positively regulated by the σ^{28} flagella specific sigma factor (FliA) and negatively regulated by an anti-sigma factor, FlgM (Macnab, 1996; Macnab, 2003). However, the transcriptional hierarchy of polar flagella genes shows more heterogeneity than peritrichous flagella genes. Polar flagella genes of *Gammaproteobacteria,* such as *V. cholerae, V. parahaemolyticus* and *Pseudomonas aeruginosa,* are divided into four transcriptional levels. In *V. cholerae,* at the top of the hierarchy is the σ^{54}-associated transcriptional activator FlrA (FlaK of *V. parahaemolyticus* and FleQ of *P. aeruginosa*). This regulator activates class II σ^{54}-dependent promoters encoding structural components of the flagella export system and the basal body, the σ^{28} flagellar-specific sigma factor and the two component signal-transducing system FlrBC (FlaLM of *V. parahaemolyticus* and FleSR in *P. aeruginosa).* The FlrC regulator activates class III σ^{54}-dependent promoters encoding the basal body components, the hook and the flagellin proteins essential for motility. The sigma factor σ^{28} activates transcription of class IV genes encoding the motor genes and non-essential flagellin subunits (Prouty *et al.,* 2001; Dasgupta *et al.,* 2003; Correa *et al.,* 2005).

Three different polar flagella transcriptional models were described in *Alphaproteobacteria*: *Sinorhizobium meliloti, Rhodobacter sphaeroides* and *Caulobacter crescentus* models. *S. meliloti* has three transcriptional levels: class I comprises two LuxR-type regulators, *visN* and *visR;* class II comprises flagellar assembly (class IIA) and motor (class IIB) genes, controlled by VisNR; and class III comprises flagellin and chemotaxis genes that require functional class I and class IIA genes for expression (Sourjik *et al.,* 2000). *R. sphaeroides* and *C. crescentus* have four transcriptional levels. In *R. sphaeroides,* class I comprise the σ^{54}-transcriptional activator FleQ, which control transcription of class II genes. Class II genes activate structural proteins and the second σ^{54}-transcriptional regulator FleT that, along with FleQ act together as a hetero-oligomer, to control the expression of class III genes, which results in completion of the hook-basal body complex, export of FlgM and late genes expression by a σ^{28} flagella specific factor (Poggio *et al.,* 2005). In *C. crescentus,* there is only a single class I gene, *ctrA,* which is under the control of cell cycle signals. Phosphorylation of CtrA regulates the σ^{70}-dependent promoters of class II genes, encoding the MS ring complex, the export system and the two component signal-transducing system FlbDE. Class III and class IV genes have σ^{54}-dependent promoters stimulated by FlbD. In this process, FlbD works as a key component of a genetic switch by repressing class II expression after class II expression has been activated. To obtain a hierarchic expression of class III and class IV genes, the FlbT protein destabilizes class IV transcripts before class III completion. No σ^{28}-dependent genes were found in this transcriptional model (Mangan *et al.,* 1999; Gober *et al.,* 2000; England *et al.,* 2001).

The lateral flagella system (secondary flagella) of *V. parahaemolyticus* is the best studied lateral flagella at the regulatory level (Stewart *et al.,* 2003) (Fig. 11.3). This flagella system does not show FlhDC master regulators as in the peritrichous flagella system and it is activated by the σ^{54}-associated transcriptional activator LafK, which is homologous to the *V. cholerae* FlrA polar flagella regulator. LafK controls the expression of class II genes and a lateral flagella specific σ^{28} factor, encoded by a class II gene (*fliA*$_L$) that controls the expression of class III genes. No two component signal-transducing system was found in the lateral flagella system. This transcriptional hierarchy model seems to be present in the lateral flagella of other *Gammaproteobacteria,* such as *A. hydrophila* (Canals *et al.,* 2006a), *P. profundum* and *Y. pseudotuberculosis.* Recently, Chang *et al.* proposed that the lateral flagellar regulation profile of the *Alphaproteobacteria A. brasilense* is probably more similar to *C. crescentus* polar flagella, than to *V. parahaemolyticus* and *A. hydrophila* lateral flagella systems (Chang *et al.,* 2007). In fact, a CtrA homologue was found in *A. brasilense.* In addition, the late lateral flagella genes seem to be activated by a σ^{54} factor instead of σ^{28} flagella specific factor and the σ^{54}-associated transcriptional activator FlbD is an indispensable activator of lateral flagella and a repressor of *fliF* and *flgB* of polar flagella system, playing a key component in the regulation of dual flagella systems. These results and the hypothesis that the lateral flagella system appears to have originated twice (Liu *et al.,* 2007) may suggest two transcriptional hierarchy models in *Gamma-* and *Alphaproteobacteria* lateral flagella systems.

Master bacterial regulators with single flagella systems are essential for flagella expression, but this situation seems to be different in some bacteria with dual flagella systems. For example, in *V. parahaemolyticus,* the σ^{54}-dependent polar flagella response regulator FlaK is dispensable for polar flagella expression, since the lateral flagella transcriptional activator LafK, that is essential for lateral flagella generation, compensates for its loss (Kim *et al.*, 2004) (Fig. 11.3).

The presence of two active flagella systems implies a high energetic cost for a bacterium, therefore lateral flagella synthesis is carefully regulated in response to different environmental conditions. The most studied environmental conditions known to induce transcription of lateral flagella genes are those that reduce polar flagella movement, i.e. increase in media viscosity. In some dual flagellated bacteria, such as *V. parahaemolyticus* and *A. brasilense,* the polar flagella act as a mechano-sensor by measuring viscosity and transducing the reduction of polar flagella rotation to control lateral flagella expression. In these two bacterial species, both flagella types are intimately linked and their regulation systems seem to interact, since defects in polar flagella formation or motility allow lateral flagella expression to be constitutive (McCarter *et al.*, 1988; Kawagishi *et al.*, 1995; Alexandre *et al.*, 1999). However, the molecular mechanism for sensing polar flagella inhibition and the signal transducing pathway regulating lateral flagella expression are not known. However, polar flagellum defects in *Aeromonas* spp. or *R. centenum* do not induce constitutive lateral flagella formation (Jiang *et al.*, 1998; Altarriba *et al.*, 2003; Canals *et al.*, 2006a), demonstrating that lateral flagella expression is still under its natural control, and suggesting that polar flagella do not interact with lateral flagella regulation in these species. In spite of transcriptional independence of both *R. centenum* flagella systems, lateral flagella are unable to move the cell in the absence of functional polar flagella, although they are able to rotate. This suggests that coordination of lateral flagellum rotation, such as proper formation of a flagella bundle, is defective in cells that contain a non functional polar flagellum (McClain *et al.*, 2002). Other environmental conditions were reported to affect lateral flagella formation or stability. In *V. parahaemolyticus,* lateral flagella formation is stimulated by growth in iron-depleted medium (McCarter *et al.*, 1989) and inhibited by alkaline conditions (Kimura *et al.*, 1979) or anionic detergents. Lateral flagella have also been reported to dissociate from the cell surface at high temperature in some strains (Shinoda *et al.*, 1977). In *R. centenum,* increasing growth temperature induces lateral flagella formation in liquid media (Berleman *et al.*, 2005). In *A. brasilense* Cd, but not in other *Azospirillum* species, static liquid growth stimulates lateral flagella (Madi *et al.*, 1988).

Bacteria sense and respond to environmental inputs via networks of signal transducing pathways and signal integration systems. Some of these networks either produce small molecules or specifically respond to them and modulate different bacterial components. A network of proteins containing GGDEF or EAL motif were reported to participate in detecting environmental signals and modulate the level of the second messenger, cyclic dimeric-guanosine monophosphate (c-di-GMP) (Romling *et al.*, 2005), which appears to control the transition from the motile to the biofilm community. In *V. parahaemolyticus,* two GGDEF-EAL containing proteins, ScrC and ScrG, regulate inversely the expression of two surface structures important to life on surface, i.e. lateral flagella and capsule, by modulation of the c-di-GMP pool (Fig. 11.3). ScrC is part of the *srcABC* operon and is a cytoplasmic membrane protein that contains a periplasmic-sensing domain. ScrB is a soluble periplasmic-binding protein that receives an input signal and interacts with the periplasmic domain of ScrC. ScrA is a cytoplasmic protein that resembles a pyridoxal-phosphate-dependent aminotransferase, which is required for signal transduction, although its role is unclear. ScrA and ScrB modulate the activity of the cytoplasmic GGDEF and EAL domains of ScrC, which are involved in synthesis and hydrolysis, respectively, of c-di-GMP. Thus, ScrC, in the presence of ScrA and ScrB, acts as a phosphodiesterase decreasing the level of the c-di-GMP pool which induces lateral flagella expression and prevents capsule polysaccharide transcription. The absence of these two proteins acts as a diguanylate cyclase enhancing the c-di-GMP pool, which evokes the complete opposite effect (Boles *et al.*, 2002; Ferreira *et al.*, 2008). The second *V. parahaemolyticus* GGDEF-EAL protein involved in regulation of genes relevant for surface growth is ScrG. This protein shows three highly conserved domains: PAS (signal sensor domain), GGDEF and EAL, although the GGDEF domain is rather poorly conserved and lacks the GGDEF signature. ScrG acts as a phosphodiesterase influencing, by itself, swarming and capsule polysaccharide regulation, in contrast to ScrC whose regulatory activity requires other products (ScrA and ScrB). ScrC and ScrG appear to act in the same pathway and none of them affect swimming motility (Kim *et al.*, 2007). Moreover, the loss of *scrABC* or *scrG* only reduces but does not abolish swarm differentiation, suggesting that it may play a role in mediating the switch between lateral flagella and capsule expression during surface colonization.

Quorum sensing is a bacterial cell–cell communication system shown to control a number of factors important for colonization, virulence and biofilm formation. In *V. parahaemolyticus*, the homologous quorum-sensing components OpaR, SwrT and SwrZ modulate lateral flagella expression. OpaR is the central transcriptional regulator involved in the opaque-translucent switching by upregulating capsule expression, but it also represses lateral flagella expression. However, in the absence of OpaR, the swarming phenotype remains tightly surface regulated. SwrT appears to modulate swarming by repressing SwrZ transcription (the lateral flagella repressor). Interestingly, a SwrZ mutant does not display constitutive lateral flagella, suggesting that SwrZ is not responsible for transmitting either the iron starvation or the polar flagella inhibition signal (Jaques *et al.*, 2006) (Figure 3). In *V. alginolyticus*, the LuxS quorum-sensing system regulates polar and lateral flagella biogenesis. It down-regulates *flaK* and *fliS* polar flagella genes and up-regulates *lafK* and *lafA* lateral flagella genes in semi-solid plates. Moreover, the in-frame *luxS* deletion abolishes swimming and swarming, confirming the involvement of LuxS in the regulation of *V. alginolyticus* motility (Tian *et al.*, 2008). No motility differences were found in a *V. vulnificus luxS* mutant, suggesting that LuxS-dependent motility regulation is not common throughout *Vibrio* spp. (Kim *et al.*, 2006).

In some cases, the control of gene expression by environmental factors is mediated by global regulators that affect the DNA conformation of regulatory regions, such as histone-like DNA-binding proteins. *V. parahaemolyticus trh*-positive strains express the VpaH protein, which is homologous to the histone DNA-binding protein H-NS that positively regulates the enterobacterial flhDC operon by DNA supercoiling. VpaH positively regulates lateral flagella biogenesis whereas no effect was observed in the expression of polar flagella (Park *et al.*, 2005). Another mechanism for the rapid control of critical regulatory proteins is the proteolytic degradation of master regulators. This mechanism is used by the *V. parahaemolyticus* ATP-dependent protease LonS, which inhibits the swarmer cell phenotype by degrading a transcriptional activator of lateral flagella genes and a cell division inhibitor. Therefore, LonS mutants express lateral flagella and produce elongated cells in liquid medium (Stewart *et al.*, 1997) (Fig. 11.3).

The chemotaxis signalling pathway appears to play a role in the ability of peritrichously flagellated bacteria, such as *S. enterica* serovar Typhimurium, to swarm, since mutants in the chemosensory system display few and shortened flagella when propagated on a surface (Wang *et al.*, 2005). In the dual flagellated bacterium *R. centenum*, components of the second chemotatic operon (Che_2) do not appear to affect directly or indirectly chemotaxis, but their mutation reduces polar and lateral flagella synthesis, suggesting some level of coordinated regulation between both flagella types. The presence of the *fliC*$_L$ flagella gene transcript in *che2* disrupted mutants indicates little or no disturbance of flagella gene expression, implicating the Che2 transducing cascade in the post-transcriptional or post-translational regulation of lateral and polar flagella (Berleman *et al.*, 2005). Some dual flagellated bacteria, like *V. parahaemolyticus* and *V. alginolyticus*, have a unique set of *che* genes and therefore a unique small soluble protein CheY, which in its active phosphorylated form is the response regulator that binds to the flagella motor to promote switching. In *V. alginolyticus*, it was reported that substitutions in CheY residues have different effects on the polar and lateral flagella motors, indicating that CheY interacts differently with the two motors. The lateral motor rotates exclusively counterclockwise and is slowed down by CheY, whereas the polar motor turns both counterclockwise and clockwise, and CheY controls its rotational direction (Kojima *et al.*, 2007).

Lateral flagella post-translational modifications: glycosylation

Protein glycosylation has been long recognized as an important posttranslational modification in eukaryotic cells, whereas bacterial protein glycosylation is uncommon. Recently, several surface-layer proteins, pilins, adhesins, flagellins and secreted proteins of Gram-negative bacteria were shown to possess two types of glycosidic linkage: *N*- and *O*-glycosides (Benz *et al.*, 2002; Power *et al.*, 2003; Schmidt *et al.*, 2003). In most cases, the genes responsible for glycosylation are located close to the genes encoding the target protein. Several studies have shown the role of glycosylation in infection and interference with the immune response (Upreti *et al.*, 2003; Szymanski *et al.*, 2005).

Although no evidence for flagellin glycosylation in the *Enterobacteriaceae* has been reported, polar flagella from different Gram-negative bacteria appear to be glycosylated in *O*-linkage to serine/threonine residues (Logan, 2006). Thus, in *P. aeruginosa* strain PAK, a glycan consisting of up to 11 monosaccharide units was attached through a rhamnose residue to two polar flagellin glycosylation sites located in the central protein domain (Arora *et al.*, 2001; Schirm *et al.*, 2004). *P. aeruginosa* polar flagella glycosylation plays an important role in the proinflammatory response (Arora *et al.*, 2005), and, in *P. syringae,* it is responsible for host plant recognition and hypersensitivity reactions (Takeuchi *et al.*, 2003). Flagellins subunits of *Campylobacter* and *Helicobacter* are heavily glycosylated with a pseudaminic acid and, in contrast to *Pseudomonas,* this sugar is required for the proper development of the flagella and motility (Thibault *et al.*, 2001). *C. jejuni* flagella glycosylation contributes to gut colonization (Szymanski *et al.*, 2002). In mesophilic *Aeromonas,* polar and lateral flagella of *A. hydrophila* and *A. caviae* are glycosylated (Rabaan *et al.*, 2001; Gavin *et al.*, 2002). Mass spectrometry analysis of *A. caviae* UU51 purified polar flagellin subunits show that up to six sites on FlaA and seven sites on the FlaB polar flagellin monomers are glycosylated with a derivative of pseudaminic acid by an O-glycoside linkage (Schirm *et al.*, 2005). However, the number of sites and the glycan nature of lateral flagellin remain to be established. In *A. hydrophila* AH-3 and *A. caviae* Sch3N, six genes were reported to be involved in pseudaminic acid biosynthesis: *flmA, flmB, maf2, neuA, flmD* and *neuB.* These genes are homologous to *Campylobacter* glycosylation genes involved in polar and lateral flagella assembly (Power *et al.*, 2002). In *A. caviae* Sch3N, all genes are arranged in a unique locus and are also involved in LPS O-antigen expression (Gryllos *et al.*, 2001). In contrast, these genes are distributed in two different chromosomal regions in *A. hydrophila* AH-3: *flmA-flmB* locus and *maf2-neuA-flmD-neuB* locus that is adjacent to the polar flagella region 2 (Canals *et al.*, 2007). Moreover, two *A. hydrophila* genes (*maf-1* and 5), whose products are homologous to *H. pylori* and *C. jejuni* Maf proteins (Power *et al.*, 2002), appear to be involved in specific polar and lateral flagella glycosylation, respectively (Canals *et al.*, 2006a, Canals *et al.*, 2006b). To date, little is known about the role of lateral flagellin glycosylation in *Aeromonas* virulence.

Role of lateral flagella in virulence

The prerequisite event for any infection is the encounter of pathogenic bacteria with the target tissues; therefore many pathogenic bacteria produce flagellar structures that allow bacteria to swim. Swimming, combined with chemotaxis, enables a fine tuned access of pathogens to their target on mucosal tissues, but the flagellum also contributes in adhesion to epithelial cells (Otteman *et al.*, 1997; Josenhans *et al.*, 2002) and is an important stimulator of the TLR5-dependent immune response (Ramos *et al.*, 2004). Adhesion and colonization of the target cells, after initial attachment, are the most critical steps in the infection process. Swarming motility allows rapid colonization of the surface before irreversible adsorption, contributing to the infection process, as reported in *P. mirabilis* urinary-tract infections (Mobley *et al.*, 1995). Surface colonization usually involves biofilm formation that is an accumulation of microorganisms adhered on a surface embedded in their produced polysaccharide matrix (Costerton *et al.*, 1999). Bacteria in biofilms express more virulence factors as a result of gene activation through quorum sensing or gene transfer (Greenberg, 1999), and are more resistant to host defences and antimicrobial agents than they are as free-floating planktonic cells.

Lateral flagella involvement in the first steps of the infection process was reported in some dual flagellated water-borne bacteria and in a nitrogen-fixing rhizobacterium. The waterborne bacteria *V. parahaemolyticus*, *A. hydrophila* and *A. caviae* are involved in different animal diseases, as well as in gastroenteritis, wound infection and septicaemia in humans. Lateral flagella of these three bacteria link bacteria on agar and on cell surfaces and contribute to microcolony formation and allow bacteria to adhere more firmly (Belas *et al.*, 1982a; Kirov *et al.*, 2002). *V. parahaemolyticus* lateral flagella have been shown to attach to chitinaceous plankton in a temperature, salinity and pH dependent manner, by a mechanism probably different from that used by the polar flagellum (Belas *et al.*, 1982b). Furthermore, the analysis of *V. parahaemolyticus vapH* or *lafTU* mutants, unable to form or to move lateral flagella, respectively, showed that both had a 50% reduction in their adherence to HeLa cells and about 60% lower ability to form biofilms than the wild type. This reduction in cell adherence and biofilm formation phenotypes demonstrate the importance of lateral flagella in *V. parahaemolyticus* virulence (Park *et al.*, 2005). In mesophilic *Aeromonas*, at least 50% of strains commonly associated with diarrhoeal illness produce lateral flagella, such as *A. hydrophila* AH-3 and *A. caviae* Sch3N. In these strains, lateral flagella facilitate adherence to, and probably invasion of, the epithelial cell line HEp-2 (Gavín *et al.*, 2002; Gavín *et al.*, 2003; Canals *et al.*, 2006a), as well as the intestinal cell lines Henle INT 407 and Caco-2 (Kirov *et al.*, 2004), suggesting that lateral flagella may be a virulence determinant for enteropathogenic strains. Lateral flagella also facilitate biofilm formation on abiotic surfaces, such as plastic microtiter plates and borosilicate glass tubes (Gavín *et al.*, 2003; Kirov *et al.*, 2004). In nitrogen-fixing rhizobacteria, the efficient colonization of the plant rhizosphere is an important factor for being competitive in the soil. Swimming through the water spaces and chemotaxis enable bacteria to move towards plant roots (Zhulin *et al.*, 1993), but this motility is limited by soil moisture (Bashan, 1986). In *A. brasilense*, the first step in the plant root colonization seems to be a functional polar flagellum (Croes *et al.*, 1993), whereas lateral flagella enable the bacteria to move along the root and are thought to be important for long-term colonization (Moens *et al.*, 1995).

Conclusions

Motility is widespread throughout the prokaryotes and a wide variety of structures are able to confer motility in different circumstances. The bacterial flagellum, the most common and best studied prokaryotic motility structure, allows bacteria to swim and swarm. Many bacteria, such as *Salmonella* and *Proteus*, have a single flagella system that accommodates both swimming and swarming motility. Only a few *Alpha-*, *Beta-* and *Gammaproteobacteria* have dual flagella systems that modulate in different environmental circumstances. To date, all bacteria with functional dual flagella systems have a constitutive polar flagellum and an inducible lateral flagella system that is expressed in highly viscous media or on surfaces. Complete lateral flagella systems have also been found in bacteria with constitutive peritrichous flagella, but their functionality is not proven. The origin of lateral flagella systems seems to have occurred twice from the duplication or horizontal transfer of primary flagella systems: once in the *Alphaproteobacteria* and the other in the common ancestor of the *Beta-* and *Gammaproteobacteria*. Biosynthesis and motility of two flagella systems results in a high metabolic cost for the bacteria; thus, lateral flagella expression is highly regulated by multiple sensory inputs which activate different regulatory systems, such as quorum-sensing and GGDEF-EAL proteins that modulate the bacterial c-di-GMP pool. Lateral flagella synthesis can be controlled at the transcriptional, translational or post-translational level. In *V. parahaemolyticus* lateral flagella and capsule polysaccharide synthesis are inversely regulated. In pathogenic bacteria, the lateral flagella and the swarmer cell phenotype contribute to host cell adhesion and biofilm formation. Further investigations into lateral flagella regulation, glycosylation, host cell interaction and proinflammatory action are needed to understand their pathogenic relevance.

Acknowledgements

This work was supported by Plan Nacional de I + D and FIS grants (Ministerio de Educación, Ciencia y Deporte and Ministerio de Sanidad, Spain) and from Generalitat de Catalunya.

References

Aldridge, P., and Hughes, K.T. (2002). Regulation of flagellar assembly. Curr. Opin. Microbiol. *5*, 160–165.

Alexandre, G., Rohr, R., and Bally, R. (1999). A phase variant of *Azospirillum lipoferum* lacks a polar flagellum and constitutively expresses mechanosensing lateral flagella. Appl. Environ. Microbiol. *65*, 4701–4704.

Altarriba, A., Merino, S., Gavín, R., Canals, R., Rabaan, A., Shaw, J.G., and Tomás, J.M. (2003). A polar flagella operon (*flg*) of *Aeromonas hydrophila* contains genes required for lateral flagella expression. Microb. Pathog. *34*, 249–259.

Atsumi, T., McCarter, L.L., and Imae, Y. (1992). Polar and lateral flagellar motors of marine *Vibrio* are driven by different ion-motive forces. Nature *355*, 182–184.

Arora, S.K., Bangera, M., Lory, S., and Ramphal, R. (2001) A genomic island in *Pseudomonas aeruginosa* carries the determinants of flagellin glycosylation. Proc. Natl. Acad. Sci. U.S.A. *98*, 9342–9347.

Arora, S.K., Neely, A.N., Blair, B., Lory, S., and Ramphal, R. (2005). Role of motility and flagellin glycosylation in the pathogenesis of *Pseudomonas aeruginosa*. Burn Wound Infections. Infect. Immun. *73*, 4395–4398.

Bardy, S.L., Ng, S.Y.M., and Jarrell, K.F. (2003) Prokaryotic motility structures. Microbiology *149*, 295–304.

Bashan, Y. (1986). Migration of the rhizosphere bacteria *Azospirillum brasilense* and *Pseudomonas fluorescens* towards wheat roots in the soil. J. Gen. Microbiol. *132*, 3407–3414.

Belas, M.R., and Colwell, R.R. (1982a). Scanning electron microscope observation of the swarming phenomenon of *Vibrio parahaemolyticus*. J. Bacteriol. *150*, 956–959.

Belas, M.R., and Colwell R.R. (1982b). Adsorption kinetics of laterally and polarly flagellated *Vibrio*. J. Bacteriol. *151*, 1568–1580.

Benz, I., and Schmidt, M.A. (2002). Never say never again: protein glycosylation in pathogenic bacteria. Mol. Microbiol. *45*, 267–276.

Berleman, J.E., and Bauer, C.E. (2005). A *che*-like signal transduction cascade involved in controlling flagella biosynthesis in *Rhodospirillum centenum*. Mol. Microbiol. *55*, 1390–1402.

Boles, B.R., and McCarter, L.L. (2002). *Vibrio parahaemolyticus scrABC*, a novel operon affecting swarming and capsular polysaccharide regulation. J. Bacteriol. *184*, 5946–5954.

Burnens, A.P., Stanley, J., Sack, R., Hunziker, P., Brodard, I., and Nicolet, J. (1997). The flagellin N-methylase gene *fliB* and an adjacent serovar-specific IS200 element in *Salmonella typhimurium*. Microbiology *143*, 1539–1547.

Canals, R., Altarriba, M., Vilches, S., Horsburgh, G., Shaw, J.G., Tomás, J.M., and Merino, S. (2006a). Analysis of the lateral flagellar gene system of *Aeromonas hydrophila* AH-3. J. Bacteriol. *188*, 852–862.

Canals, R., Ramirez, S., Vilches, S., Horsburgh, G., Shaw, J.G., Tomás, J.M., and Merino, S. (2006b). Polar flagellum biogenesis in *Aeromonas hydrophila*. J. Bacteriol. *188*, 542–555.

Canals, R., Vilches, S., Wilhelms, M., Shaw, J.G., Merino, S., and Tomas, J.M. (2007). Non-structural flagella genes affecting both polar and lateral flagella-mediated motility in *Aeromonas hydrophila*. Microbiology *153*, 1165–1175.

Chang, Y., Tang, T., Li, J-L. (2007) Isolation of a flagellar operon in *Azospirillum brasilense* and functional analysis of FlbD. Res. Microbiol. *158*, 521–528.

Correa, N.E., and Klose, K.E. (2005) Characterization of enhancer binding by the *Vibrio cholerae* flagellar regulatory protein FlrC. J. Bacteriol. *187*, 3158–3170.

Costerton, J.W., Stewart, P.S., and Greenberg, E.P. (1999). Bacterial biofilms: a common cause of persistent infections. Science *284*, 1318–1322.

Croes, C.L., Moens, S., Van Bastelaere, E., Vanderleyden, J., and Michiels, K.W. (1993). The polar flagellum mediates *Azospirillum brasilense* adsorption to wheat roots. J. Gen. Microbiol. *139*, 960–967.

Dasgupta, N., Wolfgang, M.C., Goodman, A.L., Arora, S.K., Jyot, J., Lory, S., and Ramphal, R. (2003). A four-tiered transcriptional regulatory circuit controls flagellar biogenesis in *Pseudomonas aeruginosa*. Mol. Microbiol. *50*, 809–824.

England, J.C., and Gober, J.W. (2001) Cell cycle control of cell morphogenesis in *Caulobacter*. Curr. Opin. Microbiol. *4*, 674–680.

Ferreira, R.B.R., Antunes, L.C.M., Greenberg, E.P, and McCarter, L.L. (2008). *Vibrio parahaemolyticus* ScrC modulates cyclic dimeric GMP regulation of gene expression relevant to growth on surfaces. J. Bacteriol. *190*, 851–860.

Gavín, R., Rabaan, A.A., Merino, S., Tomás, J.M., Gryllos, I., and Shaw, J.G. (2002). Lateral flagella of *Aeromonas* species are essential for epithelial cell adherence and biofilm formation. Mol. Microbiol. *43*, 383–397.

Gavín, R., Merino, S., Altarriba, M., Canals, R., Shaw, J.G., and Tomás, J.M. (2003). Lateral flagella are required for increased cell adherence, invasion and biofilm formation by *Aeromonas* spp. FEMS Microbiol. Lett. *224*, 77–83.

Gober, J.W., and England, J.C. (2000). Regulation of flagellum biosynthesis and motility in *Caulobacter*. In *Prokaryotic Development*. Y.V. Brun, and L.J. Schimkets, ed. (Washington D.C., USA: American Society of Microbiology Press), pp. 319–340.

Greenberg, E.P. 1999. Quorum sensing in Gram-negative bacteria. ASM News *63*, 371–377.

Gryllos, I., Shaw, J.G., Gavin, R., Merino, S., and Tomás, J.M. (2001). Role of *flm* locus in mesophilic *Aeromonas* species adherence. Infect. Immun. *69*, 65–74.

Hall, P.G., and Krieg, N.R. (1983). Swarming of *Azospirillum brasilense* on solid media. Can. J. Microbiol. *29*, 1592–1594.

Harshey, R.M. (1994). Bees aren't the only ones: swarming in Gram-negative bacteria. Mol. Microbiol. *13*, 389–394.

Harshey, R.M., and Matsuyama, T. (1994). Dimorphic transition in *Escherichia coli* and *Salmonella typhimurium*: surface-induced differentiation into hyperflagellate swarmer cells. Proc. Natl. Acad. Sci. U.S.A. *92*, 8631–8635.

Henrichsen, J. (1972). Bacterial surface translocation: a survey and a classification. Bacteriol. Rev. *36*, 478–503.

Inoue, K., Kosako, Y., Suzuky, K., and Shimada, T. (1991). Peritrichous flagellation in *Plesiomonas shigelloides* strains. Jpn. J. Med. Sci. Biol. *44*, 141–146.

Jaques, S., and McCarter, L.L. (2006). Three new regulators of swarming in *Vibrio parahaemolyticus*. J. Bacteriol. *188*, 2625–2635.

Jarrell, K.F., Bayley, D.P., and Kostyukova, A.S. (1996). The archaeal flagellum: a unique motility structure. J. Bacteriol. *178*, 5057–5064.

Jiang, Z.Y., and Bauer, C.E. (1997). Analysis of a chemotaxis operon from *Rhodospirillum centenum*. J. Bacteriol. *179*, 5712–5719.

Jiang, Z.Y., Rushing, B.G., Bai, Y., Gest, H., and Bauer, C.E. (1998) Isolation of *Rhodospirillum centenum* mutants defective in phototactic colony motility by transposon mutagenesis. J. Bacteriol. *180*, 1248–1255.

Josenhans, C., and Suerbaum, S. (2002). The role of motility as a virulence factor in bacteria. Int. J. Med. Microbiol. *29*, 605–614.

Kanbe, M., Yagasaki, J., Zehner, S., Göttfert, M., and Aizawa, S-I. (2007). Characterization of two sets of subpolar flagella in *Bradyrhizobium japonicum*. J. Bacteriol. *189*, 1083–1089.

Kaneko, T., Nakamura, Y., Sato, S., Minamisawa, K., Uchiumi, T., Sasamoto, S., Watanabe, A., Idesawa, K., Iriguchi, M., Kawashima, K., Kohara, M., Matsumoto, M., Shimpo, S., Tsuruoka, H., Wada, T., Yamada, M., and Tabata M. (2002). Complete genomic sequence of nitrogen-fixing symbiotic bacterium *Bradyrhizobium japonicum* USDA110. DNA Res. *9*, 189–197.

Karlinshev, A.V., Linton, D., Gregson, N.A., and Wren, B.W. (2002). A novel paralogous gene family involved in phase-variable flagella-mediated motility in *Campylobacter jejuni*. Microbiology *148*, 473–480.

Kawagishi, I., Maekawa, Y., Atsumi, T., Homma, M., and Imae, Y. (1995). Isolation of the polar and lateral flagellum-defective mutants in *Vibrio alginolyticus* and identification of their flagellar driving energy sources. J. Bacteriol. *177*, 5158–5160.

Khammas, K.M., Ageron, E., Grimont, P.A.D., and Kaiser, P. (1989). *Azospirillum irakense* sp. nov., a nitrogen-fixing bacterium associated with rice roots and rhizosphere soils. Res. Microbiol. *140*, 679–693.

Kim, M.Y., Park, R.Y., Choi, M.H., Sun, H.Y., Kim, C.M., Kim, S.Y., Rhee, J.H., and Shin, S.H. (2006). Swarming differentiation of *Vibrio vulnificus* downregulates the expression of the *vvhBA* hemolysin gene via the LuxS quorum-sensing system. J. Microbiol. *44*, 226–232.

Kim, Y-K, and McCarter, L.L. (2007). ScrG, a GGDEF-EAL Protein, participates in regulating swarming and sticking in *Vibrio parahaemolyticus* J. Bacteriol.*189*, 4094–4107.

Kim, Y.K., and McCarter, L.L. (2004) Cross-regulation in *Vibrio parahaemolyticus*: compensatory activation of polar flagellar genes by the lateral flagellar regulator LafK. J. Bacteriol. *186*, 4014–4018.

Kimura, K., Tateiri, S., and Iida, H. (1979). Effects of pH of the medium on flagellation of *Vibrio parahaemolyticus*. Appl. Environ. Microbiol. 37, 1248–1249.

Kirov, S.M., Tassell, B.C., Semmler, A.B.T., O'Donovan, L.A., Rabaan, A.A., and Shaw, J.G. (2002). Lateral flagella and swarming motility in *Aeromonas* species. J. Bacteriol. *184*, 547–555.

Kirov, S.M. (2003). Bacteria that express lateral flagella enable dissection of the multifunctional roles of flagella in pathogenesis. FEMS Microbiol. Lett. *224*, 151–159.

Kirov, S.M., Castrisios, M., and Shaw, J,G. (2004). *Aeromonas* flagella (polar and lateral) are enterocyte adhesins that contribute to biofilm formation on surface. Infect. Immun. *72*, 1939–1945.

Kojima, M., Kubo, R., Yakushi, T., Homma, M., and Kawagishi, I. (2007). The bidirectional polar and unidirectional lateral flagellar motors of *Vibrio alginolyticus* are controlledby a single CheY species. Mol. Microbiol. *64*, 57–67.

Liu, R., and Ochman, H. (2007a) Origins of flagellar gene operons and secondary flagellar systems. J. Bacteriol. *189*, 7098–7104.

Liu, R., and Ochman, H. (2007b) Stepwise formation of the bacterial flagellar system. Proc. Natl. Acad. Sci. USA *104*, 7116–7121.

Logan, S.M. (2006) Flagellar glycosylation: a new component of the motility repertoire? Microbiology *152*, 1249–1262.

Macnab, R.M. (1996). Flagella and motility. In: *Escherichia coli* and *Salmonella typhimurium*: Cellular and Molecular Biology. F.C. Neidhardt, R. Curtiss III, C.A. Gross, J.L. Ingraham, E.C.C. Lin, K.B. Low, B. Magasanik, W. Reznikoff, M. Riley, M. Schaechter and H.E. Umbarger, eds (Washington D.C., USA: American Society of Microbiology Press), pp. 123–145.

Macnab, R.M. (1999). The bacterial flagellum: reversible rotary propellor and type III export apparatus. J. Bacteriol. *181*, 7149–7153.

Macnab, R.M. (2003). How bacteria assemble flagella. Annu. Rev. Microbiol. *57*, 77–100.

Madi, L., Kessel, M., Sadovnik, E., and Henis, Y. (1988). Electron microscopic studies of aggregation and pellicle formation in *Azospirillum* spp. Plant Soil *109*, 115–121.

Mangan, E.K., Malakoot,i J., Caballero, A., Anderson, P., Ely, B., Gober, J.W. (1999). FlbT couples flagellum assembly to gene expression in *Caulobacter crescentus*. J. Bacteriol. *181*, 6160–6170.

Manson, M.D., Armitage, J.P., Hoch, J.A., and Macnab, R.M. (1998). Bacterial locomotion and signal transduction. J. Bacteriol. *180*, 1009–1022.

Martinez, A., Torello, S., and Kolter, R. (1999). Sliding motility in mycobacteria. J. Bacteriol. *181*, 7331–7338.

Mattick, J.S. (2002). Type IV pili and twitching motility. Annu. Rev. Microbiol. *56*, 289–314.

McCarter, L.L., Hilmen, M., and Silvermanm, M. (1988). Flagellar dynamometer controls swarmer cell differentiation of *V. parahaemolyticus*. Cell *54*, 345–351.

McCarter, L.L., and Silverman, M. (1989). Iron regulation of swarmer cell differentiation of *Vibrio parahaemolyticus*. J. Bacteriol. *171*, 731–736.

McCarter, L.L., and Silverman, M. (1990). Surface-induced swarmer cell differentiation of *Vibrio parahaemolyticus*. Mol. Microbiol. *4*, 1057–1062.

McCarter, L.L. (1999). The multiple identities of *Vibrio parahaemolyticus*. J. Mol. Microbiol. Biotechnol. *1*, 51–57.

McCarter, L.L. (2004). Dual flagellar systems enable motility under different circumstances. J. Mol. Microbiol. Biotechnol. *7*, 18–29.

McCarter, L.L. (2006). Regulation of flagella. Curr. Opin. Microbiol. *9*, 1–7.

McClain, J., Rollo, D.R., Rushing, B.G., and Bauer, C.E. (2002). *Rhodospirillum centenum* utilizes separate motor and switch components to control lateral and polar flagellum rotation. J. Bacteriol. *184*, 2429–2438.

Merino, S., Gavín, R., Vilches, S., Shaw, J.G., and Tomas J.M.(2003). A colonization factor (production of lateral flagella) of mesophilic *Aeromonas* spp. is inactive in *Aeromonas salmonicida* strains. Appl. Environ. Microbiol. *69*, 663–667.

Merino, S., Shaw, J.G., and Tomas, J.M. (2006) Bacterial lateral flagella: an inducible flagellar system. FEMS Microbiol. Lett. *263*, 127–135.

Mobley, H.L., and Belas, R. (1995). Swarming and pathogenicity of *Proteus mirabilis* in the urinary tract. Trends Microbiol. *3*, 280–284.

Moens, S., Michiels, K., Keijer, V., Van Leuven, F., and Vanderleyden, J. (1995). Cloning, sequencing, and phenotypic analysis of *laf1*, encoding the flagellin of the lateral flagella of *Azospirillum brasilense* Sp7. J. Bacteriol. *177*, 5419–5426.

Okabe, M., Yakushi, T., Kojima, M., and Homma, M. (2002). MotX and MotY, specific components of the sodium-driven flagellar motor, colocalize to the outer membrane in *Vibrio alginolyticus*. Mol. Microbiol. *46*, 125–134.

Otteman, K.M., and Miller, J.F. (1997). Roles for motility in bacterial-host interactions. Mol. Microbiol. *24*, 1109–1117.

O'Rouke, J., Lee, A., and Fox, J.G. (1992). An ultrastructural study of *Helicobacter mustelae* and evidence of a specific association with gastric mucosa. J. Med. Microbiol. *36*, 420–427.

Park, K-S., Arita, M., Iida, T., and Honda, T. (2005). *vpa*H, a gene encoding a novel histone-like nucleoid structure-like protein that was possibly horizontally acquired, regulates the biogenesis of lateral flagella in *trh*-positive *Vibrio parahaemolyticus* TH3996. Infect. Immun. *73*, 5754–5761.

Pereira, M., Parente, J.A., Bataus, L.A., Cardoso, D.D., Soares, R.B., and Soares, C.M.(2004). Chemotaxis and flagellar genes of *Chromobacterium violaceum*. Genet. Mol. Res. *3*, 92–101.

Poggio, S., Osorio, A., Dreyfus, G., and Camarena, L. (2005). The flagellar hierarchy of *Rhodobacter sphaeroides* is controlled by the concerted action of two enhancer-binding proteins. Mol. Microbiol. *58*, 969–983.

Power, P.M., and Jennings, M.P. (2003) The genetics of glycosylation in Gram-negative bacteria. FEMS Microbiol. Lett. *218*, 211–222.

Prouty, M.G., Correa, N.E., and Klose, K.E. (2001). The novel sigma54 and sigma28 dependent flagellar gene transcription hierarchy of *Vibrio cholerae*. Mol. Microbiol. *39*, 1595–1609.

Rabaan, A., Gryllos, I., Tomás, J.M., and Shaw, J.G. (2001). Motility and the polar flagellum are required for *Aeromonas caviae* adherence to HEp-2 cells. Infect. Immun. *69*, 4257–4267.

Ragatz, L., Jiang, Z-Y., Gest, H., and Bauer, C. (1994). Macroscopic phototactic behavior of the purple photosynthetic bacterium *Rhodospirillum centenum*. Arch. Microbiol. *163*, 1–6.

Ramos, H.C., Rumbo, M., and Sirard, J.C. (2004). Bacterial flagellins: mediators of pathogenicity and host immune responses in mucosa. Trends Microbiol. *12*, 500–517.

Ren, C.P., Beatson, S.A., Parkhill, J., and Pallen, M.J. (2005) The Flag-2 locus, an ancestral gene cluster, is potentially associated with a novel flagellar system from *Escherichia coli*. J. Bacteriol. *187*, 1430–1440.

Romling, U., Gomelsky, M., and Galperin, M.Y. (2005) C-di- GMP: the dawning of a novel bacterial signalling system. Mol. Microbiol. *57*, 629–639.

Scheludko, A.V., Katsy, E.I., Ostudin, N.A., Gringauz, O.K., and Panasenko, V.I. (1998). Novel classes of *Azospirillum brasilense* mutants with defects in the assembly and functioning of polar and lateral flagella. Mol. Gen. Mikrobiol. Virusol. *4*, 33–37.

Schirm, M., Arora, S.K., Verma, A., Vinogradov, E., Thibault, P., Ramphal, R., and Logan, S.M. (2004) Structural and genetic characterization of glycosylation of type a flagellin in *Pseudomonas aeruginosa*. J. Bacteriol. *186*, 2523–2531.

Schirm, M., Schoenhofen, I.C., Logan, S.M., Waldron, K.C., and Thibault, P. (2005). Identification of unusual bacterial glycosylation by tandem mass spectrometry analyses of intact proteins. Anal. Chem. *77*, 7774–7782.

Schmidt, M.A., Riley, L.W., and Benz, I. (2003). Sweet new world: glycoproteins in bacterial pathogens. Trends Microbiol. *11*, 554–561.

Schoenhals, G.J., Kihara, M., and Macnab, R.M. (1998). Translation of the flagellar gene *fliO* of *Salmonella typhimurium* from putative tandem starts. J. Bacteriol. *180*, 2936–2942.

Shimada, T., Sakazaki, R., and Suzuki, K. (1985) Peritrichous flagella in mesophilic strains of *Aeromonas*. Jpn. J. Med. Sci. Biol. *38*, 141–145.

Shinoda, S., and Okamoto, K.K. (1977). Formation and function of *Vibrio parahaemolyticus* lateral flagella. J. Bacteriol. *129*, 1266–1271.

Shinoda, S., Yakiyama, I., Yasui, S., Kim, Y., Ono, B., and Nakagami, S. (1992) Lateral flagella of vibrios: serological classification and genetical similarity. Microbiol. Immunol. *36*, 303–309.

Sourjik, V., Muschler, P., Scharf, B., and Schmitt, R. (2000) VisN and VisR are global regulators of chemotaxis, flagellar, and motility genes in *Sinorhizobium* (*Rhizobium*) *meliloti*. J. Bacteriol. *182*, 782–788.

Soutourina, O.A., and Bertin, P.N. (2003). Regulation cascade of flagellar expression in Gram-negative bacteria. FEMS Microbiol. Rev. *27*, 505–523.

Stewart, B.J., Enos-Berlage, J.L., and McCarter, L.L. (1997). The *lonS* gene regulates swarmer cell differentiation of *Vibrio parahaemolyticus*. J. Bacteriol. *179*, 107–114.

Stewart, B.J., and McCarter, L.L. (2003). Lateral flagellar gene system of *Vibrio parahaemolyticus*. J. Bacteriol.*185*, 4508–4518.

Szymanski, C.M., Burr, D.H., and Guerry, P. (2002). *Campylobacter* protein glycosylation affects host cell interactions. Infect. Immun. *70*, 2242–2244.

Szymanski, C.M., and Wren, B.W. (2005). Protein glycosylation in bacterial mucosal pathogens. Nature Rev. Microbiol. 3, 225–237.

Tian, Y., Wang, Q., Liu, Q., Ma, Y., Cao, X., Guan, L., and Zhang, Y. (2008). Involvement of LuxS in the regulation of motility and flagella biogenesis in *Vibrio alginolyticus*. Biosci. Biotechnol. Biochem. 72, 1–9.

Takeuchi, K., Taguchi, F., Inagaki, Y., Toyoda, K., Shiraishi, T., and Ichinose, Y. (2003). Flagellin glycosylation island in *Pseudomonas syringae* pv. Glycinea and its role in host specificity. J. Bacteriol. *185*, 6658–6665.

Tarrand, J.J., Krieg, N.R., and Döbereiner, J. (1978). A taxonomic study of the *Spirillum lipoferum* group with description of a new genus, *Azospirillum* gen. nov. and two species *Azospirillum lipoferum* (Beijerinck) comb. nov. and *Azospirillum brasilense* sp. nov. Can. J. Microbiol. *24*, 967–980.

Thomas, N.A., Bardy, S.L., and Jarrell, K.F. (2001). The archaeal flagellum: a different kind of prokaryotic motility structure. FEMS Microbiol. Rev. *25*, 147–174.

Thibault, P., Logan, S.M., Kelly, J.F., Brisson, J-R., Ewing, C.P., Trust, T.J., and Guerry. P. (2001) Identification of the carbohydrate moieties and glycosylation motifs in *Campylobacter jejuni* flagellin. J. Biol. Chem. *276*, 34862–34870.

Ulitzur, S. (1975). The mechanism of swarming of *Vibrio alginolyticus*. Arch. Microbiol. *104*, 67–71.

Upreti, R.K., Kumar, M., and Shankar, V. (2003). Bacterial glycoproteins: Functions, biosynthesis and applications. Proteomics 3, 363–379.

Wang, Q., Suzuki, A., Mariconda, S., Porwollik, S., and Harshey, R.M. (2005) Sensing wetness: a new role for the bacterial flagellum. EMBO J. *24*, 1–9.

Yun-Kyeong, K., and McCarter, L.L. (2000). Analysis of the polar flagellar gene system of *Vibrio parahaemolyticus*. J. Bacteriol. *182*, 3693–3704.

Zhulin, I.B., and Armitage, J.P. (1993). Motility, chemokinesis, and methylation-independent chemotaxis in *Azospirillum brasilense*. J. Bacteriol. *175*, 952–958.

The Bacterial Flagellum as a Surface Display and Expression Tool

12

Katariina Majander, Lena Anton, Riikka Kylväjä, and Benita Westerlund-Wikström

Abstract

The complex and self-assembling bacterial flagellum was earlier considered solely as a motility organelle used by bacteria and archaea for swimming towards attractants. In recent years, additional functions for the flagellum have been reported, such as adhesive properties, involvement in biofilm formation, sensor of environmental wetness, and importantly, a molecule recognized by components of the innate immune system of eukaryotic cells. This review will summarize the basis and use of the flagellum in bacterial surface display, as an apparatus for extracellular secretion, as a vaccine component, and in expression techniques. Targeting of foreign polypeptides to the bacterial surface is based on fusion by genetic methods of the heterologous polypeptide to a surface-localized carrier protein. Subunits of the flagellum, i.e. the major subunit FliC and the tip-localized FliD, can be used for surface display of foreign polypeptides in, for example, basic research, construction of random peptide libraries, vaccinology, and for diagnostic purposes. Surface-localized flagellar subunits are translocated to the extracellular milieu by the flagellar secretion apparatus, which can be utilized for high-yield export of foreign polypeptides into the bacterial growth medium. The flagellar subunit is, due to its potency as an immunomodulator, being explored as a carrier of antigens in vaccine design.

Introduction

Bacterial molecular surface display techniques facilitate functional or antigenic presentation of heterologous polypeptides on the surface of the microbial cell. A variety of such methods have been developed and applied in basic research and vaccinology, giving alternatives or complements to more established expression techniques. The use of surface display methods for diagnostic purposes, in construction of biocatalysts, bioadsorbents, random peptide libraries, and in other biotechnological approaches are examples of more recent technologies. Targeting of foreign polypeptides to the bacterial cell surface is based on fusion of the heterologous polypeptide to a surface-localized carrier protein of choice by genetic methods.

Various surface-localized protein types have been used as carrier proteins. In Gram-negative bacteria, outer membrane proteins, lipoproteins, and subunits of fimbriae and flagella have been applied, and in Gram-positive bacteria, surface proteins carrying the sortase-dependent cell wall-anchoring motif have most frequently been used for surface display purposes (Samuelson *et al.*, 2002; Lee *et al.*, 2003; Marraffini *et al.*, 2006). Reported surface display applications frequently involve well-known epitopes of viruses, bacterial adhesins, toxins, or enzymes, 10 to 40 amino acid residues in length, and only in a few cases have bacterial polypeptides, up to 300 residues in length, been displayed on bacterial cells. For further details on different bacterial surface display techniques, the reader is referred to review articles (Samuelson *et al.*, 2002; Lee *et al* 2003; Daugherty, 2007). Expression technologies which facilitate export of foreign proteins to the extracellular growth medium or into eukaryotic cells are gaining increasing interest. These methods rely on fusions to genes encoding efficiently secreted carrier proteins or, on use of fusions to secretion signals or motifs of well-characterized protein secretion pathways. In Gram-positive bacteria, which lack a periplasmic space, the secretion is in principle more easily achieved by simple fusion of the polypeptide to be secreted with an efficient translocating signal. In practice, however, secretion of foreign peptides in Gram-positive bacteria can be complicated due to the fact that genetic tools are more restricted and genetic manipulation of many Gram-positive bacteria is more complicated than in *Escherichia coli* and *Salmonella*. In Gram-negative bacteria, mainly the type I and the type III secretion pathways have been applied in extracellular export of foreign proteins (Mergulhao *et al.*, 2005; Panthel *et al.*, 2008). Substrates of these secretion routes do not contain SecA-dependent N-terminal signal sequences, and their secretion-competence resides in the C-terminus (type I secretion) or in short N-terminal conformational motifs. Extracellular secretion methods are currently being developed for applications such as high-level expression of diagnostically important proteins, injection of therapeutically important proteins into target cells, and use of live bacteria for vaccination or other therapeutic purposes. In this review we will summarize the basis and the use of flagellum display as well as flagellum-based secretion and expression techniques.

Structure and assembly of the flagellar filament

The flagellum is the motility organelle of bacteria and represents a complex self-assembling apparatus. The flagellar structure, assembly and regulation processes have been most extensively studied in *Salmonella* and *E. coli*, and can be used as a model for flagella of other species. One should, however, keep in mind that the content and organization of the flagella-encoding genes as well as the flagellar structure differ among bacteria (Beatson *et al.*, 2006; Galkin *et al.*, 2008; Maruyama *et al.*, 2008).

In Gram-negative bacteria, the flagellum traverses the inner membrane, the periplasm and the outer membrane and protrudes from the bacterial cell. It is composed of more than 20 different proteins and is built up of three main components: the basal body and the secretion apparatus, the hook and the filament (Fig. 12.1A). Flagellar assembly is strictly regulated by hierarchical expression of the genes encoding flagellar components and proceeds, in a highly ordered manner, from the components of the basal body and the secretion apparatus towards the hook and the filament (reviewed by Chevance and Hughes, 2008).

The flagellar filament can consist of up to 20,000 copies of the FliC (flagellin) subunit, which are arranged in a helical manner, and build up a 20-nm-wide and up to 15 μm long, thin cylinder with a 2-nm-wide hollow central channel (Fig. 12.1B). The filament is connected to the curved hook, which is composed of 120 FlgE proteins arranged in a helical manner (Samatey *et al.*, 2004), and the basal body via the junction proteins FliK and FliL. The flagellar cap at the distal end of the filament is composed of five FliD molecules, which form a pentameric stool-like structure at the tip of the filament (Fig. 12.1C). The cap has a central role in flagellar polymerization (Yonekura *et al.*, 2000; Maki-Yonekura *et al.*, 2003). When the hook-basal body complex has been synthesized, FliD proteins are attached to FliK and FliL at the tip of the hook and the nascent FliC subunits insert between the junction proteins and the cap. Deletion of *fliD* results in a non-flagellated phenotype, which is unable to polymerize FliC into filaments and secretes unassembled flagellin monomers into the growth medium (Ikeda *et al.*, 1987). The flagellar building blocks are exported to their destination by the flagellar-specific Type III secretion system (T3SS) in an unfolded or partially folded form (recently reviewed by Chevance and Hughes, 2008). The initial entry of export substrates to the T3SS is energized by ATP hydrolysis of an ATPase complex and the successive translocation process is driven by proton motive force (Minamino and Namba, 2008; Paul *et al.*, 2008). The reader is referred to excellent reviews (Macnab, 2004; Chevance and Hughes, 2008) for more detailed information on flagellar structure, regulation and assembly.

N- and C-terminal amino acid sequences of bacterial flagellins are conserved whereas the central region of FliC is highly variable (Beatson *et al.*, 2006). Based on the three-dimensional structure of N- and C-terminally truncated

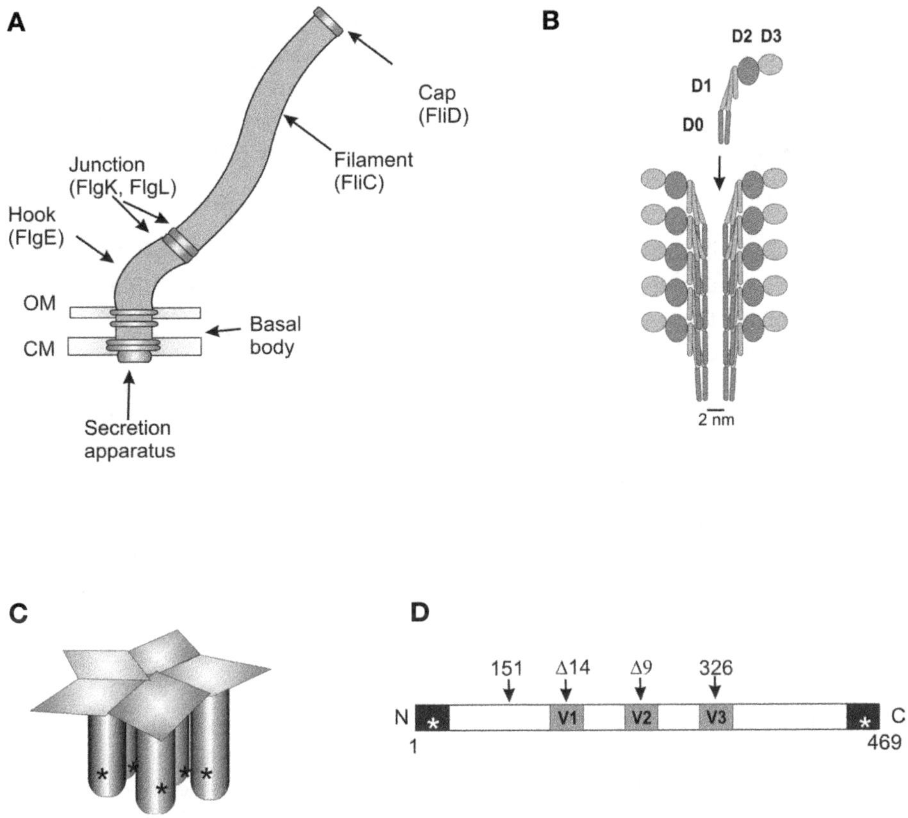

Figure 12.1 The flagellum of *Salmonella*. **(A)** Structure of the flagellum. **(B)** Schematic presentation of a longitudinal section of the flagellum showing at the top an individual FliC molecule with the domains D0-D3 indicated. D0 and D1 form the flagellar core and are crucial for flagellar assembly. D2 and D3 are surface exposed; D3 is highly variable and indispensable for polymerization of flagella. The flagellins are arranged in a helical manner and form a 2 nm-wide hollow central channel in the flagellum. **(C)** Schematic presentation the flagellar cap based on the published three-dimensional image of *Salmonella* FliD. The leg domain of each FliD molecule (indicated with stars) contains conserved, structurally disordered N- and C-termini. The plate-like structure is surface exposed, oriented outwards from the distal tip of the filament, and contains variable regions. **(D)** Schematic presentation of *E. coli* FliD and regions applicable in surface display. Conserved N- and C- termini are indicated with stars, and variable regions (V1, V2 and V3) with grey bars. The numbers refer to amino acids in FliD, Δ14 and Δ9 refer to regions that tolerate deletions and insertions, whereas 151 and 326 are positions in FliD not suitable for surface display.

Salmonella enterica serovar Typhimurium (from now on called *S.* Typhimurium) flagellin and the atomic modelling of the flagellar filament, it has been concluded that FliC has an overall shape of the Greek letter 'Γ' and consists of four linearly connected domains, which are arranged from the inside to the outside of the filament (Fig. 12.1B) (Samatey *et al.*, 2001; Yonekura *et al.*, 2003). The N-terminus of FliC starts in domain D0, the polypeptide goes through D1, D2 and reaches the D3 domain, the polypeptide then turns and comes back through D2 and D1, and ends with the C-terminus in domain D0. The D3 domain is surface exposed and corresponds to the highly variable central region to which antibody response is directed (Frankel *et al.*, 1989), D2 is also on the surface of the filament, whereas the conserved domains D0 and D1 compose the core of the flagellum, interact with neighbouring flagellins, and are required for filament assembly. Diversity in flagellin structure was recently demonstrated by the determination of the crystal structure of an N- and C-terminally truncated *Sphingomonas* sp. surface-bound flagellin. The domains involved in filament assembly highly resemble those of *Salmonella* flagellin, whereas the central part of *Sphingomonas* FliC significantly differs from the D2 and D3 domains of *Salmonella* FliC (Maruyama *et al.*, 2008).

Functions of the bacterial flagellum

Flagella in bacterial motility and fitness

The flagellum has a multitude of biological functions and effects on the animal host. The flagellum was earlier only considered as a motility organelle used by many bacteria and archaea for swimming towards nutrients or other attractants. During recent years the flagellum has been assigned additional, novel functions. In pathogenic bacteria, the flagellum is often an important factor rendering the pathogen more capable of moving towards the site of colonization (Wright *et al.*, 2005; Lane *et al.*, 2007) and in some cases, for example in *Bacillus cereus, E. coli* and *Pseudomonas aeruginosa,* the flagellum functions as an adhesin mediating bacterial adhesion to host cell surfaces (Scharfman *et al.*, 2001; Tasteyre *et al.*, 2001; Girón *et al.*, 2002; Erdem *et al.*, 2007). The flagellum is involved in early stages of biofilm formation (reviewed by Prüss *et al.*, 2006) and in *S.* Typhimurium the flagellum functions as a sensor for external wetness and regulates, in addition to its own biogenesis, other physiological functions (Wang *et al.*, 2005). In addition, the flagellum has in various bacteria, for example *Campylobacter jejuni, E. coli, S.* Typhimurium, and *Yersinia enterocolitica,* been shown to function as an export apparatus for virulence-associated non-flagellar and even heterologous proteins (Young *et al.*, 1999; Konkel *et al.*, 2004; Lee and Galán, 2004; Majander *et al.*, 2005a).

Role of flagella in host immunity

The flagellum has traditionally been used in immunization experiments due to its advantageous effects as an immunoadjuvant. More recently, both the adaptive and the innate immunity have been identified as underlying biological reasons for the immunopotency of flagella. The adaptive immune system generates antibodies against the variable D3 domain of flagellin (Frankel *et al.*, 1989), and responds to MHC class II-presented FliC peptides from various domains of the protein (Cookson and Bevan, 1997; Alaniz *et al.*, 2006). The innate immune system involves transmembrane proteins called Toll-like receptors (TLRs), which recognize microbe-specific pathogen associated molecular patterns (PAMPs) in the extracellular environment. As a result of the recognition, TLRs induce expression of proinflammatory components and antimicrobial defence molecules in the host cell (Akira and Takeda, 2004); mechanisms of importance for efficient immunity. The flagellin-specific TLR5 molecule recognizes 10–14 amino acid long sequences located in the N- and C-termini of the flagellin D1 domain (Smith *et al.*, 2003; Murthy *et al.*, 2004). TLR5 is activated by monomeric flagellins from a wide variety of bacteria, also by the flagellin of *Salmonella* that is frequently used in vaccine development. Two clades of flagellated bacteria, which include for example the human pathogens *C. jejuni, Helicobacter pylori,* and *Bartonella bacilliformis,* possess unusual flagellin sequences that permit TLR5 evasion while preserving bacterial motility (Andersen-Nissen *et al.*, 2005). In the eukaryotic cytoplasm, monomeric flagellin is recognized by two other components of the innate immune system called Ipaf and Bircle/Naip5, which indirectly enhance cytokine gene expression (Franchi *et al.*, 2006; Miao *et al.*, 2006; Ren *et al.*, 2006).

Flagellum surface display

Background and principles

After Kuwajima and coworkers in 1988 did the pioneering work in flagellum display, by fusing an antigenic determinant of hen egg-white lysozyme to the central region of *E. coli* flagellin (Kuwajima *et al.*, 1988), a number of reports on the flagellum as a surface carrier molecule have appeared. Foreign polypeptides can be displayed on the surface of the flagellum as fusions to the major filament subunit FliC or to the capping protein FliD.

In FliC-based flagellum display the foreign polypeptides are fused, by genetic methods, into the variable, surface-exposed D3 domain of FliC of, for example, *E. coli* and *Salmonella.* In the H7 serovariant of *E. coli* FliC (585 amino acids in length) up to 187 amino acid residues can be deleted from the D3 region without loss of flagellar polymerization or function. Complementation of a Δ*fliC* mutation in the *E. coli* host strain with the hybrid *fliC* gene results in expression of thousands of copies of the fusion flagellin along the flagellar filament (Westerlund-Wikström *et al.*, 1997).

Amino acid comparison of FliD molecules, combined with structural data of the pentameric FliD cap, have revealed that FliD, like FliC, contains conserved as well as less conserved regions. The conserved N- and C- terminal regions are located in the leg-like domains of the cap, which are oriented inwards at the distal end of the flagellar filament. Some of the variable regions are located in the plate-like structure of the cap, are surface exposed, and can be partially deleted and used for insertion of foreign polypeptides by genetic fusions into FliD (Fig. 12.1D) (Majander *et al.*, 2005b). For surface display, hybrid *fliD* is used to complement a Δ*fliD* deletion in the *E. coli* host to create a pentameric flagellar tip -localized display system.

Bacterial FliC and FliD sequences generally lack cysteine residues (Wilson and Beveridge, 1993) and the flagellar assembly machinery bypasses the bacterial periplasmic space (Macnab, 2004; Chevance and Hughes, 2008) where disulphide bonds are formed. Thus, flagellins and capping proteins do not contain disulphide bonds, and as a consequence, such bonds are also lacking in FliC and FliD fusion proteins, a fact that must be considered when foreign epitopes are presented on the flagellum.

Flagellum display of random peptide libraries

In 1995, Lu and coworkers (Lu *et al.*, 1995) developed the FliC-based flagellum display technique into an alternative to phage display. In this application called FliTrx, thioredoxin of *E. coli* was fused to the D3 domain of *E. coli* FliC. The active site of thioredoxin consists of a short disulphide-bonded loop known to be permissive for insertion of a wide variety of peptide sequences and enabling expression of conformationally constrained peptides (LaVallie *et al.*, 2003). A random peptide library of 1.77×10^8 individual dodecapeptides was expressed in the thioredoxin active-site loop in FliC. The functionality of the library was demonstrated by taking the clones of the random peptide library through a repeated biopanning procedure using monoclonal antibodies against human interleukins and macrophage colony stimulating factor as immobilized target molecules. The inserts of the enriched clones, which had bound strongly to the antibodies, were sequenced and demonstrated to have distinct consensus binding sequences for each antibody. The FliTrx random peptide library technology has subsequently been applied in identification of a number of adhesive dodecapeptides reacting mainly with eukaryotic proteins (see Table 12.1).

Compared to phage display, the FliTrx system has several advantages: i) cells are self-replicative, whereas phages need a two-step replication consisting of infection of host cells and subsequent purification, ii) expression of peptides as fusions to thioredoxin in FliC provides peptides with constrained conformations, and iii) the peptides identified as important for binding by the biopanning procedure can be expressed as fusions directly to thioredoxin. Such fusion proteins can be expressed at high concentrations and facilitate structure predictions (Li *et al.*, 2008).

Applications of FliC- and FliD-based flagellum display

Conventional methods for expression and structural analysis of bacterial adhesins are frequently based on a single copy of adhesive peptide per fusion protein, which may be insufficient, e. g. in the case of Gram-negative bacterial adhesins often showing low affinity to their receptors. Expression of adhesive peptides in thousands of intimately associated copies along the flagellar filament in FliC creates a high-affinity display system; hence, flagellum display offers an alternative tool for the analysis of adhesin-receptor interactions. The FliC-based display in *E. coli* has been applied in adhesion research in several ways: i) bacterial adhesive polypeptides up to 302 amino acids in length have been displayed in a functional, adhesive form along the flagellar filament as fusions to FliC, ii) hybrid flagella have been used in mapping of adhesive epitopes in bacterial adhesins, iii) it has been demonstrated that immunization with hybrid flagella raises antibodies that can be used in inhibition of bacterial adhesion, and iv) hybrid flagella carrying adhesive epitopes have been used to localize receptor-active regions in frozen tissue sections and cell cultures (Westerlund-Wikström *et al.*, 1997; Hynönen *et al.*, 2002).

Table 12.1 Examples of peptide-ligand interactions identified with the FliTrx system

Immobilized target	Aim or application	Identified consensus sequence or amino acid repeats	Reference
Matrigel-invading and tumour-derived endothelial cells	Identification of tumour vasculature binding peptides for targeted tumour therapy	ArgArgLeu and ArgLysLeu	Brown *et al.*, 2000
Proliferating cell nuclear antigen (PCNA)	Identification of PCNA-binding peptides involved in DNA replication, DNA repair and cell cycle control	Known PCNA-binding motif and a novel motif present in the N-terminus of human DNA polymerase σ	Xu *et al.*, 2001
Cu_20 and ZnO oxides	Identification of peptides binding to semiconductive metals oxides for understanding of metal oxide binding	ArgXXArg[a] for metal oxide binding, ArgArg for binding of Cu_20, and ArgLys for ZnO binding	Thai *et al.*, 2004
Human prostate carcinoma PC-3 cells	Identification of tumour-specific peptides for therapeutic use	AspArgGlyAlaArgArgLeuPheGlyArgIleGlu	Zitzmann *et al.*, 2005
Human umbilicalcord hyaluronan (HA)	Identification of HA-binding peptides	Lys/Arg-x_7-Lys/Arg[b], Lys/Arg-x_5-Lys/Arg, and ArgArg	Amemiya *et al.*, 2005
Normal liver L02 cells and hepatoma HepG2 cells	Identification of tumour-specific peptides for liver cancer therapy	IleAlaValAlaProGlyTrpLeuTrpGluGluGlu, LysGluLeuCysGluLeuAspSerLeuLeuArgIle, IleArgGluLeuTyrSerTyrAspAspAspPheGly	Li *et al.*, 2008

a X, any amino acid

b x, any non-acidic amino acid

The FliC-based display technique has been further developed into a double hybrid display system, in which two different polypeptides can be expressed in separate flagellin subunits in the same flagellar filament on the bacterial cell, as demonstrated by Western blotting, immuno-electron microscopy and adhesion experiments (Tanskanen *et al.*, 2000). The FliC-based double-hybrid display technique has recently been combined with the FliD-based display method into a trihybrid technique, in which foreign polypeptides are expressed along the flagellum and at the filament tip (Majander *et al.*, 2005b). See Fig. 12.2 for examples of FliC- and FliD-based surface display.

The use of a Gram-positive bacterium in flagellum display was recently described for the first time (Crampton *et al.*, 2007). The alkaliphilic, thermotolerant bacterium *Bacillus halodurans* Alk36, which constitutively overexpresses flagellin, was manipulated for optimal FliC-based surface display. The chromosomal *hag* gene (corresponds to *fliC* in *Salmonella*) of the strain was deleted by site-specific mutagenesis and the deletion was complemented with the flagellin gene expressed from its own promoter in an *E. coli-Bacillus* thermosensitive shuttle vector. Optimal sites for surface display in the cloned *hag* gene were screened by insertion of in-frame polylinkers at various sites, and one permissive site in a region corresponding to the variable region of *E. coli* flagellin was identified, that restored motility in the non-motile host strain. The permissive site in FliC was utilized to construct Ni^{2+}-ion-binding hybrid flagella carrying a 13-mer polyhistidine peptide, and hybrid flagella carrying the gp120 epitope of HIV-1 subtype C (29 amino acid residues) that reacted in Western blot with anti-HIV

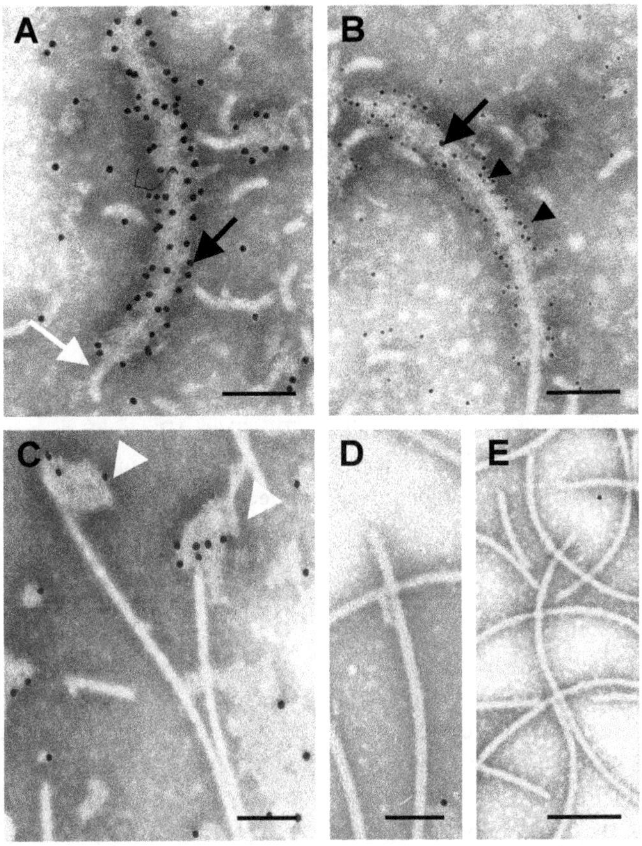

Figure 12.2 Immunoelectron microscope analysis of hybrid flagella detached from the *E. coli* surface. **A** FliC-display of an epithelial-cell binding N-terminal fragment (ΔSlpaA, 150 amino acids) of the S-layer protein SlpA from *Lactobacillus brevis*. The flagella were incubated with anti-SlpA antibodies and Protein A-conjugated gold particles (10 nm in diameter). **B** Double hybrid FliC-display of ΔSlpA and a fragment of the YadA adhesin (ΔYadA, amino acids 84 to 385) from *Y. enterocolitica*. ΔSlpA, detected with anti-SlpA antibodies and Protein A–gold (10 nm diameter), and ΔYadA, detected with anti-YadA monoclonal antibodies and gold-conjugated anti-mouse antibodies (5 nm in diameter), are expressed in different FliC molecules within the same flagellar filament. **C** FliD-display of the fibronectin-binding D3 polypeptide (39 amino acids) of the FnBPA protein of *S. aureus* at the tip of flagella. The tip-localization of the hybrid FliD was shown by addition of fibronectin and detection of bound fibronectin with anti-fibronectin antibodies and Protein A–gold. **D** Control double hybrid flagella (FliC::ΔSlpA and FliC::ΔYadA carrying wild-type FliD at the tip. The flagella have been incubated with fibronectin, antibodies and Protein A–gold as described in **C**. **E** Control flagella carrying wild-type FliC and FliD incubated with anti-SlpA antibodies and Protein A–gold as described in **A**. Note the small gold particles and a large gold particle in **B** as well as the tip-localized mass of fibronectin in **C** (large gold particles). Note also the lack of small and large gold particles in **D** and **E**. The bar is 200 nm in panel **E**, the bars represent 100 nm in all other panels. Arrows indicate large gold particles (anti-SlpA antibodies), small arrow heads indicate small gold particles (anti-YadA antibodies), and large arrowheads refer to fibronectin bound to the hybrid FliD at the tip of flagella in **C**. The white arrow shows the bare hook of a flagellum in **A**.

peptide antibodies. This study demonstrates that *Bacillus*-based flagellum display can also be used in a range of applications.

Flagellum display in vaccine development

Flagellum display is an appealing choice in vaccine design due to the cheap and unlimited supply of the bacterial vaccine strain, use of whole bacteria as live vaccine, and the possibility to simultaneously present polypeptides in FliC and FliD. An additional important aspect is the strong immunogenic character and the immunoadjuvant effect of the flagellum, a reason why FliC-based display is one of the expression methods which have been tested in laboratory animals to generate novel recombinant vaccines. A variety of short, antigenic B-cell epitopes have been displayed as fusions to mainly *Salmonella* FliC. Many of the short epitopes, 10–30 amino acid residues in length, mainly of viral and bacterial proteins, or toxins have been expressed in attenuated *Salmonella* strains proven in phase I clinical trials in humans to be safe and effective as vaccines (see, for example, Dilts *et al.*, 2000). Numerous studies have shown that immunization with hybrid flagella induces immune responses in laboratory animals against the epitope inserted in FliC, but the efficiency and type of immune response as well as the reactivity of the obtained antibodies with the antigen have varied substantially. Depending on the immunization route, the antibodies have in some cases reacted *in vitro* with recombinant proteins carrying the foreign epitope or even with the native protein from which the epitope originated, whereas in other cases, the antibodies have failed to recognize the corresponding epitope on the native protein. Apparently, the outcome of immunization with hybrid flagellar constructs varies depending on factors such as the type of flagellin, the laboratory animal, the immunization procedure as well as the structure of the epitope that was used. The protective capacity of the immune response obtained after vaccination with hybrid flagella has only in a few cases been verified in animals. Systematic studies to define the optimal length, copy number of epitopes per FliC to be displayed, and the most efficient administration routes in animals for evoking an optimal immune response should be performed, since existing results are contradictory. For examples of applications of flagellum display see Table 12.2 and for a more detailed review on immunization trials with hybrid flagella, see Westerlund-Wikström (2000).

Flagellum as a secretion and expression tool

Background

Bacteria have evolved several protein export mechanisms for translocation of nascent proteins from the cytoplasm into various cell compartments: the cytoplasmic membrane, the periplasmic space, the outer membrane or the cell wall; or into external sites in host cells or the growth medium. In *E. coli*, at least eight common pathways of protein translocation have been described to date. The pathways differ from each other regarding the nature and localization of a signal sequence for secretion, the secretion route, and the destination of the secreted substrate (deKeyzer *et al.*, 2003; Sauer *et al.*, 2004; Cianciotto, 2005; Holland *et al.*, 2005; Lee *et al.*, 2006; Dautin and Bernstein, 2007).

Flagellar secretion as a Type III secretion system

The Type III secretion systems (T3SS) found in human and plant pathogenic bacteria, translocate effector proteins from the cytoplasm to the exterior of the bacterial cell and into eukaryotic target cells without involvement of periplasmic intermediates (Galán and Wolf-Watz, 2006). They are Sec-bypass routes lacking the cleavable leader peptide typical of SecA-YEG-dependent proteins (deKeyzer *et al.*, 2003) and the nature of the Type III secretion signal remains controversial. An mRNA signal leading to cotranslational secretion of the nascent secretion substrate was originally proposed (Anderson and Schneewind, 1999) although later the hypothesis of an N-terminal structural region of the protein to be secreted as a translocating signal has become more widely accepted (Lloyd *et al.*, 2001). Export of the substrate proteins by T3SS is dependent on contact with the host cell (Galán and Wolf-Watz, 2006).

The flagellar secretion system, responsible for the synthesis of the flagellar filament, is considered to be a variant or an evolutionary progenitor of the pathogenic T3SS based on the homology between the components of the secretion machineries of the two systems. Pathogenic T3SS include secretion-related chaperones of different roles that bind to N-terminal structural motifs of their substrates (Galán and Wolf-Watz, 2006). The flagellar T3SS involve chaperones, which prevent proteolysis of the substrate, localize the substrate to the secretion machinery, and function as regulators of gene expression. Differently from the pathogenic T3SS chaperones, the flagellar secretion chaperones recognize C-terminal parts of the proteins to be secreted (Evdokimov *et al.*, 2003; Galán and Wolf-Watz, 2006; Chevance and Hughes, 2008). The specificity of T3SS

Table 12.2 Examples of the flagellum in surface display and expression applications

Carrier protein	Heterologous polypeptide	Application	References
Surface display			
Salmonella FliC	Virus, toxin, enzyme epitopes	Immunization, vaccine design	Newton *et al.*, 1989; McEwen *et al.*, 1992; Luna *et al.*, 1997; Guillobel *et al.*, 2000
E. coli FliC	Virus, toxin, bacterial, and enzyme epitopes	Immunization, vaccine design	Kuwajima *et al.*, 1988; Westerlund-Wikström *et al.*, 1997;Westerlund-Wikström, 2000
	Fragments of bacterial adhesins	Mapping of adhesive regions, localization of receptor-active domains, double hybrid display	Westerlund-Wikström *et al.*, 1997; Tanskanen *et al.*, 2000; Hynönen *et al.*, 2002; Majander *et al.*, 2005b
	Thioredoxin	FliTrx for random peptide libraries	Lu *et al.*, 1995
B. halodurans FliC	Polyhistidine,epitope of HIV virus	Ni^{2+}-binding, Western blot	Crampton *et al.*, 2007
E. coli FliD	Bacterial adhesive epitopes	Flagellar tip-localized display	Majander *et al.*, 2005b
E. coli FliC and FliD	Bacterial adhesive epitopes	Trihybrid display for vaccine design	Majander *et al.*, 2005b
FliC-based secretion			
E. coli flagellum	Gfp, bacterial proteins	High-level extracellular export	Majander *et al.*, 2005a
Salmonella flagellum	Complement control protein 2 module	Extracellular export	Végh *et al.*, 2006
FliC fusion peptides			
Salmonella FliC	*Listeria monocytogenes* proteins, epitopes of West Nile virus and influenza hemagglutinin	Vaccine design for innate immunity -mediated response	Huleatt *et al.*, 2007; McDonald *et al.*, 2007; Song *et al.*, 2008
Flagellin fragment			
Salmonella ΔFliC		Agonist against radiation	Burdelya *et al.*, 2008

chaperones apparently is not very strict, since there are several reports from various bacterial species, including *C. jejuni*, *S.* Typhimurium, and *Y. enterocolitica*, on secretion of virulence-associated non-flagellar proteins by the flagellar T3SS (Young *et al.*, 1999; Konkel *et al.*, 2004; Lee and Galán, 2004) and of export of flagellar proteins by *Salmonella* T3SS (Sun *et al.*, 2007).

Extracellular secretion of foreign proteins

The flagellum has been applied as machinery for extracellular export of heterologous proteins in *E. coli* (Majander *et al.*, 2005a). The *fliD* and *fliC* genes were deleted from the chromosome of the *E. coli* K12 host strain (Blattner *et al.*, 1997) and the *fliC* deletion was complemented with cloned gene fragments, which encoded fusion proteins containing C-terminally truncated FliC sequences. Fusion proteins carrying the N-terminal FliC residues 1–20, 1–80, 1–130 and 1–183 were all secreted into the growth medium. The eukaryotic green fluorescent protein fused to FliC residues 1–20 (FliC1–20Gfp) was one example of efficiently secreted FliC fusion proteins (Fig. 12.3). The hypothesis that translocation of FliC monomers from the cytoplasm may depend also on sequences in the untranslated DNA region upstream of the *Salmonella fliC* gene (Karlinsey *et al.*, 2000) was tested in *E. coli*. The gene, encoding the heterologous polypeptide of interest, was inserted between a 173-bp untranslated DNA fragment upstream of *E. coli fliC* (5′UTR) and a transcriptional terminator downstream of *fliC* (3′UTR). Complementation of the *fliC* deletion in the Δ*fliC*Δ*fliD E. coli* host with such hybrid gene fragments facilitated extracellular secretion of foreign proteins. The model polypeptides from *C. jejuni*, *Staphylococcus aureus*, *Streptococcus pneumoniae*, 115 to 434 amino acid residues in length, were secreted into the growth medium at

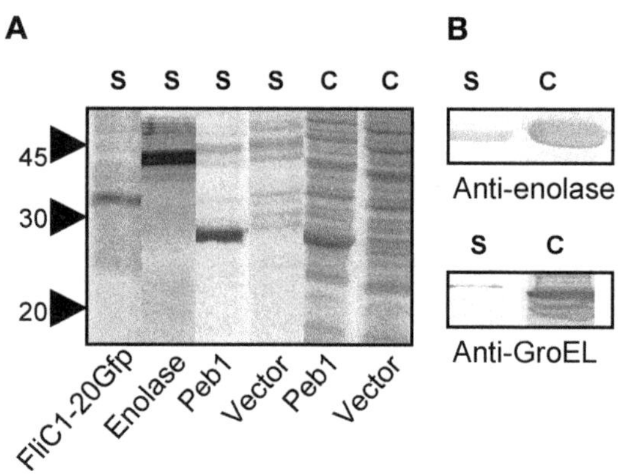

Figure 12.3 Examples of foreign proteins exported into the bacterial growth medium by the *E. coli* Δ*fliCfliD* host. **(A)** SDS-PAGE analysis of the extracellularly exported proteins FliC1–20Gfp, enolase from *S. pneumoniae*, and Peb1 from *C. jejuni*. S refers to 1 ml precipitated cell-free supernatant, and C cells from 100 µl bacterial culture. Vector indicates samples from cells carrying an empty vector. **(B)** Western blot analysis with anti-enolase and anti-GroEL antibodies to show the specificity and lack of cell lysis in the samples. S denotes 100 µl precipitated cell-free supernatant, C cells from 100 µl bacterial culture. Molecular size markers in kDa are indicated in A.

concentrations ranging from 1 to 15 mg/l (Fig. 12.3) and represented >50% of total secreted protein (Majander *et al.*, 2005a). C-terminal affinity tags, for example the polyhistidine tag and the hydrophobic FLAG™ tag, genetically fused to the proteins to be exported, did not affect the secretion efficiency. This facilitates a convenient strategy for affinity purification of proteins directly from cell-free growth medium. Rabbit antibodies that were raised for diagnostic purposes against the extracellularly exported, C-terminally His-tagged, affinity-purified campylobacterial Peb1 protein reacted strongly and specifically in Western blots with the homologous protein in cell lysates of *C. jejuni*. The results demonstrated that the extracellular export technique is convenient for the production of immunization antigens. Finally, the *fliC* promoter, which was included in the 5′UTR region used in the study of Majander and coworkers is very efficient and an alternative in conventional expression of heterologous proteins in *E. coli*.

Another study applied the flagellin of *S.* Typhimurium for secretion of foreign proteins into the bacterial growth medium (Végh *et al.*, 2006). Using terminally truncated flagellins the authors showed that, in *Salmonella*, the N-terminal 13 residues were not essential for export whereas the flagellin residues 26–47 were critical for extracellular transport of FliC hybrids, a region partially overlapping with the *E. coli* FliC region identified earlier (Majander *et al.*, 2005a). The effect of the 5′UTR region of *fliC* alone on flagellum-based secretion of heterologous proteins in *Salmonella* was not analysed, thus making it difficult on the basis of these results to draw a conclusion on whether the motifs directing proteins to the flagellar export route are similar or slightly different in *E. coli* and *S.* Typhimurium. For examples of flagellum-based extracellular protein export, see Table 12.2.

Flagellin as a vaccine carrier

Currently, when flagellin has been recognized as a TLR5 ligand positively affecting the signalling pathways that induce immune and inflammatory genes, efficient FliC-based vaccine antigens are being generated in a much more focused manner than before. See Table 12.2 for examples. Reports show that physically linking the vaccine antigen to flagellin generates a more potent vaccine than simple mixing of antigen and flagellin (Huleatt *et al.*, 2006; McDonald *et al.*, 2007; Huleatt *et al.*, 2008; Song *et al.*, 2008). In a recent example, the antigenic EIII domain (100 amino acid residues) of West Nile virus envelope protein was genetically fused to truncated *Salmonella* flagellin lacking the variable D3 domain (ΔFliC). The fusion protein expressed in and purified from *E. coli* cells was shown to stimulate cytokine production in a TLR5-dependent manner in reporter cells *in vitro* and to elicit strong IgG response to West Nile virus when administered without adjuvant to mice. The dependence on TLR5-mediated recognition was verified by demonstration of only a weak antigen-specific response in TLR5-deficient mice (McDonald *et al.*, 2007). The report on efficacious recombinant influenza vaccines is another excellent example of current flagellin-based vaccine design (Song *et al.*, 2008). The hemagglutinin (HA) of influenza virus rapidly changes antigenically to evade the host immune response and a regular update of vaccine strains is required. To improve the efficiency of vaccine production and to enhance the immunopotency of influenza vaccines, flagellin fusions have been constructed that carry the protective head domain of influenza virus HA at the C-terminus of *Salmonella* flagellin. The HA-FliC fusion protein purified from *E. coli* was shown to be TLR5-active *in vitro*, recognized in Western blots by convalescent mouse sera, and reactive in ELISA with neutralizing monoclonal antibodies, thus indicating that the HA epitope was displayed in flagellin in a conformationally correct manner. Challenge of BALB/c mice, immunized with the HA-FliC protein and showing HA-specific IgG response, with $1xLD_{90}$ of a mouse-adapted influenza virus administered intranasally, clearly demonstrated that the flagellin-based recombinant vaccine induced HA-specific immune response and protected the immunized mice from lethal challenge of influenza virus A. The procedure was successfully repeated with the currently circulating strain of influenza A virus, demonstrating that the technique can be applied on different HA molecules. Flagellin-based vaccines have also been generated using other epitopes, for example listeriolysin and p60 protein of *Listeria monocytogenes* (Huleatt *et al.*, 2007). It should be stressed that TLR5 recognizes monomeric flagellin but not FliC polymerized into intact flagella, a fact to be considered when flagellum-based display systems are being developed for vaccine applications. On the other hand, the hybrid-FliC vaccines based on monomeric flagellin can be efficiently expressed in cost-efficient *E. coli* fermentor systems, which is an apparent strength.

Conclusions and future trends

Flagellum display

Compared to surface display techniques based on outer membrane proteins or lipoproteins, the advantages of the flagellum-based display methods are (i) the possibility to detach flagella from the bacterial surface and use them as soluble reagents; (ii) presentation of the foreign polypeptide in thousands of copies along the flagellar filament, which makes the system multivalent in comparison to monovalent systems like the outer-membrane protein display; (iii) simultaneous display of different polypeptides in the same flagellar filament; (iv) simultaneous FliD- and FliC-based display; and (v) functional display of polypeptides up to 300 residues in length. Similar to flagellum display, the fimbriae-based display techniques allow multivalent display along the fimbrial filament and facilitate a choice between presentation of the foreign peptide along the filament as fusions to the fimbrillin or as fusion to specific molecules at the tip of the fimbriae (Klemm and Schembri, 2000). A disadvantage of the fimbriae-based display, compared to flagellum-based techniques, is the fact that only short peptides can be tolerated without disturbing the polymerization of fimbrial filaments. Obvious disadvantages of flagellum display are the lack of disulphide formation in the displayed polypeptides and the fact that only molecules narrow enough to pass through the 2-nm wide hollow core of the flagellum are exported. If formation of disulphide bonds is a requirement, an alternative carrier to the flagellum must be considered, such as outer membrane proteins and autotransporters, which are translocated to the surface via the periplasm and facilitate the formation of disulphide bonds in the polypeptide to be displayed.

Hybrid flagella are well suited, and have previously been applied, in identification of peptide–ligand interactions and as recombinant antigens for vaccine design, powerful applications that undoubtedly will be used also in the future. Reports also exist which describe functional presentation of metal-binding peptides on the flagellum (Thai *et al.*, 2004; Crampton *et al.*, 2007). Bacteria carrying such hybrid flagella could be applied in bioremediation, i.e. the purification of contaminated soil or water by biological means; an environmentally oriented future trend of FliC display.

The trihybrid flagellum-based display, i.e. the simultaneous presentation of different peptides in FliC and FliD, could be a very potent technique in vaccine development. The hybrid cap at the tip of the filament could carry a specific adhesin for targeting the hybrid flagellum to a desired epithelial site. Individual flagellins of the same hybrid flagellar filament could carry different antigens or antigenic epitopes, against which immune response is required. This approach would offer whole bacteria carrying a targeting molecule (displayed in FliD), several antigens (presented in different FliC molecules) and the adjuvant in one package, which in addition is cost-effective to produce. An additional future application is the intracellular delivery of hybrid flagella, for example for therapeutic purposes. Peptides identified as motifs for entry into eukaryotic cells (Tashner *et al.*, 2002; Ryu *et al.*, 2003; Vitte and Jalinot, 2008) could be fused to FliD for intracellular delivery of several different effector molecules expressed as flagellin fusions along the flagellar filament. In harsh applications where full-length flagellar filaments may be broken, or in applications where FliC presence is a disadvantage due to its strong immunological properties, another future trend could be hook-based display. Unpublished results by Majander and coworkers (2005b) have indicated that hybrid FliD can be displayed at the tip of the hook in a Δ*fliC*Δ*fliD* host strain.

Flagellum-based secretion and expression

In Gram-negative bacteria, the versatile Type I secretion system (T1SS) is a competitive method to flagellum-based secretion, and it has been used to export substrates of variable length and from different sources. The drawback of the T1SS-based technology, compared to the flagellum-based one, is the significantly lower concentrations of extracellularly exported heterologous proteins (Lee *et al.*, 2003). Various pathogenic T3SSs have also been applied in export of foreign polypeptides. These applications mainly rely on fusions to central regions of T3SS substrates and have been developed for injection of proteins into eukaryotic cells rather than for secretion into the growth medium, due to the reported contact-dependent secretion of the pathogenic T3SSs.

A current topic is the use of monomeric flagellin as an immunopotent TLR5-recognized carrier for polypeptides in vaccine design, since the FliC hybrid proteins are often highly immunogenic, and have been shown to protect laboratory animals against viral challenge (Song *et al.*, 2008). A competitive way of producing low-cost, efficient vaccine antigens would be the export of FliC fusion polypeptides directly into the bacterial growth medium using the techniques for extracellular export of foreign proteins in a Δ*fliCfliD* host as described in *E. coli* (Majander *et al.*, 2005a) and in *S.* Typhimurium (Végh *et al.*, 2006). Currently, FliC hybrid vaccine candidates are produced without affinity tags by conventional techniques and purified from the cytoplasm of the bacteria. Sun and coworkers (2007) have shown that *S.* Typhimurium flagellin is injected into the cytosol of *Salmonella*-infected human macrophages by a mechanism independent of the flagellar T3SS but involving a *Salmonella* T3SS upregulated during the early stages of *Salmonella* infection. The T3SS of *S.* Typhimurium has also successfully been used for *in vivo* antigen delivery of vaccine antigens (Rüssman *et al.*, 1998) and injection of therapeutic cancer vaccines into the cytosol of professional and nonprofessional antigen presenting cells (Nishikawa *et al.*, 2006). Based on these reports, a technique for direct delivery into the host cell cytosol of a polypeptide of interest, fused to the N- and C-termini of FliC as receptors for innate immunity, could be developed for therapeutic use or vaccine design.

Ionizing radiation is an effective therapy for certain forms of cancer, but it destroys tumour cells as well as bone marrow cells and cells of the gastrointestinal tract due to programmed cell death, or apoptosis. Tumour cells block apoptosis by activation of a transcription factor NF-κβ, and cancer therapies based on inhibition of NF-κβ have been developed. It has been demonstrated that the survival of normal cells after radiation can be improved by activation of NF-κβ with flagellin (Burdelya *et al.*, 2008). A 328-mer *Salmonella enterica* serovar *Dublin* flagellin derivative was constructed by deletion of the variable domain (corresponding to residues 177–401) and insertion of a 16-mer flexible linker at the site of the deletion. The ΔFliC derivative was recognized by TLR5 and activated NF-κβ *in vitro*. The ΔFliC polypeptide, expressed and purified from the bacterial cytoplasm, was injected into mice and rhesus monkeys before exposure to the ionizing radiation, and the results showed that the injected animals were more likely to survive exposure to lethal radiation than uninjected control animals. This type of flagellin derivatives could in the future be tested also in other conditions with excess apoptosis (Bhattacharjee, 2008).

A further future application of flagellum-based extracellular secretion technology could be the construction of whole-genome random secretion libraries of important pathogenic bacteria. Depending on the bacterial species used for generation of the library and on the application, the random polypeptides secreted into the growth medium of the library clones could be used in a variety of different analyses, for example in screening of the immunoproteome of the pathogen, in analysis of immunomodulators, or in identification and characterization of adhesion- or invasion-associated polypeptides. Due to the uncomplicated technologies involved in flagellum-based surface display and extracellular protein export, these techniques are competitive methods in basic research as well as in biotechnological applications.

Web resources

GenBank accession numbers

E. coli strain MG1655 (AE000285), *E. coli* $FliC_{MG1655}$ (NP_416433), *E. coli* $FliC_{H7}$ (L07388), *E. coli* $FliD_{MG1655}$ (NP_416434), crystal structure of *Salmonella* FliC protofilament (1IO1), crystal structure of *Salmonella* FliC (1UCU), crystal structure of *Sphingomonas* FliC protofilament (2ZBI), crystal structure of flagellar FliS chaperone in complex with its cognate binding partner (1ORY).

Acknowledgements

We thank Prof. T. K. Korhonen for valuable comments on the manuscript. Electron microscopy was carried out at the Electron Microscopy Unit, Institute of Biotechnology, University of Helsinki. This work was supported by the Academy of Finland (grant number 110716) and the Viikki Graduate School in Biosciences.

References

Akira, S., and Takeda, K. (2004) Toll-like receptor signalling. Nat. Rev. Immunol. *4*, 499–511.

Alaniz, R.C., Cummings, L.A., Bergman, M.A., Rassoulian-Barrett, S.L., and Cookson, B.T. (2006) *Salmonella typhimurium* coordinately regulates FliC location and reduces dendritic cell activation and antigen presentation to CD4+ T cells. J. Immunol. *177*, 3983–3993.

Amemiya, K., Nakatani, T., Saito, A., Suzuki, A., and Munakata, H. (2005). Hyaluronan-binding motif identified by panning a random peptide display library. Biochim. Biophys. Acta *1724*, 94–99.

Andersen-Nissen, E., Smith, K.D., Strobe, K.L., Barrett, S.L., Cookson, B.T., Logan, S.M., and Aderem, A. (2005). Evasion of Toll-like receptor 5 by flagellated bacteria. Proc. Natl. Acad. Sci. U.S.A. *102*, 9247–9252.

Anderson, D.M., and Schneewind, O. (1999). *Yersinia enterocolitica* type III secretion: an mRNA signal that couples translation and secretion of YopQ. Mol. Microbiol. *31*, 1139–1148.

Beatson, S.A., Minamino, T., and Pallen, M.J. (2006). Variation in bacterial flagellins: from sequence to structure. Trends Microbiol. *14*, 151–155.

Bhattacharjee, Y. (2008). Drug bestows radiation resistance on mice and monkeys. Science *320*, 163.

Blattner, F.R., Plunkett, G. III, Bloch, C.A., Perna, N.T., Burland, V., Riley, M., Collado-Vides, J., Glasner, J.D., Rode, C.K., Mayhew, G.F., Gregor, J., Davis, N.W., Kirkpatrick, H.A., Goeden, M.A., Rose, D.J., Mau, B., and Shao, Y. (1997). The complete genome sequence of *Escherichia coli* K12. Science *277*, 1453–1463.

Brown, C.K., Modzelewski, R.A., Johnson, C.S. and Wong, M.K. (2000). A novel approach for the identification of unique tumor vasculature binding peptides using an *E. coli* peptide display library. Ann. Surg. Oncol. *7*, 743–749.

Burdelya, L.G., Krivokrysenko, V.I., Tallant, T.C., Strom, E., Gleiberman, A.S., Gupta, D., Kurnasov, O.V., Fort, F.L., Osterman, A.L., Didonato, J.A., Feinstein E., and Gudkov, A.V. (2008). An agonist of Toll Like Receptor 5 has radioprotective activity in mouse and primate models. Science *320*, 226–230.

Chevance, F.F., and Hughes, K.T. (2008). Coordinating assembly of a bacterial macromolecular machine. Nat. Rev. Microbiol. *6*, 455–465.

Cianciotto, N.P. (2005). Type II secretion: a protein secretion system for all seasons. Trends Microbiol. *13*, 581–588.

Cookson, B.T., and Bevan, M.J. (1997). Identification of a natural T cell epitope presented by *Salmonella*-infected macrophages and recognized by T cells from orally immunized mice. J. Immunol. *158*, 4310–4319.

Crampton, M., Berger, E., Reid, S. and Louw, M. (2007). The development of a flagellin surface display expression system in a moderate thermophile, *Bacillus halodurans* Alk36. Appl. Microbiol. Biotechnol. *75*, 599–607.

Daugherty, P.S. (2007). Protein engineering with bacterial display. Curr. Opin. Struct. Biol. *17*, 474–480.

Dautin, N., and Bernstein, H.D. (2007). Protein secretion in Gram-negative bacteria via the autotransporter pathway. Annu. Rev. Microbiol. *61*, 89–112.

de Keyzer, J., van der Does, C., and Driessen, A.J. (2003). The bacterial translocase: a dynamic protein channel complex. Cell Mol. Life Sci. *60*, 2034–2052.

Dilts, D.A., Riesenfeld-Orn, I., Fulginiti, J.P., Ekwall, E., Granert, C., Nonenmacher, J., Brey, R.N., Cryz, S.J., Karlsson, K., Bergman, K., Thompson, T., Hu, B., Brückner, A.H., and Lindberg, A.A. (2000). Phase I clinical trials of *aroA aroD* and *aroA aroD htrA* attenuated *S. typhi* vaccines; effect of formulation on safety and immunogenicity. Vaccine *18*, 1473–1484.

Erdem, A.L., Avelino, F., Xicohtencatl-Cortes, J., and Girón, J.A. (2007). Host protein binding and adhesive properties of H6 and H7 flagella of attaching and effacing *Escherichia coli*. J. Bacteriol. *189*, 7426–35.

Evdokimov, A.G., Phan, J., Tropea, J.E., Routzahn, K.M., Peters, H.K., Pokross, M., and Waugh, D.S. (2003). Similar modes of polypeptide recognition by export chaperones in flagellar biosynthesis and type III secretion. Nat. Struct. Biol. *10*, 789–793.

Franchi, L., Amer, A., Body-Malapel, M., Kanneganti, T.D., Ozoren, N., Jagirdar, R., Inohara, N., Vandenabeele, P., Bertin, J., Coyle, A., Grant, E.P., and Núñez, G. (2006). Cytosolic flagellin requires Ipaf for activation of caspase-1 and interleukin 1β in salmonella-infected macrophages Nat. Immunol. *7*, 576–582.

Frankel, G., Newton, S.M., Schoolnik, G.K., and Stocker, B.A. (1989). Intragenic recombination in a flagellin gene: characterization of the *H1-j* gene of *Salmonella typhi*. EMBO J. *8*, 3149–3152.

Galán, J.E., and Wolf-Watz, H. (2006). Protein delivery into eukaryotic cells by type III secretion machines. Nature *444*, 567–573.

Galkin, V.E., Yu, X., Bielnicki, J., Heuser, J., Ewing, C.P., Guerry, P., and Egelman, E.H. (2008). Divergence of quaternary structures among bacterial flagellar filaments. Science *320*, 382–385.

Girón, J.A., Torres, A.G., Freer, E., and Kaper, J.B. (2002). The flagella of enteropathogenic *Escherichia coli* mediate adherence to epithelial cells. Mol. Microbiol. *2*, 361–379.

Guillobel, H.C., Carinhanha, J.I., Cárdenas, L., Clements, J.D., Almeida, D.F., and Ferreira, L.C.S. (2000). Adjuvant activity of a nontoxic mutant of *Escherichia coli* heat-labile enterotoxin on systemic and mucosal immune responses elicited against a heterologous antigen carried by a live *Salmonella enterica* serovar Typhimurium vaccine strain. Infect. Immun. *68*, 4349–4353.

Holland, I.B., Schmitt, L., and Young, J. (2005). Type 1 protein secretion in bacteria, the ABC-transporter dependent pathway. Mol. Membr. Biol. *22*, 29–39.

Huleatt, J.W., Jacobs, A.R., Tang, J., Desai, P., Kopp, E.B., Huang, Y., Song, L., Nakaar, V., and Powell, T.J. (2007). Vaccination with recombinant fusion proteins incorporating Toll-like receptor ligands induces rapid cellular and humoral immunity. Vaccine *25*, 763–75.

Huleatt, J.W., Nakaar, V., Desai, P., Huang, Y., Hewitt, D., Jacobs, A., Tang, J., McDonald, W., Song, L., Evans, R.K., Umlauf, S., Tussey, L., and Powell, T.J. (2008). Potent immunogenicity and efficacy of a universal influenza vaccine candidate comprising a recombinant fusion protein linking influenza M2e to the TLR5 ligand flagellin. Vaccine *26*, 201 14.

Hynönen, U., Westerlund-Wikström, B., Palva, A., and Korhonen, T.K. (2002). Identification by flagellum display of an epithelial cell- and fibronectin-binding function in the SlpA surface protein of *Lactobacillus brevis*. J. Bacteriol. *184*, 3360–3367.

Ikeda, T., Homma, M., Iino, T., Asakura, S., and Kamiya, R. (1987). Localization and stoichiometry of hook-associated proteins within *Salmonella typhimurium* flagella. J. Bacteriol. *169*, 1168–1173.

Karlinsey, J.E., Lonner, J., Brown, K.L., and Hughes, K.T. (2000). Translation/secretion coupling by type III secretion systems. *Cell 102*, 487–497

Klemm, P., and Schembri, M.A. (2000). Fimbrial surface display systems in bacteria: from vaccines to random libraries. Microbiology *12*, 3025–3032.

Konkel, M.E. Klena, J.D., Rivera-Amill, V., Monteville, M.R., Biswas, D., Raphael, B., and Mickelson, J. (2004). Secretion of virulence proteins from *Campylobacter jejuni* is dependent on a functional flagellar export apparatus. J. Bacteriol. *186*, 3296–3303.

Kuwajima, G., Asaka, J.-I., Fujiwara, T., Nakano, K., and Kondoh, E. (1988). Presentation of an antigenic determinant from hen egg-white lysozyme on the flagellar filament of *Escherichia coli*. Bio/Technology *6*, 1080–1083.

Lane, M.C., Alteri, C.J., Smith, S.N., and Mobley, H.L. (2007). Expression of flagella is coincident with uropathogenic *Escherichia coli* ascension to the upper urinary tract. Proc. Natl. Acad. Sci. U.S.A. *104*, 16669–16674.

LaVallie, E.R., DiBlasio-Smith, E.A., Collins-Racie, L.A., Lu, Z., and McCoy, J.M. (2003). Thioredoxin and related proteins as multifunctional fusion tags for soluble expression in *E. coli*. Methods Mol. Biol. *205*, 119–140.

Lee, S.Y., Choi, J.H., and Xu, Z. (2003). Microbial cell-surface display. Trends Biotechnol. *21*, 45–52.

Lee, S.H. and Galán, J.A. (2004). *Salmonella* type III secretion-associated chaperones confer secretion-pathway specificity. Mol. Microbiol. *51*, 483–495.

Lee, P.A., Tullman-Ercek, D., and Georgiou, G. (2006). The bacterial twin-arginine translocation pathway. Annu. Rev. Microbiol. *60*, 373–395.

Li, W., Lei, P., Yu, B., Wu, S., Peng, J., Zhao, X., Zhu, H., Kirschfink, M., and Shen, G. (2008). Screening and identification of a novel target specific for hepatoma cell line HepG2 from the FliTrx bacterial peptide library. Acta Biochim. Biophys. Sin. (Shanghai) *40*, 443–451.

Lloyd, S.A., Norman, M., Rosqvist, R., and Wolf-Watz, H. (2001). *Yersinia* YopE is targeted for type III secretion by N-terminal, not mRNA, signals. Mol. Microbiol. *39*, 520–531.

Lu, Z., Murray, K.S., Van Cleave, V., LaVallie, E.R., Stahl, M.L., and McCoy, J.M. (1995). Expression of thioredoxin random peptide libraries on the *Escherichia coli* cell surface as functional fusions to flagellin: a system designed for exploring protein–protein interactions. Bio/Technology *13*, 366–372.

Luna, M.G., Martins, M.M., Newton, S.M.C., Costa, S.O.P., Almeida, D.F., and Ferreira, L.C.S. (1997). Cloning and expression of colonization factor antigen I (CFA/I) epitopes of enterotoxigenic *Escherichia coli* (ETEC) in *Salmonella* flagellin. Res. Microbiol. *148*, 217–228.

Macnab, R.M. (2004). Type III flagellar protein export and flagellar assembly. Biochim. Biophys. Acta *1694*, 207–217.

Majander, K., Anton, L., Antikainen, J., LÅng, H., Brummer, M., Korhonen, T.K., and Westerlund-Wikström, B. (2005a). Extracellular secretion of polypeptides using a modified *Escherichia coli* flagellar secretion apparatus. Nat. Biotechnol. *23*, 475–481.

Majander, K., Korhonen, T.K., and Westerlund-Wikström, B. (2005b). Simultaneous display of multiple foreign peptides in the FliD capping and FliC filament proteins of the *Escherichia coli* flagellum. Appl. Environ. Microbiol. *71*, 4263–4268.

Maki-Yonekura, S., Yonekura, K., and Namba, K. (2003). Domain movements of HAP2 in the cap-filament complex formation and growth process of the bacterial flagellum. Proc. Natl. Acad. Sci. U.S.A. *100*, 15528–15533.

Marraffini, L.A., Dedent, A.C., and Schneewind, O. (2006). Sortases and the art of anchoring proteins to the envelopes of Gram-positive bacteria. Microbiol. Mol. Biol. Rev. *70*, 192–221.

Maruyama, Y., Momma, M., Mikami, B., Hashimoto, W., Murata, K. (2008). Crystal structure of a novel bacterial cell-surface flagellin binding to a polysaccharide. Biochemistry *47*, 1393–1402.

McDonald, W.F, Huleatt, J.W., Foellmer, H.G., Hewitt, D., Tang, J., Desai, P., Price, A., Jacobs, A., Takahashi, V.N., Huang, Y., Nakaar, V., Alexopoulou, L., Fikrig, E., and Powell T.J. (2007). A West Nile virus recombinant protein vaccine that coactivates innate and adaptive immunity. J. Infect. Dis. *195*, 1607–17.

McEwen, J., Levi, R., Horwitz, R.J., and Arnon, R. (1992). Synthetic recombinant vaccine expressing haemagglutinin epitope in *Salmonella* flagellin leads to partial protection in mice. Vaccine *10*, 405–411.

Mergulhão, F.J., Summers, D.K., and Monteiro, G.A. (2005). Recombinant protein secretion in *Escherichia coli*. Biotechnol. Adv. *23*, 177–202.

Miao, E.A., Alpuche-Aranda, C.M., Dors, M., Clark, A.E., Bader, M.W., Miller, S.I., and Aderem, A. (2006). Cytoplasmic flagellin activates caspase-1 and secretion of interleukin 1β via Ipaf. Nat. Immunol. *7*, 569–575.

Minamino, T., and Namba, K. (2008). Distinct roles of the FliI ATPase and proton motive force in bacterial flagellar protein export. Nature *451*, 485–488.

Murthy, K.D., Deb, A., Goonesekera, S., Szabó, C., and Salzman, A.L. (2004). Identification of conserved domains in *Salmonella muenchen* flagellin that are essential for its ability to activate TLR5 and to induce an inflammatory response *in vitro*. J. Biol. Chem. *279*, 5667–5675.

Newton, S.M.C., Jacob, C.O., and Stocker, B.A.D. (1989). Immune response to cholera toxin epitope inserted in *Salmonella* flagellin. Science *244*, 70–72.

Nishikawa, H., Sato, E., Briones, G., Chen, L.M., Matsuo, M., Nagata, Y., Ritter, G., Jäger, E., Nomura, H., Kondo, S., Tawara, I., Kato, T., Shiku, H., Old, L.J., Galán, J.E., and Gnjatic, S. (2006). *In vivo* antigen delivery by a *Salmonella typhimurium* type III secretion system for therapeutic cancer vaccines. J. Clin. Invest. *116*, 1946–1954.

Panthel, K., Meinel, K.M., Sevil Domènech, V.E., Trülzsch, K., and Rüssmann, H. (2008). *Salmonella* type III-mediated heterologous antigen delivery: a versatile oral vaccination strategy to induce cellular immunity against infectious agents and tumors. Int. J. Med. Microbiol. *298*, 99–103.

Paul, K., Erhardt, M., Hirano, T., Blair, D.F., and Hughes, K.T. (2008). Energy source of flagellar type III secretion. Nature *451*, 489–492.

Prüss, B.M., Besemann, C., Denton, A., and Wolfe, A.J. (2006). A complex transcription network controls the early stages of biofilm development by *Escherichia coli*. J. Bacteriol. *188*, 3731–3739.

Ren, T., Zamboni, D.S., Roy, C.R., Dietrich, W.F., and Vance, R.E. (2006). Flagellin-deficient *Legionella* mutants evade caspase-1- and Naip5-mediated macrophage immunity. PLoS Pathog. 2, e18.

Rüssmann, H., Shams, H., Poblete, F., Fu, Y., Galán, J.E., and Donis, R.O. (1998). Delivery of epitopes by the *Salmonella* type III secretion system for vaccine development. Science *281*, 565–568.

Ryu, J., Han, K., Park, J., and Choi, S.Y. (2003). Enhanced uptake of a heterologous protein with an HIV-1 Tat protein transduction domains (PTD) at both termini. Mol. Cells *16*, 385–391.

Samatey, F.A., Imada, K., Nagashima, S., Vonderviszt, F., Kumasaka, T., Yamamoto, M., and Namba, K. (2001). Structure of the bacterial flagellar protofilament and implications for a switch for supercoiling. Nature *410*, 331–337.

Samatey, F.A., Matsunami, H., Imada, K., Nagashima, S., Shaikh, T.R., Thomas, D.R., Chen, J.Z., Derosier, D.J., Kitao, A., and Namba, K. (2004). Structure of the bacterial flagellar hook and implication for the molecular universal joint mechanism. Nature *431*, 1062–1068.

Samuelson, P., Gunneriusson, E., Nygren, P.A., and StÅhl, S. (2002). Display of proteins on bacteria. J. Biotechnol. *96*, 129–154.

Sauer, F.G., Remaut, H., Hultgren, S.J., and Waksman, G. (2004). Fiber assembly by the chaperone-usher pathway. Biochim. Biophys. Acta *1694*, 259–267.

Scharfman, A., Arora, S.K., Delmotte, P., Van Brussel, E., Mazurier, J., Ramphal, R., and Roussel, P. (2001). Recognition of Lewis x derivatives present on mucins by flagellar components of *Pseudomonas aeruginosa*. Infect. Immun. *69*, 5243–5248.

Smith, K.D., Andersen-Nissen, E., Hayashi, F., Strobe, K., Bergman, M.A., Barrett, S.L., Cookson, B.T., and Aderem, A. (2003). Toll-like receptor 5 recognizes a conserved site on flagellin required for protofilament formation and bacterial motility. Nat. Immunol. *4*, 1247–1253.

Song, L., Nakaar, V., Kavita, U., Price, A., Huleatt, J., Tang, J., Jacobs, A., Liu, G., Huang, Y., Desai, P., Maksymiuk, G., Takahashi, V., Umlauf, S., Reiserova, L., Bell, R., Li, H., Zhang, Y., McDonald, W.F., Powell, T.J., and Tussey, L. (2008). Efficacious recombinant influenza vaccines produced by high yield bacterial expression: a solution to global pandemic and seasonal needs. PLoS ONE. 3, e2257.

Sun, Y.H., Rolán, H.G., and Tsolis, R.M. (2007). Injection of flagellin into the host cell cytosol by *Salmonella enterica* serotype Typhimurium. J. Biol. Chem. *282*, 33897–33901.

Tanskanen, J., Korhonen, T.K., and Westerlund-Wikström, B. (2000). Flagellar filaments carrying two foreign adhesive peptides. Appl. Environ. Microbiol. *66*, 4152–4156.

Taschner, S., Meinke, A., von Gabain, A., and Boyd, A.P. (2002). Selection of peptide entry motifs by bacterial surface display. Biochem. J. *367*, 393–402.

Tasteyre, A., Barc, M.-C., Collignon, A., Boureau, H., and Karjalainen, T. (2001). Role of FliC and FliD flagellar proteins of *Clostridium difficile* in adherence and gut colonization. Infect. Immun. *69*, 7937–7940.

Thai, C.K., Dai, H., Sastry, M.S., Sarikaya, M., Schwartz, D.T., and Baneyx, F. (2004). Identification and characterization of Cu(2)O- and ZnO-binding polypeptides by *Escherichia coli* cell surface display: toward an understanding of metal oxide binding. Biotechnol. Bioeng. *87*, 129–37.

Végh, B.M., Gál, P., Dobó, J., Závodszky, P., and Vonderviszt, F. (2006). Localization of the flagellum-specific secretion signal in *Salmonella* flagellin. Biochem. Biophys. Res. Commun. *345*, 93–98.

Vitte, A.L., and Jalinot, P. (2008). Intracellular delivery of peptides via association with ubiquitin or SUMO-1 coupled to protein transduction domains. BMC Biotechnol. *8*, 24.

Wang, Q., Suzuki, A., Mariconda, S., Porwollik, S., and Harshey, R.M. (2005). Sensing wetness: a new role for the bacterial flagellum. EMBO J. *24*, 2034–2042.

Westerlund-Wikström, B. (2000). Peptide display on bacterial flagella: principles and applications. Int. J. Med. Microbiol. *290*, 223–230.

Westerlund-Wikström, B., Tanskanen, J., Virkola, R., Hacker, J., Lindberg, M., Skurnik, M., and Korhonen, T.K. (1997). Functional expression of adhesive peptides as fusions to *Escherichia coli* flagellin. Prot. Engin. *10*, 1319–1326.

Wilson, D.R., and Beveridge, T.J. (1993). Bacterial flagellar filaments and their component flagellins. Can. J. Microbiol. *39*, 451–472.

Wright , K.J., Seed, P.C., and Hultgren, S.J. (2005). Uropathogenic *Escherichia coli* flagella aid in efficient urinary tract colonization. Infect. Immun. *73*, 7657–68.

Xu, H., Zhang, P., Liu, L., and Lee, M.Y. (2001). A novel PCNA-binding motif identified by the panning of a random peptide display library. Biochemistry *40*, 4512–4520.

Yonekura, K., Maki, S., Morgan,D., DeRosier, G.D.J., Vonderviszt, F., Imada, K., and Namba, K. (2000). The bacterial flagellar cap as the rotary promotor of flagellin self-assembly. Science *290*, 2148–2152.

Yonekura, K., Maki-Yonekura, S., and Namba, K. (2003). Complete atomic model of the bacterial flagellar filament by electron cryomicroscopy. Nature *424*, 643–650.

Young, G.M., Schmiel, D.H., and Miller, V.L. (1999). A new pathway for the secretion of virulence factors by bacteria: The flagellar export apparatus functions as a protein-secretion system. Proc. Natl. Acad. Sci. U.S.A. *96*, 6456–6461.

Zitzmann, S., Krämer, S., Mier, W., Mahmut, M., Fleig, J., Altmann, A., Eisenhut, M., and Haberkorn, U. (2005). Identification of a new prostate-specific cyclic peptide with the bacterial FliTrx system. J. Nucl. Med. *46*, 782–785.

Origin and Evolution of the Bacterial Flagellar System 13

Renyi Liu

Abstract

Elucidating the origin and evolution of complex systems such as the bacterial flagellum is one of the major challenges of evolutionary studies. Taking advantages of whole genome sequences from hundreds of bacterial genomes, recent studies have shown that the bacterial flagellum evolved from a simple secretion system to a well-honed structure, and various evolutionary forces have contributed to the stepwise formation process, including gene duplication, horizontal gene transfer, and gene loss. Whereas structural genes arose before the divergence of major bacterial lineages, auxiliary and regulatory genes are recruited in a lineage-specific fashion. Although the physical structure of the flagellum is conserved across bacterial phyla, the composition, organization, and operon structure of flagellar genes are under constant modification. Additional genome sequences and better functional characterization of genetic components will help illustrate details of the evolutionary process in the future.

Introduction

The bacterial flagellum is a well-honed organelle that gives bacteria the ability to move around in liquid or semi-solid media and seek better living conditions. Intensive studies in the past several decades revealed that the bacterial flagellum is a complex system consisting of at least six components, including a basal body, a motor, a switch, a hook, a filament, and an export apparatus (Macnab, 2003). Adding regulatory and chemotaxis genes, the total number of flagellar genes in the model bacterium *E. coli* is at least 50. Just like other complex systems such as the eye and the heart, the bacterial flagellum poses a major challenge to modern evolutionary theory: if it can be demonstrated that such complex systems could not have evolved through numerous small steps, Darwin's theory of evolution would break down (Darwin, 1859).

In addition, flagellar systems display extensive diversity among different groups of bacteria (Pallen and Matzke, 2006). Bacterial flagella not only appear in different locations on the cell surface (polar or peritrichous) and in different numbers, but also differ in form and function. For example, unlike the model organism *Salmonella enterica* serovar Typhimurium, spirochaetes possess flagella that do not penetrate the outer membrane and stay inside the periplasm (Charon and Goldstein, 2002). Some bacteria, such as *Vibrio parahaemolyticus*, have two distinct flagellar systems (one sheathed polar flagellum and numerous unsheathed lateral flagella) that are responsible for movements in different environments (swimming in liquid and swarming on solid surface) and use different motive forces (proton and sodium) (McCarter, 2004). How has the bacterial flagellar system originated and evolved to achieve this high level of complexity and diversity? This article intends to summarize recent advancements towards answering this question and to provide a couple of pointers to future research.

A simple start

Several lines of evidence suggested that the flagellar system originated from a system similar to

F_0F_1-ATP synthase. Both systems are membrane-bound, rotary machines using proton flow as the power source, and some of their components share significant similarity. For example, FliI is the ATPase of the export apparatus of the flagellar system and it shares significant homology to the α and β subunits of F_0F_1-ATP synthase in both primary protein sequence and crystal structure (Imada *et al.*, 2007). Another flagellar protein in the export apparatus, FliH, represents fusions of domains homologous to the second-stalk components (the b and δ subunits) of F_0F_1-ATP synthase (Pallen *et al.*, 2006).

In the early stage of evolution, the primary function of the bacterial flagellar system was likely to be a simple proton-driven secretion system (Musgrave, 2004). The flagellar motor proteins MotA/B have been shown to be homologous to the TolQ/R and ExbB/D proteins in the proton-driven, cross-membrane transport systems Tol-Pal and TonB, respectively (Cascales *et al.*, 2001). Secretion remains a critical part of function of extant flagella as protein components are transported by the export apparatus through the hollow structure of the basal body and then used for the assembly of the structure that lies outside of the outer-membrane (hook and filament) (Macnab, 2004). Other non-flagellar proteins including virulence factors may also be secreted by the flagellar system (Young *et al.*, 1999). Furthermore, the sequence and structural homologies between the flagellar system and the type III secretion system (T3SS) were well established (Nguyen *et al.*, 2000; Blocker *et al.*, 2003), indicating their common origin. Because T3SS is restricted in some lineages of Gram-negative bacteria whereas the flagellar system has much broader phylogenetic distribution – including some Gram-positive bacteria – the flagellar system appears to have originated earlier in the evolutionary history and the T3SS is derived (Saier, 2004).

Formation of the structure components

In order to illustrate the evolutionary history of the bacterial flagellar system, it is critical to find out how and when each component has originated and been integrated together and what evolutionary forces have contributed to this process. With the advance of DNA sequencing technology, complete genome sequences from all major lineages have become available and it offers an unprecedented opportunity to trace the evolutionary history of each gene involved in the structure and assembly of bacterial flagella. A particularly effective approach is phylogenetic profiling, where the presence and absence of genes within a phylogenetic tree are recorded and evolutionary events are deduced (Pellegrini *et al.*, 1999). Applying this technique to the flagellar genes, it is clear that some flagellar genes arose quite early and other genes are much younger (Campillos *et al.*, 2006; Liu and Ochman, 2007b). Based on the profiling results, a core set of 24 flagellar genes were defined as they are present in all major bacterial phyla considered and thus likely originated before the divergence of these lineages (Fig. 13.1; Liu and Ochman, 2007b). This core set coincides with the set of genes whose protein products form the physical structure of the flagellum.

What are the major evolutionary mechanisms behind the origination of the core set of structure genes? Early studies revealed that some of the rod and hook proteins shared significant similarity (Homma *et al.*, 1990), suggesting that gene duplication might have played a role in their origination. When core proteins from more genomes are compared to all proteins in the same genome, the best hits are usually to other core proteins. Together they form a network of sequence similarity. In addition, multiple copies of individual flagellar genes are found in many genomes, and duplication of the whole flagellar gene set is not uncommon. Therefore, gene duplication and diversification is a major force behind the origination of all core proteins (Liu and Ochman, 2007b).

Although horizontal gene transfer (HGT) is a major mechanism for bacteria to gain new genetic material from distantly related organisms and new abilities to adapt to specific niches (Ochman *et al.*, 2000), it played only a limited role in the evolution of the bacterial flagellum (Liu and Ochman, 2007b), as in the case of the archaeal flagellum (Desmond *et al.*, 2007). Most of the core flagellar genes have originated before the divergence of the major bacterial lineages and show congruent evolutionary histories with one another. Nevertheless, cases of HGT of flagellar

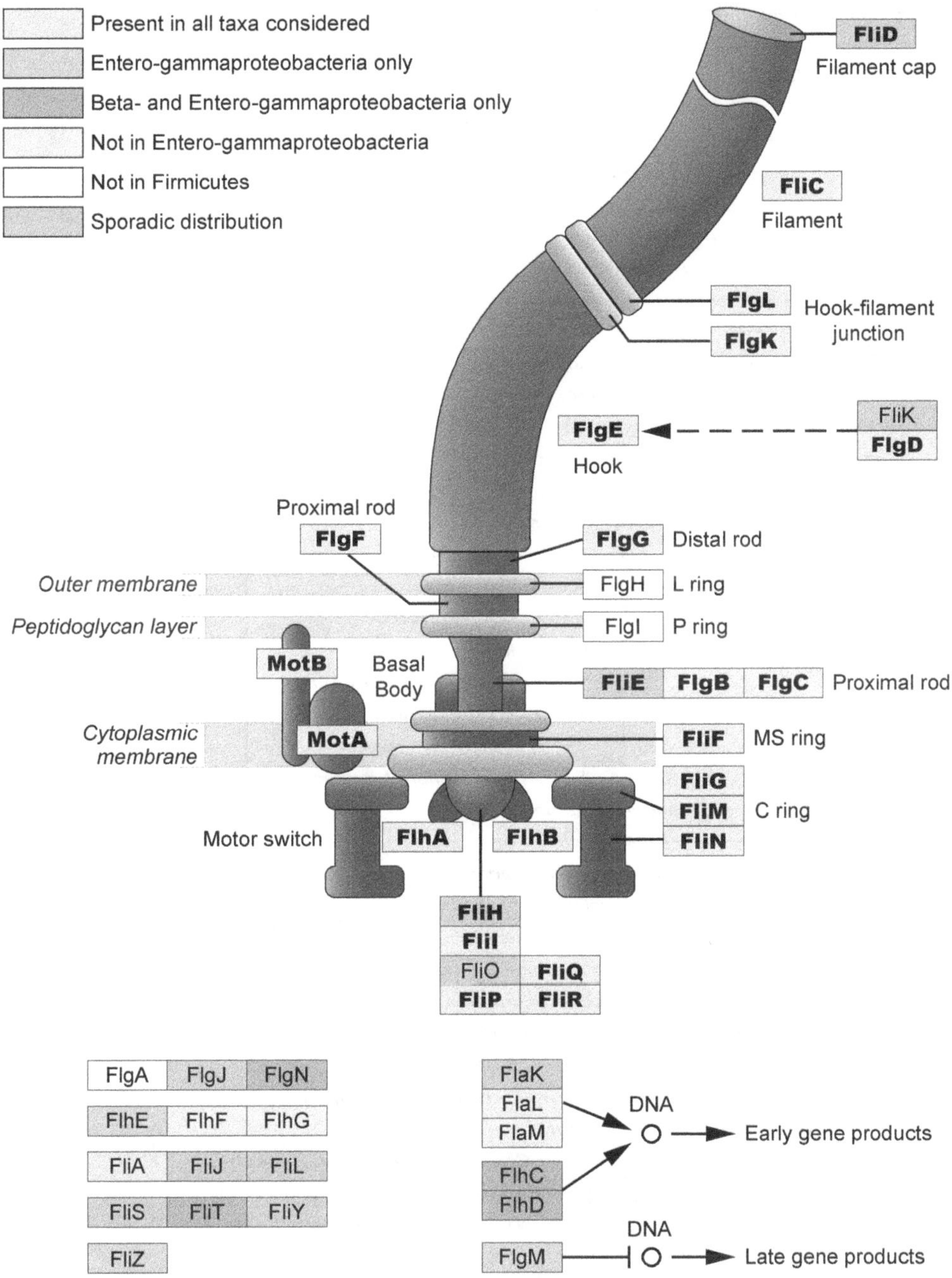

Figure 13.1 Phylogenetic profiling results of bacterial flagellar proteins. Core proteins that are present in all major bacterial phyla considered are designated in bold. Reprinted with permission from Liu and Ochman (2007b). Copyright (2007) National Academy of Sciences, USA.

genes can be found. Examples include the transfer of the primary (and only) flagellar system in *Zymomonas mobilis* (Fig. 13.2; Liu and Ochman, 2007b) and the secondary flagellar system in *Rhodobacter sphaeroides* (Poggio *et al.*, 2007). In addition, HGT of the entire chemotaxis system between bacteria and archaea has been proposed (Faguy and Jarrell, 1999). Although the flagella in bacteria and archaea are distinct in their composition and assembly (Jarrell *et al.*, 1996), their chemotaxis systems are very similar in terms of sequence and gene organization.

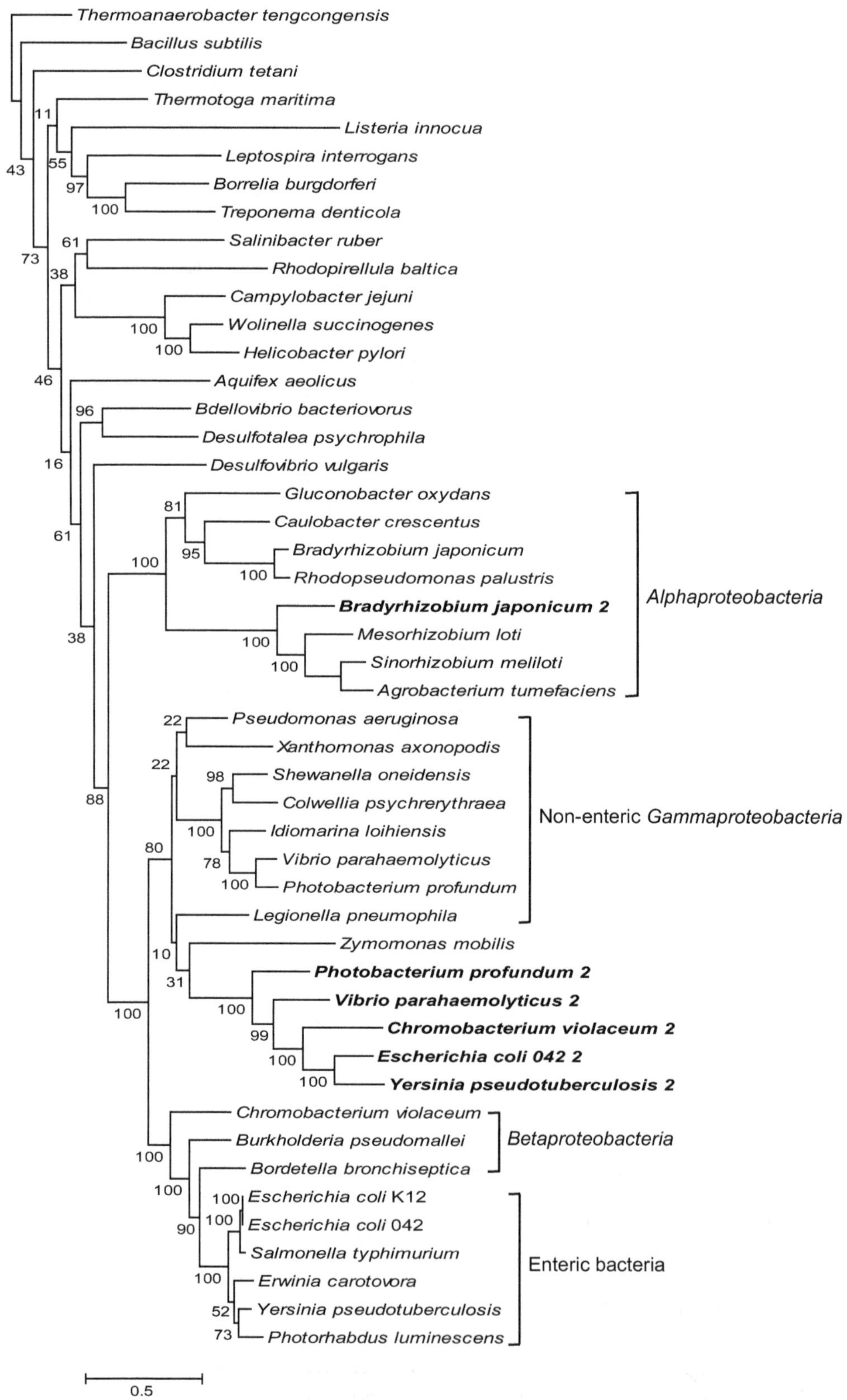

Figure 13.2 Phylogenetic tree of primary and secondary flagellar systems based on the concatenated alignment of 14 flagellar proteins. The secondary systems are labelled with ‘2’ after the species names and are in bold.

Recruitment of regulatory genes

Among known flagellar genes, about half encode the structural components of the bacterial flagellum, and the rest encode auxiliary proteins such as chaperones, regulatory proteins that control the flagellar assembly process and proper function, or chemotaxis proteins that are responsible for detection and transduction of environmental stimuli. The phylogenetic distribution of auxiliary, regulatory, and chemotaxis genes are usually sporadic (Fig. 13.1; Pallen and Matzke, 2006; Liu and Ochman, 2007b), indicating that they may have been recruited in specific lineages and thus have different evolutionary histories from those of structural genes.

The assembly of the bacterial flagellum is a sequential process that proceeds from inside of the cell to the exterior, with the filament being the last structure to be assembled (Macnab, 2003). This process requires the coupling of temporal gene expression to specific stages of the assembly process (Ferris and Minamino, 2006; Chevance and Hughes, 2008). Therefore, in flagellated bacteria there exists a regulation network to ensure the sequential expression of flagellar genes. However, the regulation cascades are quite different among genomes in terms of regulators involved and regulatory interactions (Soutourina and Bertin, 2003; McCarter, 2006). For example, the regulation hierarchy in *E. coli* has three tiers and is controlled by the master regulators FlhD and FlhC, whereas in another Gammaproteobacterium, *Pseudomonas aeruginosa*, it is a four-tiered network controlled by FleQ (McCarter, 2006). Most regulator genes are lineage specific and are likely late additions after the divergence of bacterial lineages. For instance, the master regulators *flhD* and *flhC* are not found beyond enteric bacteria. The diversity of flagellar regulator genes is consistent with the general pattern revealed by genome-wide analyses: bacterial regulator networks are extremely flexible and transcription factors evolved much faster than their target genes (Babu *et al.*, 2006; Lozada-Chavez *et al.*, 2006).

Evolution of gene organization

Although flagellar genes usually form several tight clusters on chromosomes, the size, gene content, and gene order of clusters vary greatly among bacterial lineages (Liu and Ochman, 2007a), suggesting that the flagellar gene organization is quite dynamic. Because operons organize multiple genes into a single transcription unit and coordinate the expression of functionally related genes, it is particularly informative to investigate the evolution of flagellar gene operons. Using 10 multigene operons in *E. coli* as reference, the occurrence and organization of member genes in each operon were compared among different bacterial lineages and major events of operon formation, expansion, and disruption were inferred. The general trend is that flagellar gene operons expanded from small, usually two-gene units to larger gene complexes by recruiting new genes and joining clusters. However, in some lineages such as *Epsilon-* and *Alphaproteobacteria*, operon structures are highly disrupted through dissociation and rearrangement of member genes (Liu and Ochman, 2007a). These results indicate that destruction of operons may be selectively neutral and that the formation and maintenance of flagellar operons might be driven by horizontal gene transfer (Lawrence and Roth, 1996), co-duplication (Reams and Neidle, 2004), or co-regulation (Price *et al.*, 2005) of neighbour genes.

Origin of secondary systems

In addition to the primary flagellar system, some bacteria possess a second flagellar gene set (McCarter, 2004; Kanbe *et al.*, 2007). These two gene sets seldom overlap with each other on chromosomes, and they are often responsible for different types of movements and appear to be under distinct expression regulations. Does the secondary system have a similar evolutionary history as the primary system in the same genome? Phylogenetic analysis of secondary flagellar systems in five proteobacterial genomes suggests that this is not the case. In the phylogenetic tree of concatenated flagellar proteins, the secondary system in the alphaproteobacterium *Bradyrhizobium japonicum* clusters with the primary systems of three other *Alphaproteobacteria* (*Mesorhizobium loti*, *Sinorhizobium meliloti*, and *Agrobacterium tumefaciens*), whereas the secondary systems of four *Gammaproteobacteria* (*E. coli* 042, *Photobacterium profundum*, *Vibrio parahaemolyticus*, and *Yersinia pseudotuberculosis*), and one *Betaproteobacterium* (*Chromobacterium violaceum*) cluster with the primary systems of non-enteric *Gammaproteobacteria* (Fig. 13.2). Therefore, secondary flagellar systems originated at least twice, once in the alphaproteobacterial lineage and another in the common ancestor of *Gamma-* and *Betaproteobacteria* (Liu and Ochman, 2007a). Horizontal transfer of the whole gene set appears to be the mechanism of origination of the secondary system (Liu and Ochman, 2007a; Poggio *et al.*, 2007).

Degradation of flagellar genes

The gene content in bacterial genomes is under constant change and the flagellar gene set is no exception. Like other genes, flagellar genes are always under mutation pressure. In certain environments where possessing flagella does not enhance the fitness of bacteria, flagellar genes become redundant and are subject to deletion. A whole flagellar gene set can be removed in a short time, as suggested by the presence of the Flag-2 locus in some strains of *E. coli* and *Yersinia enterocolitica* but its absence in others (Ren *et al.*, 2005; Bresolin *et al.*, 2008). All flagellar genes at the Flag-2 locus reside next to each other and are likely to have been deleted simultaneously, as indicated by the remnant boundary genes in *E. coli* K12 (Ren *et al.*, 2005).

Degradation of flagellar genes also occurred on a gene-by-gene basis. Individual flagellar genes first become pseudogenes and are then removed from the genome. On one hand, pesudogenes can be found in the genomes of bacteria with functional flagella, as exemplified by the *flgE* gene in *Thermotoga maritima* and *flhD* gene in *Yersinia pestis* CO92 (Liu and Ochman, unpublished data); on the other hand, some flagellar genes are still retained in nonmotile bacteria such as endosymbiotic bacteria *Buchnera aphidicola* (Maezawa *et al.*, 2006) and *Wigglesworthia glossinidia* (Akman *et al.*, 2002). Comparative analysis shows that flagellar genes have been differentially lost in endosymbiotic bacteria of insects: whereas filament and hook proteins are lost in most strains, proteins involved in the export apparatus are retained (Toft and Fares, 2008). The retained flagellar genes are actually transcribed and translated to form hook–basal-body complexes on the cell surface of *Buchnera aphidicola* (Maezawa *et al.*, 2006), indicating that the primary function of the export apparatus has likely been changed from motility to protein exporter.

Conclusions

With well-characterized structure and genetic components, the bacterial flagellar system offers a great opportunity for studying the evolution of complex systems. Comparative analyses based on whole genome sequence data from representative bacterial species show that the bacterial flagellum originated before the divergence of major bacterial lineages and have been under constant change, driven by various evolutionary forces, including gene duplication, horizontal gene transfer, and gene loss. Therefore, the bacterial flagellar system is subject to the same evolutionary mechanisms as other biological entities, whether simple or complex, and there are no conceptual obstacles in understanding its evolution.

Future trends

We have moved a step closer to completely understanding the billion-year evolution process of the bacterial flagellum. However, further evolutionary details on how and when each component was added and integrated to build extant functional flagella remain to be elucidated. In order to understand these details, it is necessary to obtain better functional characterization of genetic components involved in diverse flagellar systems and to sample many more bacterial genomes. Despite decades of molecular studies, our understanding on the genetic components of the bacterial flagellar system is far from complete, as indicated by two recent genome scale studies that revealed quite a few new genetic loci and protein–protein interactions involved in bacterial motility (Girgis *et al.*, 2007; Rajagopala *et al.*, 2007). Exploration of more genomes, including those from uncultured organisms from environmental samples (Tringe and Rubin, 2005), and functional characterization of specific flagellar genes may help us discover the intermediate forms and functional transitions during the evolution of the flagellum. We can also use phylogenetic methods to reconstruct the ancestral forms of flagellar genes from extant genes, examine their sequence and structure properties, and even synthesize and investigate their functions and their interactions with other proteins directly using various experimental approaches (Pallen and Matzke, 2006). Together, these approaches will certainly give a much clearer picture of the evolution of the bacterial flagellar system.

References

Akman, L., Yamashita, A., Watanabe, H., Oshima, K., Shiba, T., Hattori, M., and Aksoy, S. (2002). Genome sequence of the endocellular obligate symbiont of tsetse flies, *Wigglesworthia glossinidia*. Nat. Genet. *32*, 402–407.

Babu, M.M., Teichmann, S.A., and Aravind, L. (2006). Evolutionary dynamics of prokaryotic transcriptional regulatory networks. J. Mol. Biol. *358*, 614–633.

Blocker, A., Komoriya, K., and Aizawa, S. (2003). Type III secretion systems and bacterial flagella: Insights into their function from structural similarities. Proc. Natl. Acad. Sci. U.S.A. *100*, 3027–3030.

Bresolin, G.N., Trek, J., Scherer, S., and Fuchs, T.M. (2008). Presence of a functional flagellar cluster Flag-2 and low-temperature expression of flagellar genes in *Yersinia enterocolitica* W22703. Microbiology *154*, 196–206.

Campillos, M., von Mering, C., Jensen, L.J., and Bork, P. (2006). Identification and analysis of evolutionarily cohesive functional modules in protein networks. Genome Res. *16*, 374–382.

Cascales, E., Lloubes, R., and Sturgis, J.N. (2001). The ToIQ-ToIR proteins energize ToIA and share homologueies with the flagellar motor proteins MotA-MotB. Mol. Microbiol. *42*, 795–807.

Charon, N.W., and Goldstein, S.F. (2002). Genetics of motility and chemotaxis of a fascinating group of bacteria: The spirochetes. Annu. Rev. Genet. *36*, 47–73.

Chevance, F.F.V., and Hughes, K.T. (2008). Coordinating assembly of a bacterial macromolecular machine. Nat. Rev. Microbiol. *6*, 455–465.

Darwin, C. (1859). The origin of species by means of natural selection, or the perservation of favoured races in the struggle for life (London, John Murray).

Desmond, E., Brochier-Armanet, C., and Gribaldo, S. (2007). Phylogenomics of the archaeal flagellum: rare horizontal gene transfer in a unique motility structure. BMC Evol. Biol. *7*, 106.

Faguy, D.M., and Jarrell, K.F. (1999). A twisted tale: the origin and evolution of motility and chemotaxis in prokaryotes. Microbiology *145*, 279–281.

Ferris, H.U., and Minamino, T. (2006). Flipping the switch: bringing order to flagellar assembly. Trends Microbiol. *14*, 519–526.

Girgis, H.S., Liu, Y.C., Ryu, W.S., and Tavazoie, S. (2007). A comprehensive genetic characterization of bacterial motility. PLoS Genet. *3*, 1644–1660.

Homma, M., Derosier, D.J., and Macnab, R.M. (1990). Flagellar hook and hook-associated proteins of *Salmonella typhimurium* and their relationship to other axial components of the flagellum. J. Mol. Biol. *213*, 819–832.

Imada, K., Minamino, T., Tahara, A., and Namba, K. (2007). Structural similarity between the flagellar type III ATPase FliI and F_1 ATPase subunits. Proc. Natl. Acad. Sci. U.S.A. *104*, 485–490.

Jarrell, K.F., Bayley, D.P., and Kostyukova, A.S. (1996). The archaeal flagellum: A unique motility structure. J. Bacteriol. *178*, 5057–5064.

Kanbe, M., Yagasaki, J., Zehner, S., Goettfert, M., and Aizawa, S.I. (2007). Characterization of two sets of subpolar flagella in *Bradyrhizobium japonicum*. J. Bacteriol. *189*, 1083–1089.

Lawrence, J.G., and Roth, J.R. (1996). Selfish operons: Horizontal transfer may drive the evolution of gene clusters. Genetics *143*, 1843–1860.

Liu, R., and Ochman, H. (2007a). Origins of flagellar gene operons and secondary flagellar systems. J. Bacteriol. *189*, 7098–7104.

Liu, R., and Ochman, H. (2007b). Stepwise formation of the bacterial flagellar system. Proc. Natl. Acad. Sci. U.S.A. *104*, 7116–7121.

Lozada-Chavez, I., Janga, S.C., and Collado-Vides, J. (2006). Bacterial regulatory networks are extremely flexible in evolution. Nucleic Acids Res. *34*, 3434–3445.

Macnab, R.M. (2003). How bacteria assemble flagella. Annu. Rev. Microbiol. *57*, 77–100.

Macnab, R.M. (2004). Type III flagellar protein export and flagellar assembly. Biochim. Biophys. Acta *1694*, 207–217.

Maezawa, K., Shigenobu, S., Taniguchi, H., Kubo, T., Aizawa, S., and Morioka, M. (2006). Hundreds of flagellar basal bodies cover the cell surface of the endosymbiotic bacterium *Buchnera aphidicola* sp strain APS. J. Bacteriol. *188*, 6539–6543.

McCarter, L.L. (2004). Dual flagellar systems enable motility under different circumstances. J. Mol. Microbiol. Biotechnol. *7*, 18–29.

McCarter, L.L. (2006). Regulation of flagella. Curr. Opin. Microbiol. *9*, 180–186.

Musgrave, I. (2004). Evolution of the bacterial flagellum. In Why intelligent design fails : a scientific critique of the new creationism, M. Young, and T. Edis, eds. (New Brunswick, N.J., Rutgers University Press), pp. 72–84.

Nguyen, L., Paulsen, I.T., Tchieu, J., Hueck, C.J., and Saier, M.H. (2000). Phylogenetic analyses of the constituents of type III protein secretion systems. J. Mol. Microbiol. Biotechnol. *2*, 125–144.

Ochman, H., Lawrence, J.G., and Groisman, E.A. (2000). Lateral gene transfer and the nature of bacterial innovation. Nature *405*, 299–304.

Pallen, M.J., Bailey, C.M., and Beatson, S.A. (2006). Evolutionary links between FliH/YscL-like proteins from bacterial type III secretion systems and second-stalk components of the F_0F_1 and vacuolar ATPases. Protein Sci. *15*, 935–940.

Pallen, M.J., and Matzke, N.J. (2006). From The Origin of Species to the origin of bacterial flagella. Nat. Rev. Microbiol. *4*, 784–790.

Pellegrini, M., Marcotte, E.M., Thompson, M.J., Eisenberg, D., and Yeates, T.O. (1999). Assigning protein functions by comparative genome analysis: Protein phylogenetic profiles. Proc. Natl. Acad. Sci. U.S.A. *96*, 4285–4288.

Poggio, S., Abreu-Goodger, C., Fabela, S., Osorio, A., Dreyfus, G., Vinuesa, P., and Camarena, L. (2007).

A complete set of flagellar genes acquired by horizontal transfer coexists with the endogenous flagellar system in *Rhodobacter sphaeroides*. J. Bacteriol. *189*, 3208–3216.

Price, M.N., Huang, K.H., Arkin, A.P., and Alm, E.J. (2005). Operon formation is driven by co-regulation and not by horizontal gene transfer. Genome Res. *15*, 809–819.

Rajagopala, S.V., Titz, B., Goll, J., Parrish, J.R., Wohlbold, K., McKevitt, M.T., Palzkill, T., Mori, H., Finley, R.L., and Uetz, P. (2007). The protein network of bacterial motility. Mol. Syst. Biol. *3*, 128.

Reams, A.B., and Neidle, E.L. (2004). Selection for gene clustering by tandem duplication. Annu. Rev. Microbiol. *58*, 119–142.

Ren, C.P., Beatson, S.A., Parkhill, J., and Pallen, M.J. (2005). The Flag-2 locus, an ancestral gene cluster, is potentially associated with a novel flagellar system from *Escherichia coli*. J. Bacteriol. *187*, 1430–1440.

Saier, M.H. (2004). Evolution of bacterial type III protein secretion systems. Trends Microbiol. *12*, 113–115.

Soutourina, O.A., and Bertin, P.N. (2003). Regulation cascade of flagellar expression in Gram-negative bacteria. FEMS Microbiol. Rev. *27*, 505–523.

Toft, C., and Fares, M.A. (2008). The evolution of the flagellar assembly pathway in endosymbiotic bacterial genomes. Mol. Biol. Evol. *25*, 2069–2076.

Tringe, S.G., and Rubin, E.M. (2005). Metagenomics: DNA sequencing of environmental samples. Nat. Rev. Genet. *6*, 805–814.

Young, G.M., Schmiel, D.H., and Miller, V.L. (1999). A new pathway for the secretion of virulence factors by bacteria: The flagellar export apparatus functions as a protein-secretion system. Proc. Natl. Acad. Sci. U.S.A. *96*, 6456–6461.

Archaeal Flagella and Pili

14

Ken F. Jarrell, David J. VanDyke and John Wu

Abstract

Archaeal flagella and pili are unique cellular appendages that are distinct from their bacterial namesakes. The better studied of the two structures is the archaeal flagellum. While it resembles the bacterial flagellum in terms of being a reversible rotating organelle responsible for swimming motility, its composition, structure and likely mode of assembly are all very different. Archaeal flagella have a unique structure which lacks a central channel. Similar to bacterial Type IV pilins, the component flagellins are made with Class 3 signal peptides and they are processed by a Type IV prepilin peptidase-like enzyme. The flagellins are typically modified by the addition of *N*-linked glycans which are necessary for proper assembly and/or function. The study of archaeal pili is extremely limited. In *Methanococcus maripaludis,* the structures are formed from Type IV pilin-like proteins, again possessing Class 3 signal peptides processed by a Type IV prepilin peptidase-like enzyme. However, the structure of the assembled pilus does not resemble any known bacterial pilus type and, unlike Type IV pili, has a central channel of a diameter similar to that observed in bacterial flagella.

Introduction

Prokaryotic cells have been shown to possess numerous cell surface appendages that perform a variety of tasks in the organisms' environmental niche. While flagella and a variety of pili have been well studied in Bacteria, investigations into surface structures in the other prokaryotic Domain, Archaea, have been, in comparison, very limited. This is despite the fact that the Archaea as a group have several unique surface structures, such as hami and cannulae, as well as unusual flagella and pili that have been shown to be widely distributed among the Domain and possessing features quite distinct from their bacterial namesakes (Ng *et al.*, 2008).

Study of archaeal appendages like flagella and pili has led to numerous findings related to motility but also to more widespread and understudied archaeal themes, including adhesion, biofilm formation, protein secretion and posttranslational modifications such as signal peptide removal and *N*-linked glycosylation (Pohlschroder *et al.*, 2005; Ng *et al.*, 2006; Ng *et al.*, 2007; Yurist-Doutsch *et al.*, 2008). Studies of archaeal model systems, often from extremophiles, continue to showcase the seemingly endless variations that living organisms can employ to reach a common end, such as a rotating motility structure or a pilus (Trachtenberg and Cohen-Krausz, 2006; Cohen-Krausz and Trachtenberg, 2008; Wang *et al.*, 2008). The recent advances in the development of genetic tools for several diverse archaeal model organisms like *Halobacterium* (Soppa, 2006), *Methanococcus* (Moore and Leigh, 2005) and *Sulfolobus* (Berkner *et al.*, 2007; Berkner and Lipps, 2008) have provided researchers the opportunity to probe the structure, assembly and genetics of archaeal flagella and pili to a degree that was not possible even five years ago.

To date, flagella mediated motility is the only mechanism of movement reported in Archaea, save for the floating movement afforded, in certain Archaea, by gas vesicles (Pfeifer *et al.*, 2002).

The recent interest in pili in various archaeal species may soon lead to the discovery of surface mediated movement in this domain, akin to the twitching motility afforded in bacteria by Type IV pili, or to some of the other functions conferred by various bacterial pili. In this chapter, we will review the current information available on the structure, assembly and genetics of archaeal flagella and summarize the limited but increasing data available on archaeal pili.

Distribution of flagellation in Archaea

Flagellation is a trait that is distributed widely throughout the Archaea domain, with many genera of both Crenarchaeota and Euryarchaeota possessing these motility structures (Thomas *et al.*, 2001a). Flagellated archaea include species from a variety of the most extreme habitats in which life has been found, and include hyperthermophiles (e.g. *Pyrococcus*), extreme halophiles (e.g. *Halobacterium*), methanogens (e.g. *Methanococcus*), thermoacidophiles (e.g. *Sulfolobus*) and haloalkaliphiles (e.g. *Natrialba*). Recent phylogenomic analysis of the archaeal flagella system suggested that a possible ancient horizontal gene transfer event of the entire flagella locus occurred between the Euryarchaeota and the Crenarchaeota (Desmond *et al.*, 2007). The presence of flagella on archaea inhabiting such extreme environments demonstrates vividly the extraordinary stability of these structures, in comparison to what is typically observed in the case of bacteria flagella. In addition, as in bacteria, many variations in the number and arrangement of flagella on the cell surface have been observed in Archaea and in all cases, save *Sulfolobus* (Szabo *et al.*, 2007a), the filaments are composed of multiple flagellin proteins (Gerl and Sumper, 1988; Kalmokoff and Jarrell, 1991; Serganova *et al.*, 2002; Jarrell *et al.*, 2007). An abbreviated list of selected archaea and their flagellation is presented in Table 14.1. An electron micrograph of one of the better studied archaea with regards to flagellation, *Methanococcus voltae*, is presented in Fig. 14.1A.

Bacterial flagella are known to be involved in more functions than just swimming and swarming motility (Moens and Vanderleyden, 1996). As the study of archaeal flagella expands, this also appears to be the case in this domain. Recent work on *Pyrococcus furiosus* flagella has indicated that they are involved in forming cable-like connections between cells and in adherence to a variety of surfaces, enabling the cells to form a biofilm on these surfaces under certain physiological conditions (Nather *et al.*, 2006). Such interactions may be a prelude to genetic exchange. More recently, flagella have also been demonstrated to be involved in the cell to cell interaction between *Pyrococcus* and another species, *Methanopyrus*, in a bispecies biofilm (Schopf *et al.*, 2008).

Flagellar structure

Archaeal flagella, like those of bacteria (Chevance and Hughes, 2008), are considered to have three distinct parts. The most distal is the long filament region, which forms the propeller blade of the organelle, followed by a short curved hook region and, finally, an anchoring complex that attaches the flagellum to the cell envelope (Fig. 14.1B). Motility is achieved through the rotation of the flagellum, as in bacteria, although the nature of the motor and its protein composition are unknown (Marwan *et al.*, 1991; Nutsch *et al.*, 2003). With the arrival of the genome sequencing age, it became very apparent that homologues of genes involved in the structure, assembly and function of bacterial flagella could not be identified in any archaeal genome and thus the archaeal flagellum was a truly unique structure (Faguy and Jarrell, 1999; Nutsch *et al.*, 2005).

Table 14.1 Distribution of flagella in selected archaea from various habitats

Organism	Flagella distribution	Flagellin GI
Hyperthermophiles		
Pyrococcus furiosus	Large bundle of flagella	18976709
		18976710

Table 14.1 Continued

Organism	Flagella distribution	Flagellin GI
Thermococcus kodakaraensis	Polar tufts	57639973
		57639974
		57639975
		57639976
		57639977
Archaeoglobus fulgidus	Monopolar polytrichous	11498659
		11498660
Thermoacidophiles		
Sulfolobus solfataricus	One or more flagella	15899081
Thermoplasma volcanium	Monopolar monotrichous	13542257
		13541438
Methanogens		
Methanococcus maripaludis	Large bundle of flagella	45359229
		45359230
		45359231
Methanospirillum hungatei	Polar tufts	88604365
		88604366
Methanococcoides burtonii	Single monotrichous	68185311
		68185312
		68185254
Extreme halophiles		
Halobacterium salinarum	Monopolar or bipolar	15790079
	bundles (5–10 flagella)	15790080
		15790081
		15790120
		15790121
		15790252
Haloalkalophiles		
Natronomonas pharaonis	Polar tuft	76557552
		76557553
		76557554
Natrialba magadii	Polar tuft	8250514
		8250515
		8250516
		8250517

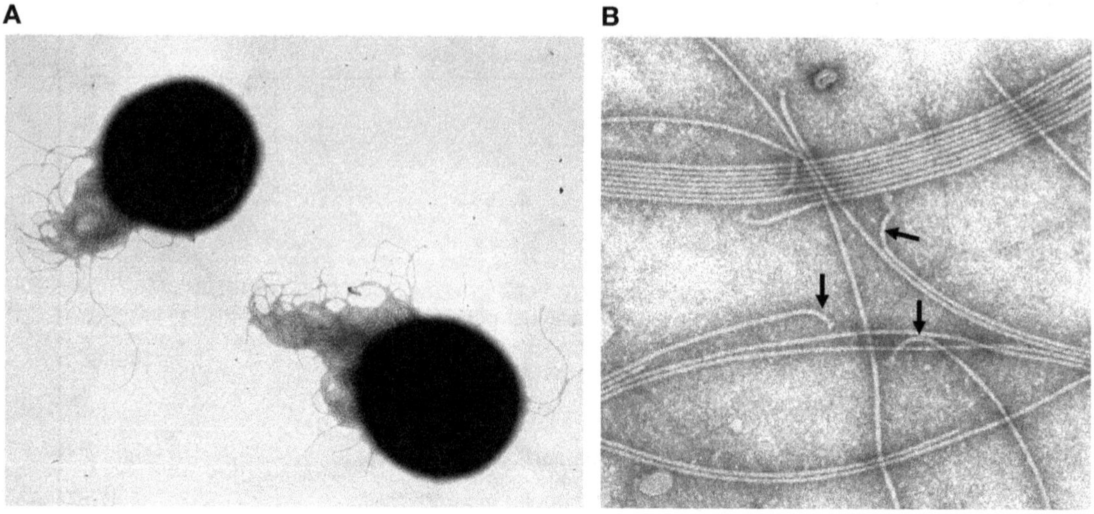

Figure 14.1 **A** Electron micrograph of two *Methanococcus voltae* cells showing tuft of dozens of flagella. Cells stained with 2% phosphotungstic acid (pH 7.0). **B** Detergent-extracted flagella from *M. voltae* showing filament, hook (arrow) and knoblike anchoring structure. Courtesy of S.I. Aizawa.

Filament

Electron microscopic observations of flagellated archaeal species indicated early on that archaeal flagella are typically much thinner (at 10–13 nm) (Black *et al.*, 1979; Cruden *et al.*, 1989; Southam *et al.*, 1990; Jarrell *et al.*, 2007) than the bacterial counterpart (typically 20 nm) (Jones and Aizawa, 1991). While rarely reported in archaea, the structure of the filament in *Halobacterium salinarum* was shown to be a right-handed helix (Alam and Oesterhelt, 1984; Cohen-Krausz and Trachtenberg, 2002) rather than the left-handed flagella bundle observed in the majority of bacterial flagella (Asakura *et al.*, 1966). Recent observations of cryo-negative stained filaments of two, only distantly related, archaea, *Halobacterium* and *Sulfolobus,* revealed a conservation of structure symmetry, which is distinct from the bacterial flagellum (Cohen-Krausz and Trachtenberg, 2002; Trachtenberg *et al.*, 2005; Cohen-Krausz and Trachtenberg, 2008). Given the evolutionary distance between *Halobacterium* and *Sulfolobus,* it is believed that this unique archaeal filament structure is likely conserved across the domain (Cohen-Krausz and Trachtenberg, 2008).

A critically important difference between bacterial and archaeal flagellar filaments was highlighted when it was observed that the archaeal flagellum was missing a central channel large enough to accommodate passage of subunits (Cohen-Krausz and Trachtenberg, 2008). In bacteria, a 2-nm channel is needed for assembly (Yonekura *et al.*, 2003); newly added subunits, made without cleaved signal peptides, travel via the flagellum-specific type III secretion system through the channel and assemble at the distal end of the growing flagella underneath the capping structure (Yonekura *et al.*, 2002; Macnab, 2003; Macnab, 2004; Chevance and Hughes, 2008). The lack of a channel is persuasive evidence for the argument that the archaeal flagellum must be assembled differently, most likely where incoming subunits are added at the base, as in Type IV pili systems (Jarrell *et al.*, 1996b).

Distinction between bacterial and archaeal flagella is seen further in post-translational modifications of the flagellin subunits. These may contribute to the needed stability of archaeal flagella in certain extreme environments. Whereas bacterial flagellins are rarely glycosylated, many archaea have been reported to possess glycosylated flagellins (Eichler and Adams, 2005). Thus far, glycosylated bacterial flagellins are only found to have glycans attached via an *O*-linkage (Logan, 2006). In contrast to bacteria, *H. salinarum*, *M. voltae* and *M. maripaludis* are known to contain *N*-linked glycans attached to the flagellins (Wieland *et al.*, 1985; Sumper, 1987; Voisin *et al.*, 2005). *N*-linked glycans are attached to asparagine residues that are part of an Asn-X-Ser/Thr sequon (Abu-Qarn and Eichler, 2007). Analysis of archaeal flagellins has indicated such sequons are present, usually in multiples, in the overwhelming majority of flagellins, suggesting that the *N*-linked glycosylation modification is an extremely common one (Ng *et al.*, 2006) (see glycosylation section below). Archaeal flagellins also contain a signal peptide sequence at its N-terminus, drawing similarities to bacterial Type IV pili rather than to bacterial flagella (Ng *et al.*, 2006; Ng *et al.*, 2007). This signal peptide must be cleaved in order for the flagellins to be properly assembled into the flagellar filament (Bardy and Jarrell, 2003) (see processing of flagellins section below).

The composition of the flagellar filaments also reveals differences from the bacterial system. Although there are many exceptions, it is common for bacterial filaments to be composed of multiple copies of only a single flagellin protein. Archaeal flagella, however, are almost always composed of multiple flagellin proteins (between 2 and 6), with the only known exception being *Sulfolobus* species (Ng *et al.*, 2006; Jarrell *et al.*, 2007; Szabo *et al.*, 2007a). It is also clear that the various flagellins perform specialized tasks since, in the few cases studied genetically, individual flagellin genes cannot be deleted without major or minor effects on flagellar assembly, structure and /or function (Jarrell *et al.*, 1996a; Tarasov *et al.*, 2000; Chaban *et al.*, 2007).

Anchoring structure

In bacteria, the flagellum is anchored to the cell through a basal body complex. In Gram-negative bacteria, the basal body consists of a rod and series of stacked rings that spans the outer and inner membrane, while passing through the periplasmic space and protruding into the cytoplasm. The basal body also contains the proteins that make up a motor switch complex and Type III secretion apparatus that are involved in flagellar function and assembly, respectively (Macnab, 2003). In contrast, little is known about the components of the archaeal anchoring structure. Extracted archaeal flagella revealed little more than a knob-like basal body, typically lacking the well defined ring structures found in bacteria (Kalmokoff *et al.*, 1988; Kupper *et al.*, 1994). Since homologues to bacterial basal body genes are not found within any archaeal genome, the attachment mechanism of the archaeal flagellum to the cell is likely unique and distinct from that of bacteria. Indeed, many flagellated archaea have the common archaeal cell wall type that consists solely of a protein or glycoprotein S-layer overlying the cytoplasmic membrane, therefore, lacking the murein layer that the flagellar stator attaches to in the bacterial system. The mechanism of archaeal flagellar attachment and the cell wall components involved have yet to be determined. Kupper *et al.* (Kupper *et al.*, 1994), however, were able to isolate flagella from *H. salinarum* that were inserted into a polar cap. Those authors were not able to determine whether this polar cap was part of the cell wall, membrane or subcytoplasmic membrane. However, in thin sections of *H. salinarum*, a plateau-like intracellular structure (the discoid lamellar structure) that lies below the cytoplasmic membrane underneath the flagella insertion sites, has been reported (Metlina, 2004) and may be the polar cap structure. The function of this structure may be for additional anchoring of flagella in certain archaea, as predicted many years ago (Alam and Oesterhelt, 1984). Currently, the components of the polar cap have not been identified.

Archaea are tactic organisms and they can respond to a variety of stimuli, including chemicals (chemotaxis) and light (phototaxis), by changing the rotation of the flagella to swim in a different direction (Rudolph *et al.*, 1995; Rudolph and Oesterhelt, 1996; Nutsch *et al.*, 2003; Nutsch *et al.*, 2005). In bacteria, a change in the direction of flagellar rotation is mediated by the interaction of the chemotaxis system with the flagellar switch. This occurs through the binding of the CheY response regulator, in its phosphorylated form, to the switch protein, FliM (Baker *et al.*, 2006). Archaea generally have a complete bacterial-like chemotaxis system with the proteins necessary for signal recognition and transduction, adaptation, excitation and signal removal (Rudolph *et al.*, 1996; Rudolph and Oesterhelt, 1996; Jarrell *et al.*, 2007). There are clear CheY homologues but, as yet, no FliM equivalent has been reported in any archaeon. To date, it is still a mystery as to how the chemotaxis system interacts with the flagella in archaea to elicit a chemotactic response to changing environmental stimuli.

Hook

In both archaea and bacteria, the hook region serves to connect the filament to the basal body. Electron microscopy revealed a distinct curved hook in many archaeal species, including *Methanococcus voltae* (Fig. 14.1B), *Methanospirillum hungatei* and *Halobacterium salinarum* (Cruden *et al.*, 1989; Bardy *et al.*, 2002; Beznosov *et al.*, 2007). Studies in *M. voltae, M. maripaludis* and *H. salinarum* led to the identification of a single 'flagellin' as being responsible for the hook region (Bardy *et al.*, 2002; Beznosov *et al.*, 2007; Chaban *et al.*, 2007) as mutations in these specific flagellins results in flagellar filaments possessing the characteristic knob-like structure without an intervening curved region (see Flagellar genes section below). Interestingly, these hookless flagella still allow the cells to swim and, in semi-swarm media, these mutants were able to migrate as far as the wild-type cells (Chaban *et al.*, 2007). Electron microscopic characterization of the archaeal hook found the size (72–265 nm) to be variable (Bardy *et al.*, 2002). This property is distinct from the bacterial hook region which is much more consistent in length: 55 nm in *Salmonella* (Hirano *et al.*, 1994).

Flagellar genes

Genes encoding at least some of the structural components of the archaeal flagellum, as well as ones presumed to be involved in assembly, have been identified so far in only one major genetic locus (Fig. 14.2). This locus can contain up to 13 genes arranged in tandem and includes genes encoding multiple flagellin proteins (*flaA* and/or *flaB*) followed by other conserved genes that are essential for flagellation (*flaC–flaJ*) (Bardy *et al.*, 2004; Jarrell *et al.*, 2007). All genes located within the operon are believed to be transcribed together (Nagahisa *et al.*, 1999; Thomas *et al.*, 2002), except where the orientation of some genes is in the opposite direction, as in *H. salinarum* (Patenge *et al.*, 2001). Variations in the flagellar locus have been observed across archaeal species with organisms found to encode differing patterns of flagellin structural proteins or to contain only a subset of the other conserved flagellar associated genes. These differences are most extreme between euryarchaeota species and crenarchaeota species (Ng *et al.*, 2006). Residing usually outside of the flagellar gene locus is a signal peptidase (*flaK*) needed to process the flagellins as well as a number of *agl* (archaeal glycosylation) genes that are required for the assembly and attachment of an *N*-linked glycan to the flagellin proteins (Bardy and Jarrell, 2002; Bardy and Jarrell, 2003; Chaban *et al.*, 2006; Shams-Eldin *et al.*, 2008). The post-translational modifications carried out by these additional elements are essential for the assembly of a functional archaeal flagellum (see below).

Since many essential proteins are still unaccounted for, including motor and switch components and ones involved in anchoring the flagellum to the envelope, it is believed that at least one additional major locus involved in flagella assembly and function is yet to be discovered.

Flagellins

Almost all archaeal genomes contain more than one flagellin gene and each flagellin is believed to be serving a specific function within the flagellar structure. Earlier work in *M. voltae* determined that FlaA, a separately transcribed and minor structural flagellin, was not required for flagella assembly but its presence was necessary for optimal function of the flagella (Jarrell *et al.*, 1996a). *M. maripaludis* encodes three flagellin proteins (FlaB1, FlaB2, and FlaB3). Mutational studies conducted in this organism have shown that deletion of either *flaB1* or *flaB2* resulted in cells that were nonflagellated and, thus, completely non-motile (Chaban *et al.*, 2007). Deletion of *flaB3* resulted in cells that still possessed flagella but had unusual swimming motility patterns, compared to wild-type cells. Electron microscopic analysis of flagella isolated from Δ*flaB3* mutant cells demonstrated that the flagella were lacking the hook portion at the base of the flagellum. These findings suggest that the flagellin proteins FlaB1 and FlaB2 make up the filament structure while FlaB3 comprises the hook portion. The latter observation supports the data previously obtained in *M. voltae* in which FlaB3 was enriched in 'stubby' flagella, which contained mainly cell proximal flagella components (Bardy *et al.*, 2002). This arrangement is unique to archaea as bacterial flagellar systems use flagellins to make up the filament and a very different protein to make up the hook structure. Studies conducted in *H. salinarum* have

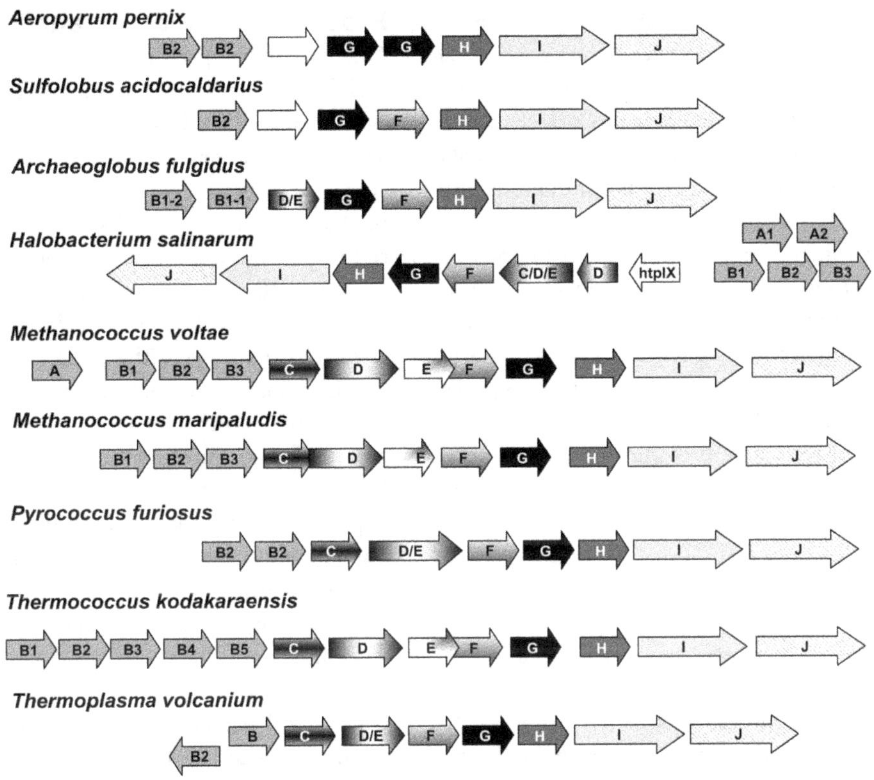

Figure 14.2 The flagella gene region from a variety of selected archaea. Similar shading indicates homologues among the different families. In *H. salinarum*, the B flagellins are located next to the *fla* accessory genes while the A flagellins are located elsewhere on the chromosome.

also demonstrated that multiple flagellins exist to carry out specific roles. In this organism, there are two flagellin gene clusters, encoding FlgA or FlgB flagellin proteins. It is believed that the FlgA proteins form the main component of the filament and are incorporated at the initial stages of assembly while the FlgB flagellins are located near the base and may be involved in anchoring the structure (Tarasov *et al.*, 2000). Here, the hook region is also believed to be composed of one specific flagellin (Beznosov *et al.*, 2007).

A key novel feature of the archaeal flagellins is their noted similarity to Type IV pilins. Early analysis indicated a conserved, very hydrophobic N-terminus that had sequence similarity to Type IV pilins (Faguy *et al.*, 1994). In addition, the flagellins are synthesized as preproteins with Class 3 signal peptides that possess many similar features to those found on Type IV pilins. In both the archaeal flagellin and Type IV pilin case, the preproteins are cleaved by a specific peptidase of the Type IV prepilin peptidase type (see Processing of the preflagellins section, below).

flaHIJ

Of the other genes encoded within flagellar operons, the most conserved are the *flaHIJ* clusters. They are always present in the immediate vicinity of the archaeal flagellins on the genome and are believed to participate in flagellum assembly. Disruption or deletion of these genes results in non-flagellated cells (Patenge *et al.*, 2001; Thomas *et al.*, 2002; Chaban *et al.*, 2007; Szabo *et al.*, 2007a). FlaI bears similarity to ATPases involved in bacterial Type IV pili extension and retraction as it contains a Walker box involved in NTP-binding (Bayley and Jarrell, 1998; Thomas *et al.*, 2002; Peabody *et al.*, 2003). FlaI possesses ATPase activity and is believed to form a membrane bound hexameric ring dependent on Mg^{2+} and ATP to shift from open to closed conformations (Albers and Driessen, 2005; Yamagata and Tainer, 2007).

Relatively little is known about the products encoded by the conserved accessory genes *flaH* and *flaJ*. FlaH contains a Walker box A but it does not have significant similarity to any bacterial proteins. Whether it actually has ATPase activity has yet to be determined. FlaJ has numerous predicted transmembrane domains and is assumed to be an integral membrane protein interacting with the other conserved components at the base of the flagellum. It is homologous to a conserved membrane protein, that recruits the ATPase to the membrane, involved in assembly of Type IV pili (Peabody *et al.*, 2003; Craig and Li, 2008). Together, FlaHIJ may provide the platform and the energy for incorporation of new subunits into the growing structure.

flaCDEFG

Archaeal flagellar operons have various other genes situated between the flagellins and the conserved *flaHIJ* genes. These genes (*flaCDEFG*) do not share significant sequence similarity to any bacterial genes and have yet to be found in any archaeal genome that does not encode flagellin proteins, suggesting that *flaCDEFG* are specific for the archaeal flagellar structure and/or mode of assembly. In certain archaea, such as *Halobacterium* and *Methanococcoides,* individual genes encoding FlaC, FlaD and FlaE are absent and instead one large gene with sequence similarity to all of *flaC, flaD* and *flaE* is found (Ng *et al.*, 2006). This seems to indicate that, in those organisms that possess separate genes for the three proteins, FlaC, FlaD and FlaE likely interact with each other. The importance of the *flaC–J* genes has been demonstrated in mutational studies conducted in *M. maripaludis* as the deletions of *flaC, flaF,* and *flaG* resulted in non-flagellated cells (Chaban *et al.*, 2007). For unknown reasons, the deletion of *flaD* and *flaE* could not be achieved in this study. Strangely though, many of these genes, especially *flaC, flaD* and *flaE* are not found in flagellated crenarchaeote species, making the role of these genes in flagellation in euryarchaeotes even more of a mystery. Indeed, the roles that any of these genes are serving, along with explanations for the presence or absence of a specific gene across archaeal species, are currently unknown.

Processing of the preflagellins

One of the most prominent unique features of the archaeal flagella system compared to the bacterial counterpart is the fact that the archaeal flagellins are made as preproteins which need to have their signal peptides removed before incorporation of the protein into the flagellar filament. Bacterial flagellins are never made with cleaved signal peptides. The presence of signal peptides on archaeal flagellins was first noted in *Methanococcus voltae* flagellins when the gene cluster encoding the flagellins was sequenced and compared to the N-terminal amino acid sequence of the flagellins obtained from purified flagella (Kalmokoff and Jarrell, 1991). The presence of Class 3 signal peptides, as first reported in bacterial Type IV pilins, was soon shown to be a feature of all known archaeal flagellins.

The Class 3 signal peptides found on archaeal flagellins are typically short, often 12 amino acids or less, and almost all end with a highly conserved basic amino acid (lysine or arginine) at the −2 position relative to the cleavage site and a −1 position that is almost invariably a glycine (Ng *et al.*, 2006). In addition, the +3 position found in the mature protein is in all cases a glycine residue. There is, in contrast, a large variability in the amino acids found at the +1 and +2 positions. Site-directed mutagenesis of the *M. voltae* FlaB2 flagellin was conducted to replace the highly conserved amino acid positions to investigate their importance in the cleavage reaction (Thomas *et al.*, 2001b). Such investigations demonstrated the essential nature of a basic amino acid at the −2 position and a glycine at −1. Changing the −1 glycine to a variety of amino acids resulted in no cleavage of the substrate; the sole exception among the amino acid substitutions tested was alanine which resulted in poor cleavage. Interestingly, similar results regarding the −1 glycine were reported for the cleavage of the Class 3 signal peptide from Type IV prepilins (Strom and Lory, 1991). The study by Thomas *et al.* (Thomas *et al.*, 2001b) also demonstrated that valine could not substitute for the glycine at +3 in FlaB2 and that the absolute conservation of that position may indeed be critical for recognition of the flagellins by the responsible signal peptidase. Similar studies have also been conducted on the substrates for the *Sulfolobus* signal peptidase, which also processes sugar binding proteins and pilins (Albers *et al.*, 2003; Szabo *et al.*, 2006). Again the importance of glycine at the -1 position was clearly shown (Albers *et al.*, 2003), although alanine was much more readily accepted than in the case in *M. voltae*. In general, other substitutions showed that the *Sulfolobus* enzyme was more relaxed in its requirements for cleavage than the *Methanococcus* enzyme, a fact already gleaned from its extended range of substrates.

Based on the similarity of archaeal flagellins to Type IV pilins and the presence of similar Class 3 signal peptides, a search for an archaeal enzyme similar to bacterial Type IV prepilin peptidases was started. The enzyme was found and studied in both *Methanococcus* (Bardy and Jarrell, 2002; Bardy and Jarrell, 2003) and *Sulfolobus* (Albers *et al.*, 2003). The methanogen enzyme was termed the preflagellin peptidase, FlaK, in analogy to the bacterial counterpart while in *Sulfolobus* the enzyme was termed PibD (protein involved in biogenesis of prepilin-like proteins) and demonstrated to have a number of substrates in addition to the preflagellins. An *in vitro* assay, based on that developed to study processing of Type IV pilins, was used to show that the cloned FlaK expressed in *E. coli* was indeed the enzyme responsible for the cleavage of preflagellins in methanococci (Bayley and Jarrell, 1999; Correia and Jarrell, 2000). The archaeal enzymes belong to the same COG 1989 as bacterial Type IV prepilin peptidases. These enzymes are a novel group of aspartic acid proteases and site directed mutagenesis was used to show that two aspartic acid residues, ones that align with aspartic acid residues previously shown to be necessary for catalytic activity in bacterial Type IV prepilin peptidases, were also necessary for catalysis by FlaK (Bardy and Jarrell, 2003) and for PibD (Szabo *et al.*, 2006). This indicates that the bacterial and archaeal enzymes likely have an identical mechanism of action.

Knockouts of *flaK* in *M. voltae* and *M. maripaludis* showed that the flagellins were no longer processed and that the cells were nonflagellated (Bardy and Jarrell, 2003; Chaban *et al.*, 2007). This indicates that removal of the signal peptides is a critical step in the assembly of the flagellar filaments. In some archaea, especially methanogens, FlaK may be dedicated to only processing proteins for the flagellar system. Even pilins in *M. maripaludis*, which have very similar signal peptides to flagellins, are processed by a separate Type IV prepilin peptidase-like enzyme (Szabo *et al.*, 2007b). In *Sulfolobus*, however, it is clear that the PibD enzyme has a much wider range of substrates, including flagellins, sugar binding proteins and pilins (Albers *et al.*, 2003). All of these substrates have Class 3 signal peptides and all may form pili-like structures emanating from the cell surface (Albers *et al.*, 2006; Ng *et al.*, 2008). Archaeal Type IV prepilin peptidase-like enzymes have only been studied in two archaeal species and two very different strategies have emerged: separate enzymes within one organism for processing of subunits possessing Class 3 signal peptides but going to different structures (pili and flagella), as in *Methanococcus*, and one enzyme which processes all the Class 3 signal peptide-containing proteins, as in *Sulfolobus*. Since potential substrates containing Class 3 signal peptides appear to be widespread in archaea (Szabo *et al.*, 2007b), determination of the most common strategy will have to await investigation of these signal peptidases in other archaeal genera.

Glycosylation

A critical step in the assembly of the archaeal flagellum is the glycosylation of flagellin proteins prior to their incorporation into the structure. Glycosylation, a posttranslational modification that involves the covalent linkage of glycans to target proteins, is often critical for proper protein structure and function. This is true for the archaeal flagellar system, as an intact flagellin glycan is required for flagellar assembly and/or optimal functioning (Chaban *et al.*, 2006; VanDyke *et al.*, 2008).

Glycosylation of flagellin proteins has been demonstrated in many archaeal species mainly based on glycan detecting stains or inhibitors that affect glycosylation (Eichler and Adams, 2005) but only in *Methanococcus* and *Halobacterium* have the covalent attachment and the structures of the glycans been determined (Wieland *et al.*, 1985; Voisin *et al.*, 2005). The most compelling evidence demonstrating the importance of flagellin glycosylation comes from studies conducted in the methanogens *M. voltae and M. maripaludis*. *M. voltae* flagellins are known to possess a trisaccharide comprising *N*-acetylglucosamine, diacetylated glucuronic acid and a modified mannuronic acid (Voisin *et al.*, 2005). This glycan is attached at multiple sites on each flagellin, specifically to asparagine residues residing within an Asn-Xaa-Ser/Thr recognition sequence (where Xaa represents any amino acid except proline), thus, classifying this structure as an *N*-linked glycan. The glycan found on *M. maripaludis* flagellin proteins has been determined to be a slightly larger *N*-linked structure, a tetrasaccharide, consisting of *N*-acetylgalactosamine, di-acetylated glucuronic acid, a modified mannuronic acid, and a novel terminal sugar which is a diglycoside of aldulose (VanDyke *et al.*, 2009). The presence of *N*-linked sugars on these flagellin proteins exemplifies the uniqueness of this archaeal structure, as studies on flagellin proteins from various bacterial species have found that in the cases where glycosylation does occur, the glycans are limited to *O*-glycosidic linkages (Logan, 2006). This can most likely be attributed to the different ways by which archaea and bacteria are proposed to assemble their flagellar appendages.

N-linked glycosylation is an ordered process utilizing an array of proteins involved in sugar biosynthesis, assembly, translocation across cellular membranes and, finally, transfer of the completed glycan to the target protein. The process is well studied in eukaryotes (Weerapana and Imperiali, 2006) and bacteria (especially in *Campylobacter* (Szymanski and Wren, 2005)) and progress has been significant in recent years in archaea (Yurist-Doutsch *et al.*, 2008), especially in *Haloferax* (Abu-Qarn and Eichler, 2006; Abu-Qarn *et al.*, 2007; Abu-Qarn and Eichler, 2007; Abu-Qarn *et al.*, 2008) and *Methanococcus* (Chaban *et al.*, 2006, 2009; Shams-Eldin *et al.*, 2008; VanDyke *et al.*, 2008). Studies conducted in *M. voltae* have identified a number of components involved in flagellin *N*-glycosylation encoded by *agl* (archaeal glycosylation) genes. Flagellin glycosylation in archaea is thought to begin at the cytosolic face of the cytoplasmic membrane. A lipid molecule, dolichol phosphate [as used in eukaryotes but different from the undecaprenol lipid carrier used by bacteria (Yurist-Doutsch *et al.*, 2008)], is embedded in the membrane and serves as a carrier for glycan assembly prior to its transfer to the target protein. In *M. voltae* glycan synthesis, the trisaccharide is assembled using three specific glycosyltransferases, each catalysing the addition of one of the three monosaccharides. The additions of the *N*-acetylglucosamine, glucuronic acid and mannuronic acid derivatives are carried out by the glycosyltransferases AglH, AglC/AglK, and AglA, respectively (Chaban *et al.*, 2006, 2009; Shams-Eldin *et al.*, 2008). In the case of the glycosyltransferase believed to be involved in the transfer of the first sugar, AglH, a genetic knockout could not be generated. However, the archaeal enzyme was shown to be able to successfully complement a conditional lethal mutation in the *agl7* gene (*N*-acetylglucosamine-1-phosphate transferase) in yeast which catalyses the same reaction (Shams-Eldin *et al.*, 2008). Once the trisaccharide is completed, it is believed to be transported across the cytoplasmic membrane to the extracellular face. Although glycan transporters have been proposed for eukaryotic and bacterial *N*-linked glycosylation systems (Weerapana and Imperiali, 2006), the archaeal equivalent has yet to be identified, which could be a reflection of the unique nature of the archaeal membrane. Upon reorientation to the extracellular face of the cytoplasmic membrane, the glycan is then transferred to specific asparagine residues on the target flagellin using an oligosaccharyltransferase (AglB), an enzyme containing a specific catalytic motif (WWDXG) also found in its eukaryotic and bacterial counterparts (Chaban *et al.*, 2006). As *M. maripaludis* flagellins are known to possess a tetrasaccharide, glycan assembly is believed to require four specific glycosyltransferases. In this organism, the attachment of the second, third, and terminal sugars are carried out by the gene products of *MMP1079, MMP1080,* and *MMP1088*, respectively (VanDyke *et al.*, 2009). Since the linking sugar in *M. maripaludis* is different from that determined in the *M. voltae* glycan (*N*-acetylgalactosamine vs *N*-acetylglucosamine), the *M. maripaludis* homologue of the *M. voltae* AglH is probably not the glycosyltransfersae involved in the transfer of the first sugar and its identity remains unknown. The *M. maripaludis* AglH homologue appears to be essential as repeated attempts to delete it have failed. Its essential role may lie in the biosynthesis of cofactor B (Namboori and Graham, 2008) or perhaps in glycolipid biosynthesis.

Early reports suggested an important role for glycosylation in flagella assembly. Studies of a *H. salinarum* mutant that overexpressed flagella showed that the flagellins were shifted to a lower molecular weight compared to wild-type, suggesting that the flagellins were undermodified and that this resulted in imperfectly integrated flagella in the cell envelope and their release into the media (Wieland *et al.*, 1985). Growth of *Methanococcus deltae* in the presence of bacitracin, a known inhibitor of glycosylation, led to the appearance of lower molecular weight forms of flagellin, likely underglycosylated ones (Bayley *et al.*, 1993). Electron microscopy revealed the absence of normal flagella in those cells with the hypoglycosylated flagellins. More recently, with the advent of improved genetic tools, directed targeting of suspected glycosylation genes has been achieved. Mutational studies conducted in *M. voltae* and *M. maripaludis* have now shed some light on the necessity of this glycan for proper flagellar assembly and function. The deletion of a specific glycosyltransferase from the pathway results in the termination of the glycan chain at that step, resulting in a truncated glycan. Interestingly, the proposed membrane transporter (flippase), as well as the oligosaccharyltransferase (AglB), still recognize these partial glycan structures and attach the truncated forms to the flagellin proteins as evidenced by shifts in molecular mass detected by immunoblotting as well as by mass spectrometry analysis (Chaban *et al.*, 2006). Thus, it is possible to study the effects of partial or no flagellin glycan attachment on flagellar assembly and efficient motility. In *M. maripaludis*, an *aglB* deletion mutant, harbouring flagellins that do not possess any portion of the *N*-glycan, is not able to assemble flagella and is, not surprisingly, non-motile as evidence by both light microscopy observations and semiswarm plate analysis (VanDyke *et al.*, 2009). Identical results were observed in a *M. voltae* mutant inactivated for *aglB* (Chaban *et al.*, 2006). Deletion of the glycosyltransferase responsible for carrying out the addition of the second sugar of the *M. maripaludis* glycan (MMP1079) also resulted in nonflagellated cells. Deletions of the third and fourth glycosyltransferases (MMP1080, MMP1088) results in flagellins with two or three sugars attached, respectively. Interestingly, these mutants are still able to assemble flagella but measurements of motility by semi-swarm plates indicated that the presence of the two terminal sugars was still a requirement for optimal motility as the Δ*MMP1080* mutant was only slightly motile and Δ*MMP1088* cells had moderate motility compared to wild-type cells. Thus, it seems that attachment of a glycan composed of a minimum of two sugars is needed for flagellin assembly into flagellar filaments but attachment of the entire glycan is necessary for the flagellum to operate optimally in *M. maripaludis* (VanDyke *et al.*, 2009). Studies conducted in *M. voltae* have shown similar results with attachment of a glycan with two of the three sugars being required to assemble flagellar filaments but the complete trisaccharide being necessary for optimal motility (Chaban *et al.*, 2006).

Proposed assembly model for archaeal flagella

Several lines of evidence point strongly to a unique mechanism of assembly of archaeal flagella when compared to the bacterial counterpart. While there are still some puzzling aspects to the assembly of the bacterial flagellum that are yet to be fully understood, the basic mechanism is well known (Chevance and Hughes, 2008; Jarrell and McBride, 2008). A type III secretion system located at the base of the MS ring is responsible for delivering the component subunits for the rod, hook and filament into the hollow growing organelle. The subunits are made without signal peptides and pass through the narrow hollow centre of the growing structure. In the case of the flagellins, they assemble under the capping protein at the distal end of the structure, which can be many cell body lengths from the surface. In archaea, the flagellins appear more like Type IV pilins with their short signal peptides and the associated Type IV prepilin peptidase-like enzyme required for processing. The glycosylation so far observed on flagellins in archaea is *N*-linked, rather than the *O*-linked that is sometimes found in bacterial flagellins. Given the membrane-associated nature of the *N*-linked process and the necessity for flagellins to be accessed by the preflagellin peptidase, it would seem the logical explanation is that the flagellins are membrane embedded at some point prior to their incorporation into the filament. Theoretically, this could all occur within a secretion apparatus located in the membrane at the base of the structure but the lack of a channel in the archaeal flagellar filament large enough to accommodate the passage of the flagellin subunits indicates that a bacterial type assembly mechanism is not occurring in archaea. We have proposed for many years, based on the similarity of the archaeal flagella system to Type IV pili, that archaea flagella are assembled by incorporation of subunits at the base of the structure, in direct contrast to bacterial flagella (Jarrell *et al.*, 1996b; Jarrell and McBride, 2008).

Incorporating the latest genetic data, as well as information on posttranslational processing, a hypothetical model for the assembly of archaeal flagella, using *Methanococcus maripaludis* as the paradigm, is presented in Fig. 14.3. Flagellins are synthesized as preproteins in the cytoplasm and likely delivered by chaperones (Polosina *et al.*, 1998) to the cytoplasmic membrane. It is here that two key steps occur. One is the attachment of the *N*-linked glycan by AglB (Chaban *et al.*, 2006) and the other is the removal of the signal peptide by FlaK (Bardy and Jarrell, 2003). Either of these two steps can occur without the other and in what order the steps naturally occur in the cells is unknown. Since all archaeal flagellin systems have FlaHIJ, and FlaIJ are homologous to Type IV proteins involved in assembly of Type IV pili, it is believed the flagellin subunits are incorporated into the archaeal structure using the ATPase activity demonstrated for FlaI, with FlaJ as a membrane assembly site. FlaH, also a potential ATPase, may possibly be involved in the assembly or in the actual rotation of the flagella. Roles for the other Fla accessory proteins are unknown although all tested are required for flagella assembly (Chaban *et al.*, 2007). The modified flagellins are

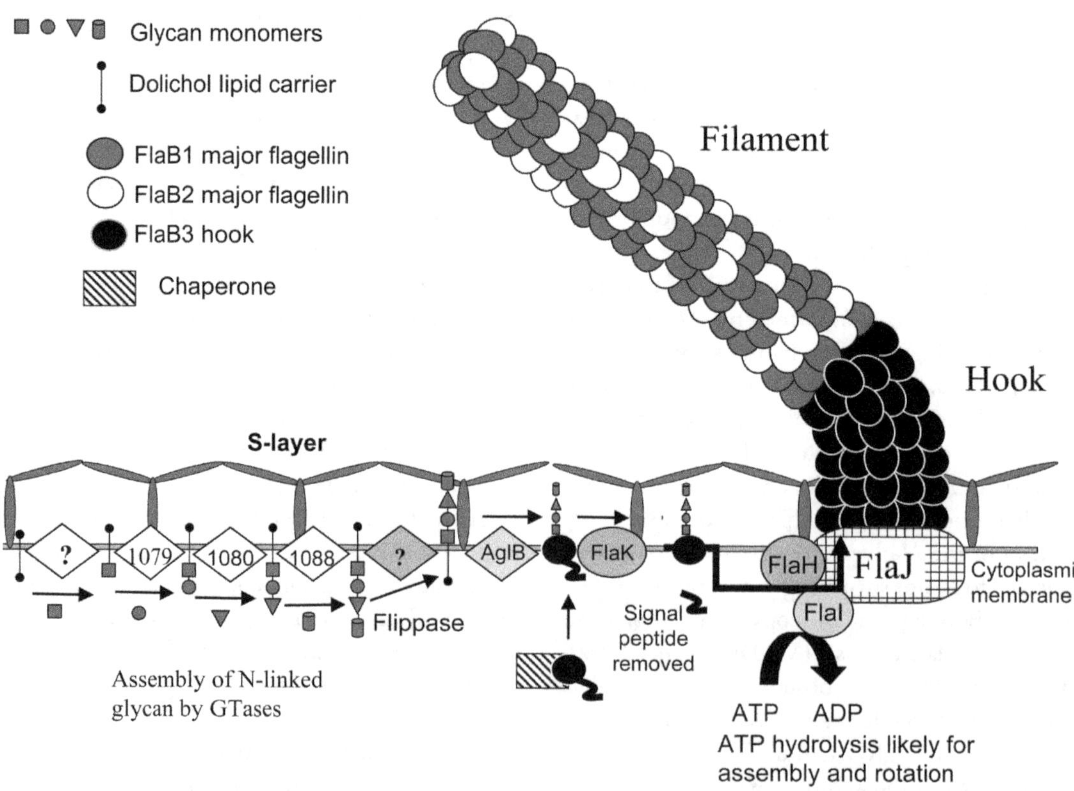

Figure 14.3 Proposed model of assembly for archaeal flagella, using *Methanococcus maripaludis* as the paradigm. See text for details.

then incorporated at the base of the structure, presumably, since their passage through the filament is not possible as it is not a hollow structure. While there seems to be no alternative given the structural data provided by Trachtenberg's group (Cohen-Krausz and Trachtenberg, 2002; Trachtenberg *et al.*, 2005: Cohen-Krausz and Trachtenberg, 2008), no polarity of growth experiments have yet been performed to show the site of flagellin incorporation. If this model of assembly is accurate, then the filament would be assembled prior to the hook, which does appear to be the case since deletion of *flaB3* leads to hookless filaments (Chaban *et al.*, 2007). The much more variable length of the archaeal hook compared to the bacterial one may be a reflection of it being the terminal step in the assembly. In *M. maripaludis* the distribution of the other two flagellins (FlaB1 and FlaB2) within the filament is not known nor whether one flagellin is preferentially needed to begin the assembly.

Comparison of archaeal flagella to bacterial flagella and Type IV pili

Prokaryotic cells have developed a wide variety of mechanisms to move in their environment (Bardy *et al.*, 2003; Jarrell and McBride, 2008). Two well studied structures involved in motility in bacteria are flagella, which can be used for swimming and swarming and Type IV pili which are used for surface translocation, called twitching. Bacterial flagella are reversible, rotating structures in which the major structural proteins, the flagellins, are translocated via a specialized type III secretion system located at the base of the flagellum. The flagellins are made without cleaved signal peptides and are sent through the hollow centre of the flagellum to incorporate at the distal end under the cap protein (Chevance and Hughes, 2008). Type IV pili mediate twitching by extension of the pili, attachment onto a surface and subsequent retraction of the pilus which drags the cell along the surface (Mattick, 2002; Burrows, 2005). The major structural subunits, the pilins, are made as prepilins, with unusual, Class 3 signal peptides which are cleaved by a specialized signal peptidase called the prepilin peptidase prior to incorporation into the structure (Strom *et al.*, 1993). The pilus does not have a central channel large enough to accommodate the passage of subunits and incorporation (for extension) and removal (for retraction) of the subunits occurs at the base (Craig and Li, 2008). Two separate ATPases are involved, one for extension and the other for retraction (Burrows, 2005; Craig and Li, 2008). A conserved membrane protein that recruits the ATPase to the membrane is also found in these systems and it is believed that the ATPases and membrane protein form a platform for pilus assembly (Craig and Li, 2008).

Thus far, motility in archaea has been limited to swimming by flagella although pili clusters, including Type IV pili clusters have been recently reported (Szabo *et al.*, 2007b). Unexpectedly, initial studies of archaeal flagella revealed that the flagellins in this domain were made as preproteins with Class 3 signal peptides that resembled those of Type IV pilins. The associated prepilin peptidase-like enzyme was also found to be necessary to remove the signal peptide. With the abundance of sequenced archaeal genomes, it became clear that the flagellin genes in archaea were universally accompanied by the neighbouring genes, *flaIJ*, which are homologues of the genes encoding an ATPase and conserved membrane protein seen in Type IV pili systems. Analysis of archaeal genomes also revealed the complete lack of any bacterial flagella-like genes, such as ring, rod and motor proteins (Faguy and Jarrell, 1999). Structural work on archaeal flagella showed the lack of a significant central channel so key to bacterial flagellum assembly. These studies have also indicated that the archaeal flagella share structural features with Type IV pili to the exclusion of bacterial flagella. All of these observations lead to the hypothesis that archaeal flagella assemble like Type IV pili, with preproteins cleaved by specialized signal peptidases with subsequent incorporation of the subunits at the base, but then function like bacterial flagella, i.e. by a reversible rotating mechanism.

A summary of the comparison of the archaeal flagella system to bacterial flagella and Type IV pili is presented in Table 14.2.

Archaeal pili

While data has been accumulating on the unusual flagella of archaea for decades, the information available for other surface appendages, especially

Table 14.2 Comparison of archaeal flagella to bacterial flagella and Type IV pili

Bacterial flagella	Archaeal flagella	Type IV pili
• Usually single flagellin with some exceptions • No sequence similarities between bacterial and archaeal flagellins	• Multiple flagellins with exception of *Sulfolobus* • Archaeal flagellins have N-terminal sequence similarity to Type IV pilins, no similarity to bacterial flagellins	• Single major pilin • Several minor pilins • N-terminal sequence similarities with archaeal flagellins
• Flagellins rarely posttranslationally modified • If glycosylated, O-linked • May require glycosylation for flagellin assembly	• Flagellins typically glycosylated • Glycosylation N-linked • Minimum amount of glycosylation required for flagellin assembly • Full glycan needed for optimal activity	• Pilins may be glycosylated • Glycosylation O-linked
• Flagellins do not have cleaved N-terminal signal peptides	• Preflagellins have cleaved N-terminal signal peptides • Cleaved by FlaK/PibB of the Type IV prepilin peptidase family	• Prepilins have cleaved N-terminal signal peptides • Cleaved by prepilin peptidase
• Rod and ring containing basal body • No homologues of known archaeal *fla* genes	• Knob-like anchoring structure • May have sub-cytoplasmic membrane anchoring site • No ring, rod, stator or motor gene homologues	• No anchoring structure observed
• No homologues to Type IV pili genes in bacterial flagella system	• FlaI and FlaJ are homologues to Type IV pili ATPase and membrane protein	• Homologues to archaeal flagella ATPase and membrane protein
• Hook region • Well-defined hook region length of about 55 nm	• Curved hook region composed of one 'flagellin' • Great variability in hook length (72–265 nm).	• No hook
• ~20 nm diameter • 2-nm channel allowing for passage of flagellin subunits	• ~10–13 nm diameter • Channel too narrow for subunits to pass through	• ~5–7 nm diameter • Channel too narrow for proteins to pass through
• Swimming and swarming motility	• Swimming and swarming motility	• Surface twitching motility
• Rotating filament • Switching of rotation • Rotation driven by pmf	• Rotating filament • Switching of rotation • rotation driven by ATP • Motor unidentified	• Movement by pilus extension/retraction through subunit addition/removal at base
• Growth at distal end • type 3 secretion system	• Presumed growth at proximal end (base)	• Growth at proximal end (base)
• Associated chemotaxis system • CheY-P interacts with FliM	• Associated bacterial-like chemotaxis system • CheY-P but no FliM	• Often associated chemotaxis system

pili, is scant. Pili were reported on archaeal cells via electron microscopy observations as early as the 1970's (Weiss, 1973; Jones *et al.*, 1977; Doddema *et al.*, 1979). Pili, in bacteria, are categorized into different classes (for example Type I, Type IV, sex pili, curli and Gram-positive pili) where function, structure and mode of assembly differ (Chapman, 2006; Babic *et al.*, 2008; Barnhart and Chapman, 2006; Craig and Li, 2008; Mandlik *et al.*, 2008; Remaut *et al.*, 2008). With study of archaeal pili in its infancy, information regarding function, structure and the genetic locus involved is still limited; however, several research groups have begun to focus on these extracellular appendages in order to answer these questions. Specifically, data is now appearing on various aspects of pili in *Methanococcus*, *Sulfolobus* and *Methanothermobacter*.

In the cases of *Methanococcus* and *Sulfolobus* there is clear evidence of a Type IV pilus-like locus (Albers and Driessen, 2005; Froels *et al.*, 2007; Ng *et al.*, 2008). There are 2–3 pilin genes encoding proteins with Class 3 signal peptides that are cleaved by a Type IV prepilin peptidase-like enzyme leaving mature proteins with a hydrophobic N-terminus (Szabo *et al.*, 2007b). The pilins are somewhat smaller than Type IV pilins in bacteria, at about 130 amino acids (without the signal peptide). In the case of two of the three putative pilins of *M. maripaludis*, there is a +5 glutamic acid that is also almost universally found in bacterial Type IV pilins, however, these pilins lack the two cysteines that are also highly conserved amongst Type IV pilins.

The study on *M. maripaludis* pilins and pilin processing is particularly interesting (Szabo *et al.*, 2007b). Although flagellins and pilins have similar signal peptides, the pilin locus has a prepilin peptidase-like enzyme EppA, in addition to the prepilin peptidase-like enzyme involved in flagellin processing (FlaK) located elsewhere on the chromosome. The preflagellin peptidase in *M. maripaludis* was found to be specific in cleaving only the flagellin proteins while EppA was specific for cleavage of the signal peptides from the pilin-like proteins. Swapping four amino acid regions surrounding the cleavage site between flagellins and pilins led to remarkable results. While EppA could not process flagellins, replacing the –2 to +2 amino acids of a flagellin with those of a pilin led to processing by EppA. FlaK was also able to process this flagellin-pilin hybrid. It was suggested that the key requirement needed for processing by EppA was a glutamine at position +1. In the comparable experiment, where the –2 to +2 of pilins was replaced with those of a flagellin, neither EppA nor FlaK was able to cleave the hybrid molecule. This was consistent with the need for a +1 glutamine for EppA cleavage, since the +1 position in the flagellin is alanine. Unlike the case with EppA, FlaK cleavage was not dependent solely on the amino acids within the –2 to +2 area around the cleavage site. However, previous studies of the cleavage requirements for FlaK in *M. voltae* had already determined that the absolutely conserved +3 glycine found in all archaeal flagellins was necessary (Thomas *et al.*, 2001b). This position in the pilins is not glycine, but serine. It may very well be that FlaK, which can accommodate a variety of different amino acids at the +1 and +2 positions, could cleave the prepilins if the prepilins had a single amino acid change to make the +3 position a glycine. This is clearly a different situation than in *Sulfolobus* where a single enzyme, PibD, cleaves signal peptides from all Class 3 signal peptide-containing proteins (flagellins, pilins and sugar binding proteins) (Ng *et al.* 2008).

Pili isolated from *M. maripaludis* were shown to be composed primarily of a single protein species with a molecular mass of approximately 17 kDa. Furthermore, the initial predictions of these genes as being involved in the formation of the pili observed on *M. maripaludis* cells was confirmed when in-frame deletions of two of the putative pilins lead to completely nonpiliated cells (Fig. 14.4). Pili could be restored to the cell surface of the mutants when the deleted pilin gene was complemented back on a plasmid (Ng *et al.*, 2008).

Considering the similarity of the *M. maripaludis* pilin genes to those encoding bacterial Type IV pilins, including their processing by a Type IV prepilin peptidase-like enzyme, it was surprising when electron cryo-microscopy of purified *M. maripaludis* pili revealed a structure that does not look like Type IV pili of bacteria or indeed like any bacterial pilus (Wang *et al.*, 2008). The archaeal pili are heterogeneous with respect to their helical symmetry, with two different packing arrangements observed, remarkably, within the same filaments. It is not yet determined whether the two subunit packing formations identified are due to the presence of two different major pilin proteins, most likely MMP0236 and MMP0237. Interestingly, whereas archaeal flagella lack a central channel, the archaeal pilus was found to possess a narrow hollow channel of similar diameter to that seen for *Salmonella* flagella (Yonekura *et al.*, 2003). Thus, paradoxically, archaeal pili could

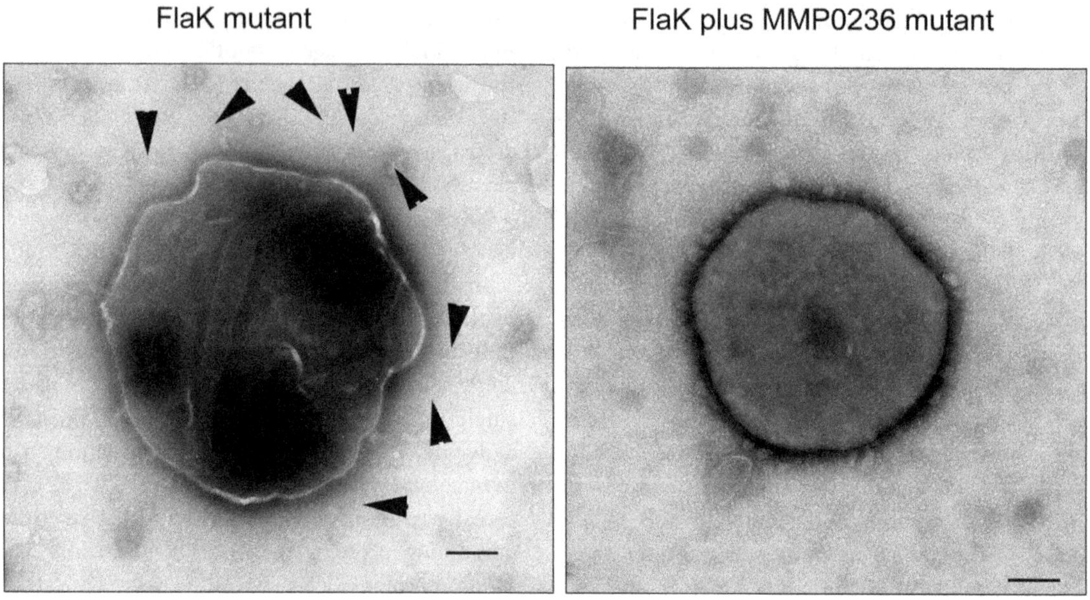

Figure 14.4 Electron micrograph of *flaK* deletion of *Methanococcus maripaludis* showing several pili (arrows) but no flagella and a double mutant with a deletion of both *flaK* and a pilin gene *MMP0236*. The double mutant is lacking both flagella and pili. Cells stained with 2% phosphotungstic acid (pH 7.0). Courtesy of S.I. Aizawa. Mutant generated by Sandy Ng.

share a mode of assembly similar to bacterial flagella while archaeal flagella seem to assemble in a similar fashion to bacterial Type IV pili.

Recently, a gene affecting both flagella and pili has been identified in *M. maripaludis* (VanDyke *et al.*, 2008). The gene, MMP0350, annotated as an acetyltransferase, is responsible for a step in the biosynthesis of sugars comprising the *N*-linked glycan attached to the flagellin proteins. Not unexpectedly, this glycosylation mutant was nonflagellated, but surprisingly, electron microscopy also revealed that these mutants were mainly nonpiliated. Electron micrographs of the mutant culture showed that while the cells were underpiliated compared to a culture with an intact MMP0350 gene, pili filaments were still assembled but they did not remain attached to the cell surface. Instead, the pili were found in the culture supernatant. This strongly suggested that the problem with the pili was distinct from the effect of the deletion on flagella. While this deletion likely resulted in underglycosylated flagellins that could not be assembled into filaments, the pili clearly could be assembled but not anchored to the cell. It is currently unknown whether the pilins themselves are glycosylated or not and if so, whether the glycan is *O*- or *N*-linked. It may be that the major pilin subunits are unaffected by the presence or absence of glycan but that a minor anchoring pilin subunit needs a glycosylation modification to function properly.

Reports of a pilus operon were also presented in *S. solfataricus*, where exposure to UV light induced cellular aggregation via pili formation (Froels *et al.*, 2007; Ng *et al.*, 2008). Mutants missing a putative pili secretion ATPase failed to aggregate after UV exposure while overexpression of suspected pilin genes in non-piliated mutants led to long, irregular pili (Fröls *et al.*, 2008). Since enhanced genetic exchange also occurs upon UV exposure, it has been proposed that the pili-induced aggregation may be responsible for increased DNA exchange via conjugation.

Also among the recent reports on archaeal pili is the characterization of pili from *Methanothermobacter thermoautotrophicus*. The purified pili were shown to consist of one major protein, a 16 kDa glycoprotein encoded by the *mth60* gene (Thoma *et al.*, 2008). Antibodies to the pilus subunit expressed in *E. coli* were able to detach adhering cells from various surfaces. In addition, when cells were grown on solid surfaces, they produced many more pili than cells in liquid culture. This is the first evidence showing a function for archaeal pili, namely their use as adhesions. Unfortunately, analyses of these interesting structures will be limited by the lack of a tractable genetic system in this organism.

Discussion and future trends

The study of archaeal flagella and pili has lead to novel findings concerning how prokaryotes can assemble structures in different ways and yet still perform the same function.

What is known about archaeal flagella raises many questions. What is the nature of the motor and the actual driving force behind rotation? It appears to be unique to archaea. If archaeal flagella actually do assemble at the base, how do they regenerate new flagella that are sheared off? Without murein in archaea, how is the flagellum anchored to the envelope? Since a 'flagellin' appears to form the hook in *Halobacterium* and *Methanococcus*, what does *Sulfolobus* do for its hook since it is an exception with only a single flagellin? Is there even conclusive evidence for hook structures in *Sulfolobus*? What is distinctive about the FlaB3 flagellin, compared to the FlaB1 and FlaB2 flagellins in *Methanococcus* that allows it to play the role of hook protein? What is the role of the submembraneous presumed anchoring structure for flagella observed in several archaea? How does the sensory system, with classical bacterial components including CheY, interact with the flagellum switch complex which lacks a FliM homologue?

As for the investigations of pili in archaea, the data is so scarce that the questions greatly outnumber the data accumulated. Of particular fascination is whether there are multiple types of pili with distinctive roles in archaea, as are found in bacteria. What might the range of functions be and could there be roles for pili in archaea, especially the extremophilic archaea, not seen in bacteria? Why would *Methanococcus* utilize two distinctive Type IV prepilin peptidase-like enzymes, one for flagellins and one for pilins whereas *Sulfolobus* employs a single enzyme? Is this some form of regulatory control?

No doubt future investigations will focus on these and other aspects of these fascinating organelles.

Acknowledgements

Research in the Jarrell lab has been supported by the Natural Sciences and Engineering Research Council of Canada. We are grateful to all the previous graduate students and collaborators who have contributed significantly to the work from the Jarrell laboratory presented here.

References

Abu-Qarn, M., and Eichler, J. (2007). An analysis of amino acid sequences surrounding archaeal glycoprotein sequons. Archaea *2*, 73–81.

Abu-Qarn, M., and Eichler, J. (2006). Protein N-glycosylation in Archaea: defining *Haloferax volcanii* genes involved in S-layer glycoprotein glycosylation. Mol. Microbiol. *61*, 511–525.

Abu-Qarn, M., Giordano, A., Battaglia, F., Trauner, A., Hitchen, P.G., Morris, H.R., Dell, A., and Eichler, J. (2008). Identification of AglE, a second glycosyltransferase involved in N-glycosylation of the *Haloferax volcanii* S-layer glycoprotein. J. Bacteriol. *190*, 3140–3146.

Abu-Qarn, M., Yurist-Doutsch, S., Giordano, A., Trauner, A., Morris, H.R., Hitchen, P., Medalia, O., Dell, A., and Eichler, J. (2007). *Haloferax volcanii* AglB and AglD are involved in N-glycosylation of the S-layer glycoprotein and proper assembly of the surface layer. J. Mol. Biol. *374*, 1224–1236.

Alam, M., and Oesterhelt, D. (1984). Morphology, function and isolation of halobacterial flagella. J. Mol. Biol. *176*, 459–475.

Albers, S.V., and Driessen, A.J.M. (2005). Analysis of ATPases of putative secretion operons in the thermoacidophilic archaeon *Sulfolobus solfataricus*. Microbiology *151*, 763–773.

Albers, S.V., Szabo, Z., and Driessen, A.J.M. (2006). Protein secretion in the Archaea: multiple paths towards a unique cell surface. Nat. Rev. Microbiol. *4*, 537–547.

Albers, S.V., Szabo, Z., and Driessen, A.J.M. (2003). Archaeal homologue of bacterial Type IV prepilin signal peptidases with broad substrate specificity. J. Bacteriol. *185*, 3918–3925.

Asakura, S., Eguchi, G., and Iino, T. (1966). *Salmonella* flagella: *in vitro* reconstruction and over-all shapes of flagellar filaments. J. Mol. Biol. *16*, 302–316.

Babic, A., Lindner, A.B., Vulic, M., Stewart, E.J., and Radman, M. (2008). Direct visualization of horizontal gene transfer. Science *319*, 1533–1536.

Baker, M.D., Wolanin, P.M., and Stock, J.B. (2006). Systems biology of bacterial chemotaxis. Curr. Opin. Microbiol. *9*, 187–192.

Bardy, S.L., and Jarrell, K.F. (2003). Cleavage of preflagellins by an aspartic acid signal peptidase is essential for flagellation in the archaeon *Methanococcus voltae*. Mol. Microbiol. *50*, 1339–1347.

Bardy, S.L., and Jarrell, K.F. (2002). FlaK of the archaeon *Methanococcus maripaludis* possesses preflagellin peptidase activity. FEMS Microbiol. Lett. *208*, 53–59.

Bardy, S.L., Mori, T., Komoriya, K., Aizawa, S., and Jarrell, K.F. (2002). Identification and localization of flagellins FlaA and FlaB3 within flagella of *Methanococcus voltae*. J. Bacteriol. *184*, 5223–5233.

Bardy, S.L., Ng, S.Y., and Jarrell, K.F. (2004). Recent advances in the structure and assembly of the archaeal flagellum. J. Mol. Microbiol. Biotechnol. *7*, 41–51.

Bardy, S.L., Ng, S.Y., and Jarrell, K.F. (2003). Prokaryotic motility structures. Microbiology *149*, 295–304.

Barnhart, M.M., and Chapman, M.R. (2006). Curli biogenesis and function. Annu. Rev. Microbiol. *60*, 131–147.

Bayley, D.P., and Jarrell, K.F. (1999). Overexpression of *Methanococcus voltae* flagellin subunits in *Escherichia coli* and *Pseudomonas aeruginosa*: a source of archaeal preflagellin. J. Bacteriol. *181*, 4146–4153.

Bayley, D.P., and Jarrell, K.F. (1998). Further evidence to suggest that archaeal flagella are related to bacterial Type IV pili. J. Mol. Evol. *46*, 370–373.

Bayley, D.P., Kalmokoff, M.L., and Jarrell, K.F. (1993). Effect of bacitracin on flagellar assembly and presumed glycosylation of the flagellins of *Methanococcus deltae*. Arch. Microbiol. *160*, 179–185.

Berkner, S., Grogan, D., Albers, S.V., and Lipps, G. (2007). Small multicopy, non-integrative shuttle vectors based on the plasmid pRN1 for *Sulfolobus acidocaldarius* and *Sulfolobus solfataricus*, model organisms of the (cren-) archaea. Nucleic Acids Res. *35*, e88.

Berkner, S., and Lipps, G. (2008). Genetic tools for *Sulfolobus* spp.: vectors and first applications. Arch. Microbiol. *190*, 217–230.

Beznosov, S.N., Pyatibratov, M.G., and Fedorov, O.V. (2007). On the multicomponent nature of *Halobacterium salinarum* flagella. Microbiology (Russ.) *76*, 435–441.

Black, F.T., Freundt, E.A., Vinther, O., and Christiansen, C. (1979). Flagellation and swimming motility of *Thermoplasma acidophilum*. J. Bacteriol. *137*, 456–460.

Burrows, L.L. (2005). Weapons of mass retraction. Mol. Microbiol. *57*, 878–888.

Chaban, B., Logan, S.M., Kelly, J. and Jarrell, K.F. (2009). AglC and AglK are involved in biosynthesis and attachment of diacetylated glucuronic acid to the N-glycan in *Methanococcus voltae*. J. Bacteriol. *191*, 187–195.

Chaban, B., Ng, S.Y., Kanbe, M., Saltzman, I., Nimmo, G., Aizawa, S.I., and Jarrell, K.F. (2007). Systematic deletion analyses of the *fla* genes in the flagella operon identify several genes essential for proper assembly and function of flagella in the archaeon, *Methanococcus maripaludis*. Mol. Microbiol. *66*, 596–609.

Chaban, B., Voisin, S., Kelly, J., Logan, S.M., and Jarrell, K.F. (2006). Identification of genes involved in the biosynthesis and attachment of *Methanococcus voltae* N-linked glycans: insight into N-linked glycosylation pathways in Archaea. Mol. Microbiol. *61*, 259–268.

Chevance, F.F., and Hughes, K.T. (2008). Coordinating assembly of a bacterial macromolecular machine. Nat. Rev. Microbiol. *6*, 455–465.

Cohen-Krausz, S., and Trachtenberg, S. (2008). The flagellar filament structure of the extreme acidothermophile *Sulfolobus shibatae* B12 suggests that archaeabacterial flagella have a unique and common symmetry and design. J. Mol. Biol. *375*, 1113–1124.

Cohen-Krausz, S., and Trachtenberg, S. (2002). The structure of the archeabacterial flagellar filament of the extreme halophile *Halobacterium salinarum* R1M1 and its relation to eubacterial flagellar filaments and Type IV pili. J. Mol. Biol. *321*, 383–395.

Correia, J.D., and Jarrell, K.F. (2000). Posttranslational processing of *Methanococcus voltae* preflagellin by preflagellin peptidases of *M. voltae* and other methanogens. J. Bacteriol. *182*, 855–858.

Craig, L., and Li, J. (2008). Type IV pili: paradoxes in form and function. Curr. Opin. Struct. Biol. *18*, 267–277.

Cruden, D., Sparling, R., and Markovetz, A.J. (1989). Isolation and ultrastructure of the flagella of *Methanococcus thermolithotrophicus* and *Methanospirillum hungatei*. Appl. Environ. Microbiol. *55*, 1414–1419.

Desmond, E., Brochier-Armanet, C., and Gribaldo, S. (2007). Phylogenomics of the archaeal flagellum: rare horizontal gene transfer in a unique motility structure. BMC Evol. Biol. *7*, 106.

Doddema, H.J., Derksen, J.W.M., and Vogels, G.D. (1979). Fimbriae and flagella of methanogenic bacteria. FEMS Microbiol. Lett. *5*, 135–138.

Eichler, J., and Adams, M.W. (2005). Posttranslational protein modification in Archaea. Microbiol. Mol. Biol. Rev. *69*, 393–425.

Faguy, D.M., and Jarrell, K.F. (1999). A twisted tale: the origin and evolution of motility and chemotaxis in prokaryotes. Microbiology *145*, 279–281.

Faguy, D.M., Jarrell, K.F., Kuzio, J., and Kalmokoff, M.L. (1994). Molecular analysis of archael flagellins: similarity to the Type IV pilin-transport superfamily widespread in bacteria. Can. J. Microbiol. *40*, 67–71.

Fröls, S., Gordon, P.M., Panlilio, M.A., Duggin, I.G., Bell, S.D., Sensen, C.W., and Schleper, C. (2007). Response of the hyperthermophilic archaeon *Sulfolobus solfataricus* to UV damage. J. Bacteriol. *189*, 8708–8718.

Fröls, S., Ajon, M., Wagner, M., Teichmann, D., Zolghadr, B., Folea, M., Boekema, E.J., Driessen, A.J.M., Schleper, C. and Albers, S.V. (2008). UV-inducible cellular aggregation of the hyperthermophilic archaeon *Sulfolobus solfataricus* is mediated by pili formation. Mol. Microbiol. *70*, 938–952.

Gerl, L., and Sumper, M. (1988). Halobacterial flagellins are encoded by a multigene family. Characterization of five flagellin genes. J. Biol. Chem. *263*, 13246–13251.

Hirano, T., Yamaguchi, S., Oosawa, K., and Aizawa, S. (1994). Roles of FliK and FlhB in determination of flagellar hook length in *Salmonella typhimurium*. J. Bacteriol. *176*, 5439–5449.

Jarrell, K.F., Bayley, D.P., Florian, V., and Klein, A. (1996a). Isolation and characterization of insertional mutations in flagellin genes in the archaeon *Methanococcus voltae*. Mol. Microbiol. *20*, 657–666.

Jarrell, K.F., Bayley, D.P., and Kostyukova, A.S. (1996b). The archaeal flagellum: a unique motility structure. J. Bacteriol. *178*, 5057–5064.

Jarrell, K.F., and McBride, M.J. (2008). The surprisingly diverse ways that prokaryotes move. Nat. Rev. Microbiol. *6*, 466–476.

Jarrell, K.F., Ng, S.Y., and Chaban, B. (2007). Flagellation and chemotaxis. In Archaea: molecular and cellular biology, Cavicchioli, R. ed., (Washington, D.C.: ASM Press) pp. 385–410.

Jones, C.J., and Aizawa, S. (1991). The bacterial flagellum and flagellar motor: structure, assembly and function. Adv. Microb. Physiol. *32,* 109–172.

Jones, J.B., Bowers, B., and Stadtman, T.C. (1977). *Methanococcus vannielii:* ultrastructure and sensitivity to detergents and antibiotics. J. Bacteriol. *130,* 1357–1363.

Kalmokoff, M.L., and Jarrell, K.F. (1991). Cloning and sequencing of a multigene family encoding the flagellins of *Methanococcus voltae*. J. Bacteriol. *173,* 7113–7125.

Kalmokoff, M.L., Jarrell, K.F., and Koval, S.F. (1988). Isolation of flagella from the archaebacterium *Methanococcus voltae* by phase separation with Triton X-114. J. Bacteriol. *170,* 1752–1758.

Kelly, J., Logan, S.L., Jarrell, K.F., VanDyke, D.J. and Vinogradov, E. (2009). A novel N-linked flagellar glycan from *Methanococcus maripaludis*. Carbohydr. Res. *344,* 648-653.

Kupper, J., Marwan, W., Typke, D., Grunberg, H., Uwer, U., Gluch, M., and Oesterhelt, D. (1994). The flagellar bundle of *Halobacterium salinarium* is inserted into a distinct polar cap structure. J. Bacteriol. *176,* 5184–5187.

Logan, S.M. (2006). Flagellar glycosylation – a new component of the motility repertoire? Microbiology *152,* 1249–1262.

Macnab, R.M. (2004). Type III flagellar protein export and flagellar assembly. Biochim. Biophys. Acta *1694,* 207–217.

Macnab, R.M. (2003). How bacteria assemble flagella. Annu. Rev. Microbiol. *57,* 77–100.

Mandlik, A., Swierczynski, A., Das, A., and Ton-That, H. (2008). Pili in Gram-positive bacteria: assembly, involvement in colonization and biofilm development. Trends Microbiol. *16,* 33–40.

Marwan, W., Alam, M., and Oesterhelt, D. (1991). Rotation and switching of the flagellar motor assembly in *Halobacterium halobium*. J. Bacteriol. *173,* 1971–1977.

Mattick, J.S. (2002). Type IV pili and twitching motility. Annu. Rev. Microbiol. *56,* 289–314.

Metlina, A.L. (2004). Bacterial and archaeal flagella as prokaryotic motility organelles. Biochemistry (Mosc) *69,* 1203–1212.

Moens, S., and Vanderleyden, J. (1996). Functions of bacterial flagella. Crit. Rev. Microbiol. *22,* 67–100.

Moore, B.C., and Leigh, J.A. (2005). Markerless mutagenesis in *Methanococcus maripaludis* demonstrates roles for alanine dehydrogenase, alanine racemase, and alanine permease. J. Bacteriol. *187,* 972–979.

Nagahisa, K., Ezaki, S., Fujiwara, S., Imanaka, T., and Takagi, M. (1999). Sequence and transcriptional studies of five clustered flagellin genes from hyperthermophilic archaeon *Pyrococcus kodakaraensis* KOD1. FEMS Microbiol. Lett. *178,* 183–190.

Namboori, S.C., and Graham, D.E. (2008). Acetamido sugar biosynthesis in the Euryarchaea. J. Bacteriol. *190,* 2987–2996.

Nather, D.J., Rachel, R., Wanner, G., and Wirth, R. (2006). Flagella of *Pyrococcus furiosus*: multifunctional organelles, made for swimming, adhesion to various surfaces, and cell-cell contacts. J. Bacteriol. *188,* 6915–6923.

Ng, S.Y.M., Zolghadr, B., Driessen, A.J.M., Albers, S.V., and Jarrell, K.F. (2008). Cell surface structures of Archaea. J. Bacteriol. *190,* 6039–6047.

Ng, S.Y., Chaban, B., and Jarrell, K.F. (2006). Archaeal flagella, bacterial flagella and Type IV pili: a comparison of genes and posttranslational modifications. J. Mol. Microbiol. Biotechnol. *11,* 167–191.

Ng, S.Y., Chaban, B., VanDyke, D.J., and Jarrell, K.F. (2007). Archaeal signal peptidases. Microbiology *153,* 305–314.

Nutsch, T., Marwan, W., Oesterhelt, D., and Gilles, E.D. (2003). Signal processing and flagellar motor switching during phototaxis of *Halobacterium salinarum*. Genome Res. *13,* 2406–2412.

Nutsch, T., Oesterhelt, D., Gilles, E.D., and Marwan, W. (2005). A quantitative model of the switch cycle of an archaeal flagellar motor and its sensory control. Biophys. J. *89,* 2307–2323.

Patenge, N., Berendes, A., Engelhardt, H., Schuster, S.C., and Oesterhelt, D. (2001). The *fla* gene cluster is involved in the biogenesis of flagella in *Halobacterium salinarum*. Mol. Microbiol. *41,* 653–663.

Peabody, C.R., Chung, Y.J., Yen, M.R., Vidal-Ingigliardi, D., Pugsley, A.P., and Saier, M.H.,Jr. (2003). Type II protein secretion and its relationship to bacterial Type IV pili and archaeal flagella. Microbiology *149,* 3051–3072.

Pfeifer, F., Gregor, D., Hofacker, A., Plosser, P., and Zimmermann, P. (2002). Regulation of gas vesicle formation in halophilic archaea. J. Mol. Microbiol. Biotechnol. *4,* 175–181.

Pohlschroder, M., Gimenez, M.I., and Jarrell, K.F. (2005). Protein transport in Archaea: Sec and twin arginine translocation pathways. Curr. Opin. Microbiol. *8,* 713–719.

Polosina, Y.Y., Jarrell, K.F., Fedorov, O.V., and Kostyukova, A.S. (1998). Nucleoside diphosphate kinase from haloalkaliphilic archaeon *Natronobacterium magadii*: purification and characterization. Extremophiles *2,* 333–338.

Remaut, H., Tang, C., Henderson, N.S., Pinkner, J.S., Wang, T., Hultgren, S.J., Thanassi, D.G., Waksman, G., and Li, H. (2008). Fiber formation across the bacterial outer membrane by the chaperone/usher pathway. Cell *133,* 640–652.

Rudolph, J., Nordmann, B., Storch, K.F., Gruenberg, H., Rodewald, K., and Oesterhelt, D. (1996). A family of halobacterial transducer proteins. FEMS Microbiol. Lett. *139,* 161–168.

Rudolph, J., and Oesterhelt, D. (1996). Deletion analysis of the *che* operon in the archaeon *Halobacterium salinarium*. J. Mol. Biol. *258,* 548–554.

Rudolph, J., Tolliday, N., Schmitt, C., Schuster, S.C., and Oesterhelt, D. (1995). Phosphorylation in halobacterial signal transduction. EMBO J. *14,* 4249–4257.

Schopf, S., Wanner, G., Rachel, R., and Wirth, R. (2008). An archaeal bi-species biofilm formed by *Pyrococcus furiosus* and *Methanopyrus kandleri*. Arch. Microbiol. *190*, 371–377.

Serganova, I., Ksenzenko, V., Serganov, A., Meshcheryakova, I., Pyatibratov, M., Vakhrusheva, O., Metlina, A., and Fedorov, O. (2002). Sequencing of flagellin genes from *Natrialba magadii* provides new insight into evolutionary aspects of archaeal flagellins. J. Bacteriol. *184*, 318–322.

Shams-Eldin, H., Chaban, B., Niehus, S., Schwarz, R.T., and Jarrell, K.F. (2008). Identification of the archaeal *alg7* gene homologue (encoding N-acetylglucosamine-1-phosphate transferase) of the N-linked glycosylation system by cross-domain complementation in *Saccharomyces cerevisiae*. J. Bacteriol. *190*, 2217–2220.

Soppa, J. (2006). From genomes to function: haloarchaea as model organisms. Microbiology *152*, 585–590.

Southam, G., Kalmokoff, M.L., Jarrell, K.F., Koval, S.F., and Beveridge, T.J. (1990). Isolation, characterization, and cellular insertion of the flagella from two strains of the archaebacterium *Methanospirillum hungatei*. J. Bacteriol. *172*, 3221–3228.

Strom, M.S., and Lory, S. (1991). Amino acid substitutions in pilin of *Pseudomonas aeruginosa*. Effect on leader peptide cleavage, amino-terminal methylation, and pilus assembly. J. Biol. Chem. *266*, 1656–1664.

Strom, M.S., Nunn, D.N., and Lory, S. (1993). A single bifunctional enzyme, PilD, catalyzes cleavage and N-methylation of proteins belonging to the Type IV pilin family. Proc. Natl. Acad. Sci. U.S. A. *90*, 2404–2408.

Sumper, M. (1987). Halobacterial glycoprotein biosynthesis. Biochim. Biophys. Acta *906*, 69–79.

Szabo, Z., Albers, S.V., and Driessen, A.J.M. (2006). Active-site residues in the Type IV prepilin peptidase homologue PibD from the archaeon *Sulfolobus solfataricus*. J. Bacteriol. *188*, 1437–1443.

Szabo, Z., Sani, M., Groeneveld, M., Zolghadr, B., Schelert, J., Albers, S.V., Blum, P., Boekema, E.J., and Driessen, A.J.M. (2007a). Flagellar motility and structure in the hyperthermoacidophilic archaeon *Sulfolobus solfataricus*. J. Bacteriol. *189*, 4305–4309.

Szabo, Z., Stahl, A.O., Albers, S.V., Kissinger, J.C., Driessen, A.J.M., and Pohlschroder, M. (2007b). Identification of diverse archaeal proteins with class III signal peptides cleaved by distinct archaeal prepilin peptidases. J. Bacteriol. *189*, 772–778.

Szymanski, C.M., and Wren, B.W. (2005). Protein glycosylation in bacterial mucosal pathogens. Nat. Rev. Microbiol. *3*, 225–237.

Tarasov, V.Y., Pyatibratov, M.G., Tang, S.L., Dyall-Smith, M., and Fedorov, O.V. (2000). Role of flagellins from A and B loci in flagella formation of *Halobacterium salinarum*. Mol. Microbiol. *35*, 69–78.

Thoma, C., Frank, M., Rachel, R., Schmid, S., Nather, D., Wanner, G., and Wirth, R. (2008). The Mth60-fimbriae of *Methanothermobacter thermoautotrophicus* are functional adhesins. Environ. Microbiol. *10*, 2785–2795.

Thomas, N.A., Bardy, S.L., and Jarrell, K.F. (2001a). The archaeal flagellum: a different kind of prokaryotic motility structure. FEMS Microbiol. Rev. *25*, 147–174.

Thomas, N.A., Chao, E.D., and Jarrell, K.F. (2001b). Identification of amino acids in the leader peptide of *Methanococcus voltae* preflagellin that are important in posttranslational processing. Arch. Microbiol. *175*, 263–269.

Thomas, N.A., Mueller, S., Klein, A., and Jarrell, K.F. (2002). Mutants in *flaI* and *flaJ* of the archaeon *Methanococcus voltae* are deficient in flagellum assembly. Mol. Microbiol. *46*, 879–887.

Trachtenberg, S., and Cohen-Krausz, S. (2006). The archaeabacterial flagellar filament: a bacterial propeller with a pilus-like structure. J. Mol. Microbiol. Biotechnol. *11*, 208–220.

Trachtenberg, S., Galkin, V.E., and Egelman, E.H. (2005). Refining the structure of the *Halobacterium salinarum* flagellar filament using the iterative helical real space reconstruction method: insights into polymorphism. J. Mol. Biol. *346*, 665–676.

VanDyke, D.J., Wu, J., Logan, S.L., Kelly, J.F., Mizuno, S., Aizawa, S.-I. and Jarrell, K.F. (2009). Identification of genes involved in the assembly and attachment of a novel N-linked tetrasaccharide important for motility in the archaeon *Methanococcus maripaludis*. Mol. Microbiol. DOI:10.1111/j.1365-2958.2009.06671.x

VanDyke, D.J., Wu, J., Ng, S.Y., Kanbe, M., Chaban, B., Aizawa, S.I., and Jarrell, K.F. (2008). Identification of putative acetyltransferase gene, MMP0350, which affects proper assembly of both flagella and pili in the archaeon *Methanococcus maripaludis*. J. Bacteriol. *190*, 5300–5307.

Voisin, S., Houliston, R.S., Kelly, J., Brisson, J.R., Watson, D., Bardy, S.L., Jarrell, K.F., and Logan, S.M. (2005). Identification and characterization of the unique N-linked glycan common to the flagellins and S-layer glycoprotein of *Methanococcus voltae*. J. Biol. Chem. *280*, 16586–16593.

Wang, Y.A., Yu, X., Ng, S.Y.M., Jarrell, K.F., and Egelman, E.H. (2008). The structure of an archaeal pilus. J. Mol. Biol. *381*, 456–466.

Weerapana, E., and Imperiali, B. (2006). Asparagine-linked protein glycosylation: from eukaryotic to prokaryotic systems. Glycobiology *16*, 91R-101R.

Weiss, R.L. (1973). Attachment of bacteria to sulfur in extreme environments. J. Gen. Microbiol. *77*, 501–507.

Wieland, F., Paul, G., and Sumper, M. (1985). Halobacterial flagellins are sulfated glycoproteins. J. Biol. Chem. *260*, 15180–15185.

Yamagata, A., and Tainer, J.A. (2007). Hexameric structures of the archaeal secretion ATPase GspE and implications for a universal secretion mechanism. EMBO J. *26*, 878–890.

Yonekura, K., Maki-Yonekura, S., and Namba, K. (2003). Complete atomic model of the bacterial flagellar filament by electron cryomicroscopy. Nature *424*, 643–650.

Yonekura, K., Maki-Yonekura, S., and Namba, K. (2002). Growth mechanism of the bacterial flagellar filament. Res. Microbiol. *153*, 191–197.

Yurist-Doutsch, S., Chaban, B., VanDyke, D.J., Jarrell, K.F., and Eichler, J. (2008). Sweet to the extreme: protein glycosylation in Archaea. Mol. Microbiol. *68*, 1079–1084.

Index

F

G

H

I

L

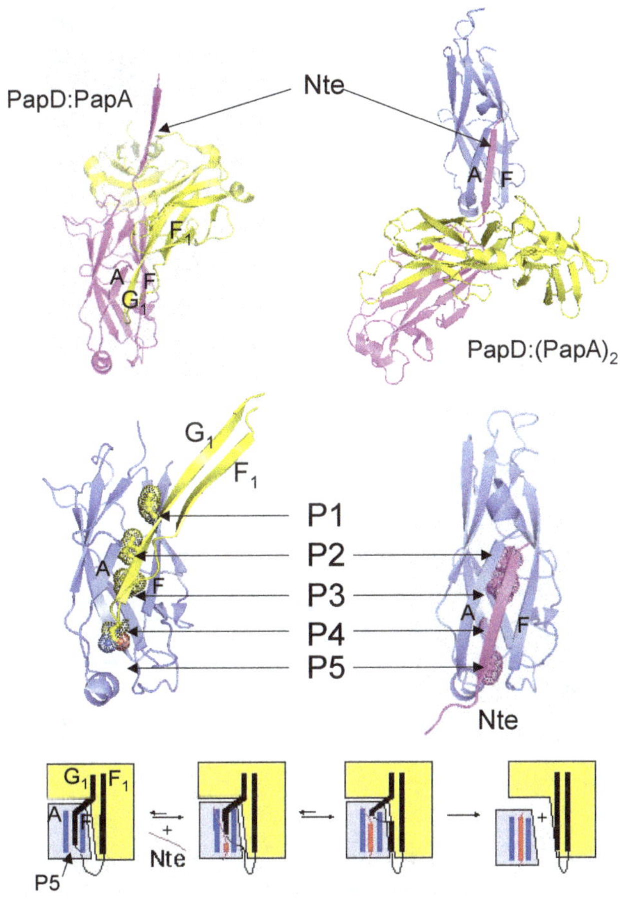

Plate 4.4 Donor-strand exchange and subunit–subunit interactions. **Top left and right**: ribbon diagram of the P pilis PapD–PapA chaperone–subunit (coloured yellow–magenta; PDB entry 2UY6) and PapD–PapA–PapA chaperone–subunit–subunit complexes (coloured yellow–magenta–blue), respectively. **Middle left and right**: ribbon diagram of the PapA subunit in donor-strand complementation and donor-strand exchange with the chaperone G_1 strand and PapA N-terminal extension (Nte), respectively. Labels indicate the P1, P2, P3, P4 and P5 sites and the subunit A and F strands. **Bottom**: schematic representation of the concerted donor-strand exchange mechanism. The diagrams show the F_1-G_1 β-hairpin in the chaperone (coloured yellow) and A and F strands in the subunit (coloured blue). The attacking Nte is coloured red.

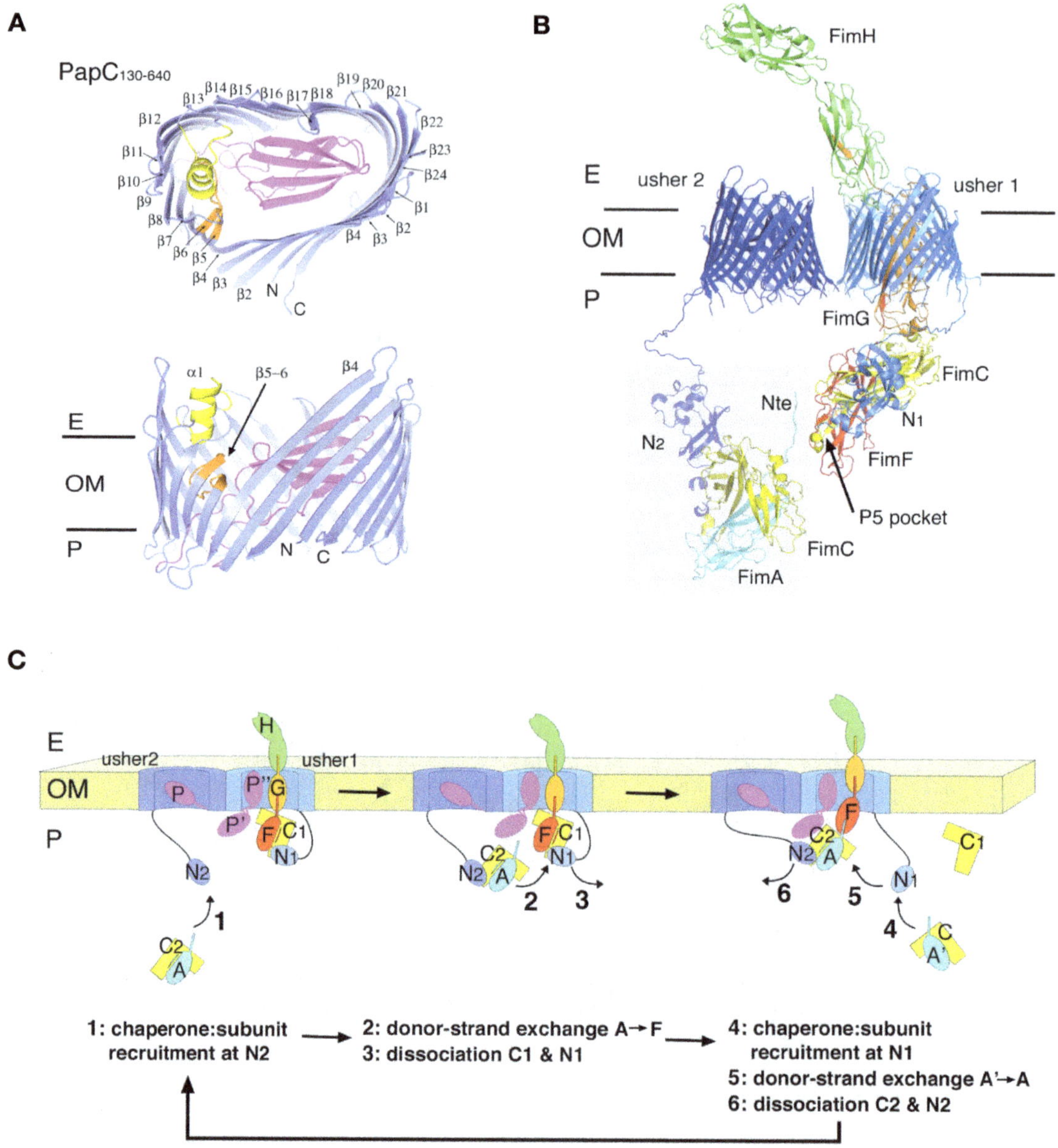

Plate 4.5 Pilus assembly at the usher. **A** Ribbon diagram of the P pilus usher (PapC) translocation pore (PDB entry 2VQI), view from the extracellular side (upper panel) or within the membrane (lower panel) (extracellular space, the outer membrane and periplasm are labelled E, OM, P, respectively). The 24-stranded β-barrel, β5–6 hairpin and central plug domain are coloured blue, orange and magenta, respectively. **B** Atomic model of the usher-bound 3-subunit intermediate of the UPEC Type 1 pilus (FimD usher 1 and 2, the FimH adhesin, FimG and FimF subunits and FimC chaperone are coloured blue, dark blue, green, orange, red and yellow, respectively) (Remaut *et al.*, 2008). Shaded in grey, a new chaperone–subunit complex (FimC–FimA, coloured yellow–cyan) bound to the N-terminal domain of usher pore 2 is modelled in. Labels indicate the P5 initiation site for donor-strand exchange on the penultimate, chaperone-bound subunit. **C** Schematic diagram of fibre assembly at the usher.

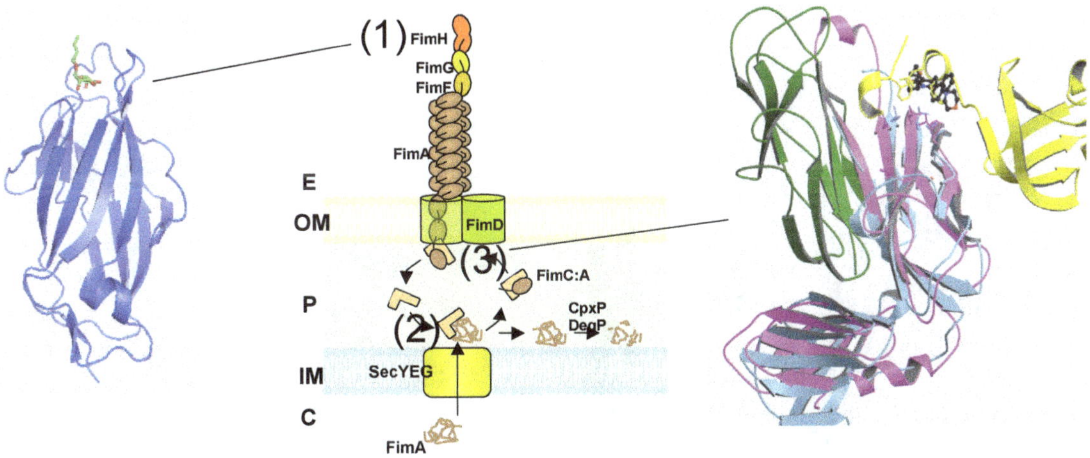

Plate 4.6 Chemical antagonist of CU assembly and receptor recognition. **Left**: ribbon diagram of the FimH adhesin domain bound to the synthetic ligand butyl α-D-mannose (in stick representation). **Middle**: schematic diagram of chaperone–usher pilus assembly pathway. The diagram indicates three areas of interest in CU antagonist development (1) receptor binding, (2) chaperone–subunit interaction and subunit folding and (3) chaperone–subunit recruitment to the usher. **Right**: ribbon diagram of the P pilus chaperone PapD (coloured blue) bound to a pilicide (in stick representation) superimposed onto the Type 1 pilus FimC–FimH chaperone–adhesin complex (coloured magenta–green) bound to the N-terminal domain of the FimD usher (coloured yellow). A patch of conserved hydrophobic residues on the usher/pilicide binding surface are shown in stick representation.

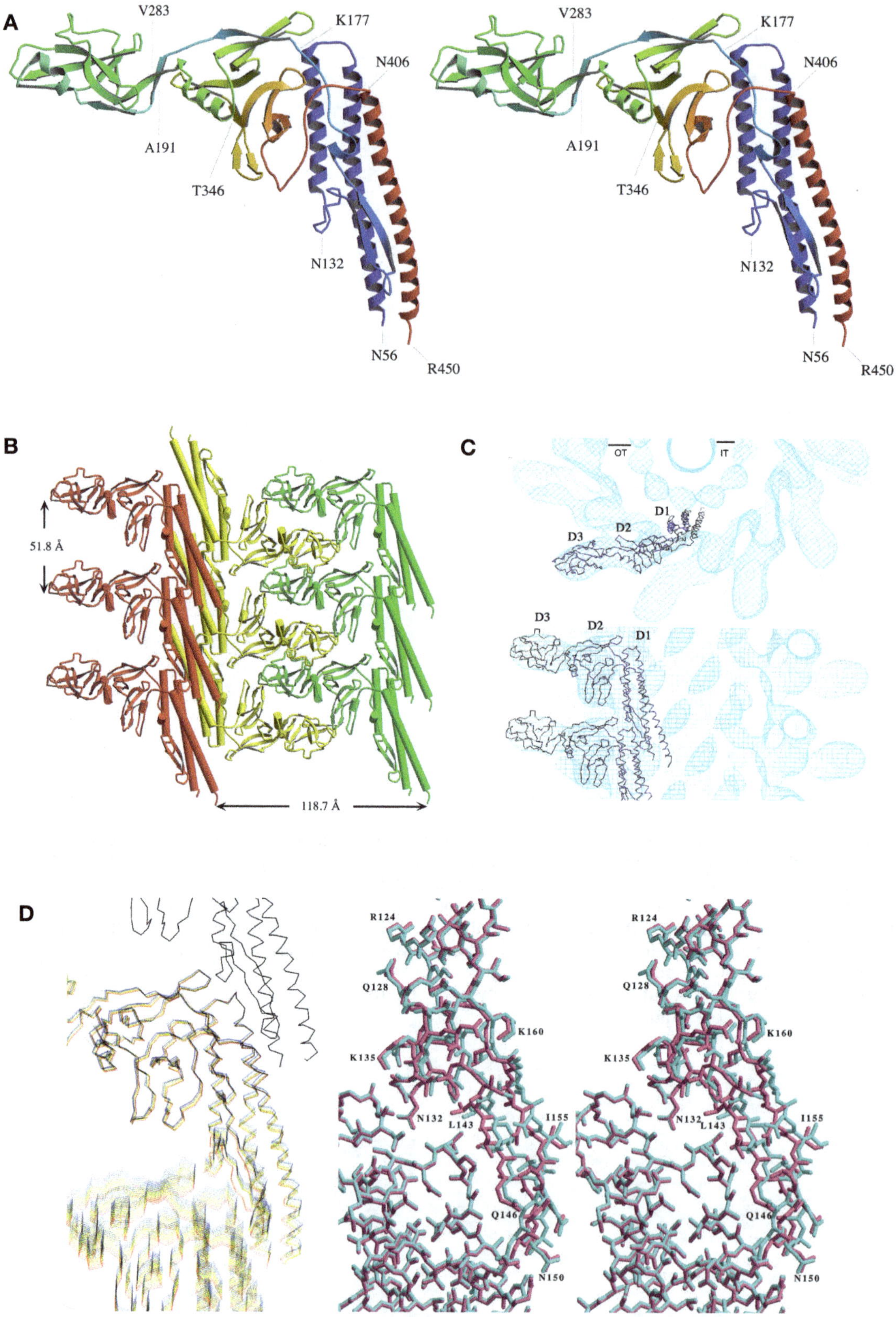
A
V283
K177
N406
A191
T346
N132
N56
R450
V283
K177
N406
A191
T346
N132
N56
R450
B
51.8 Å
118.7 Å
C
OT
IT
D1
D2
D3
D3
D2
D1
D
R124
Q128
K160
K135
N132
L143
I155
Q146
N150
R124
Q128
K160
K135
N132
L143
I155
Q146
N150

Plate 9.3 (opposite) Model of the bacterial flagellar filament. **A** Stereo diagram of the Cα backbone trace of F41 with the three domains, D1, D2 and D3. The chain is colour coded from blue to red from the N to the C terminus. **B** Crystal packing of F41 in the a-c plane viewed down the b axis. Each array of the F41 molecules along the 'a' axis, which has been identified as the protofilament of the flagellar filament, has a different colour. The repeat distances are labelled. Antiparallel packing of the protofilaments by the P21 symmetry of the crystal is clearly shown. **C** Docking of a single protofilament model into the density map of the filament that was obtained by electron cryomicroscopy. The top panel shows an end-on view from the top, with a 50-Å-thick cross-section of the map. The map is painted where the density is higher than the contour level. Bottom panel shows a side view representation of the docking, with a 50-Å-thick longitudinal section of the map. The resolution of the density map was limited to 20 Å for easy evaluation of the model fitting. D0, D1, D2 and D3 indicate the domains of flagellin; IT and OT represent the inner- and outer-tube regions, respectively. **D** Superimposition of 12 sampled stages of the simulated extension. Domains D1 and D2 of three F41 molecules are shown (bottom half of the top, whole of the middle, and top half of the bottom) and the 12 stages are colour coded from blue to red. Each step of the extension was 0.1 Å, but every five steps is sampled and displayed here; therefore, the bottom subunits in different stages are equally separated by 0.5 Å. The whole β-hairpin at the bottom of domain D1 of the middle subunit shows a discrete jump in its conformation at an extension from 4.5 to 5.0 Å, whereas other parts show only gradual deformation of an elastic nature. On the right panel is a close-up view of the axial contact region in stereo, including the β-hairpin showing the conformational switch. The structures at two sampled stages, at an extension of 4.5 Å in cyan and of 4.7 Å in dark pink, are superimposed to show the conformational switch of the β-hairpin, of which the start and the end are marked by yellow arrowheads. Reproduced from Samatey *et al.* (2001) Nature *410*, 331–337.

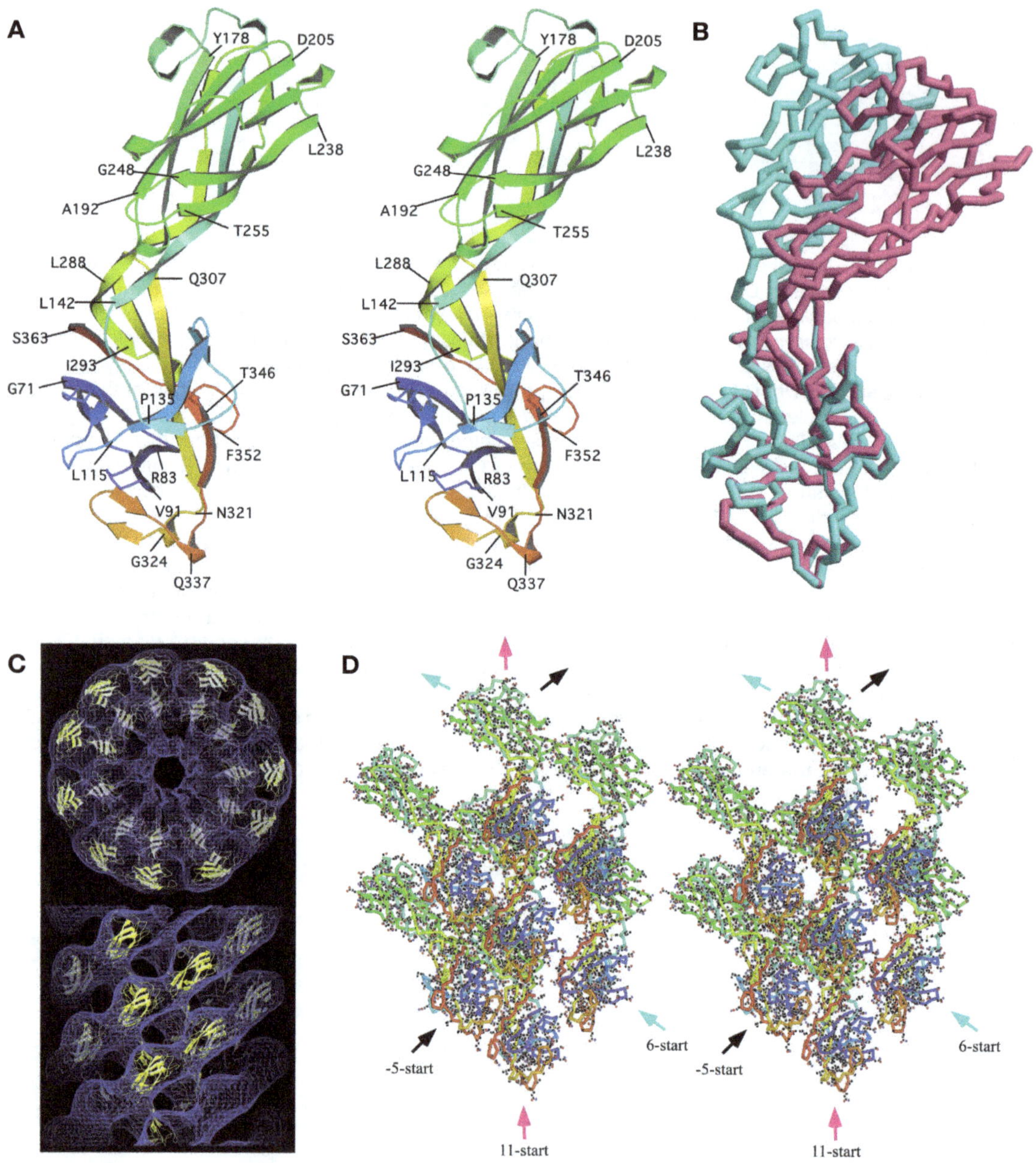

Plate 9.5 Atomic model of the bacterial flagellar straight hook. **A** Stereo view of the Cα backbone trace of FlgE31. The chain is colour-coded from blue to red, going through the rainbow colours, from the N terminus to the C terminus. The model is oriented with domain D1 at the bottom and D2 at the top. **B** Relative domain rearrangement necessary for the docking of FlgE31 into the density map of the straight hook obtained by electron microscopy. The cyan model represents the crystal structure; the dark pink model represent the refined model in the hook. The molecules were aligned respective to the domain D1. **C** Docking of the atomic model of FlgE31 into the outer two domains of the hook. On the top panel shows an end-on view and the bottom panel a side view. In the side view, slightly more than half of the hook density on the back is trimmed off for clarity. **D** Stereo view of three protofilaments in the hook with seven subunits viewed from the inside. Atomic models are represented with the Cα backbone in stick form and side chains in ball-and-stick form, and colour-coded as in **A**. Arrows in three different colours indicate the directions of three representative helical lines: the bottom three for D1 domains and the top three for D2 domains. Reproduced from Samatey *et al.* (2004) Nature *431*, 1062–1068.

Plate 9.6 (overleaf) Model of the bacterial flagellar coiled hook. **A** Atomic model of the coiled hook with a schematic diagram of the basal body spanning the outer membrane and the cytoplasmic membrane. This coiled hook model is part of a supercoiled polyhook with a helical pitch of 950 Å and a diameter of 350 Å. **B** Magnified image of the coiled hook with the innermost and outermost protofilaments on the left and right, respectively. The inner core domains formed from both terminal chains and the central channel are represented by dotted grey lines. **C** Intermolecular packing arrangements of D2 domains on the inner side (left panel) and on the outer side (right panel) of the coiled hook surface. Domain D1 is coloured in grey. Domain D2 is colour-coded from blue to red, going through the rainbow colours, from the N-terminus to the C-terminus. **D** Simulated extension and compression of hook protofilament. Three subunits were used, and domain D2 at the top subunit was moved upwards, whereas domain D1 of the bottom subunit was moved downwards. Protofilament models at five different stages at 5-Å intervals are superimposed with different colours: dark blue, light blue, green, yellow and red, from the most compressed to the most extended state. The left panel shows the whole three subunits while the right panel shows a magnified view of the upper half. D2 domains at the top and D1 domains at the bottom are at equal intervals (2.5 Å) along the vertical axis. It should be noted that the top corner of the triangular loop of domain D1 of the top subunit (indicated by arrows for the most compressed and the most extended, respectively) shows stepwise movements (a jump from dark blue to light blue, very small movements from light blue to green and to yellow, and another jump to red), whereas domain D2 of the middle subunit stays more or less in the same position. This indicates that the triangular loop of domain D1 and the surface of domain D2 can have distinct side-chain bonding partners depending on the state of extension or compression. Reproduced from Samatey *et al.* (2004) Nature *431*, 1062–1068.

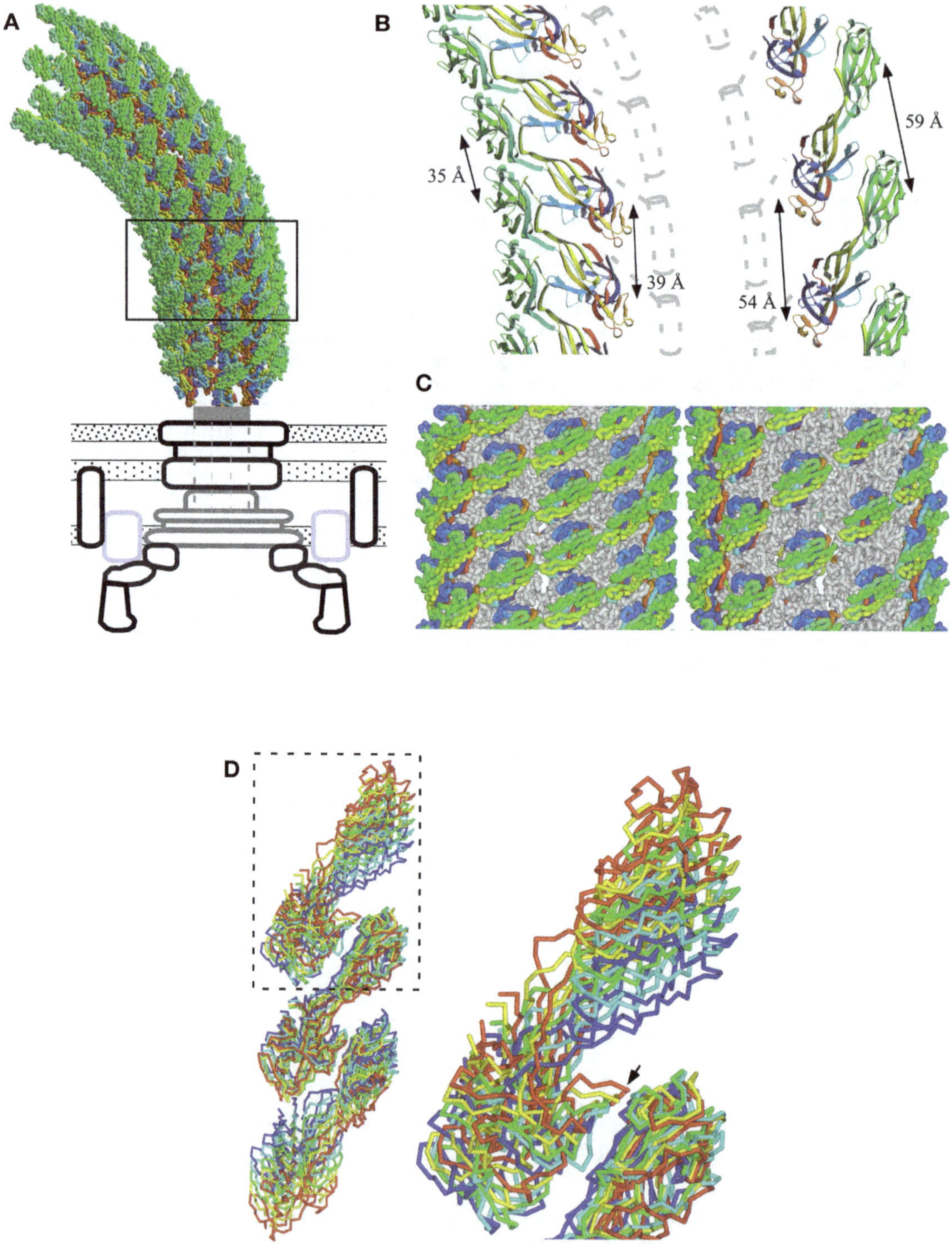
A
B
35 Å
39 Å
54 Å
59 Å
C
D

www.ingramcontent.com/pod-product-compliance
Lightning Source LLC
LaVergne TN
LVHW062230200926
848484LV00006B/1511